Lecture Notes in Physics

Volume 917

The Lecture Notes in Physics

The series Lecture Notes in Physics (LNP), founded in 1969, reports new developments in physics research and teaching-quickly and informally, but with a high quality and the explicit aim to summarize and communicate current knowledge in an accessible way. Books published in this series are conceived as bridging material between advanced graduate textbooks and the forefront of research and to serve three purposes:

- to be a compact and modern up-to-date source of reference on a well-defined topic
- to serve as an accessible introduction to the field to postgraduate students and nonspecialist researchers from related areas
- to be a source of advanced teaching material for specialized seminars, courses and schools

Both monographs and multi-author volumes will be considered for publication. Edited volumes should, however, consist of a very limited number of contributions only. Proceedings will not be considered for LNP.

Volumes published in LNP are disseminated both in print and in electronic formats, the electronic archive being available at springerlink.com. The series content is indexed, abstracted and referenced by many abstracting and information services, bibliographic networks, subscription agencies, library networks, and consortia.

Proposals should be sent to a member of the Editorial Board, or directly to the managing editor at Springer:

Christian Caron
Springer Heidelberg
Physics Editorial Department I
Tiergartenstrasse 17
69121 Heidelberg/Germany
christian.caron@springer.com

More information about this series at http://www.springer.com/series/5304

Peter R. Lang · Yi Liu

Editors

Soft Matter at Aqueous Interfaces

 Springer

Editors
Peter R. Lang
Forschungszentrum Juelich GmbH
Jülich
Germany

Yi Liu
Forschungszentrum Juelich GmbH
Jülich
Germany

ISSN 0075-8450   ISSN 1616-6361   (electronic)
Lecture Notes in Physics
ISBN 978-3-319-24500-3   ISBN 978-3-319-24502-7   (eBook)
DOI 10.1007/978-3-319-24502-7

Library of Congress Control Number: 2015950444

Springer Cham Heidelberg New York Dordrecht London

Printed on acid-free paper

Springer International Publishing AG Switzerland is part of Springer Science+Business Media
(www.springer.com)

Preface

Aqueous interfaces, by which we mean interfaces between an aqueous phase and a solid, another liquid or a gaseous phase, are ubiquitous in daily life, technological applications and biological systems. The properties of these interfaces are of crucial importance for a wide variety of processes, products and biological functions, such as the formulation of personal care and food products, paints and coatings, microfluidic and lab-on-a-chip applications, cell membranes, and lung surfactant. Accordingly, there is a considerable amount of scientific activity on the subject in academia and in industry. However, research and training in this field appear to be distributed over a broad variety of disciplines, ranging from theoretical physics, over engineering science through to biophysics and biology. Consequently, a great deal of excellent work is performed, but progress may be hampered by a lack of awareness of the available knowledge in other disciplines.

As exemplified by the more general field of soft matter science, where a similar situation was prevalent until roughly 25 years ago, a broad interdisciplinary approach will certainly be beneficial for the scientific understanding of aqueous interfaces and for the design of systems with desirable properties. Thus, in 2010 the consortium of SOMATAI was convinced it was the right time to train young researchers in the field of aqueous interfaces to acquire a high degree of expertise in their original discipline in addition to gaining the necessary knowledge and scientific contacts to tackle problems using a broad interdisciplinary approach. After two attempts, the application for a Marie Sklodowska Curie Intitial Training network was granted by the European Commission, which was the starting point of SOMATAI's research and training activities.

As part of SOMATAI's training programme the summer school "Soft Matter at Aqueous Interfaces" was held in Berlin, hosting forty young researchers from all over the world. Besides a series of research papers presented by experts in relevant fields and some lectures dedicated to the specifics of industrial research, the main objective of this school was to provide lectures and tutorials, covering a wide range of topics, from the fundamental text-book physics of fluid interfaces, to advanced experimental and theoretical methods applied in ongoing research. In this book we are collecting the lecture notes of the latter courses.

The contents cover a wide variety of topics in two ways. On the one hand, they span the range of knowledge levels from the basics of physical chemistry to state of the art experimental and theoretical methods, and on the other the diverse range of the scientific fields involved include electro chemistry and corrosion protection right through to colloidal hydrodynamics. This variety offers a fascinating spectrum of information for the newcomer to the field, regardless of whether they are young researchers starting their first project, or experienced scientists intending to broaden their scope of activities. However, despite the editors' best efforts, it became impossible to merge all the contributions into a monograph-style text book. Rather, the individual chapters should be regarded as stand-alone entry points to the challenging research field of soft matter at aqueous interfaces.

The programme of the school was complemented by a series of research papers, some lectures dedicated to the specifics of industrial research, and one session on the ethics of science, which further demonstrated the school's versatility. Regrettably, transcripts of these presentations can not be included in this book. Therefore we especially want to express our gratitude to:

- Kitty van Gruijthuijsen, Firmenich Perfumery—SC, Genève Switzerland
- Katja Hübel Max Planck Institut für Eisenforschung, Düsseldorf, Germany
- Hanne Juling, BASF HR, Ludwigshafen, Germany
- Katharina Kreth, BASF Coatings GmbH, Münster, Germany
- Dominique Langevin, Univerité Paris Sud, Paris, France
- Donald Bruce, Edinethics Ltd, Edinburgh, United Kingdom
- Gerhard Ertl, Fritz-Haber-Institut, Berlin, Germany
- Patrick Keil, BASF Coatings GmbH, Münster, Germany
- Willem Norde, Wageningen UR, Wageningen, The Netherlands
- Gert Strobl, Universität Freiburg, Freiburg, Germany
- Jan Vermant, ETH Zürich, Zürich, Switzerland

for their contributions, which were highly appreciated by the participants.

On behalf of all participants and lecturers we gratefully acknowledge the financial support of the European Commission through the Seventh Framework Programme for research, technological development and demonstration under grant agreements no. 316866 SOMATAI and no. 262348 ESMI as well as Forschungszentrum Jülich.

Finally we would like to wholeheartedly thank the staff of the Hotel Aquino in Berlin for their hospitality and assistance while hosting the school, and Ulrike Nägele whose dedication was of inestimable value for the success of the event.

Jülich   Peter R. Lang
June 2015   Yi Liu

Contents

Part I
Introductions and Basics

Chapter 1
Introduction to Soft Matter

Neus Vilanova and Ilja Karina Voets

Abstract In this introductory chapter we introduce the basic features and building blocks of soft matter focusing in particular on the role of interfaces. As the fundamentals of the behaviour of particles, surfactants, and polymers at interfaces are described, several classical physico-chemical concepts are introduced. A very brief historical overview of the fields of colloid and polymer science is given. Practical applications of soft matter science in areas like personal care, food technology, biology and materials science are outlined at the end of the chapter.

1.1 What Is Soft Matter?

Soft matter is a steadily growing field of science and technology dealing with soft materials, that is, materials that are readily deformable by a small energy input on the order of the thermal energy. Building blocks of soft matter include man-made and naturally occurring small molecules, colloids, polymers, surfactants, liquid crystals, and nanoparticles. Function often arises upon assembly of these units into hierarchical structures in complex architectures such as micelles, vesicles, fibrils, (micro)emulsion droplets, thin films, gels, glasses, and crystals. As the terminology suggests, the key unifying feature is unmistakably that all materials are macroscopically 'soft'—unlike metals for example—due to weak interactions between the various species which we term 'surface forces'. These interactions are at room temperature comparable to thermal energy, which makes that soft materials are highly adaptive: the properties of soft materials are easily tuneable by small variations in the local environment (see boxed text by the 'founding father of soft matter', Nobel laureate Prof. Pierre-Gilles de Gennes). Relevant time and length scales span orders of magnitude from 10^{-9} s for atomic motion to $>10^9$ s for the

N. Vilanova · I.K. Voets (✉)
Macromolecular and Organic Chemistry, Physical Chemistry & Institute for Complex
Molecular Systems, Eindhoven University of Technology, Eindhoven, The Netherlands
e-mail: i.voets@tue.nl

© Springer International Publishing Switzerland 2016
P.R. Lang and Y. Liu (eds.), *Soft Matter at Aqueous Interfaces*,
Lecture Notes in Physics 917, DOI 10.1007/978-3-319-24502-7_1

reconfiguration dynamics of glasses and from 10^{-10^-} to 10^{-9} m for the dimensions of atoms and small molecules up to $>10^{-3}$ m for granular matter and macroscopic materials. Interfaces play a key role since many soft materials are 'colloids' composed of multiple phases—such as a liquid and a solid—separated by an interface.

'What do we mean by soft matter? Americans prefer to call it "complex fluids". This is a rather ugly name, which tends to discourage the young students. But it does indeed bring in two of the major features:

(1) *Complexity*. We may, in a certain primitive sense, say that modern biology has proceeded from studies on simple model systems (bacteria) to complex multicellular organisms (plants, invertebrates, vertebrates…). Similarly, from the explosion of atomic physics in the first half of this century, one of the outgrowths is soft matter, based on polymers, surfactants, liquid crystals, and also on colloidal grains.

(2) *Flexibility*. I like to explain this through one early polymer experiment, which has been initiated by the Indians of the Amazon basin: they collected the sap from the hevea tree, put it on their foot, let it "dry" for a short time. And, behold, they have a *boot*. From a microscopic point of view, the starting point is a set of independent, flexible polymer chains. The oxygen from the air builds in a few bridges between the chains, and this brings in a spectacular change: we shift from a liquid to a network structure which can resist tension —what we now call a *rubber* (in French: caoutchouc, a direct transcription of the Indian word). What is striking in this experiment, is the fact that a very mild chemical action has induced a drastic change in mechanical properties: a typical feature of soft matter [1].'

Pierre-Gilles de Gennes, Nobel lecture 'Soft Matter', 1991.

1.2   Building Blocks of Soft Matter

'Science may be described as the art of systematic over-simplification— the art of discerning what we may with advantage omit.'
Karl Popper in *The Open Universe: An Argument for Indeterminism*, 1992

As diverse as the plethora of soft materials is, soft matter scientists strive to elucidate generic features to describe their formation, structure, dynamics, and mechanics. Chemical details are lumped into a 'coarse grained' description of the materials in an attempt to rationalize the observable behaviour in terms of generic characteristics such as solvophilicity, surface-activity, thermal motion, connectivity, and morphology. Clearly, this approach contrasts sharply with the central paradigm in chemistry and biology which emphasizes the impact of chemical details. And with good reason, examples evidencing the relevance of tiny details are ubiquitous. A single point mutation in a protein that changes only one amino acid in a chain of over hundred monomers may completely abolish its function (Fig. 1.1). Isotopic substitution of a single hydrogen to deuterium constituting the tiniest chemical change in a molecule, may alter its solution behaviour entirely. This apparent contradiction between specificity and generality is the central challenge of soft matter science: how to identify unifying concepts to describe the behaviour of materials wherein specific interactions are at play? How to gain insight from coarse grained approaches and apply it to the very subjects whose chemical and biological details have been ignored?

A reasonable starting point from which unifying concepts can be developed is to classify soft materials based on one or more shared characteristic features. An obvious choice is to distinguish based on constituents (polymers, surfactants, colloids) as shown in Table 1.1, yet it is less straightforward than it appears at first

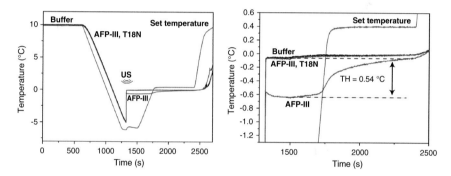

Fig. 1.1 A single point mutation from threonine to asparagine (T18 N) completely abolishes the activity of a recombinantly expressed type III antifreeze protein (AFP) from ocean pout (rQAE) as determined in a sonocrystallization assay of thermal hysteresis activity. Thermal hysteresis is the difference between freezing and melting temperature, ΔT, induced by adsorption of AFPs at specific ice crystal planes. While wildtype rQAE generates $\Delta T = 0.54$ °C, T18 N exhibits freezing and melting at the same temperature as the buffer, i.e., $\Delta T = 0.0$ °C. Reprinted with permission from ref [2]

Table 1.1 An overview of building blocks of soft materials

Building block	Examples
Small molecules	Organogelators, discotics (disc-like), calamitics (rod-like)
Surfactants	Phospholipids, sodium dodecyl sulphate (SDS), alkyl ethoxylates (C_mE_n)
Liquid crystals	Amphiphiles (lyotropic), cholesterylmyristate (thermotropic)
Nanoparticles	CdS quantum dots,
Polymers	Poly(tetrafluoroethylene), DNA, poly(dimethylsiloxane), hyaluronic acid
Colloids	Silica spheres, polymer latexes, virus capsids

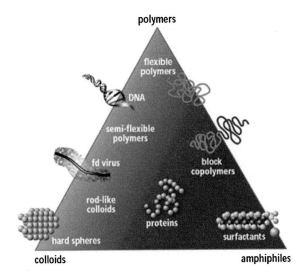

Fig. 1.2 The 'soft matter triangle' as proposed by Gompper, Dhont, and Richter [3] displays the multitude of molecules and materials that span the gap between the archetypical soft matter building blocks—colloids, polymers, and amphiphiles. The abscissa and ordinate can be viewed as gradients in amphiphilicity and elongation or flexibility, respectively

sight. It has been suggested that the building blocks of soft materials rather resemble a continuum of molecules and materials bracketed by the three archetypical examples of spherical colloids, polymers, and surfactants as shown in Fig. 1.2. [3]. Moreover, the behaviour displayed is often strongly dependent on external factors, especially of those materials found close to the midpoint of the axes, such as proteins for example. Most globular enzymes adopt a highly ordered three-dimensional structure known as their native structure in aqueous solution, resembling classical spherical colloids in many respects. But, at high temperatures, in organic solvents, and in the presence of large amounts of denaturants such as urea, proteins unfold and display behaviour characteristic of (semi-)flexible polymers [4]. In fact, much of the pioneering polymer physics experiments have been performed on unfolded proteins, since these are abundant and have a well-defined molecular weight.

An alternative classification originates from colloid science and is based on both energetic and structural considerations. Herein, colloidal materials are first categorized as reversible (equilibrium) or irreversible (metastable) colloids, thereafter they are assigned into subcategories based on the nature of the constituent phases (Table 1.2). Note that some of these colloids—e.g., solid suspensions—are typically hard rather than soft materials.

Table 1.2 Overview of reversible and irreversible colloids including their classification

Reversible colloids			
Type		Examples	
Polymer solution		Fruit juice, synovial fluid	
Network colloids		Sephadex, superabsorbent	
Association colloids		Microemulsions, micelles, vesicles, liposomes	
Irreversible colloids			
Dispersed phase	Continuous phase	Name	Examples
Liquid	Gas	Liquid aerosol	Fog, hairspray
Solid	Gas	Solid aerosol	Smoke, dust
Gas	Liquid	Foam	Beer foam, froth, detergent foam
Liquid	Liquid	Emulsion	Milk, mayonnaise, rubber, shampoo
Solid	Liquid	Sol, dispersion, suspension, paste	Paints, toothpaste, blood, ink
Gas	Solid	Solid foam	Polyurethane foam, bread, Styrofoam
Liquid	Solid	Solid emulsion	Ice cream, tarmac
Solid	Solid	Solid suspension	Wood, opal, pearl, bone, pigmented plastic

1.2.1   Colloids

'Our freedom to doubt was born out of a struggle against authority in the early days of science. It was a very deep and strong struggle: permit us to question—to doubt—to not be sure. I think that it is important that we do not forget this struggle and thus perhaps lose what we have gained.'

Richard Feynman in *The Value of Science*, address to the National Academy of Sciences, 1955

The colloidal domain is loosely defined as heterogeneous systems composed of at least two phases of which one has at least one dimension in the range from a few nanometers to a few tens of micrometers (Fig. 1.3). Colloid and soft matter science are clearly intricately connected, one might say virtually congruent with the exception of hard colloidal materials—such as wood and pigmented plastic—and soft materials composed of either very small or large species, such as sub-nanometer particles and granular matter. Often however, the term colloids is used in a more restrictive fashion to denote the building blocks of dispersions, such as e.g. spherical silica beads, polymeric latexes and poly(isopropyl acrylamide) microgels.

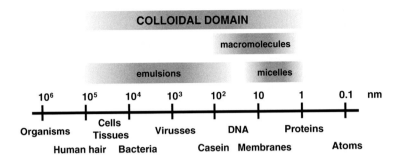

Fig. 1.3 An overview of the typical length scales and representative examples of materials within the 'colloidal domain'

The first studies on inorganic colloids date back to 1840s when the Italian toxicologist Francesco Selmi published on 'pseudo-solutions' of silver chloride, Prussian blue, and sulfur. Some years later Faraday studied colloidal gold, but the term 'colloid'—meaning glue-like—was not coined until another 10 years later in 1861 by Graham. He found that colloids could not pass through membranes. Thus, colloids could be separated from other species in solution utilizing membranes, a process which he referred to as 'dialysis'. After some decades of alternating sparse and intense investigations, colloid science ultimately became firmly established as a scientific discipline around the turn of the century. In 1903 Siedentopf and Zsigmondy developed the ultramicroscope which enabled direct visualization of molecules and colloidal particles, settling the long dispute on the discontinuous structure of matter and the existence of molecules.

Jean-Baptiste Perrin (1926 Nobel Prize in Physics) realized that both the structure and dynamics of colloidal suspensions could be examined by microscopy. In 1908 Perrin and his student Chaudesaigues carried out a series of experiments and verified Albert Einstein's (1921 Nobel Prize in Physics) theory that the irregular motion of pollen observed in 1821 by the English botanist Robert Brown originates from random collisions with liquid molecules [5, 6]. In the same landmark paper, Perrin estimated Avogadro's number from the equilibrium distribution of colloids in a dispersion wherein sedimentation due to gravity is opposed by thermal energy favouring redistribution [5]. In the 1930s Hamaker studied the origin of the attractive interaction between colloids in suspension and in the 1940s Derjaguin, Landau, Verwey and Overbeek developed a powerful and coherent theoretical framework to describe the pair potential between electrostatically stabilized colloids which became known as the 'DLVO theory'.

Perrin's idea to use colloidal particles as a model for atoms still resonates throughout the colloid community today. Indeed, the dimensions and time scales associated with the colloidal domain make these systems an ideal testing ground for atomic and molecular models, particularly when individual building blocks can be visualized in real-time and real-space. Direct imaging of the three-dimensional structure of micron-sized colloidal suspensions by confocal microscopy became

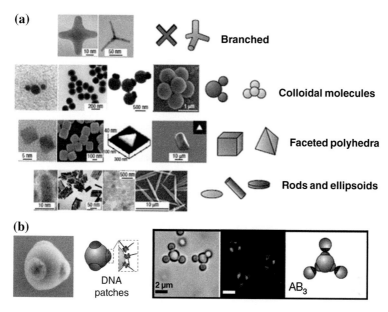

Fig. 1.4 a Examples of asymmetric colloids used as building blocks for soft materials [15] and **b** micrographs of particles with shape anisotropy. As the patches are covered with DNA, they associate with other particles with complementary DNA strands, yielding supracolloidal AB_3 molecules [14]

possible in the 1990s shortly after the seminal paper by Pusey and van Megen on the phase diagram of colloidal hard spheres [7], when the synthesis of monodisperse, micron-sized fluorescent silica colloids was first realized. Several breakthrough publications on the structure and dynamics of hard sphere colloids appeared [8–11]. More recently, complex colloids—either patchy, asymmetric in shape, or both (Fig. 1.4)—have been prepared in reasonable quantities which are now used as models for molecules and polymers [12–14] aiming e.g. to expand the realm of attainable crystal lattices to non-close packed structures [15, 16].

1.2.2 Polymers

'Polymers, Italians in miniature: messy, but very flexible.'
Roberto Piazza in *Soft Matter—The stuff that dreams are made of,* 2010

The distinction between the fields of polymer and colloid science emerged in the 1920s when the German chemist Staudinger (1953 Nobel Prize in Chemistry) first proposed in a landmark paper that polymers such as starch, cellulose, proteins, and rubber are long chains of short repeating units—monomers– linked by covalent bonds [17]. In 1924 he posed a precise definition: 'For such colloid particles, in which the molecule is identical with the primary particle, and in which the individual atoms of this colloid molecule are linked together by covalent bonds, we propose for better differentiation the name macromolecule' [18]. It was a controversial thought at the time; many leading chemists—including Fischer, Wieland—believed that the measured high molecular weights originated from aggregation of small molecules into larger clusters held together by non-covalent bonds (see boxed text [19]).

Since the pioneering studies of Staudinger and others, the field of polymer science has continued to flourish. Academic interest was matched with growing industrial efforts as the use of plastics and synthetic fibres steadily increased. Various exceptional contributions to further polymer synthesis—including Ziegler-Natta catalysis and living polymerization methods—enable contemporary polymer scientists to prepare a multitude of macromolecules of varying composition, topology, and functionality (Fig. 1.5). Interestingly, polymers in which monomers are held together by non-covalent bonds as suggested by Wieland— supramolecular polymers, also known as reversible or living polymers—have nowadays also found widespread academic and industrial interest as they combine excellent materials properties with good processability due to low melt viscosities [20–26].

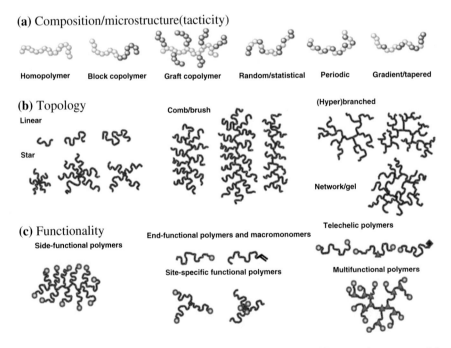

(a) Composition/microstructure(tacticity)

Homopolymer   Block copolymer   Graft copolymer   Random/statistical   Periodic   Gradient/tapered

(b) Topology

Linear

Star

Comb/brush

(Hyper)branched

Network/gel

(c) Functionality

Side-functional polymers

End-functional polymers and macromonomers

Site-specific functional polymers

Telechelic polymers

Multifunctional polymers

Fig. 1.5 Overview of polymeric architectures varying in composition or microstructure (**a**), topology (**b**), and functionality (**c**). Adapted from [36]

'Dear colleague,

Abandon your idea of large molecules, organic molecules with molecular weights exceeding 5000 do not exist. Purify your products such as rubber, they will crystallize and turn out to be low molecular weight compounds.'

Heinrich Wieland to Hermann Staudinger, from 'Arbeitserinnerungen', 1961

Major contributions by the physicists Flory and de Gennes have been adopted throughout the field and continue to inspire many soft matter scientists today. For example, the behaviour of binary and ternary mixtures, such as a single or two polymers in a solvent, can be rationalized within the framework of Flory-Huggins theory. Nowadays, this provides an effective handle to achieve a better understanding of the structure and performance of a great variety of materials ranging from organic solar cells to the cytoplasm of living cells [27–29]. Scaling approaches to describe the conformation of macromolecules in solution and at interfaces introduced originally by de Gennes find widespread application in the description of polymer assembly, responsive polymer brushes, and the behaviour of polymer melts [30–35].

1.2.3 Surfactants

Whereas the fields of polymer and colloid science grew increasingly distinct since the early 20th century, surfactant science has remained rather closely connected to colloid science. Surface-active agents known as surfactants or amphiphiles are composed of a hydrophobic tail and a hydrophilic head group (Fig. 1.6). They tend to adsorb at interfaces and form association colloids, such as micelles, above a threshold concentration which is referred to as the critical micellar concentration (CMC). At concentrations much higher than the CMC, a phase transition from micelles to lyotropic liquid crystalline phases takes place. A powerful method to predict the morphology of either the micelles or liquid crystals formed is based on the effective surfactant geometry in the aggregate, defined as the critical packing parameter, P_c, which is the ratio between the space occupied by the surfactant head and tail, respectively [37]. Spherical micelles form for $P_c < 1/3$, while (inverse) cylindrical (or worm-like) micelles, lamellar or bicontinuous liquid crystals and vesicles require larger values of P_c (Fig. 1.6).

Both adsorption and association are associated with a significant free energy gain. The solubility of surfactants in bulk phases (aqueous and organic solvents) is limited due to their amphiphilicity. Since surfactants are composed of a solvophilic and a solvophobic part, poising them towards interfaces limits the exposure of solvophobic moieties to the bulk liquid phase. Numerous types of surfactants have been prepared and characterized; we generally distinguish based on the headgroup type into three classes: ionics, non-ionics, and zwitterionics or amphoterics (Table 1.3).

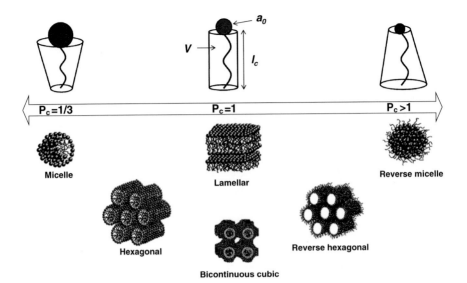

Fig. 1.6 The surfactant architecture can be described by a single dimensionless parameter referred to as the critical or molecular packing parameter $P_c = v/a_0 l_c$. It is the ratio between the volume occupied by the hydrophobic surfactant tail, v, and the interfacial area occupied by the surfactant head group, a_0, multiplied by the length of the hydrophobic tail, l_c. Surfactant self-assembly leads to a range of different structures depending on P_c. Adapted from [38]

Table 1.3 Selected examples of common amphiphiles

Type	Example	Chemical structure
Ionic	Sodium bis (2-ethylhexyl) sulfosuccinate (Aerosol OT, AOT)	
Non-ionic	Alkyl ethoxylate	
Zwitterionic	Amidobetaine	

1.3   What Makes Soft Matter 'Soft'?

From the above it has become clear that soft matter encompasses essentially all materials with polyatomic species (surfactants, polymers, colloids) that display behaviour intermediate between that of simple liquids and simple solids. Now that we know what soft matter is made off, we may understand what makes these materials soft, and perhaps more importantly, whether it matters.

Consider as a qualitative measure of the yield stress σ_y that opposes flow, the interaction energy density (U/R^3) of a material, which is given by the interaction energy U between its subunits with a radius R. While the interaction energy between atoms far exceeds the thermal energy, the 'strength' of the surface forces acting between the mesoscopic subunits of soft materials are comparable to thermal energy ($k_B T$). Consequently, U is relatively small, while R is large, such that the interaction energy density is low, which makes the materials soft.

Exercise 1
Estimate the yield stress of a colloidal crystal of silica beads of $R = 250$ nm and an atomic solid ($R = 1$ Å) with a bond energy of 100 $k_B T$.

Does soft matter? The answer is simple: definitely. Soft materials are deformable, enabling passage through narrow channels, like red blood cells through blood vessels [39]. When spheres fail to pack into crystalline lattices at high volume fraction and instead the liquid 'freezes' into a glass without apparent change in its structure, the resultant solid is strong if the original spheres are soft [40]. Local injection and sustained release of drugs embedded in hydrogels is facilitated by flow enhancement under load [41–43]. Such 'shear-thinning' is a characteristic feature of many soft materials and often vital for their performance.

1.4 Soft Matter and Interfaces

The importance of interfaces in soft matter is evident once we realize that many soft materials are colloids in the broad sense of the term, that is, mesoscopically heterogeneous systems in which one phase is finely dispersed in another phase. Given the small dimensions of the dispersed phase—typically 10^{-9}–10^{-4} m—such systems are associated with a large interfacial area.

Exercise 2
Compute the interfacial area of a 0.1 % dispersion of 2 nm CdS quantum dots and compare it to the surface of a soccer field (6500 m^2).

Clearly, interfaces must play a dominant role in soft matter. Indeed, minimization of surface free area and the associated excessive Gibbs energy are a key driving force for structural evolution and fluctuations. The interactions operating between individual components of soft materials are termed surface forces, not because they are volume- or medium-independent, but because surfaces play a crucial role. Screened electrostatic interactions in aqueous solutions originate predominantly from ionization of surface groups and ion adsorption at interfaces. The addition of polymers to a colloidal suspension may enhance colloidal stability if dense 'polymer brushes' form, whereas destabilization occurs if the polymers tend to avoid the surface instead (known as 'depletion' interaction).

1.4.1 Colloids at Interfaces

While adsorption of polymers and surfactants at interfaces has received widespread attention since decades, particle-laden interfaces only recently regained great interest throughout the soft matter community when research into Pickering emulsions was re-popularized [44, 45]. Applied interest for interface stabilization by particles is also plentiful especially in the technology areas of flotation, emulsion stabilization, cosmetics, pharmaceutics, and waste water treatment [46]. Amphiphilicity is not the key driving force for adsorption of particles. Instead, we can understand the tendency to adhere from the perspective of wetting. Thus, let us start with an explanation of some fundamental considerations to comprehend the adhesion of colloids at interfaces. The equilibrium balance of in-plane forces acting on a liquid drop at a solid substrate is described in Young's equation (Eq. (1.1)).

$$\gamma_{LS} + \gamma_{LV}\cos\theta = \gamma_{VS} \tag{1.1}$$

Each of the three tensions γ_{LS}, γ_{LV}, and γ_{VS} acts on the three-phase contact line of a sessile liquid droplet (L) at the solid substrate (S) in contact with vapour (V) to *minimize* the corresponding LS, LV, and VS contact area (Fig. 1.7). Similarly, we may use Young's equation for two immiscible liquids, such as oil and water, in contact with a solid surface, (Fig. 1.8).

$$\gamma_{AS} + \gamma_{AB}\cos\theta = \gamma_{BS} \tag{1.2}$$

With Young's equation in hand, we may use the contact angle θ to distinguish between four different types of 'wetting' behaviour: perfect wetting ($\theta = 0°$), partial wetting ($0° < \theta < 90°$), partial non-wetting ($90° < \theta < 180°$), and perfect non-wetting ($\theta = 180°$) as outlined in Fig. 1.9.

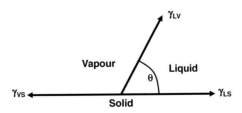

Fig. 1.7 Equilibrium balance of in-plane forces acting on a liquid drop at a solid substrate

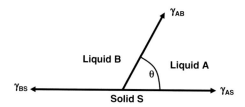

Fig. 1.8 Surface tensions for two immiscible liquids on a solid substrate

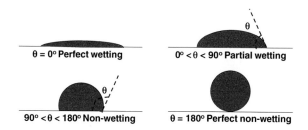

Fig. 1.9 Overview of four different types of wetting behaviour of a sessile liquid droplet on a flat solid substrate in the presence of vapour

It turns out that all colloids bind to liquid-gas interfaces as long as the contact angle is not zero [47], because the work required to remove the particle from the interface ΔG far exceeds k_BT (Eq. 1.3). Due to gravity and buoyancy, larger particles (> 100 μm) may deform the liquid interface and detach (Fig. 1.10a).

$$\Delta G = \pi R^2 \gamma_{LV}(\cos\theta - 1)^2 \tag{1.3}$$

Exercise 3
Compute the work in units of thermal energy required to remove a hydrophilic colloid ($\theta = 85°$) of 500 nm radius from the surface of an air bubble in water surface ($\gamma_{LV} = 0.072$ Jm^{-2}).

Adsorption of colloids at interfaces has been exploited extensively to stabilize emulsions, hereafter referred to as Pickering emulsions. Stabilization of oil-in-water (O/W) emulsions via particles requires partial wetting ($\theta < 90°$), whilst W/O emulsions require non-wetting ($\theta > 90°$) as shown in Fig. 1.10b. In other words, the phase that preferentially wets the colloids gives the external phase. The stability of the formed emulsions depends on several factors: size, composition, and concentration of the colloids, their contact angle and the interactions between them [45]. Strong adsorption of colloids at the oil-water interface of the emulsion droplets acts as an efficient mechanical barrier against droplet coalescence. A wide variety of

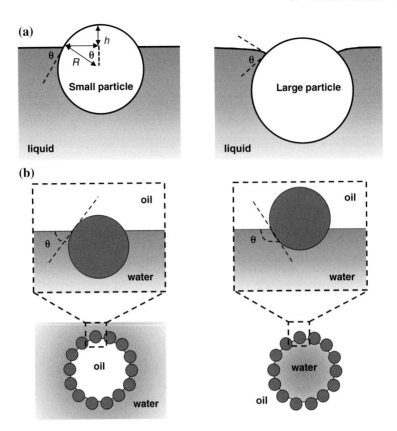

Fig. 1.10 a Small (*left*) and large (*right*) spherical particles adsorbed at a liquid–gas interface. Large particles may deform the liquid interface, while small particles do not. Adapted from [48]. **b** The position of colloids at an oil-water interface of emulsion droplets depends on their wettability: O/W emulsions form for contact angles less than 90° (*left*), while W/O emulsions require contact angles greater than 90° (*right*)

solid colloids has been used to stabilize O/W and W/O emulsions, including iron oxide, hydroxides, metal sulfates, silica, clays, carbon or polymer particles. Interestingly, the use of colloids as emulsion stabilizers may offer additional advantages beyond stability, as they may confer additional features to the system such as magnetic [49] or catalytic properties [50].

1.4.2 Polymers at Interfaces

Steric stabilization is one of the key approaches to stabilize colloids in suspension, often with polymers as key player. Polymer decorated interfaces can be realized through various means, either chemically via preformed polymers ('grafting to') and in situ polymerization ('grafting from'), or physically, via adsorption. Functionalization via chemical routes comprises tethered, i.e., end-attached chains which need not to have an affinity with the surface. By contrast, polymer layers formed via adsorption require a sufficiently favourable polymer-substrate interaction, or alternatively, sufficiently unfavourable substrate-solvent interaction.

Effective steric stabilization of colloidal dispersions via polymers is achieved only if the polymer layer is sufficiently thick and dense, which nowadays still represents a serious challenge. This is because, what makes a polymer layer an effective barrier against aggregation, impedes attachment of additional polymer chains to a partially covered interface: steric hindrance.

To understand the conformation of an adsorbed or tethered macromolecule, let us first take a closer look at the size and shape of a single polymer far away from an interface. The conformation of a polymer chain in solution is a balance between conformational entropy or elasticity (F_{el}) and osmotic interactions (F_{os}). The former is affected by the differential affinity of the monomers for themselves and for solvent molecules which may be described in terms of the Flory-Huggins interaction parameter χ. It is related to the binary excluded-volume parameter v_0 (representative of 'solvent quality') according to $v_0 = 1 - 2\chi$. Intuitively, it is easy to understand why χ largely determines the balance between the tendency of the chains to extend for entropic reasons and the tendency to contract to minimize energetically unfavourable ($\chi > 0.5$) and maximize favourable ($\chi < 0.5$) interactions.

Consider the well-known Flory scaling law $R_g \sim N^{3/5} v_0^{1/5}$, which gives the radius of gyration (R_g) for a homopolymer chain of N monomers dispersed in a good solvent. How is this affected by the presence of a surface? Naturally, we now need to account for monomer-surface interactions.

At low grafting densities, the conformation of covalently coupled chains remains largely but not fully unperturbed if surface affinity is low ('mushroom' regime). By contrast, if the chain has a high surface affinity ($\chi < 0.5$), it adopts a pancake-like conformation under good solvent conditions, which is a 2D self-avoiding walk (Fig. 1.11). Therefore, a balance is established between the energetic gain due to monomer-surface contacts and an entropic loss as the chain flattens upon attachment.

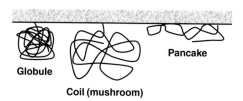

Fig. 1.11 Schematic representation of three possible chain conformations of tethered polymers at low grafting density. From *left* to *right* a non-adsorbing polymer forms a globule in a bad solvent and a coil in a good solvent ('mushroom' regime), while a pancake conformation persists in a good solvent for chains with a finite surface affinity

Polymers that are not tethered but merely adsorbed, require a finite favourable monomer-substrate interaction. A threshold surface affinity χ_c separates a regime of χ values for which adsorption is negligible from a regime where the bound fraction is high. The longer the polymer chain the sharper the rise in surface coverage around χ_c [51]. In case of adsorption, chain attachment may occur at any point along the polymer backbone and typically only a fraction of the total amount of monomers is adsorbed at any point in time. Loops are the detached sections of the polymer chain bracketed by adsorbed monomers (trains), what remains are tails (free ends) (Fig. 1.12). The balance between the adsorption and the solvation of the polymer at interfaces can be optimized by using block copolymers, in which one sort of segments can be strongly adsorbed and poorly solvated at the surface, while the other kind of segments has the opposite behaviour by means of adopting a strongly solvent-swollen conformation.

At elevated densities, adjacent tethered polymer chains interact and eventually 'brush'-like conformations start to appear (Fig. 1.13). Predictions of the thickness of polymer brushes by scaling theories, self-consistent-field (SCF) and numerical models agree well. It can be shown that the brush height, δ, of linear brushes scales linearly with chain length N according to

$$\delta = N\sigma^{1/3} \qquad (1.4)$$

with the grafting density, σ. Note that this is much stronger than the scaling $R \propto N^{3/5}$ for a good polymer in solution. Various groups have tested the predications for polymer brush thickness experimentally by e.g. small angle neutron reflectivity and found good agreement with theory [52].

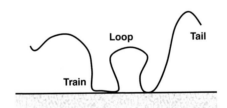

Fig. 1.12 Schematic representation of the three sections of an adsorbed polymer chain. Loops are detached sections of the polymer backbone bracketed by adsorbed monomers (trains). The dangling, non-adsorbed chain ends are referred to as tails

Fig. 1.13 As the grafting density of polymer chains at an interface increases from left to right we observe a 'mushroom', 'overlapping mushroom', and finally at high densities a 'brush' regime

In summary, isolated polymer chains at interfaces tend to adopt a swollen coil-like conformation for low surface affinity at temperatures above the theta temperature ('mushroom' regime), which flatten into a 'pancake-like' conformation as surface affinity increases and collapse into a 'globule' as solvent quality deteriorates. Both, chemical and physical attachment, generate polymer brushes at sufficiently high surface coverage. Many experimental systems approach but do not enter this regime, as the polymer chains are either too short or surface coverage is too low. This is unfortunate, since the polymer interface in this 'quasi-brush' or 'pseudo-brush' regime provides suboptimal stabilization against coagulation and antifouling. Considerable effort is therefore devoted to develop innovative approaches to realize high grafting densities [32, 53].

1.4.3 Surfactants at Interfaces

The adsorption of surfactants at liquid interfaces can be probed by a variety of methods. Most common perhaps is to record the decrease in surface tension as a function of surfactant concentration (Fig. 1.14). The surface tension of a pure liquid is the result of an imbalance in attractive forces between the constituent molecules at interfaces. At low surfactant concentrations, the surface tension (γ_{LV}) decreases as surfactant molecules adsorb onto the interface. Consequently, the surface excess of surfactant Γ_{surf} increases until Γ_{max}, at which the available surface is saturated by a surfactant monolayer. Above this concentration threshold, additional surfactant molecules assemble into association colloids—micelles—in the bulk phase.

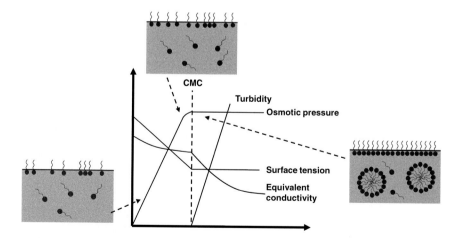

Fig. 1.14 The critical micellar concentration, CMC, of surfactants can be determined from measurements of surface tension, osmotic pressure, turbidity, and equivalent conductivity as a function of surfactant concentration. Scheme of the surfactant behavior in a liquid-gas interface below and above the CMC is also depicted

This critical micellar concentration, CMC, is evidenced by a marked change of slope in the surface tension versus concentration profile. Above the CMC, γ_{LV} remains constant. Other methods of CMC determination rely on a similarly clear break point in the dependence of an experimentally observable physical property (such as turbidity, osmotic pressure, and equivalent conductivity) on surfactant concentration.

To compare the effect of different surfactants we need to quantify the adsorbed amount and the concomitant reduction in surface tension. Note that "the interface" is not a plane, rather it is a region of few molecules in thickness, where the molecular concentration changes, see e.g. the chapter B.4 by G. H. Findenegg. Let us know look at the surface excess Γ defined as the (solvent, surfactant...) concentration in a surface plane in the interfacial region relative to the concentration in a similar plane in bulk. To facilitate computation of the adsorption isotherms that relate Γ and γ, Gibbs defined a mathematical plane within the interfacial region, such that the solvent excess in this plane is exactly zero (depicted as x' in Fig. 1.15). This became known as the 'Gibbs dividing surface'. Using this as well as several other assumptions, Gibbs derived an equation to relate surface tension to the surfactant surface excess. For non-ionic surfactants the Gibbs equation is

$$d\gamma = -\Gamma RTd\ln c, \tag{1.5}$$

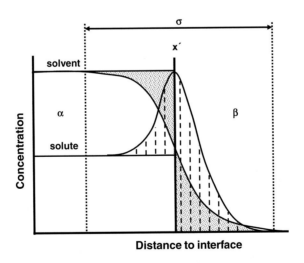

Fig. 1.15 Schematic representation of the interfacial region (bracketed by dotted lines), with width σ, separating two bulk phases, a liquid phase α and vapour, β. The Gibbs dividing surface located at x' is the plane in which the solvent excess concentration becomes zero. The solvent and surfactant surface excess concentrations, Γ, as defined in Gibbs' approach are given by the difference in the shaded areas on each side of the plane

while for 1:1 dissociating compounds (such as SDS) without added electrolyte, it is

$$d\gamma = -\Gamma 2RTd \ln c \tag{1.6}$$

with c the molar surfactant concentration in the bulk.

Surfactants can now be compared in terms of their *efficiency* and *effectiveness*. By efficiency we mean the surfactant concentration required to reduce the surface tension by a specific amount (typically 20 mNm^{-1}), which is given by the Frumkin adsorption equation

$$\gamma_0 - \gamma = \pi = -2.303RT\Gamma_m \log\left(1 - \frac{\Gamma}{\Gamma_m}\right) \tag{1.7}$$

with Γ_m the surface excess concentration at surface saturation. For convenience we introduce pC_{20} as a quantitative measure, which is the negative logarithm of the surfactant concentration (in accordance with the definition of pH) required to reduce γ by 20 mNm^{-1}. At this surface tension, surface is nearly saturated, therefore it is the minimum concentration needed to produce maximum adsorption at the interface. A larger pC_{20} indicates more efficient adsorption. Typically, pC_{20} increases as the length of the surfactant tail increases, tail branching is limited, and the head group remains uncharged. By surfactant effectiveness we mean the maximally attainable surface tension reduction irrespective of surfactant concentration. As this is typically reached at the CMC, this is the main determinant of effectiveness.

Exercise 4

Demonstrate that the interface is close to saturation for pC_{20} at 25°C and a maximum surface excess of 3×10^{-6} molm^{-2}.

The ability of surfactants to decrease the surface tension plays a crucial role in the emulsion field. Since emulsions are *metastable* colloids, they tend to phase separate into bulk oil and water phases. Addition of surfactants into the system decreases, on the one hand, the required energy to form the emulsion (to break the dispersed phase into small droplets) and, on the other hand, also promotes emulsion stability mainly by introduction of mechanical, steric and/or electrostatic barriers that prevent direct contact between droplets. Note that using two instead of one surfactant species is reported to further enhance emulsion stability, probably due to synergetic effects as well as to an increase in interfacial elasticity [54]. Surfactants are hence extremely important for emulsion preparation, stability and structure. The geometry of the surfactant is the primary determinant of the emulsion type. Surfactants with a hydrophilic head larger than the hydrophobic tail are considered hydrophilic. These have small critical packing parameters, P_c, and curve the interfaces around the oil resulting in O/W emulsions. Differently, when the surfactant tail is larger than the head, the surfactant is lipophilic resulting in oil-continuous emulsions.

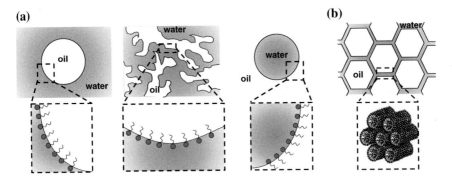

Fig. 1.16 a Schematic representation of O/W, bicontinous and W/O microemulsions (from *left* to *right*) emphasizing the orientation of the surfactants at their interfaces. **b** An O/W highly concentrated emulsion with a hexagonal liquid crystal as continuous phase. Droplets of highly concentrated emulsions are polyhedral as the maximal packing of monodisperse droplets is exceeded

Importantly, surfactants may form association colloids, such as liquid crystals, micelles, and microemulsions. Contrasting markedly with (macro)emulsions, microemulsions are *equilibrium* ternary systems composed of oil, water and surfactants molecules at the oil/water interface. Alike emulsions, their microstructure may be W/O or O/W but also bicontinuous, depending on the geometry of the surfactant (Fig. 1.16a). Unlike emulsions however, microemulsions are thermodynamically stable isotropic liquid solutions with extremely low surface tensions, which leads to remarkably small droplets (1–50 nm) as compared to typical droplet sizes of emulsions. As a consequence of their equilibrium nature, microemulsions form spontaneously once appropriate conditions are selected (temperature, volume fraction of the phases, ionic strength, …). Interestingly, nanostructured colloidal dispersions can be generated fairly easily when liquid crystalline mesophases are used instead of simple dispersed or continuous phases. The novelty of using crystalline mesophases as a dispersed phase lies on their hierarchical structure, namely with hydrophilic and hydrophobic domains. This may find use as drug delivery systems, since they enable the solubilization of both, hydrophobic and hydrophilic active molecules at the same time [55, 56]. Nanostructured continuous phases have mainly been exploited with emulsions with an internal volume phase larger than 74 %—known as highly concentrated emulsions—as exemplified in Fig. 1.16b [57]. The aim of this strategy is to enhance stability [58] or to template the preparation of solid foams with meso/macroporosity [59].

1.5 Soft Matter Science and Technology

The field of soft matter is a multidisciplinary arena where chemists, (chemical) engineers, physicists, biologists, and mathematicians interact to tackle open questions enabling the development of new technologies, some of which address key societal issues such as sustainable energy, public health, and food safety. The following sections highlight a few examples which may give you a taste for the scope and practical applications of soft matter science.

Self-organized cytoplasm. Soft matter is as central to lifeless materials as it is to living systems. Every cell in our body is a marvellous example of a highly functional, hierarchically structured soft material with fascinating properties. Not surprisingly, many soft matter scientists actively contribute to the growing understanding of key aspects of life such as replication, folding, differentiation, signal transduction, active transport, motility, self-healing, tissue re-modelling and the origin of life. To name but a few examples, the consequences of cellular crowding have been cast in terms of depletion interactions, the origin of non-membrane bound organelles has been related to associative and segregative phase separation in multicomponent mixtures, osmotic pressure has been demonstrated to passively eject a large portion of bacteriophage DNA during host infection [60, 61], single point mutations in eye lens proteins were found to reduce the solubility and raise the critical point of liquid-liquid phase separation with possible implications for cold cataract [62, 63], and the cortical actin layer has been described as an active wetting film [64].

Personal care. Control over the wetting properties, colloidal stability, and viscoelastic properties of multicomponent mixtures, in particular emulsions, are key objectives of soft matter scientists active in the field of personal care, detergency, and pharmaceutical formulations. The differences between interface stabilization by surfactants, polymers, and colloids are increasingly exploited to retain functionality of active ingredients without compromising stability [65–67]. Chemical and physical approaches are routinely combined particularly when evaporation of volatile components is a considerable risk [68, 69]. Various soft materials originally developed out of scientific curiosity have already found their way into commercial products [70–72].

Food technology. Soft matter approaches to understand food are numerous. Stabilization of emulsions via interfacial protein layers or fat crystals has attracted great interest, as well as complex coacervation for encapsulation in protein/polysaccharide mixtures, and arrested phase separation as a route to gelation [73–79]. Recently, the focus is shifting from systematic studies of macroscopic properties (e.g., interfacial tension, viscosity, turbidity, and emulsion stability, shelf-life) to the relation between structure, dynamics, and kinetics at various length- and timescales in bulk and at interfaces. Moreover, the link between physico-chemical properties and performance (taste, texture, satiating effect, and shelf-life) is increasingly investigated, as well as (partial) replacement of animal proteins by plant proteins [80, 81].

Encapsulation. Protection of active ingredients against chemical and enzymatic degradation, against aggregation, and against premature release during storage, processing, or upon diffusive transport through the body are key objectives of encapsulation strategies for soft materials. Targeted delivery and controlled release of both hydrophobic and hydrophilic drugs via tailored nano-carriers is amongst the most widely and intensely investigated topics. Interestingly, the design strategies for optimal encapsulation and delivery vehicles may be cast in rather general terms.

Crucial herein—apart from toxicity, renal clearance, endosomal escape, and other biomedical concerns- is the thin line between sufficiently strong interactions allowing for efficient encapsulation during storage, transport, and processing and not too strong interactions, which would compromise prompt release upon delivery to the target location [82–84]. Insight into the impact of the morphology and mechanical properties of the carriers is steadily growing [85, 86]. Amongst the most intensely investigated systems for transport of hydrophobic compounds are lipids and polymeric amphiphiles, while electrostatically driven complexation gains ground as an effective means to package hydrophilic and in particular charged moieties in solution and in hydrogels [87, 88].

References

1. P.G. de Gennes, Soft Matter. Rev. Mod. Phys. **64**(3), 645–648 (1992)
2. K. Meister et al., *Observation of Ice-like Water Layers at an Aqueous Protein Surface* (submitted, 2014)
3. G. Gompper, J.K.G. Dhont, D. Richter, A unified view of soft matter systems? Eur. Phys. J. E **26**(1–2), 1–2 (2008)
4. I.K. Voets et al., DMSO-Induced Denaturation of Hen Egg White Lysozyme. J. Phys. Chem. B **114**(16), 11875–11883 (2010)
5. J. Perrin, *Brownian Movement and Molecular Reality* (Taylor and Francis, London, 1910)
6. A. Einstein, Über die von der molekularkinetischen Theorie der Wärme geforderte Bewegung von in ruhenden Flüssigkeiten suspendierten Teilchen. Ann. Phys. **322**(8), 549–560 (1905)
7. P.N. Pusey, W. Vanmegen, Phase-behavior of concentrated suspensions of nearly hard colloidal spheres. Nature **320**(6060), 340–342 (1986)
8. A. Van Blaaderen et al., 3-Dimensional imaging of submicrometer colloidal particles in concentrated suspensions using confocal scanning laser microscopy. Langmuir **8**(6), 1514–1517 (1992)
9. W.K. Kegel, A. van Blaaderen, Direct observation of dynamical heterogeneities in colloidal hard-sphere suspensions. Science **287**(5451), 290–293 (2000)
10. U. Gasser et al., Real-space imaging of nucleation and growth in colloidal crystallization. Science **292**(5515), 258–262 (2001)
11. E.R. Weeks et al., Three-dimensional direct imaging of structural relaxation near the colloidal glass transition. Science **287**(5453), 627–631 (2000)
12. H.R. Vutukuri et al., Colloidal analogues of charged and uncharged polymer chains with tunable stiffness. Angewandte Chemie-International Edition **51**(45), 11249–11253 (2012)
13. S. Sacanna et al., Lock and key colloids. Nature **464**(7288), 575–578 (2010)
14. Y. Wang et al., Colloids with valence and specific directional bonding. Nature **491**(7422), 51–55 (2012)
15. S.C. Glotzer, M.J. Solomon, Anisotropy of building blocks and their assembly into complex structures. Nat. Mater. **6**(8), 557–562 (2007)
16. E. Bianchi, R. Blaak, C.N. Likos, Patchy colloids: state of the art and perspectives. Phys. Chem. Chem. Phys. **13**(14), 6397–6410 (2011)
17. H. Staudinger, Über Polymerisation. Berichte der deutschen chemischen Gesellschaft (A and B Series) **53**(6), 1073–1085 (1920)
18. H. Staudinger, Über die Konstitution des Kautschuks (6. Mitteilung). Berichte der deutschen chemischen Gesellschaft (A and B Series) **57**(7), 1203-1208 (1924)
19. R. Mulhaupt, Hermann Staudinger and the origin of macromolecular chemistry. Angewandte Chemie-International Edition **43**(9), 1054–1063 (2004)

20. T.F.A. de Greef, E.W. Meijer, Materials science—Supramolecular polymers. Nature **453** (7192), 171–173 (2008)
21. R.P. Sijbesma et al., Reversible polymers formed from self-complementary monomers using quadruple hydrogen bonding. Science **278**(5343), 1601–1604 (1997)
22. L. Brunsveld et al., Supramolecular polymers. Chem. Rev. **101**(12), 4071–4097 (2001)
23. V. Simic, L. Bouteiller, M. Jalabert, Highly cooperative formation of bis-urea based supramolecular polymers. J. Am. Chem. Soc. **125**(43), 13148–13154 (2003)
24. J.M. Lehn, Supramolecular polymer chemistry—scope and perspectives. Polym. Int. **51**(10), 825–839 (2002)
25. J.M. Lehn, Toward complex matter: supramolecular chemistry and self-organization. Proc. Natl. Acad. Sci. **99**(8), 4763 (2002)
26. B.J.B. Folmer et al., Supramolecular polymer materials: Chain extension of telechelic polymers using a reactive hydrogen-bonding synthon. Adv. Mater. **12**(12), 874–878 (2000)
27. C.F. Lee et al., Spatial organization of the cell cytoplasm by position-dependent phase separation. Phys. Rev. Lett. **111**(8) (2013)
28. J.K.J. van Duren et al., Relating the Morphology of Poly(p-phenylene vinylene)/ Methanofullerene Blends to Solar-Cell Performance. Adv. Funct. Mater. **14**(5), 425–434 (2004)
29. K. Bergfeldt, L. Piculell, P. Linse, Segregation and association in mixed polymer solutions from Flory-Huggins model calculations. J. Phys. Chem. **100**(9), 3680–3687 (1996)
30. P. Kosovan et al., Amphiphilic graft copolymers in selective solvents: molecular dynamics simulations and scaling theory. Macromolecules **42**(17), 6748–6760 (2009)
31. E.B. Zhulina, O.V. Borisov, Theory of block polymer micelles: recent advances and current challenges. Macromolecules **45**(11), 4429–4440 (2012)
32. E.P.K. Currie, W. Norde, M.A.C. Stuart, Tethered polymer chains: surface chemistry and their impact on colloidal and surface properties. Adv. Colloid Interface Sci. **100**, 205–265 (2003)
33. M.E. Cates, Reptation of living polymers—dynamics of entangled polymers in the presence of reversible chain-scission reactions. Macromolecules **20**(9), 2289–2296 (1987)
34. R.H. Colby, Structure and linear viscoelasticity of flexible polymer solutions: comparison of polyelectrolyte and neutral polymer solutions. Rheol. Acta **49**(5), 425–442 (2010)
35. M. Doi, S.F. Edwards, *The Theory of Polymer Dynamics*, vol. 73 (Oxford University Press, USA, 1988)
36. K. Matyjaszewski, N.V. Tsarevsky, Nanostructured functional materials prepared by atom transfer radical polymerization. Nat. Chem. **1**(4), 276–288 (2009)
37. J.N. Israelachvili, D.J. Mitchell, B.W. Ninham, Theory of self-assembly of hydrocarbon amphiphiles into micelles and bilayers. J. Chem. Soc.-Faraday Trans. Ii **72**, 1525–1568 (1976)
38. K. Holmberg, B. Jonsson, B. Kronberg, B. Lindman, in *Surfactants and Polymers in Aqueous Solutions* (Wiley, New York, 2002)
39. M. Abkarian, A. Viallat, Vesicles and red blood cells in shear flow. Soft Matter **4**(4), 653–657 (2008)
40. J. Mattsson et al., Soft colloids make strong glasses. Nature **462**(7269), 83–86 (2009)
41. G.M. Pawar et al., Injectable hydrogels from segmented PEG-Bisurea copolymers. Biomacromolecules **13**(12), 3966–3976 (2012)
42. R.E. Kieltyka et al., Mesoscale modulation of supramolecular ureidopyrimidinone-based Poly (ethylene glycol) transient networks in water. J. Am. Chem. Soc. **135**(30), 11159–11164 (2013)
43. A. Pape et al., Mesoscale characterization of supramolecular transient networks using SAXS and rheology. Int. J. Mol. Sci. **15**(1), 1096–1111 (2014)
44. B.P. Binks, Particles as surfactants—similarities and differences. Curr. Opin. Colloid Interface Sci. **7**(1–2), 21–41 (2002)
45. R. Aveyard, B.P. Binks, J.H. Clint, Emulsions stabilised solely by colloidal particles. Adv. Colloid Interface Sci. **100**, 503–546 (2003)
46. Y. Chevalier, M.A. Bolzinger, Emulsions stabilized with solid nanoparticles: Pickering emulsions. colloids and surfaces A-physicochemical and engineering aspects **439**, 23–34 (2013)

47. A. Scheludko, B.V. Toshev, D.T. Bojadjiev, Attachment of particles to a liquid surface (capillary theory of flotation). J. Chem. Soc.-Faraday Trans. I, 1976. **72**, 2815–2828
48. H.-J. Butt, K. Graf, M. Kappl, *Contact Angle Phenomena and Wetting, in Physics and Chemistry of Interfaces* (Wiley-VCH Verlag GmbH & Co. KGaA, 2004), pp. 118–144
49. A. Vílchez et al., Macroporous polymers obtained in highly concentrated emulsions stabilized solely with magnetic nanoparticles. Langmuir **27**(21), 13342–13352 (2011)
50. Z. Chen et al., Light controlled reversible inversion of nanophosphor-stabilized pickering emulsions for biphasic enantioselective biocatalysis. J. Am. Chem. Soc. **136**(20), 7498–7504 (2014)
51. M.A.C. Stuart, T. Cosgrove, B. Vincent, Experimental aspects of polymer adsorption at solid-solution interfaces. Adv. Colloid Interface Sci. **24**(2–3), 143–239 (1986)
52. T. Cosgrove, *Colloid Science: Principles, Methods and Applications* (Wiley, New York, 2010)
53. W.M. de Vos et al., Charge-driven and reversible assembly of ultra-dense polymer brushes: formation and antifouling properties of a zipper brush. Soft Matter **6**(11), 2499–2507 (2010)
54. D. Myers, *Surfaces, Interfaces, and Colloids, Principles and Applications* (Wiley, New York, 1999)
55. C.V. Kulkarni, R. Mezzenga, O. Glatter, Water-in-oil nanostructured emulsions: Towards the structural hierarchy of liquid crystalline materials. Soft Matter **6**(21), 5615–5624 (2010)
56. A. Yaghmur et al., Control of the internal structure of MLO-based isasomes by the addition of diglycerol monooleate and soybean phosphatidylcholine. Langmuir **22**(24), 9919–9927 (2006)
57. M.D.H.K. H. Uddin, C. Solans, Highly Concentrated Cubic Phase-Based Emulsions, in *Structure-Performance Relationships in Surfactants*, ed. by M. Dekker (Springer, New York, 2003), p. 599–626
58. C. Solans et al., Studies on macro- and microstructures of highly concentrated water-in-oil emulsions (gel emulsions). Langmuir **9**(6), 1479–1482 (1993)
59. J. Nestor et al., Facile synthesis of meso/macroporous dual materials with ordered mesopores using highly concentrated emulsions based on a cubic liquid crystal. Langmuir **29**(1), 432–440 (2013)
60. D.E. Smith et al., The bacteriophage phi 29 portal motor can package DNA against a large internal force. Nature **413**(6857), 748–752 (2001)
61. S. Tzlil et al., Forces and pressures in DNA packaging and release from viral capsids. Biophys. J. **84**(3), 1616–1627 (2003)
62. P.R. Banerjee et al., Cataract-associated mutant E107A of human gamma D-crystallin shows increased attraction to alpha-crystallin and enhanced light scattering. Proc. Natl. Acad. Sci. USA **108**(2), 574–579 (2011)
63. J.J. McManus et al., Altered phase diagram due to a single point mutation in human gamma D-crystallin. Proc. Natl. Acad. Sci. USA **104**(43), 16856–16861 (2007)
64. J.F. Joanny, et al., The actin cortex as an active wetting layer. Eur. Phys. J. E. **36**(5) (2013)
65. G. Calderó et al., Studies on controlled release of hydrophilic drugs from W/O high internal phase ratio emulsions. J. Pharm. Sci. **99**(2), 701–711 (2010)
66. G. Calderó, C. Solans, Polymeric O/W nano-emulsions obtained by the phase inversion composition (PIC) Method for biomedical nanoparticle preparation, in *Emulsion Formation and Stability* (2013), pp. 199–207
67. P. Izquierdo et al., A study on the influence of emulsion droplet size on the skin penetration of tetracaine. Skin Pharmacol. Physiol. **20**(5), 263–270 (2007)
68. S. Theisinger et al., Encapsulation of a fragrance via miniemulsion polymerization for temperature-controlled release. Macromol. Chem. Phys. **210**(6), 411–420 (2009)
69. D.L. Berthier, A. Herrmann, L. Ouali, Synthesis of hydroxypropyl cellulose derivatives modified with amphiphilic diblock copolymer side-chains for the slow release of volatile molecules. Polym. Chem. **2**(9), 2093–2101 (2011)
70. M. Philippe, *Cosmetic Use of Amphoteric Polysaccharide Compounds Containing Cationic Polymer Chain*, L'Oreal, Editor (2005)

71. J.M. Lehn, N. Giuseppone, A. Herrmann, *Imine Based Liquid Crystals for the Controlled Release of Bioactive Materials,* Firmenich, Editor (2007)
72. C. Lemoine et al., *Solid Antiperspirant and/or Deodorant Composition in the Form of a Water-in-Oil Emulsion Based on Silicone Emulsifiers and on Waxes; Method for Treating Body Odours,* L'Oreal, Editor (2010)
73. T. Harnsilawat, R. Pongsawatmanit, D.J. McClements, Stabilization of model beverage cloud emulsions using protein- polysaccharide electrostatic complexes formed at the oil-water interface. J. Agric. Food Chem. **54**(15), 5540–5547 (2006)
74. E. Dickinson, Milk protein interfacial layers and the relationship to emulsion stability and rheology. Colloids Surf. B **20**(3), 197–210 (2001)
75. D. Rousseau, Fat crystals and emulsion stability—A review. Food Res. Int. **33**(1), 3–14 (2000)
76. T.D. Dimitrova, F. Leal-Calderon, Rheological properties of highly concentrated protein-stabilized emulsions. Adv. Colloid Interface Sci. **108–109**, 49–61 (2004)
77. R. Mezzenga et al., Understanding foods as soft materials. Nat. Mater. (2005)
78. C.G. de Kruif, F. Weinbreck, R. de Vries, Complex coacervation of proteins and anionic polysaccharides. Curr. Opin. Colloid Interface Sci. **9**(5), 340–349 (2004)
79. F. Weinbreck, M. Minor, C.G. De Kruif, Microencapsulation of oils using whey protein/gum arabic coacervates. J. Microencapsul. **21**(6), 667–679 (2004)
80. M.B. Munk et al., Stability of whippable oil-in-water emulsions: Effect of monoglycerides on crystallization of palm kernel oil. Food Res. Int. **54**(2), 1738–1745 (2013)
81. L. Day, Proteins from land plants—Potential resources for human nutrition and food security. Trends Food Sci. Technol. **32**(1), 25–42 (2013)
82. F. Caruso et al., Enzyme encapsulation in layer-by-layer engineered polymer multilayer capsules. Langmuir **16**(4), 1485–1488 (2000)
83. Y. Wang, F. Caruso, Mesoporous silica spheres as supports for enzyme immobilization and encapsulation. Chem. Mater. **17**(5), 953–961 (2005)
84. X. Qiu et al., Studies on the drug release properties of polysaccharide multilayers encapsulated ibuprofen microparticles. Langmuir **17**(17), 5375–5380 (2001)
85. C.S. Peyratout, L. Dähne, Tailor-made polyelectrolyte microcapsules: From multilayers to smart containers. Angewandte Chemie—International Edition **43**(29), 3762–3783 (2004)
86. N. Vilanova et al., Fabrication of novel silicone capsules with tunable mechanical properties by microfluidic techniques. ACS Appl. Mater. Interfaces **5**(11), 5247–5252 (2013)
87. H.N. Yow, A.F. Routh, Formation of liquid core-polymer shell microcapsules. Soft Matter **2**(11), 940–949 (2006)
88. Y. Wang, A.S. Angelatos, F. Caruso, Template synthesis of nanostructured materials via layer-by-layer assembly. Chem. Mater. **20**(3), 848–858 (2008)

Chapter 2
Fundamentals of Electrochemistry, Corrosion and Corrosion Protection

Christian D. Fernández-Solis, Ashokanand Vimalanandan, Abdulrahman Altin, Jesus S. Mondragón-Ochoa, Katharina Kreth, Patrick Keil and Andreas Erbe

Abstract This chapter introduces the basics of electrochemistry, with a focus on electron transfer reactions. We will show that the electrode potential formed when a metal is immersed in a solution is most of the time not an equilibrium potential, but a mixed potential in a stationary state. This mixed potential formation is the basis of corrosion of metals in aqueous solutions. Organic coatings are introduced as protecting agents, and several types of coatings are discussed: classical passive coatings, and active coatings as modern developments. Three electrochemical techniques, which are commonly used to asses the protecting properties of coatings, are shortly introduced as well: linear polarisation measurements, electrochemical impedance spectroscopy, and scanning Kelvin probe measurements.

2.1 Basics of Electrochemistry

Electron transfer reactions are wide-spread in nature, e.g. in the respiratory chain, they are important technologically, e.g. in electrolysers and for metal plating, and they contribute to the degradation of materials, in corrosion processes of metals. This chapter shall serve as an introductory text into the basic concepts, with a special focus on their importance in the field of corrosion science. Electrochemistry is quite an old science, and hence a number of good textbooks are available for more detailed introductions of the fundamental concepts [1–3]. It is worth pointing out that there are practically important corrosion mechanisms which are not at all based on electrochemical reactions. Details are available in dedicated textbooks [4, 5].

C.D. Fernández-Solis · A. Vimalanandan · A. Altin · J.S. Mondragón-Ochoa · A. Erbe (✉)
Max-Planck-Institut für Eisenforschung GmbH, Max-Planck-Str. 1,
40237 Düsseldorf, Germany
e-mail: a.erbe@mpie.de

K. Kreth · P. Keil
BASF Coatings GmbH, 48165 Münster, Germany

© Springer International Publishing Switzerland 2016
P.R. Lang and Y. Liu (eds.), *Soft Matter at Aqueous Interfaces*,
Lecture Notes in Physics 917, DOI 10.1007/978-3-319-24502-7_2

2.1.1 Electrostatic Potentials at Interfaces

Interfaces, in particular aqueous interfaces, are almost always charged, e.g. by dissociation of surface groups, or adsorption of ions from solution. On conductive substrates, the charge state of an interface may be actively controlled, as will be discussed below. Around charged interfaces, a static electric field is present, which affects the distribution of ions in solution around this interface. The rather complex multibody-interactions between solvent, mobile charges on both sides of the phase boundary, and stationary atoms leads to a complicated interfacial structure, which is schematically shown in Fig. 2.1 [6]. This interface is often referred to by the misleading term "double layer" (see also Chap. 4 by G. H. Findenegg).

Closest to the interface is a structured layer containing adsorbed molecules of the solvent and other adsorbed species. The region where the electric charges of the absorbed ions are allocated is called inner Helmholtz layer; this region is at a distance d_1 (Fig. 2.1). Solvated ions can approach the interface up to a distance d_2. The region where the electric charges of the solvated ions are located is called outer Helmholtz layer. Due to thermal motion in the system, theses solvated ions are distributed in a three dimensional region ranging from the outer Helmholtz layer to the bulk. This region is often referred to as "diffuse layer". Counter-ion distribution

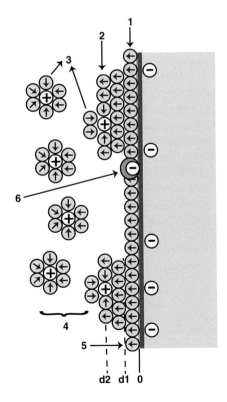

Fig. 2.1 Schematic representation of the layers at a solid/liquid interface. *1* Inner Helmholtz layer, *2* Outer Helmholtz layer, *3* Solvated ions (cations), *4* Diffuse layer, *5* Electrolyte solvent, *6* Specifically adsorbed ions. Drawing inspired by [7]

in the diffuse layer is important for some processes at a variety of interfaces (e.g., solid electrodes, biomolecules, etc.) and in technological applications as well (e.g., corrosion, paints, etc.). A simple quantitative approach to the description of the diffuse layer, balancing entropy against mean-field electrostatic attraction, is the Poisson-Boltzmann equation, which is extensively discussed—including its limitations—in [8]. Because deviations from the classic picture are found experimentally (e.g. [9]), modern conceptual works discuss in more detail solvation effects, fluctuations and ion correlation [10–14].

2.1.2 Electrochemical Potential

As an interface is charged, the work needed for a distribution of charges needs to be considered when analysing the total free enthalpy or free energy of a system. For this purpose, in charged systems, the electrochemical potential takes the role of the chemical potential.

The electrochemical potential $\bar{\mu}_i$ is defined as the mechanical work required to bring 1 mol of ions with valency z and hence charge ze from a standard state to a specified electrical potential and concentration. Thermodynamically, it is a measure of the chemical potential that takes into account the electrostatic contributions; it is expressed as energy per mole,

$$\bar{\mu}_i = \mu_i + z_i F \Phi \qquad (2.1)$$

where μ_i is the chemical potential of species i without considering the charge, F is the Faraday constant and Φ the local electrostatic potential. The first term of Eq. 2.1 takes into account the chemical potential of the species i (see Chap. 8 by R. Sigel for a thorough discussion on the chemical potential in relation to the formation of mixtures), while the second term includes the free enthalpy change brought by altering the potential Φ of the phase in which the charged species is located. Φ is also referred to as the inner potential or Galvani potential of the phase. This approach is analogous to the addition of the interface term to the chemical potential, described in detail in Chap. 4 by G. H. Findenegg.

Bringing two phases α and β with mobile charges into contact (consider metal/salt solution as an example) will result in an equilibrium where the electrochemical potentials of the two phases will equalise in a similar fashion as the chemical potentials for uncharged systems,

$$\bar{\mu}^\alpha = \bar{\mu}^\beta \qquad (2.2)$$

For simplicity, let's consider a single-valent system, for which we obtain, using Eq. 2.1,

$$\Phi^\alpha - \Phi^\beta = \frac{\mu^\beta - \mu^\alpha}{F}. \tag{2.3}$$

The result is the built-up of an electrostatic potential across the interface. The electrostatic potential difference between the two phases is the electrode potential,

$$E_{cell} = \Phi^\alpha - \Phi^\beta. \tag{2.4}$$

We need a certain reference to quantify the potential, which is an arbitrarily chosen system, for which different choices exist in the literature [15]. In analogy to the standard term in the chemical potential, to which the concentration dependence is added, we can "dump" the standard terms into a standard electrode potential E^0 for each redox system. These potentials are published in the literature typically as potential differences against the standard hydrogen electrode. They can be found e.g. in the Handbook of Chemistry and Physics [16]. The potentials are usually reported as reduction potentials, and represent the tendency of a certain species to "obtain" one or more additional electrons. The higher the potential, the higher is this tendency. The standard electrode potential is therefore a quantity similar to the free enthalpy of formation of a certain species.

The electrode potential E_{cell} is actually the potential of a single electrode. When putting two of these half-cells into electrical contact, electrons can flow through an external circuit and balance the electrode potential differences which are present. At the same time, a chemical transformation occurs: one species is reduced and one is oxidised. In one of the half cells, there will thus occur oxidation, i.e. electron "loss" of the atoms or molecules, and on the other side, there will be reduction, i.e. "gain" in electrons for a certain species. As other chemical reactions generate heat or light, redox reaction can be used to generate an electrical current, and likewise, electrical current can be used to "drive" redox reactions in a certain direction. The half cell, in which oxidation occurs, is called the "anode", and the half-cell, in which reduction occurs is the "cathode". Note that these definitions rely on the nature of the reaction that is occurring, not on the direction of the current flow or the charge. The difference in standard free enthalpy is related to the difference in standard electrode potentials to

$$\Delta G^\circ = -zFE_{cell}^\circ. \tag{2.5}$$

If $E_{cell}^\circ > 0$, the process is spontaneous, as e.g. in galvanic cells, batteries or corrosion of metals. On the other hand, if $E_{cell}^\circ < 0$, the reverse reaction is spontaneous, as e.g. in electrolytic cells.

Moving away from the standard standard state, Eq. 2.5 keeps its validity. So, at each concentration of involved species,

$$\Delta G = -zFE_{cell}. \tag{2.6}$$

For a redox reaction of the type

$$\alpha Ox + ze^- \rightarrow \gamma Red \tag{2.7}$$

the Nernst equation determines the electrode potential for an individual half-cell as

$$E = E^0 - \frac{RT}{nF} \ln \frac{[\text{Red}]^\gamma}{[\text{Ox}]^\alpha}, \tag{2.8}$$

where the activities may be approximated as concentrations in dilute solution. Here we dropped the subscript $_{\text{cell}}$ for convenience. Equipped with the Nernst equation, we can determine electrode potential differences, and hence free enthalpy differences from equilibrium, from tabulated standard electrode potentials, knowing activities/concentrations of dissolved species [1].

2.1.3 Currents Are a Measure of Reaction Rates

Electric current is the flow rate of electric charge q through a system, where t denotes the time,

$$I = \frac{dq}{dt} \tag{2.9}$$

In interface science, normalising the current in Eq. 2.9 by the interface area A is in general convenient, introducing current density $i = I/A$, i.e. current per unit area of electrode. The current through an interface is thus a convenient measure of the rate of electron transfer. If there is only a single electron transfer going on, the current density is directly related to the rate of the chemical reaction. It is possible to relate to the current through Faraday's law

$$I = \frac{dq}{dt} = \frac{nzF}{t} \tag{2.10}$$

to the total amount n of transformed substance, e.g. in a reaction like in Eq. 2.7. The magnitude of the current flowing at any potential depends on the kinetics of electron transfer. (Alternatively, in practise it often depends on the kinetics of transport to the interface. This case, is, however, not of interest in the understanding of the mechanistic aspects of reactions, which is why it is not considered here.) At any electrode potential, the measured current density is given as the sum of an anodic and a cathodic partial current

$$i = i_a + i_c. \tag{2.11}$$

Here, anodic current i_a corresponds to the current from reaction in Eq. 2.7, while the cathodic current i_c corresponds to the current of the back reaction to Eq. 2.7. In equilibrium, i.e. at the potential given by the Nernst equation, no net reaction occurs, hence no net current flows and $i_a = -i_c$, i.e. the rate of forward and back reaction are equal.

2.1.4 Equilibrium Potential and Open Circuit Potential, a Mixed Potential

If we immerse a metal plate into a solution of its salt, we expect either the metal to dissolve (according to the reverse of reaction in Eq. 2.7), or salt ions to deposit as metal, according to reaction 7, until the equilibrium concentration is reached in the solution, as given by the Nernst equation, Eq. 2.8. The electrode potential that forms at the metal/electrolyte interface after immersion of the metal into solution, the so-called open circuit potential (OCP) is hence the equilibrium potential.

However, if more than one electron transfer reaction takes place at the electrode surface, the open circuit potential is a mixed potential defined by the kinetics of all simultaneous electrochemical reactions. (Candidates are, e.g. the decomposition of the solvent, e.g. through evolution of H_2 or less likely O_2 from water, or the reduction of O_2, as typically found in corrosion.) At OCP, the anodic and cathodic current have same magnitude but opposite directions ($i_a = -i_c$), resulting in zero net current through the interface, which is not in equilibrium.

The corrosion of electrode surfaces (see also below) is explained by this theory of mixed potentials and it is based on the independence of the partial anodic and cathodic reactions. This theory states that for an electrode on which more than one electrochemical reaction takes place simultaneously [17], the measurable current-potential curves can be expressed as

$$i(E) = \sum_i |i_{a,i}(E)| - \sum_i |i_{c,i}(E)| \tag{2.12}$$

where summation is over all partial reactions.

Figure 2.2a shows the ideal situation of an electron transfer reaction, where there is an exchange current i_0 at an equilibrium potential E_{eq}, that spontaneously forms when immersing an electrode into a solution. E_{eq} is a result of the contribution of the anodic and cathodic half reactions.

In Fig. 2.2b, we can observe that at open circuit conditions, the mixed potential has an intermediate value between the equilibrium potential of two reactions, O/R and O'/R', where O stands for the oxidised and R for the reduced species. This mixed potential is determine by the kinetics of each partial reaction. (The potential dependence of the rates will be discussed in the next section.) In this case, the mixed potential corresponds to two different overpotentials, η_1 and η_2, where each η represents the difference between applied potential E and the equilibrium (Nernst)

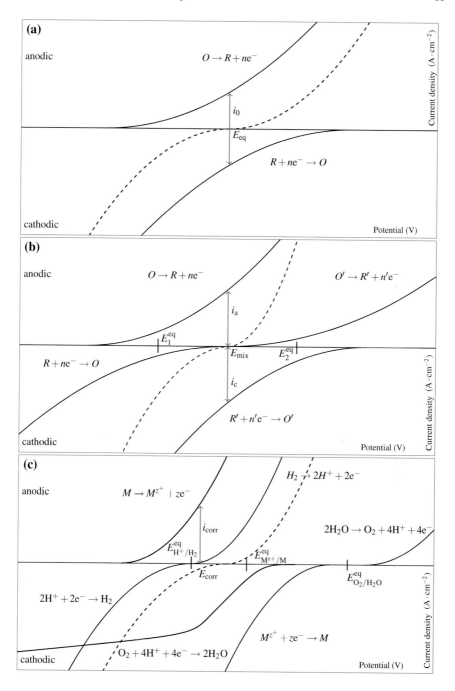

◀ **Fig. 2.2** Schematic partial current densities (ordinate) as function of electrode potentials (abscissa) for different situations. The *dashed line* shows the macroscopically observed current density-potential curve for the system. **a** Formation of equilibrium potential E_{eq} at an inert electrode with one redox process. **b** Formation of mixed potential E_{mix} at an inert electrode with two simultaneous redox processes, with formal equilibrium potentials $E_1^{(eq)}$ and $E_2^{(eq)}$. **c** Formation of corrosion potential E_{corr} by simultaneous occurrence of metal dissolution, hydrogen evolution and oxygen reduction, each of them with a given equilibrium potential. The two reactions involving oxygen and hydrogen typically show some overpotential, i.e. are not fully reversible, which is why their individual exchange current densities have been put to zero

potential E_{eq}, $\eta = E - E_{eq}$. Consequently, the forming potential cannot be related to either of the equilibrium potentials of the involved reactions [18].

In Fig. 2.2c, one of the partial electrode reaction is the dissolution of the electrode and the other half corresponds to the deposition on the electrode. However, deposition is not the only counter reaction to dissolution. Especially in corrosion, we typically have oxygen reduction as cathodic driving reaction for the anodic reaction. Further, in water we also have hydrogen evolution, which is also in some cases (e.g. in acidic solution) the important counter reaction. Consequently, the OCP is a corrosion potential different from an equilibrium potential, and the system dissolves at a rate given by the corrosion current i_{corr}.

2.1.5 Relation Between Potential and Current—Electron Transfer Reactions

The partial anodic and cathodic current densities are dependent on the concentration of the electroactive species at the site of electron transfer as typically encountered in classical chemical kinetics,

$$i_a = zF\kappa_a c_{Red} \qquad i_c = -zF\kappa_c c_{Ox} \tag{2.13}$$

The rate constants κ_a and κ_c vary with the overpotential η in first approximation in an exponential manner as

$$\kappa_a = \kappa_{a0} \exp\left(\frac{\alpha_A nF}{RT}\eta\right) \qquad \kappa_c = \kappa_{c0} \exp\left(-\frac{\alpha_C nF}{RT}\eta\right), \tag{2.14}$$

as according to the laws of thermodynamics of irreversible processes, all rates depend in first approximation exponentially on the distance of the driving force from equilibrium. α_A and α_C are constants between 0 and 1, known as the transfer coefficients for anodic and cathodic reactions, respectively. Typically, $\alpha_A \approx \alpha_C \approx 0.5$, at least on metal surfaces.

Noting the definition of the exchange current density, $i_0 = -i_c = i_a$ at $\eta = 0$ and rearranging we have the Butler-Volmer equation for the overall potential-dependence of the current density

$$i = i_0 \left[\exp\left(\frac{\alpha_A z F}{RT} \eta \right) - \exp\left(-\frac{\alpha_C z F}{RT} \eta \right) \right] \qquad (2.15)$$

The Butler-Volmer equation is a fundamental equation of electrode kinetics [19].

In practice, there are often-used limiting forms of Eq. 2.15. At sufficiently high positive overpotentials ($i_a \gg i_c$) we find Eq. 2.16 and at high negative overpotentials ($i_c \gg i_a$) Eq. 2.17,

$$\log_{10}(i) = \log_{10}(i_0) + \frac{\alpha_A z F}{\ln(10)RT} \eta \qquad (2.16)$$

$$\log_{10}(-i) = \log_{10}(i_0) - \frac{\alpha_C z F}{\ln(10)RT} \eta \qquad (2.17)$$

These equations present a simple method to determine exchange current density and a transfer coefficient from a linear relation and are known as Tafel equations. A useful plot is shown in idealised form in Fig. 2.3. From this plot, the OCP can be determined easily, and from the slopes of the linear region, the electron transfer coefficients α_A and α_C can be obtained. In the analogue plot with a linear current axis, the inverse slope around the OCP has the units of a resistance and is called the polarisation resistance R_P.

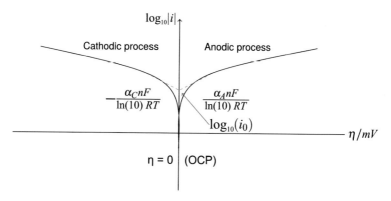

Fig. 2.3 Schematic representation of Tafel equations. On the *left side*, the cathodic process and on the *right side* the anodic process. In this particular case, the OCP equals the equilibrium potential where $i = 0$

Linear polarisation experiments, i.e. measurements of curves such as shown in Fig. 2.3, are widely used to determine corrosion currents. Corrosion currents can be transformed to corrosion rates, because oxidation at a certain current corresponds to a certain loss of material, though care must be taken when extrapolating. Experimentally, a sample is polarised in a range around its corrosion potential and the current response is measured. The slopes obtained in Fig. 2.3 for Eqs. 2.16 and 2.17 are called β_a and β_c for the anodic and cathodic processes, respectively. Extrapolating to the intercept of these linear regions, the exchange current density can be determined.

The corrosion or exchange current density i_{corr} can then be determined as [20]

$$i_{corr} = \frac{\beta_a \beta_c}{\ln(10)R_p(\beta_a + \beta_c)} \tag{2.18}$$

The validity of the analysis from the Tafel plot critically depends on the exponential relation between current and overpotential, i.e. more general in chemical kinetics the relation between driving force of a chemical reaction, and its rate constant. Such relations can be obtained by the application of transition state theory. The most famous example for electron transfer reaction is the Marcus theory, which represents an extension of the Franck-Condon principle to chemical reactions. (The Franck-Condon principle deals with the relaxation of the electron configuration after excitation. In electron transfer reactions, one goes one step further and transfers the electron.) The barrier for electron transfer reactions in this Marcus picture basically originates from the reorganisation of the solvation shell around the ions, because even simple ions change their charge during electron transfer, and the solvation shell needs to adapt to this change. The barrier in this theory is obtained as an intersection of two parabola (in the simplest case), which represent the interaction potential of the solvent with a central ion [21]. The central quantity in the theory is the solvent reorganisation energy, E_{re}. With a barrier from the Marcus theory, the relation between current and overpotential becomes

$$i \propto \left[\exp\left(\frac{zF\eta}{2RT}\right) \exp\left(\frac{(zF\eta)^2}{2E_{re}RT}\right) - \exp\left(-\frac{zF\eta}{2RT}\right) \exp\left(-\frac{(zF\eta)^2}{2E_{re}RT}\right) \right], \tag{2.19}$$

which differs from the Butler-Volmer Eq. 2.15 by the presence of a term that goes with η^2 in the exponent. For sufficiently small overpotentials, the Butler-Volmer form is obtained as limiting case, but at larger overpotentials the square term may become dominant.[1]

---

[1]An alternative approach dumps the η^2 dependence into the transfer coefficients α_c and α_a by making them potential-dependent.

2.2   Electrochemical Methods in Corrosion Science

In Sect. 2.1.4 we saw that corrosion of metals is often essentially an electrochemical phenomenon. Hence, electrochemical techniques are used to characterise corrosion processes. This chapter introduces two important experimental techniques, electrochemical impedance spectroscopy (EIS) and scanning Kelvin probe.

2.2.1   Electrochemical Impedance Spectroscopy (EIS)

In EIS, the system is excited by application of a typically small alternating perturbation of the applied electrode potential, typically in a range of frequencies f between 10^6 and 10^{-4} Hz [2]. The frequency-dependent electrical current response of the system is analysed. (An inverse variant exists, in which an alternating current is forced upon a system, and the electrode potential response is analysed. This variant shall not be discussed here.) The measured impedance can be given as a complex number, with a modulus, and a phase shift between applied potential and response current. In many cases, the frequency-dependent impedance can be described by a combination of electrical circuit elements, the so-called equivalent circuit. Therefore, the impedance of standard passive electrical circuit elements shall shortly be defined, before discussing electrochemical applications [22].

Ohm's law describes the proportionality between applied voltage U, and flowing current I, with resistance R as constant of proportionality,

$$I = \frac{1}{R}U. \tag{2.20}$$

For Ohmic conductors, this equation is valid also if an alternating voltage $U(t)$, e.g. with sinusoidal modulation with time t, is applied. If a sinusoidal voltage with a certain frequency f is applied, the corresponding alternating current has a certain amplitude and a phase shift ϕ with respect to its exciting current. The resistance in this situation is called the impedance Z,

$$Z = \frac{U(t)}{I(t)} = \frac{U_0 \sin(\omega t)}{I_0 \sin(\omega t + \phi)}, \tag{2.21}$$

with angular frequency $\omega = 2\pi f$.

If an electrical circuit consists only of a resistor, the phase shift between applied voltage and response current is zero. However, replacing the resistor with a capacitor or an inductor causes a phase shift of the impedance as schematically depicted in Fig. 2.4. Most electrochemical systems analysed by EIS can in first approximation be described by a combination of capacitors and resistors, where the so-called Randles circuit, depicted in Fig. 2.5, is the most useful circuit.

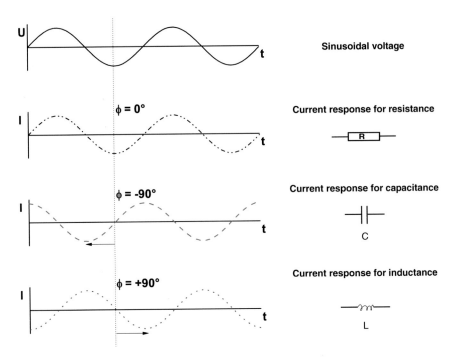

Fig. 2.4 Schematic depiction of applied sinusoidal voltage and the corresponding current responses in the case of a resistor, capacitor and inductor

As often done in spectroscopy (see also Chap. 14 by A. Erbe et al.), a complex plane representation of the voltage can be used, in which

$$U(t) = U_0 e^{i\omega t}, \qquad (2.22)$$

with $i = \sqrt{-1}$. Likewise,

$$I(t) = I_0 e^{i(\omega t - \phi)}, \qquad (2.23)$$

which allows for a simplified description of the ratio in Eq. 2.21 as

$$Z = \frac{U_0 e^{i\omega t}}{I_0 e^{i(\omega t - \phi)}} = |Z| e^{i\phi} = |Z|[\cos(\phi) + i\sin(\phi)], \qquad (2.24)$$

where $|Z| = U_0/I_0$. To analyse and interpret complex impedance measurements, a graphical depiction of the resulting spectrum is extremely useful. Two frequently used plots are Nyquist plots and Bode plots, depicted in Fig. 2.5 for the simple circuit elements. In the Nyquist plot, the real part of the impedance Re(Z) is drawn at the horizontal axis and the imaginary part IM(Z) at the vertical axis of a Cartesian coordinate system. The disadvantage of this plot is that the frequencies are not

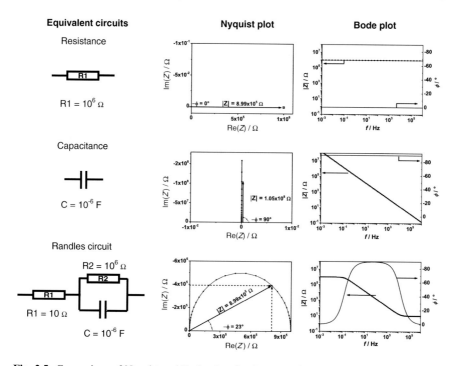

Fig. 2.5 Comparison of Nyquist and Bode plots for the respective to the equivalent circuits

directly accessible. This disadvantage is circumvented by using the Bode plot, where the modulus of impedance and the phase shift are plotted as a function of frequency. Figure 2.5 gives an overview of the Nyquist and Bode plots obtained in accordance to the electrical circuits. A further variant is the so-called Cole-Cole plot, where $\text{Re}(Z^{-1})\omega^{-1}$ is plotted against $-\text{Im}(Z^{-1})\omega^{-1}$. In this case the axes have the unit of capacitance, and hence a capacitance can directly be obtained from the plots.

All equations mentioned previously are in first approximation also valid for an electrochemical system. The voltage U is replaced by the electrode potential E and current I is replaced by the current density i. The analogy to electrical circuit elements works only if a linear relationship between potential and current density exists. As discussed in Sect. 2.1.5, this relationship is in general not linear. The actual exponential relation can be approximated by a linear relation only in a rather small electrode potential range. Thus, there will be an error in the measurement due to the fact that the current will not be sinusoidal any more. But in a very narrow region around a certain applied potential or OCP, typically ± 10 mV (depending on application between 5 mV and 30 mV root mean square) the relationship between potential and current density is approximated as linear. Several possible applications exist in electrochemistry, e.g. in-depth analysis of reaction rates for corrosion phenomena, or an analysis of the corrosion resistance and water uptake of a coating, as will be shown in Sect. 2.4.2. More advanced measurement schemes make use of the non-linearity to access detailed information on electron transfer [23].

The advantage of EIS compared to other electrochemical techniques is the fact that one can get a relaxation picture of electrochemical systems under quasi steady state conditions. The reason is that electrochemical reactions especially in corrosion science involve several electrical and ionic processes at the interface between substrate and electrolyte and that every reaction has its own relaxation time. While the rearrangement of water molecules or solvated ions in an electrolyte under potential control is an extremely fast reaction (e.g. because of the Grotthuss mechanism) with relaxation times in the range of 1 ms to 10 s, the diffusion controlled dissolution of a metal, which involves diffusion along the metal electrolyte interface, is a much slower process. Hence, the resistance of an electrolyte can be measured at high frequencies and the dissolution phenomena at lower frequencies.

Figure 2.6 shows a typical impedance spectrum of aluminium recorded in an acetate buffer at OCP. The Randles circuit introduced in Fig. 2.5 is used to extract data about the aluminium oxide. The circuit consists of a resistance simulating the electrolyte resistance R_{el}, and an RC element with R_{ox} as the oxide resistance and C_{ox} as the oxide capacitance. In the high frequency region, the electrolyte resistance dominates the spectrum. In the mid-frequency region, $|Z|$ increases due to the fact that more and more the capacitive character of the oxide is contributing to the impedance, as it can be also seen in the phase shift (note, that the current flow through a capacitor is frequency dependent—while the capacitor is short circuited at high frequencies, at lower frequencies it acts as an insulator). At lower frequencies the current flows completely along the resistances R_{el} and R_{ox} and the oxide resistance is the dominating factor. More detailed information on characterizing aluminium oxides and kinetic of growth with electrochemical techniques can be found in [24] and in the references cited therein.

Using a Randles circuit is often the starting point for building an equivalent circuit for describing and simulating more complex phenomena, like coatings. We shall return to this description and its application to protecting organic coatings in

Fig. 2.6 Electrochemical impedance spectrum of aluminium in a 0.1 M acetate buffer pH = 6 and the values received by fitting the curve with a Randles circuit

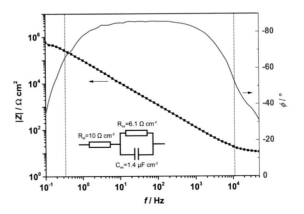

Sect. 2.4.2. Nevertheless, the usage of equivalent circuits has its limits, and an ongoing discussion concerns the validity of separating capacitive from Faradayic currents [25]. Further in-depth discussion on the use and validity of equivalent circuits can be found elsewhere [26, 27].

2.2.2 The Scanning Kelvin Probe (SKP)

While it is rather straightforward to obtain the electrode potential of a metal if the metal is immersed in electrolyte, many practical problems involve corrosion of samples in contact with an atmosphere. Electrode potentials, and to a certain extend also currents, can be obtained under atmospheric conditions by Kelvin probe measurements. Consequently, the scanning Kelvin probe (SKP) has been a valuable and versatile tool for studying corrosion problems of metals under thin electrolyte layers [28–33], and at buried interfaces beneath organic coatings [34, 35]. In principle, the Kelvin probe measures the work function [36] or, for non metallic surfaces, the surface potential of a sample. The work function itself is extremely sensitive to changes in the surface conditions and is directly influenced by e.g. adsorbed layers, charges and functional groups at the surface, surface and bulk contaminations, and for semiconducting oxides structure, doping and even imperfections [37–42].

The SKP permits a non-destructive in situ measurement of electrode potentials at buried polymer/metal interfaces with certain spatial resolution. The principle of the Kelvin probe is based on the vibrating capacitor method (see Fig. 2.7): a needle (that acts as a reference and is often referred to as "the probe") consisting of an inert corrosion-resistant metal, like Ni/Cr (80/20 wt%) alloy, is positioned approximately 50–100 μm close to the surface of the sample [43–46].

As soon as the needle and the sample are brought into electrical contact, the equilibration of the Fermi levels within the two metals will result in the formation of

Fig. 2.7 Working principle of the Kelvin probe

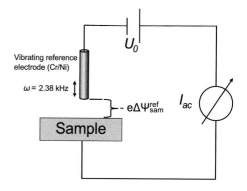

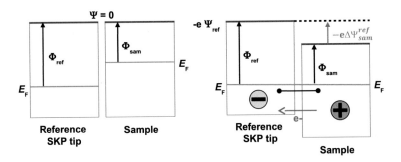

Fig. 2.8 Illustration of the Fermi level alignment of the the SKP tip and the sample. Corresponding measured Volta potential difference $\Delta\Psi^{ref}_{sample}$

a positive charge on the surface of one and a negative charge on the surface of the other metal. Thus, the needle and the sample form a capacitor and the charging of the sample with respect to the probe leads to a Volta potential difference $\Delta\Psi^{\text{ref}}_{\text{sample}}$ between the two metals (Fig. 2.8). By vibrating the needle, a current will be induced. If an external bias is applied between sample and the needle in a way that this current is compensated, surfaces will be uncharged again, the capacitance will go to zero and the original state will be obtained. Under such conditions the external bias voltage is identical to the Volta potential difference.

The work function describes the energy required to liberate an electron from an electrode's Fermi level to the vacuum level. For a metal in vacuum, the electron only has to pass the metal surface during this transfer. However, if the metal is covered by other phases, such as an aqueous electrolyte, a polymer layer or even a coated metal affected by wet deadhesion, the electron needs to pass additional interfaces. Thus, the measured Volta potential difference is strongly affected by additional potential differences across these interfaces. As a consequence, the Volta potential difference measured by the SKP is determined by the interfacial electrode potential. Therefore, changes within the interfacial structure can be detected. Depending on the different interfaces involved in the measurement, the measured Volta potential difference $\Delta\Psi^{\text{ref}}_{\text{sample}}$ can be correlated to the corrosion potential of the underlying metal E_{corr}. Practically the following cases can be distinguished. Metal surface covered by a liquid phase [47]:

$$E_{\text{corr}} = \left(\frac{W_{\text{ref}}}{F} - \chi_{\text{El}} - E^{\text{Ref}}_{1/2}\right) + \Delta\Psi^{\text{ref}}_{\text{sample}} \qquad (2.25)$$

Metal surface covered polymers with negligible dipole potentials [47]:

$$E_{\text{corr}} = \left(\frac{W_{\text{ref}}}{F} - \chi_{\text{Pol}} - E^{\text{Ref}}_{1/2}\right) + \Delta\Psi^{\text{ref}}_{\text{sample}} \qquad (2.26)$$

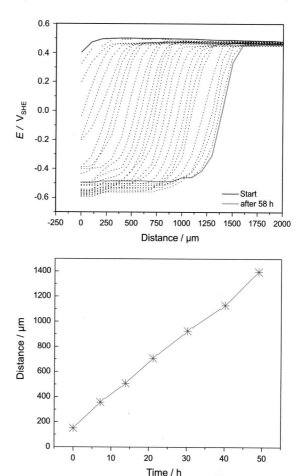

Fig. 2.9 *Top* Typical potential distribution of a polymer covered zinc substrate (as model for galvanised steel, i.e. steel covered with a metallic zinc layer) in humid air for different delamination times detected with the SKP (electrolyte in the defect: 1 M KCl). *Bottom* Time dependence of the delamination front position

Additional liquid phase between the metal and the coating (during deadhesion) [47]:

$$E_{corr} = \left(\frac{W_{ref}}{F} - \chi_{Pol} - E_{1/2}^{Ref} \right) + \Delta \Psi_{sample}^{ref} + \Delta \Phi_D \qquad (2.27)$$

where $W_{ref}, \chi_{El}, E_{1/2}^{Ref}, \chi_{Pol}, \Delta \Phi_D$ and F denote the work function of the reference (probe/needle), the surface dipole potential of the electrolyte, the half-cell potential of the reference (probe/needle), the surface dipole potential of the polymer surface, the Donnan potential of the polymer and the Faraday constant, respectively. These equations show two advantages of the SKP: (1) The measured Volta potential difference $\Delta \Psi_{sample}^{ref}$ is related to the corrosion potential in a linear manner; (2) The electrode potential of the buried interface can be determined across a dielectric medium of infinite resistance. Due to the linear nature of the equations, a calibration constant is required in order to calculate the electrode or corrosion potential from a Volta potential difference measurement. In practice, this constant can be obtained for aqueous electrolytes simply by measuring the $\Delta\Psi^{ref}_{sample}$ between the needle used for the measurements and a metal electrode (as sample) which is exposed to an electrolyte containing the metal cation in a defined concentration. Commonly, the calibration is performed against a Cu/CuSO$_4$ electrode [46].

In corrosion science, SKP has been used to study delamination kinetics and mechanisms of organically coated metals [30, 43–45, 48–50], also with control of surface chemistry [51], or morphology [52, 53]. In the following, an example for cathodic delamination will be given. In the presence of moisture and oxygen, the electrode potential of the buried coating/metal interface is changing in a well-defined manner with increasing distance from a defect. A defect can be an artificial scrape or defect, a cut edge, or a coating defect like a blister. For steel (including galvanized steel that is covered with a zinc layer), basically two potential levels characterise the SKP profiles measured during delamination. The potential of the defect, and of the potential of the intact coated area. In most cases, the potential close to the defect has a very negative value (pointing to cathodic processes), whereas far away from the defect predominantly positive potentials are observed (see Fig. 2.9). The progress of cathodic delamination is reflected by the lateral displacement of steep increase within the repeatedly measured potential profiles. The order of the delamination kinetics can be seen if the delamination front position is plotted against the exposure time: A linear relation suggests that the process of delamination is reaction controlled, whilst a hyperbolic (or square-root) relation suggests that the interfacial ion mobility is rate-determining. We will return to the discussion of the actual failure mechanisms in Sect. 2.4.3, after briefly looking at the chemical composition of the polymer coatings, which are frequently used in corrosion protection.

2.3 Types of Binder Resins for Passivating Organic Coatings

From the discussion in Sects. 2.1.4 and 2.1.5, it is clear that simple ways of protecting a metal against corrosion are to reduce the active area, to reduce the access of oxygen and finally also to reduce the access of water to the metal. Such reduction in accessibility is reached by barrier coatings, and organic coatings often are used as such. If the coating properly wets the substrate surface and shows good adhesion even in the presence of water and electrolyte, then sufficient protection can be achieved. To create that effect, the choice of an appropriate binder resin is crucial. These binder resins are generally organic polymers which can be classified with respect to their chemical structure. From an industrial point of view, the most important resins are hereby polymers made from epoxides, polyurethanes, polyesters, melamine formaldehyde resins, polyacrylates and phenolic polymers. The chemical background and the properties of these classes are described in more detail in the following paragraphs [54].

2.3.1 Epoxy Based Coatings

Resins for epoxy coatings are derived from compounds that contain one or more epoxy groups. These are derivatives of the three-membered, heterocyclic oxirane ring, which is also called ethylene oxide, epoxy ethane, 1,2-epoxide or oxacyclopropane. The methyloxirane form is also known as glycidyl group. Due to its high ring strain, the epoxy group is highly reactive and can therefore be converted with many different functional groups [55, 56].

The most basic and most frequently used epoxy resin is prepared by the reaction of bisphenol A with two moles of epichlorohydrine in alkaline medium to form the diglycidyl ether of Bisphenol A (DGEBA) according to Fig. 2.10. After reaction with the hydroxyl group, the epoxy ring is reformed by elimination of HCl, and is then available for further reactions. If it is converted again, a secondary hxdroxy group is created. By varying the ratio of Bisphenol A to epichlorohydrine, a broad variety in molecular weight can be obtained. Since the molecular weight of DGEBA is only 340 g/mol, it is available in liquid state. With a mean molecular weight of 1000 g/mol, a polymer made from the same reactants is solid at room temperature. On the one hand, low molecular weight molecules can be used to obtain thick coatings, because there is only little solvent that needs to evaporate. On the other hand, high molecular weight epoxy polymers have a higher strength, but must be dissolved in organic solvent to render them utilisable for manufacturing and the application processes [55, 56].

In most cases, epoxy polymers are cross-linked to improve the mechanical properties and the heat resistance of the coating. This process implies the usage of hardeners, which are agents that can react with the secondary hydroxyl group

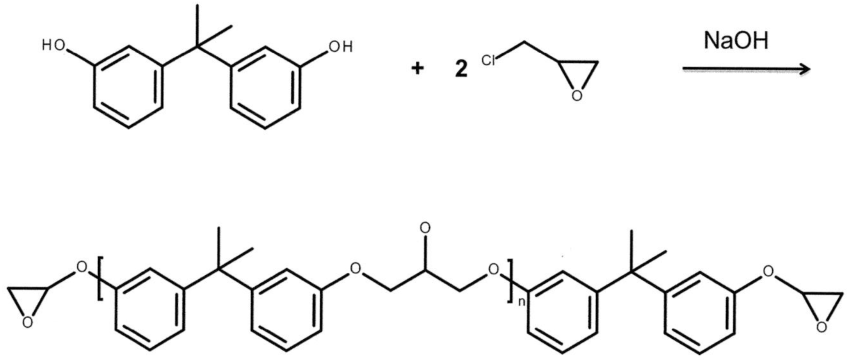

Fig. 2.10 Reaction of Bisphenol A with epichlorohydrine

and/or with the terminal epoxy groups. Examples of substances that can react with epoxy groups are amines, acids and acid anhydrides, alcohols and mercaptans. Furthermore, instead of reaction with the hardener, the terminal epoxy groups can also be used for further modification of the binder resin to tune the properties in a desired way [55, 56].

2.3.2 *Phenol Formaldehyde Resins*

Another binder resin can be obtained by the condensation reaction of phenol, or substituted phenols, with aldehydes, e.g. phenol (P) with formaldehyde (F). Products are classified as novolacs or resols. The first class is synthesized under alkaline pH conditions with an excess of F. The latter is produced in acidic medium with an excess of P. Both reactions are highly exothermic. The different architecture of the hereby achieved products is shown in Fig. 2.11. If the molar ratio of F and P is equal to one or higher (as for resol), every phenol is theoretically linked to another phenol by a methylene bridge and there are also additional aldehyde moieties available that can be used for later hardening. Having a lower amount of F, novolac only hardens with addition of a cross-linking agent. Phenol formaldehyde

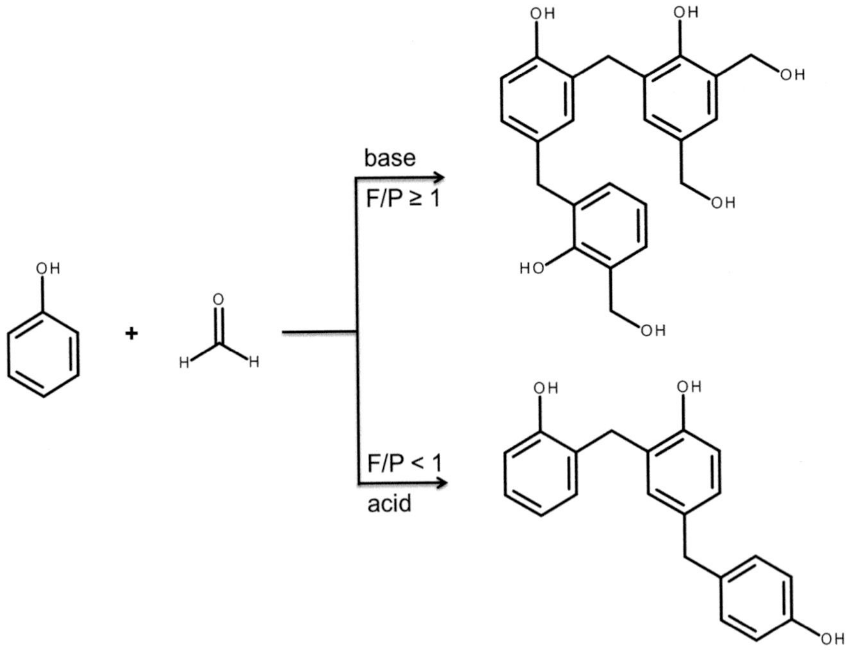

Fig. 2.11 Reaction of phenol and formaldehyde to resol and novolac

resin based coatings are utilized for high temperature applications and very humid and aggressive environments [55, 56].

2.3.3 Polyurethane Resins

Polyurethanes are materials which are traditionally formed by reaction of a compound containing two or more isocyanate groups with a polyol to form urethane bonds as shown in Fig. 2.12. The broad applicability of polyurethane coatings is due to the versatility in selection of monomeric materials from a huge list of diisocyanates, like hexamethylene diisocyanate, toluene diisocyanate or isophorone diisocyanate, and diols, macrodiols or polyols. Cross-linked polyurethanes provide hard and dense coatings with a very good chemical and moisture resistance and are the most versatile group of polymers in industry [55, 56].

2.3.4 Polyacrylate Resins

Polyacrylate resins are synthesized from ethylenically unsaturated monomers which are esterified derivatives from (meth)acrylic acid. These monomers can be equipped either hydrophic or hydrophilic and can insert special functionalities into the resin. The product is usually obtained by emulsion or solution polymerisation started by initiators that dissociate into free radicals under elevated temperature. In a chain reaction, an acrylic polymer is build-up in a very short time according to Fig. 2.13 [55, 56].

Commercially available polyacrylic resins are based on monomers like ethyl acrylate, ethyl acetate, n-butyl acrylate, methyl methacrylate, vinyl acetate, acrylic

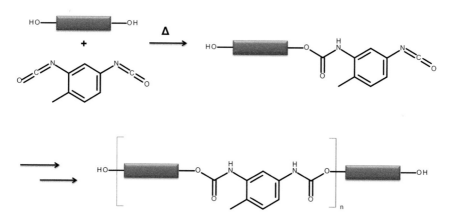

Fig. 2.12 Formation reaction of polyurethane from toluene diisocyanate and a macrodiol

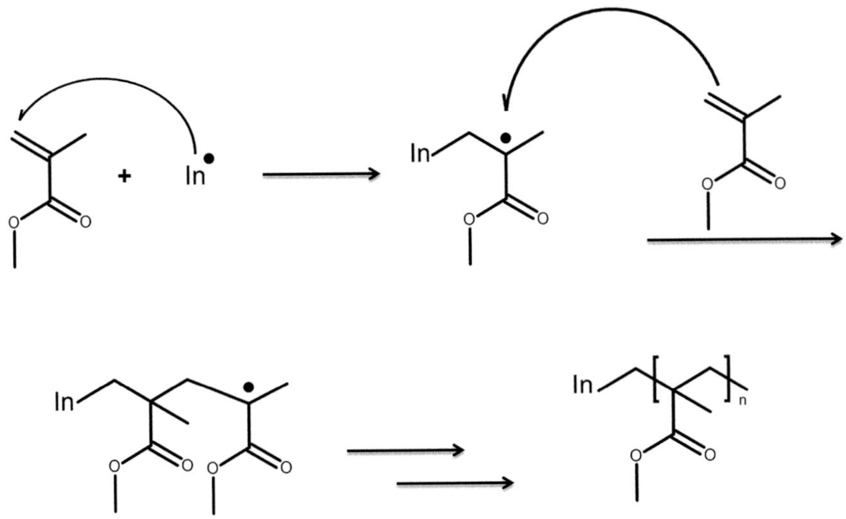

Fig. 2.13 Radical polymerisation of methyl methacrylate as example for the build-up of a polyacrylate

acid and other derivatives thereof. Furthermore, multifunctional acrylic monomers such as ethylene glycol, dimethacrylate, trymethylolpropane trimethacrylate, 1,4-butanediol dimethacrylate or neopentyl glycol dimethacrylate can be used in small amounts to act as cross-linkers. Usually, high molecular weight polymers can be created by this reaction. Chain transfer agents like mercaptans can be incorporated to control the formation of long chains [55, 56].

2.3.5 Polyester Resins

Polyesters resins are made from di-or polyols and dicarbonic acids or anhydrides via a step-wise condensation reaction according to Fig. 2.14. This procedure involves a long reaction time because the produced water has to be constantly distilled from the often high viscous reaction mixture. Appropriate monomers are phthalic anhydride, isophthalic anhydride, therephthalic anhydride, trimellithic anhydride, adipic acid, maleic anhydride, fumaric acid, glycol, 1,4-butane diol, trimethylolpropane, neopentyl glycol and many other acids or alcohols [55, 56].

Alkyd resins represent a special class of polyester resins, which contain mainly glycerin and different fatty acids combined with other dicarbonic acids and alcohols or diols. Hydroxyl functional oligo- or polyesters can also be used as building block e.g. for the synthesis of polyurethane resins. Polyesters with unsaturated double bonds can be used for later cross-linking with acrylic monomers [55, 56].

HO
OH
+
HO
OH
O
O
Δ
- H_2O
HO
O
OH
- n-1 H_2O
H
n

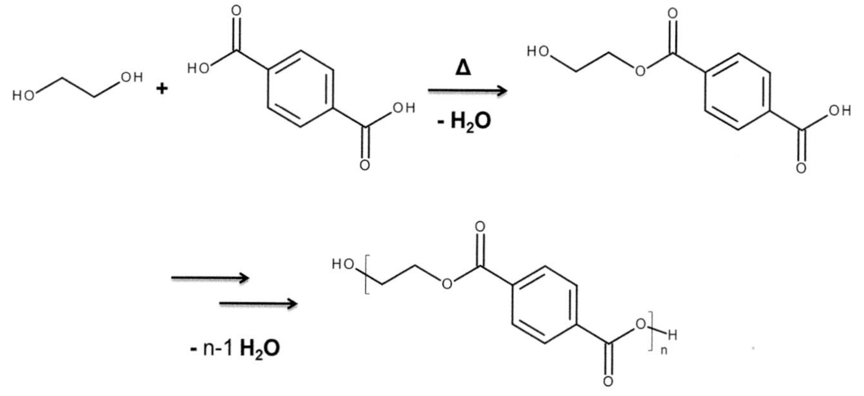

Fig. 2.14 Condensation reaction of ethyl glycol and therephthalic acid as example for polyester formation

Overall, polyesters can be designed with a broad variety of different properties and can therefore be used in various coating systems [55, 56].

2.3.6 *Melamine Formaldehyde Resins*

An important class of binders is obtained by the reaction of melamine with formaldehyde and further etherification to obtain multifunctional hardeners according to Fig. 2.15. Depending on the amount and character of modification, these show higher or lower reactivity with alcohols. A faster reaction at lower

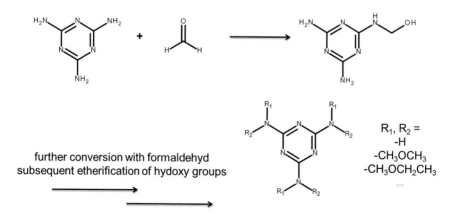

H_2N
NH_2
N
+
O
H
H
OH
further conversion with formaldehyd
subsequent etherification of hydoxy groups
R_1
R_2
R_1, R_2 =
-H
-CH_3OCH_3
-$CH_3OCH_2CH_3$
...

Fig. 2.15 Reaction of melamine and formaldehyde to form differently modified melamine formaldehyde resins

temperatures is usually achieved with more N-H-groups that are still available and shorter carbon chains within the ether segments [55, 56].

2.4 Coating Performance and Failure

2.4.1 Multilayer Coatings Are Applied in Real Coating Applications

In the automotive sector, multi-layer coatings are applied to fulfil corrosion protection standards [57]. Moreover, the whole spectrum of different and sometimes contrary properties desired for automotive coatings can be achieved and balanced only by the interplay of several layers. Usually, the top layer is represented by the clear coat. It provides not only the scratch resistance and the gloss, but protects also from all outer influences like humidity, sunlight, chemicals and natural compounds like bird droppings or tree resins. The colourful appearance and the metallic effects are delivered by the second layer from top, the base coat. Beneath that more "superficial" job, it provides adhesion between the upper and lower layer. The second-named is the filler that serves as a kind of protective shield against damage from stone chips. This flexible and thick layer also smooths the surface in preparation for the decorative layers and, to some content, provides protection against corrosion, owing to its barrier properties. The task of protection is mainly fulfilled by the lowest layer of a standard automotive coating: the cathodic electrodeposition coating (e-coat). Whereas all other layers are applied via spray application, this one is precipitated by an electrochemical approach. The car body is connected as cathode and the positively charged paint particles are attracted, thereby leading to a good surface coverage. Together with the pre-treatment and galvanization layer of the steel, the e-coat gives an appropriate protection against corrosion. The stack of these four layers is shown in Fig. 2.16, applied according to a standard procedure. Modern processes try to get rid of single layers to safe time, money and energy. Of course, the same good overall properties should be maintained [54].

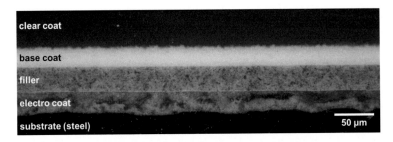

Fig. 2.16 Microscopic picture of a crosscut of a standard build-up of an automotive coating

For industrial coatings, the focus is a bit more on the side of protection than of decoration. For such applications, the clear coat and base coat functions are combined within a topcoat layer. Below that, the so called primer is responsible for corrosion resistance and adhesion to the metal. This layer contains inhibitor pigments to actively inhibit the corrosion at a defect.

2.4.2 Accessing Coating Barrier Property with EIS

All organic coatings are to some extent permeable to water and corrosive ions. The driving forces for water permeation are: (1) a concentration gradient (e.g. during immersion), (2) osmosis (e.g. from impurities at a coating/coating interface) or (3) capillary forces (e.g. defects in the coating due to inadequate curing or improper solvent evaporation). Corrosion underneath organic coatings could be basically classified in two categories: (1) Delamination and blistering of the coating due to clustering of water at the interface leading to wet deadhesion or osmotic blistering and (2) delamination caused by specific corrosion mechanisms like cathodic or anodic delamination. Often, the formation of blisters is the starting point for cathodic or anodic delamination processes. Therefore, it is important to study the barrier properties of organic coatings towards water and ions.

EIS was already introduced in Sect. 2.2.1 as a powerful tool for the electrochemical characterisation of interfaces. The final purpose of the EIS characterisation of protecting organic coatings is to obtain information about system properties, such as the presence of defects, reactivity of the interface, adhesion, and barrier properties to water. Practically speaking, EIS can quantitatively measure both the resistance of the organic coating to aqueous and/or ionic transport, and capacitances in the electrochemical cell. In this context, a resistance implies a current flow, and hence the presence of electron transfer reactions, like corrosion. Organic coatings deteriorate with time during exposure to an electrolyte. Deterioration can be monitored e.g. as a change in the capacitance of a coating.

During exposure to corrosive electrolytes, metals covered with organic coatings pass basically through four different stages of deterioration [47, 49, 50, 58]:

- Water uptake within the organic coating.
- Corrosion underneath the coating without any defect in the coating itself.
- Partially damaged coating with cracks, defects or pores reaching the metal surface.
- Partially damaged coating with corrosive undermining of the coating.

These physical and chemical processes are usually quantified by means of equivalent circuits, see also Sect. 2.2.1. Each element of the equivalent circuit depicts one constituent of the specimen that is in contact to the electrolyte [59–61]. Schematic impedance spectra with the corresponding equivalent circuits for the different stages of deterioration of metal coated with polymers exposed to corrosive electrolytes are illustrated in Fig. 2.17. The parameters in the equivalent circuits are:

Fig. 2.17 Bode plots of typical impedance response of coatings with and without defects and the corresponding equivalent circuits

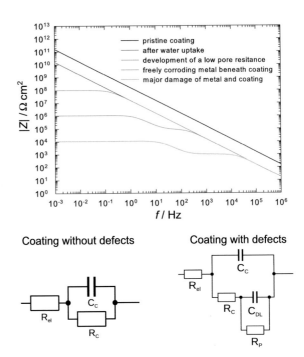

- R_{el}: Uncompensated electrolyte resistance of the electrolyte between the working and reference electrode, which is usually very low.
- C_C: The coating capacitance changes e.g. due to water uptake or swelling.
- R_C: The pore or coating resistance changes during exposure due to the penetration of the electrolyte into pores and defects of the coating.
- C_{DL}: Interfacial capacitance at the metal/electrolyte interface.
- R_P: Polarization resistance of the metal surface that is in contact with ionically conducting paths through the coating.

The two quantities that dominate the impedance during the initial stages of immersion are usually C_C and R_C: Typically, $C_C \sim 10$ nF cm^{-2} of an undamaged pristine coating (e.g. for automotive or coil coating applications) with good barrier properties. This capacitance corresponds to a modulus of the impedance at 0.1 Hz of 10^9 Ω cm^2.[2] Upon immersion, R_C is exceedingly high (10^{12} Ω cm^2 or even more), which leads to a nearly ideal capacitive response in the impedance. Capacitance decreases over orders of magnitude with increasing exposure time, leading to a horizontal plateau in the Bode plot at low frequencies. The pore resistance of a coating is directly influenced by many parameters like the crosslinking density or pigment to volume ratio, amongst others.

---

[2]In the units here, the electrode area dependence has been removed, which is critical for comparison of different coatings. As $C \propto A$, a comparison of C/A in F m^{-2} is independent of electrode area. Likewise, as $R \propto 1/A$, $R\,A$ with units of Ω m^2 is independent of electrode area.

One example how structural parameters of a coating can influence the barrier properties is shown in Fig. 2.18 [49]: The pore resistance of a pigment free UV-curable chlorinated polyester changes significantly with immersion time and degree of conversion (which is proportional to the cross-link density). Due to the formation of the polymer network R_C rises with increasing degree of conversion until the on-going polymerisation of the coating leads to internal stress within the coating [49]. Consequently the coating resistance drops for higher degree of conversion.

The water uptake changes the dielectric constant of the polymer and therefore the impedance and the capacitance of the coating: $C_C = \varepsilon\varepsilon_0 A/d$ (with ε: dielectric constant of the coating, $\varepsilon_0 = 8.8510^{-14}$ F cm^{-1}, A area of exposure, d thickness of the coating). At ambient conditions, the dielectric constant of a coating ranges between 2 and 8, whereas that of water is nearly 80. As consequence, the water uptake by the coating will lead to a significant increase in the effective dielectric constant of the coating and therefore in C_C.

A common approach for evaluating impedance spectra during water uptake is the determination of the diffusion coefficient D of water diffusion through the coating based on Fick's second law by fitting the development of the coating capacitance C_C values with time (see Fig. 2.18) [59, 62]. Theoretically, for ideal

Fig. 2.18 *Top* Typical time evolution of a coating capacitance upon immersion to 5 % NaCl. *Bottom* Dependence of coating resistance (R_C) of the differently UV-cured polyester acrylate layers after 1 h of exposure to the electrolyte (based on [49])

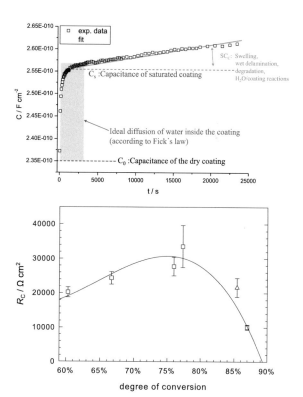

coatings plotted capacitances show an exponential increase in the initial stage of water uptake [Fickian (case-I) diffusion of water into the coating] and then result in a saturation plateau (eventually followed by another raise of inhomogeneous water accumulation due to polymer breakdown). However, water uptake curves measured on real polymeric systems (e-coating, coil coating, complete OEM lacquers, powder coatings, primer layers,) almost never show this ideal behaviour. These films tend to undergo swelling, degradation or other transport processes during longer exposure times to an electrolyte. As a result, the polymer capacitance does not reach a constant value after water saturation with time, but shows an additional increase (or decrease if the coating just undergoes a swelling transition) in the capacitance (see Fig. 2.18). Empirically, this additional increase typically shows often a linear behaviour. To quantify these influences and to improve the accuracy of the calculated diffusion coefficients, this increase can be mathematically compensated by combining Fickian Diffusion and case-II-sorption, e.g. by adding a slope coefficient to capacitance values [59]. This model consists of a Fickian-like starting period where water diffuses in interstitials/pores without interaction with polarizable groups. This is followed by a case-II-sorption period (linear part) where swelling of the polymer due to interaction with the penetrant takes place (see Fig. 2.18). Prerequisite for the applicability of this model is that the penetrant mobility is much bigger than the polymer segment mobility. This implies that the polymer relaxation controls the uptake under considerable swelling and as a result, the mass uptake proceeds linear with time, forming a sharp front.

If the knowledge of the moisture take-up rate by the coating during early stages of absorption is required, the Brasher-Kingsbury equation [63] is giving a reasonable estimate of the water volume fraction $\phi(t) = \log[C(t)/C(t = 0)]/\log(\varepsilon_w)$, where ε_w denotes the dielectric constant of water. Nevertheless, non-Fickian (or anomalous) sorption is a common behaviour for coating-water systems below the polymer glass transition temperature T_g [64]. The slow relaxation of the polymer chains due to the sorption of the water molecules is one reason for the abnormal sorption kinetic: Motions of the entire or parts of the glassy polymer chains are not sufficiently fast to completely homogenize the change of environment as water or other penetrants enter the polymer matrix (see Chap. 5 by Monteux et al. and Chap. 9 by J.-U. Sommer for an introduction into polymer dynamics). Consequently, water molecules or dissolved ions can find their way into "irregular cavities" with different intrinsic diffusional mobility. In such cases, more complex models describing the anomalous diffusion, adapting a suitable effective medium theory for the moisture in the coating, need to be applied [65, 66].

The determination of polarization resistance and interfacial capacitance for coated metals without artificial defects like a scratch is only possible if the impedance of the coating is smaller than or equal to the impedance of the interfacial reaction. For commercial coating systems one can hardly clearly separate the time constants of the interfacial reaction and the polymer film. Thus, the impedance of the interface cannot be deduced from the overall impedance of the polymer-coated metal leaving the analysis of the interface (R_P and C_{DL}) limited to very thin polymers and highly inhibited interfaces or special arrangements by placing the reference electrode directly at the interface. Another drawback are for instance non-linear effects within the coating systems or electrochemical setups. These are often overcome by using constant phase elements instead of capacitors in equivalent circuits used for fitting of data obtained from industrial coating systems, though there are intrinsic limitations in the interpretation [25, 67, 68].

2.4.3 Cathodic Delamination

Coatings with good corrosion protection properties prevent the access of hydrated ions to the coating/metal interface, acting as a barrier. At the interface, such coatings occupy metal or oxide adsorption sites. Thereby, the water activity at the interface is reduced. Although the interfacial ingress of water often weakens the adhesion of the polymer onto the metal therefore reducing the interfacial stability, such wet de-adhesion mechanisms are significantly accelerated and supported by electrochemical corrosion processes like cathodic delamination. Cathodic delamination is one of the—if not even the—fastest failure mechanisms of organically coated metals and a result of an electrochemical cell in which a separation of the anodic and cathodic reaction site takes place. It is the dominant corrosion mechanism in atmospheres with high humidity for coated metals like iron or zinc, which are covered by conductive oxide structures. Depending on the nature of the metallic substrate, the stability of the coating/metal interface against cathodic delamination is mainly determined by several properties [47, 69–71]:

1. The electron-transfer reaction at the interface between polymer and metal.
2. The redox reaction of the metallic oxide between the metal and the coating.
3. The chemical stability of the interface with respect to the reaction products of the electron transfer reaction.
4. The oxygen transport through the organic coating or defects.

Taking these four factors into account, distinct differences between the three most relevant technical surfaces must be expected:

- Steel: Basically iron, with low band-gap oxides which are highly electron conducting and stable in alkaline media.
- Galvanized Steel (Z, ZE): Basically zinc, with semiconducting oxides, which are not stable in alkaline media.
- Aluminium: High band-gap semiconducting oxides, which act as insulator and are not stable in alkaline media.

Cathodic delamination starts with randomly distributed anodes and cathodes in a blister-like defect or at a delaminated area [72]. It is a type of reaction that dominates on polymer/metal interfaces of metals covered by electron conducting or semiconducting oxides with a low band-gap in the presence of coating defects. The determining factors for cathodic delamination are the electrochemical and pH-stability of the metal oxide, the oxygen permeability and adhesion of the organic coating, the kind of electrolyte in the defect, as well as the relative humidity of the surrounding atmosphere.

As discussed in Sect. 2.2.2, the kinetics of cathodic delamination can be investigated by means of SKP measurements and basically two potential levels characterise the SKP profiles measured during delamination in the presence of oxygen and moisture. At and close to the defect, the measured potential is considerably more negative due to the anodic dissolution of the metal (see Fig. 2.19). When iron is taken as an example, oxygen will be reduced ($O_2 + 2H_2O + 4e^- \rightarrow 4OH^-$) in the defect area until the diffusion limiting current density level is reached. Iron dissolves as anodic counter reaction in the defect [69, 70].

Initial oxygen reduction also occurs at the intact coating/metal interface, but it is kinetically strongly inhibited. The intact interface is characterized by a high electronic conductivity of the oxide-covered iron surface, which allows electron transfer but due to the presence of the adhering polymer no ion transfer reactions. Therefore, the oxidation of iron oxide acts as the anodic counter reaction to the oxygen reduction. Consequently, a fast reduction of donor density within the oxide is observed, resulting in a decreasing rate of the electron transfer reaction and an anodic shift of the interfacial potential. As a consequence, above a certain anodic potential, the rate of oxygen reduction is extremely small, and no further anodic potential shift is observed. This is the actual potential as measured by the SKP in the area of an intact coating [43–45].

As discussed above, a distinct potential step between defect area and intact interface is monitored by SKP measurements. For most coatings, this sudden change in potential marks the delamination front and is the most interesting feature because at this location those processes responsible for the loss of adhesion of the coating occur [43–45]. This steep step in the potential originates from the migration of hydrated ions: The electrolyte in the defect can penetrate into the adjacent intact polymer/metal interface. Thereby the kinetic barrier for oxygen reduction at the intact interface will be bypassed and a galvanic couple between two formerly isolated electrochemical cells is established. In order to compensate the local charge (due to the oxygen reduction) at the delamination front, soluble cations of the electrolyte in the defect are electrostatically attracted. These ions have to be

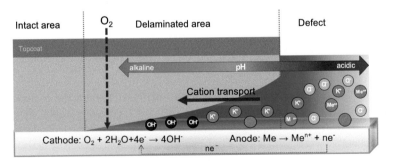

Fig. 2.19 Schematic illustration of the mechanism of cathodic delamination

transported either by migration or diffusion to the zone of oxygen reduction. Additionally, in the delamination zone, corrosive anions like chlorides are repelled by the negative charge of generated hydroxide ions.

During oxygen reduction, a strongly alkaline electrolyte is formed, with several consequences. Because interfacial iron dissolution is still inhibited, the iron surface becomes passivated. Further, the oxidative degradation of the polymer at the interface, either by the generated hydroxide ions, or by highly reactive oxygen species (in particular radicals or other intermediates evolved in the oxygen reduction reaction, see e.g. [73, 74]), results in delamination of the organic coating only by bond breaking within the organic coating near the interface. It is still an open questions which bond actually breaks [51].

The nature of cathodic delamination, especially the separation of local anode (iron dissolution within the defect) and local cathode (oxygen reduction within the delaminated area) requires a transfer of electrons as well as hydrated cations to the electrolyte front position. For many coating systems, the interfacial ion transport represents the slowest process step and therefore is rate-determining. In such cases, the delamination scales with the square root of time. However, especially in the presence of novel pre-treatments, highly cross-linked plasma polymers or e-coats, the rate determining step is reaction controlled, either by limiting the electron transfer or the oxygen reduction reaction. As a consequence, the delamination scales are linear with time. Details of analyses of the different regimes are found in the literature [51, 75].

As pointed out above, a strongly alkaline electrolyte is formed during oxygen reduction. Compared to iron or cold rolled steel, the oxides of zinc present on galvanized steel are less stable within such an alkaline environment. Although the basic mechanisms of cathodic delamination are similar to iron, additional anodic reactions in the section of interfacial ion transport are present, especially if the defect consists of iron [75–77]. On the one hand, zinc cathodically protects the underlying steel. On the other hand, this anodic process is attributed to amphoteric properties of zinc oxide. Although zinc also gets passivated in alkaline environments, $Zn(OH)_2$ tends to dissolve as zincate as the pH increases [78]. In the defect area, this is hardly of relevance for the delamination process, but within the section of interfacial ion transport, distinct oxide growth is reported for ongoing delamination at polymer/zinc interfaces by dissolution/precipitation of $Zn(OH)_2$ [75–77, 79].

The difference between cathodic delamination on iron or mild steel and corrosive delamination at zinc or galvanized surfaces can be visually observed: Within the delaminated area, mild steel remains shiny, whereas galvanized surfaces expose white colored zinc corrosion products. In summary, cathodic delamination requires electron conducting oxides, interfaces, which allow electrochemical reactions like the oxygen reduction reaction, and interfacial transport of hydrated cations. Therefore, cathodic delamination will dominate for substrates such as steel or galvanized steel at high relative humidity. At low humidity and in the presence of insulating oxides like those of magnesium or aluminium alloy grades, cathodic disbonding cannot occur and anodic driven de-adhesion reactions like filiform corrosion will prevail.

2.5 Smart Coatings for Active Corrosion Protection

The mechanism of passive corrosion protection, namely the existence of a diffuse double layer and blocking of electrochemical reactions at the interface, have been discussed in Sect. 2.4, together with the mechanism of organic coating failure. Modern coatings systems may contain layers where the onset of corrosion triggers release of corrosion inhibitors. In fundamental research, a number of triggering mechanisms are currently investigated. Examples include a change of pH, ionic strength or the change of electrode potential. Recent developments in the field of intelligent coatings for active corrosion protection will be summarised in this section. Further reading is available in current reviews [80–84].

2.5.1 Ionic Strength and pH Triggered Release

Electrochemically driven corrosion processes need—besides electron-transfer reactions—also the counterbalance of ionic charges (see Sect. 2.4.3). Therefore, the corrosion rate is faster if the metal is in contact with an electrolyte with high ion concentration at equivalent pH and temperature. Furthermore, corrosion of metals is always accompanied by local pH changes. While the pH in the vicinity of local anodes is lower due to hydrolysis of solvated metal cations, the pH at local cathodes is predominantly alkaline. Smart coatings releasing inhibitors due to the environmental ion concentration change or change in pH are actually indirectly responding to an onset of corrosion.

Interesting concepts using "ion-sensing" materials as a part of a binder coating system have been introduced. Ion-sensing materials contain naturally existing or synthesised ion-exchange materials and nano-carriers either modified or made of layer-by-layer assembled polyelectrolytes [84–89].

Ion-exchange materials can be divided into cation and anion exchange materials [90, 91]. Cation exchange materials can carry corrosion inhibiting cations, which are known to inhibit the oxygen reduction reaction (e.g. Ce^{3+}, Zn^{2+}, rare earth cations). Zeolite and bentonite clays and cross-linked polystyrene-sulphonate are the materials of choice for storing cationic inhibitors. Excellent corrosion inhibition and especially cathodic delamination inhibiting properties have been shown, if such materials have been applied as pigments in a coating matrix. The mechanism of release is due to the interaction of cations from a corrosive electrolyte with the pigments, leading to an exchange of the cations from the electrolyte with the corrosion inhibiting cation in the pigments. The corrosion inhibiting cation can form insoluble complexes with hydroxide ions, which can precipitate and block preferentially the cathodic reaction.

The predominantly investigated anion exchange pigment is layered double hydroxide compounds like hydrotalcite (HT). The structure consists of metal-hydroxide layers with excess positive charges, where anions are stored in between the layers to compensate the positive charges. The mechanism of release is similar as for the cation exchange pigments, but with the only difference that in this case aggressive anions like chloride are removed from the electrolyte and are adsorbed inside the HT, while the inhibitors are released. The use of pigments as part of a corrosion inhibiting primer coating on aluminium alloys has investigated in depth [84, 92, 93].

The advantage of ion-exchange materials is their very fast response and release of inhibitors. However, for a prolonged corrosion protection, fast releasing and slow releasing nano-carriers are needed in a combination in a coating formulation [94].

There is almost no limit in giving organic polymeric materials desired functionality for the application in corrosion protection. Very successfully applied polymer classes are polyelectrolytes, which on the one hand have been used for sensing the change in ionic-strength and on the other hand to sense also pH changes during corrosion. Three important strategies have been followed [88, 95, 96]:

1. The storage of inhibitors inside polyelectrolyte coatings.
2. Encapsulation and surface modification of hollow nano materials bearing corrosion inhibitors with polyelectrolytes.
3. Preparation of hollow polyelectroylte capsules, with corrosion inhibitors encapsulated inside the core.

The layer-by-layer polyelectrolyte assembly provides an inexpensive and robust method to build coatings that are water resistant and stable, even at high ionic strength and in acidic and basic media over a wide range of temperature, pH and solvent changes. Polyelectrolyte multilayers built by layer-by-layer assembly can be deposited on several metal substrates such as gold, stainless steel, iron, aluminium, titanium, nickel, silver, copper, and zinc.

Examples of a negatively-charged polyelectrolytes include polyelectrolytes comprising a sulfonate group ($-SO_3^-$). Examples include poly(styrenene sulfonic acid), poly(2-acrylamido-2-methyl-1-propane sulfonic acid), sulfonated poly(ether ether ketone), polycaboxilates such as poly(acrylic acid), and poly(methacrylic acid). Examples of positively-charged polyelectrolytes include polyelectrolytes comprising a quaternary ammonium group such as poly(diallyldimethylammonium chloride), poly(vinylbenzyltrimethylammonium halogenides), 2-dimethylaminoethyl metacrylate or polyelectrolytes containing a pyridinium group, such as poly (N-methylvinylpyridine).

By changing the type and number of layer-by-layer deposited polyelectrolytes, the responsiveness to corrosion stimuli can be adjusted and engineered. In [97] the surface of inhibitor filled mesoporous silica particles are modified with different combination of polyelectrolytes. While for example the combination of strong polyelectrolytes results in a polymer shell which releases the inhibitors due to the change in ionic strength, the combination of weak polyelectrolytes results in a polymer shell which is pH responsive and releases the inhibitors at alkaline or acidic pH.

Polyelectrolyte complexes have also been used in the development of self-healing matrix coatings. In this case corrosion inhibitors are incorporated as a second component in the complex. The general concept of these films is to entrap the inhibitor into the depot, between one layer and another, from which it can be released only when the corrosion starts. The inhibitor-loaded depot system is then impregnated into the standard corrosion protecting film. Corrosion is accompanied by local changes of the pH, flux of the corrosion products, corrosive species, and of the electrochemical potential. Hence, if the release properties of the inhibitor depot can be affected by any of the reactions accompanying the corrosion process, the release of the entrapped inhibitor is initiated by the corrosion process itself. Polyelectrolyte coatings and complexes are sensitive to pH changes in a local area and can influence very effectively the depot of the corrosion inhibitors and their release due to pH fluctuations during the corrosion process [88].

2.5.2 Electrode Potential Triggered Release

While the change in pH and ionic strength are an indirect indication that corrosion takes place, the change of the electrode potential of a metal is a direct, reliable and the most case-selective indicator for a corrosion process. The electrode potential of every non-inert metal is always and only decreasing when corrosion occurs.

Conducting polymers (CP) are an excellent material of choice for sensing the corrosion potential change and for storing and releasing inhibitors.

CP consist of conjugated chains containing π-electrons delocalised along the polymer backbone. In their natural form, conductive polymers are semiconducting materials that can be doped and converted into electrically conductive forms. The doping can be either oxidative or reductive, though oxidative doping is more common. Herein, some part of the polymer backbone is getting positively charged and counterbalanced by anions (which can be also corrosion inhibitors). There are generally two states of CP: non-conducting (uncharged/reduced state) and oxidized (p-doped/oxidized and the most stable state) where electrons are removed from the backbone [98]. The doping processes are usually reversible and typical conductivities can be switched between those of insulators (less than 10^{-10} S cm^{-1}) and those of metals (10^5 S cm^{-1}). Almost all of the conductive polymers used in corrosion protection fall into one of the following classes: polyanilines, polyheterocycles and poly(phenylene vinylenes).

The most widely studied of the intrinsically conducting polymers (ICPs) for corrosion protection coatings has been polyaniline (PANI). The advantages of this material are (i) easy chemical and electrochemical polymerisation, (ii) easy doping and de-doping by treatment with aqueous acids and bases and (iii) its high resistance to environmental degradation [99]. PANI is usually prepared in the emeraldine salt form by oxidative polymerisation of aniline in acidic environment. While there is general agreement that PANI performs well in preventing corrosion, the mechanism of this process is still under investigation. Several hypotheses have been suggested for the mechanism of corrosion protection using conductive polymers, specifically PANI: (i) PANI contributes to the formation of an electrical field at the metal surface, restricting the flow of electrons from metal to oxidant; (ii) PANI forms a dense, strongly adherent, low-porosity film similar to a barrier coating; and (iii) PANI causes the formation of protective layers of metal oxides on a metal surface. However, PANI possesses the important disadvantage that it is insoluble in the majority of solvents, making its processing very difficult [100].

One of the features which makes use of CP attractive for pigment coatings is the fact that corrosion inhibitor anions can be stored as counter ions in the oxidized polymer backbone. These anions can be released upon the onset of corrosion and the subsequent reduction of CP. Considerable research has been conducted on CP for corrosion protection, however there are four crucial points which have to be generally taken into consideration for designing potential triggered release systems based on CP [101–106]:

1. The efficiency of release of anionic inhibitors is based on their ionic mobility and has to compete with cation incorporation in the CP, which is usually the preferred situation.
2. Due to fast cation incorporation, electropolymerized CP films with continuous ionic networks may even enhance corrosion and can lead to a fast breakdown of the coating system. This means the direct electrochemical synthesis of PANI on a metal surface has to be ruled out for corrosion protection applications.
3. Only negatively charged molecules with non-oxidizable groups and with high solubility in acidic environment can be used as dopants for CP. However, corrosion inhibitors like mercaptobenzothiazole are not soluble in acidic environment and the thiol functional group can be easily oxidized in the oxidative environment used to synthesise conducting polymers.
4. CPs applied especially on zinc tend to react with the metal. Consequently, metal and CP are usually electronically decoupled (Fermi-level misalignment) and the CP looses its capability to sense the potential changes of the metal, due to the formation of an insulating layer between CP and zinc.

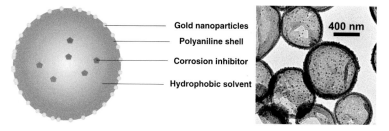

Fig. 2.20 Schematic depiction of of the release concept based on PANI capsules and the corresponding TEM micrograph of a PANI capsule synthesised via miniemulsion technique and decorated with gold nanoparticles

In a recent work [107] these challenges in using conducting polymers have been addressed, and a novel structure for storing and releasing inhibitors has been introduced (Fig. 2.20). The new structure synthesised by the miniemulsion technique comprises a redox-sensitive PANI shell and a corrosion inhibitor (3-nisa/3-nitrosalicylic acid) encapsulated in the hydrophobic (ethylenebenzene/ hexadecane) core. Furthermore, the shell is decorated by gold nanoparticles to circumvent electronic decoupling of the capsules from the metal surface (Fig. 2.20).

If the capsules are formulated with a non conducting coating and applied as a composite coating on zinc, they show a self-healing behaviour of a defect. The corrosion potential has been recorded by SKP and indicates that a scratch can be passivated by a coating containing the PANI capsules with inhibitors. On the other hand, the capsules without inhibitors and the binder coating do not show any kind of passivating effect, Fig. 2.21. Delamination performance measured by SKP unravels that after passivation of the defect the progress of cathodic delamination is completely inhibited for the coating containing the PANI capsules with inhibitor.

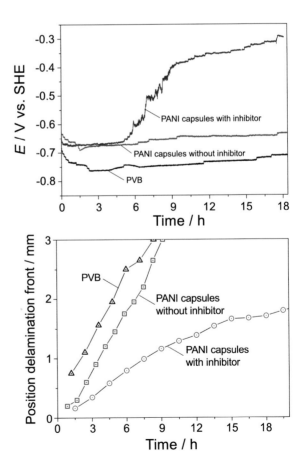

Fig. 2.21 *Top* Corrosion potential monitored by the SKP in an artificial defect down to zinc covered with 1 M KCl. *Bottom* Delamination progress over time for non-selfhealing coatings and the coating containing PANI capsules filled with inhibitor

2.6 Conclusion

Overall, relatively simple electrochemical processes lead to degradation of metals, also known as corrosion. To combat corrosion on a technological level, several rather sophisticated methods have been developed. Many of these systems involve at least in one step a soft matter system at an aqueous interface. Especially the modern, active systems of corrosion protection are still areas of active research, and a thorough understanding of polymers at aqueous interfaces is needed for the next generation of corrosion protection systems to become reality.

Acknowledgments A.A. and A.E. acknowledge support from the DFG (Deutsche Forschungsgemeinschaft) by grant number ER 601/3-1 within the Priority Program 1640 "Joining by plastic deformation". J.S.M.O. thanks the Mexican Consejo Nacional de Ciencia y Tecnología (Conacyt) for a scholarship. The authors thank Prof. M. Stratmann for continuous support and Michael Rohwerder for helpful discussions.

References

1. C.M.A. Brett, A.M.O. Brett. *Electrochemistry—Principles, Methods, and Applications* (Oxford University Press, Oxford, 1993)
2. A.J. Bard, L.R. Faulkner, *Electrochemical Methods: Fundamentals and Applications* (Wiley, New York, 2001)
3. C.H. Hamann, A. Hamnett, W. Vielstich. *Electrochemistry* (Wiley, Weinheim, 1998)
4. Helmut Kaesche, *Corrosion of Metals: Physicochemical Principles and Current Problems* (Springer, Berlin, 2003)
5. R.W. Revie (ed.), *Uhlig's Corrosion Handbook*, 3rd edn. (Wiley, Hoboken, 2011)
6. D.F. Evans, H. Wennerström. *The Colloidal Domain* (Wiley, New York, 1999)
7. Tosaka. https://en.wikipedia.org/wiki/File:Electric_double-layer_%28BMD_model%29_NT. PNG, August 2008
8. J. Lyklema, *Fundamentals of Interface and Colloid Science*, vol. 2—Solid-Liquid Interfaces (Academic Press, San Diego, 1995)
9. A. Erbe, K. Tauer, R. Sigel, Ion distribution around electrostatically stabilized polystyrene latex particles studied by ellipsometric light scattering. Langmuir **23**, 452–459 (2007)
10. R. Okamoto, A. Onuki, Charged colloids in an aqueous mixture with a salt. Phys. Rev. E **84**, 051401 (2011)
11. W. Dreyer, C. Guhlke, M. Landstorfer, A mixture theory of electrolytes containing solvation effects. Electrochem. Commun. **43**, 75–78 (2014)
12. J. Lyklema, Quest for ionion correlations in electric double layers and overcharging phenomena. Adv. Colloid Interface Sci. **147–148**, 205–213 (2009)
13. R.R. Netz, H. Orland, Beyond Poisson-Boltzmann: Fluctuation effects and correlation functions. Eur. Phys. J. E **1**, 203–214 (2000)
14. R.R. Netz, Static van der Waals interactions in electrolytes. Eur. Phys. J. E **5**, 189–205 (2001)
15. S. Trasatti, The "absolute" electrode potential—the end of the story. Electrochim. Acta **35**, 269–271 (1990)
16. W.M. Haynes (ed.), *Handbook of Chemistry and Physics* (CRC Press, Boca Raton, 2014)
17. Carl Wagner, Wilhelm Traud, Über die Deutung von Korrosionsvorgängen durch Überlagerung von elektrochemischen Teilvorgängen und über die Potentialbildung an Mischelektroden. Z. Elektrochem. Angew. Phys. Chem. **44**, 391–402 (1938)

18. C.H. Bamford, R.G. Compton (eds.), *Electrode Kinetics: Principles and Methodology* (Elsevier, Amsterdam, 1986)

19. T.J. Kemp, *Southampton Electrochemistry Group—Instrumental Methods in Electrochemistry* (Ellis Horwood, Chichester, 1985)

20. M. Stern, A.L. Geary, Electrochemical polarization: I. A theoretical analysis of the shape of polarization curves. J. Electrochem. Soc. **104**, 56–63 (1957)

21. R.A. Marcus, Electron transfer reactions in chemistry. Theory and experiment. Rev. Mod. Phys. **65**, 599–610 (1993)

22. D.D. Macdonald, Reflections on the history of electrochemical impedance spectroscopy. Electrochim. Acta **51**, 1376–1388 (2006)

23. A. Battistel, F. La Mantia, Nonlinear analysis: the intermodulated differential immittance spectroscopy. Anal. Chem. **85**, 6799–6805 (2013)

24. J.W. Schultze, A.W. Hassel, in *Encyclopedia of Electrochemistry. Passivity of Metals, Alloys and Semiconductors*, vol 4 (Wiley, Weinheim, 2007), pp. 216–235

25. S.-L. Wu, M.E. Orazem, B. Tribollet, V. Vivier, The influence of coupled faradaic and charging currents on impedance spectroscopy. Electrochim. Acta **131**, 3–12 (2014)

26. M.E. Orazem, B. Tribollet, *Electrochemical Impedance Spectroscopy* (Wiley, Hoboken, 2008)

27. E. Barsoukov, JR Macdonald, *Impedance Spectroscopy—Theory, Experiment and Applications* (Wiley, Hoboken, 2005)

28. M. Stratmann, The investigation of the corrosion properties of metals, covered with adsorbed electrolyte layers—a new experimental technique. Corros. Sci. **27**, 869–872 (1987)

29. S. Yee, R.A. Oriani, M. Stratmann, Application of a Kelvin microprobe to the corrosion of metals in humid atmospheres. J. Electrochem. Soc. **138**, 55–61 (1991)

30. K. Wapner, B. Schoenberger, M. Stratmann, G. Grundmeier, Height-regulating scanning kelvin probe for simultaneous measurement of surface topology and electrode potentials at buried polymer/metal interfaces. J. Electrochem. Soc. **152**, E114–E122 (2005)

31. M. Stratmann, H. Streckel, On the atmospheric corrosion of metals which are covered with thin electrolyte layers—I. Verification of the experimental technique. Corros. Sci. **30**, 681–696 (1990)

32. M. Stratmann, H. Streckel, On the atmospheric corrosion of metals which are covered with thin electrolyte layers—II. Experimental results. Corros. Sci. **30**, 697–714 (1990)

33. M. Stratmann, H. Streckel, K.T. Kim, S. Crockett, On the atmospheric corrosion of metals which are covered with thin electrolyte layers-III. The measurement of polarisation curves on metal surfaces which are covered by thin electrolyte layers. Corros. Sci. **30**, 715–734 (1990)

34. M. Stratmann, R. Feser, A. Leng, Corrosion protection by organic films. Electrochim. Acta **39**, 1207–1214 (1994)

35. M. Stratmann, A. Leng, W. Fürbeth, H. Streckel, H. Gehmecker, K.-H. Große-Brinkhaus, The scanning Kelvin probe; a new technique for the in situ analysis of the delamination of organic coatings. Prog. Org. Coat. **27**, 261–267 (1996)

36. H. Baumgärtner, H.D. Liess, Micro Kelvin probe for local work-function measurements. Rev. Sci. Instr. **59**, 802–805 (1988)

37. M.F. Becker, A.F. Stewart, J.A. Kardach, A.H. Guenther, Surface charging in laser damage to dielectric surfaces and thin films. Appl. Opt. **26**, 805–812 (1987)

38. D. Mao, A. Kahn, M. Marsi, G. Margaritondo, Synchrotron-radiation-induced surface photovoltage on gaas studied by contact-potential-difference measurements. Phys. Rev. B **42**, 3228–3230 (1990)

39. P. Chiaradia, J.E. Bonnet, M. Fanfoni, C. Goletti, G. Lampel, Schottky barrier and surface photovoltage induced by synchrotron radiation in GaP(110)/Ag. Phys. Rev. B **47**, 13520–13526 (1993)

40. H. Ren, H. Sinha, A. Sehgal, M.T. Nichols, G.A. Antonelli, Y. Nishi, J.L. Shohet, Surface potential due to charge accumulation during vacuum ultraviolet exposure for high-k and low-k dielectrics. Appl. Phys. Lett. **97**, 072901 (2010)

41. J.L. Lauer, J.L. Shohet, Surface potential measurements of vacuum ultraviolet irradiated Al_2O_3, Si_3N_4, and SiO_2. IEEE Trans. Plasma Sci. **33**, 248–249 (2005)
42. B. Salgin, D. Pontoni, D. Vogel, H. Schroder, P. Keil, M. Stratmann, H. Reichert, M. Rohwerder, Chemistry-dependent x-ray-induced surface charging. Phys. Chem. Chem. Phys. **16**, 22255–22261 (2014)
43. A. Leng, H. Streckel, M. Stratmann, The delamination of polymeric coatings from steel. Part 1: calibration of the Kelvin probe and basic delamination mechanism. Corros. Sci. **41**, 547–578 (1998)
44. A. Leng, H. Streckel, M. Stratmann, The delamination of polymeric coatings from steel. Part 2: first stage of delamination, effect of type and concentration of cations on delamination, chemical analysis of the interface. Corros. Sci. **41**, 579–597 (1998)
45. A. Leng, H. Streckel, K. Hofmann, M. Stratmann, The delamination of polymeric coatings from steel. Part 3: effect of the oxygen partial pressure on the delamination reaction and current distribution at the metal/polymer interface. Corros. Sci. **41**, 599–620 (1998)
46. G.S. Frankel, M. Stratmann, M. Rohwerder, A. Michalik, B. Maier, J. Dora, M. Wicinski, Potential control under thin aqueous layers using a Kelvin probe. Corros. Sci. **49**, 2021–2036 (2007)
47. G. Grundmeier, W. Schmidt, M. Stratmann, Corrosion protection by organic coatings: electrochemical mechanism and novel methods of investigation. Electrochim. Acta **45**, 2515–2533 (2000)
48. G. Williams, H.N. McMurray, Polyaniline inhibition of filiform corrosion on organic coated AA2024-T3. Electrochim. Acta **54**, 4245–4252 (2009)
49. R. Posner, P.R. Sundel, T. Bergman, P. Roose, M. Heylen, G. Grundmeier, P. Keil, UV-curable polyester acrylate coatings: barrier properties and ion transport kinetics along polymer/metal interfaces. J. Electrochem. Soc. **158**, C185–C193 (2011)
50. G. Frankel, M. Rohwerder, in *Encyclopedia of Electrochemistry*. Electrochemical Techniques for Corrosion, vol 4 (Wiley, Weinheim, Germany, 2007), pp. 687–723
51. D. Iqbal, J. Rechmann, A. Sarfraz, A. Altin, G. Genchev, A. Erbe, Synthesis of ultrathin poly (methyl methacrylate) model coatings bound via organosilanes to zinc and investigation of their delamination kinetics. ACS Appl. Mater. Interfaces **6**, 18112–18121 (2014)
52. N.W. Khun, G.S. Frankel, Effects of surface roughness, texture and polymer degradation on cathodic delamination of epoxy coated steel samples. Corros. Sci. **67**, 152–160 (2013)
53. D. Iqbal, R.S. Moirangthem, A. Bashir, A. Erbe, Study of polymer coating delamination kinetics on zinc modified with zinc oxide of different morphologies. Mater. Corros. **65**, 370–375 (2014)
54. A. Goldschmidt, H. Streitberger, *BASF-Handbuch Lackiertechnik* (Vincentz, Hannover, 2002)
55. B. Tieke, *Makromolekulare Chemie* (Wiley, Weinheim, 1997)
56. D. Braun, H. Cherdron, H. Ritter, *Praktikum der Makromolekularen Stoffe* (Wiley, Weinheim, 1999)
57. G. Grundmeier, A. Simões, in *Encyclopedia of Electrochemistry*. Corrosion Protection by Organic Coatings, vol 4 (Wiley, Weinheim, Germany, 2007), pp. 500–566
58. F. Deflorian, S. Rossi, M. Fedel, Organic coatings degradation: comparison between natural and artificial weathering. Corros. Sci. **50**, 2360–2366 (2008)
59. E.P.M. van Westing, G.M. Ferrari, J.H.W. de Wit, The determination of coating performance with impedance measurements—III. In situ determination of loss of adhesion. Corros. Sci. **36**, 979–994 (1994)
60. E.P.M. van Westing, G.M. Ferrari, J.H.W. de Wit, The determination of coating performance with impedance measurements-I. Coating polymer properties. Corros. Sci. **34**, 1511–1530 (1993)
61. J.R. Scully, S.T. Hensley, Lifetime prediction for organic coatings on steel and a magnesium alloy using electrochemical impedance methods. Corrosion **50**, 705–716 (1994)
62. F. Mansfeld, Use of electrochemical impedance spectroscopy for the study of corrosion protection by polymer coatings. J. Appl. Electrochem. **25**, 187–202 (1995)

63. D.M. Brasher, A.H. Kingsbury, Electrical measurements in the study of immersed paint coatings on metal. I. Comparison between capacitance and gravimetric methods of estimating water-uptake. J. Appl. Chem. **4**, 62–72 (1954)

64. R. Posner, K. Wapner, S. Amthor, K.J. Roschmann, G. Grundmeier, Electrochemical investigation of the coating/substrate interface stability for styrene/acrylate copolymer films applied on iron. Corros. Sci. **52**, 37–44 (2010)

65. B.R. Hinderliter, S.G. Croll, Simulation of transient electrochemical impedance spectroscopy due to water uptake or oxide growth. Electrochim. Acta **54**, 5344–5352 (2009)

66. V. La Saponara, Environmental and chemical degradation of carbon/epoxy and structural adhesive for aerospace applications: Fickian and anomalous diffusion, Arrhenius kinetics. Compos. Struct. **93**, 2180–2195 (2011)

67. B. Hirschorn, M.E. Orazem, B. Tribollet, V. Vivier, I. Frateur, M. Musiani, Determination of effective capacitance and film thickness from constant-phase-element parameters. Electrochim. Acta **55**, 6218–6227 (2010)

68. M.R.S. Abouzari, F. Berkemeier, G. Schmitz, D. Wilmer, On the physical interpretation of constant phase elements. Solid State Ionics **180**, 922–927 (2009)

69. U. Stimming, J.W. Schultze, The capacity of passivated iron electrodes and the band structure of the passive layer. Ber. Bunsenges. **80**, 1297–1302 (1976)

70. U. Stimming, J.W. Schultze, A semiconductor model of the passive layer on iron electrodes and its application to electrochemical reactions. Electrochim. Acta **24**, 859–869 (1979)

71. S. Yee, R.A. Oriani, M. Stratmann, Application of a Kelvin microprobe to the corrosion of metals in humid atmospheres. J. Electrochem. Soc. **138**, 55–61 (1991)

72. H. Leidheiser Jr., W. Wang, L. Igetoft, The mechanism for the cathodic delamination of organic coatings from a metal surface. Prog. Org. Coat. **11**, 19–40 (1983)

73. S. Nayak, P.U. Biedermann, M. Stratmann, A. Erbe, A mechanistic study of the electrochemical oxygen reduction on the model semiconductor n-Ge(100) by ATR-IR and DFT. Phys. Chem. Chem. Phys. **15**, 5771–5781 (2013)

74. S. Nayak, P.U. Biedermann, M. Stratmann, A. Erbe, In situ infrared spectroscopic investigation of intermediates in the electrochemical oxygen reduction on n-Ge (100) in alkaline perchlorate and chloride electrolyte. Electrochim. Acta **106**, 472–482 (2013)

75. W. Fürbeth, M. Stratmann, The delamination of polymeric coatings from electrogalvanised steel—a mechanistic approach. Part 1: delamination from a defect with intact zinc layer. Corros. Sci. **43**, 207–227 (2001)

76. W. Fürbeth, M. Stratmann, The delamination of polymeric coatings from electrogalvanized steel-a mechanistic approach. Part 2: delamination from a defect down to steel. Corros. Sci. **43**, 229–241 (2001)

77. W. Fürbeth, M. Stratmann, The delamination of polymeric coatings from electrogalvanized steel–a mechanistic approach.: Part 3: delamination kinetics and influence of CO_2. Corros. Sci. **43**, 243–254 (2001)

78. M. Pourbaix, *Atlas of Electrochemical Equilibria in Aqueous Solutions.* (National Association of Corrosion Engineers/Centre Belge d'Etude de la Corrosion CEBELCOR, Houston/Bruxelles, 1974)

79. D. Iqbal, A. Kostka, A. Bashir, A. Sarfraz, Y. Chen, A.D. Wieck, A. Erbe, Sequential growth of zinc oxide nanorod arrays at room temperature via a corrosion process: application in visible light photocatalysis. ACS Appl. Mater. Interfaces **6**, 18728–18734 (2014)

80. S.J. Garcia, H.R. Fischer, S. van der Zwaag, A critical appraisal of the potential of self healing polymeric coatings. Prog. Org. Coat. **72**, 211–221 (2011)

81. A.E. Hughes, I.S. Cole, T.H. Muster, R.J Varley, Designing green, self-healing coatings for metal protection. NPG Asia Mater. **2**, 143–151 (2010)

82. M.F. Montemor, Functional and smart coatings for corrosion protection: a review of recent advances. Surf. Coat. Technol. **258**, 17–37 (2014)

83. M. Zheludkevich, in *Self-Healing Materials—Fundamentals, Design Strategies, and Applications.* Self-Healing Anticorrosion Coatings, Chapter 4 (Wiley, Weinheim, 2009), pp. 101–139

84. M.L. Zheludkevich, J. Tedim, M.G.S. Ferreira, "Smart" coatings for active corrosion protection based on multi-functional micro and nanocontainers. Electrochim. Acta **82**, 314–323 (2012)
85. D.V. Andreeva, D. Fix, H. Möhwald, D.G. Shchukin, Self-healing anticorrosion coatings based on pH-sensitive polyelectrolyte/inhibitor sandwichlike nanostructures. Adv. Mater. **20**, 2789–2794 (2008)
86. B. Blaiszik, S. Kramer, S. Olugebefola, J. Moore, N. Sottos, S. White, Self-healing polymers and composites. Ann. Rev. Mater. Res. **40**, 179–211 (2010)
87. S.J. Garcia, H.R. Fischer, P.A. White, J. Mardel, Y. Gonzalez-Garcia, J.M.C. Mol, A.E. Hughes, Self-healing anticorrosive organic coating based on an encapsulated water reactive silyl ester: synthesis and proof of concept. Prog. Org. Coat. **70**, 142–149 (2011)
88. D.O. Grigoriev, K. Köhler, E. Skorb, D.G. Shchukin, H. Möhwald, Polyelectrolyte complexes as a smart depot for self-healing anticorrosion coatings. Soft Matter **5**, 1426–1432 (2009)
89. M.L. Zheludkevich, D.G. Shchukin, K.A. Yasakau, H. Möhwald, M.G.S. Ferreira, Anticorrosion coatings with self-healing effect based on nanocontainers impregnated with corrosion inhibitor. Chem. Mater. **19**, 402–411 (2007)
90. G. Williams, S. Geary, H.N. McMurray, Smart release corrosion inhibitor pigments based on organic ion-exchange resins. Corros. Sci. **57**, 139–147 (2012)
91. G. Williams, H.N. McMurray, Inhibition of filiform corrosion on organic-coated AA2024-T3 by smart-release cation and anion-exchange pigments. Electrochim. Acta **69**, 287–294 (2012)
92. R.G. Buchheit, H. Guan, S. Mahajanam, F. Wong, Active corrosion protection and corrosion sensing in chromate-free organic coatings. Prog. Org. Coat. **47**, 174–182 (2003)
93. G. Williams, H.N. McMurray, M.J. Loveridge, Inhibition of corrosion-driven organic coating disbondment on galvanised steel by smart release group II and Zn(II)-exchanged bentonite pigments. Electrochim. Acta **55**, 1740–1748 (2010)
94. M.F. Montemor, D.V. Snihirova, M.G. Taryba, S.V. Lamaka, I.A. Kartsonakis, A.C. Balaskas, G.C. Kordas, J. Tedim, A. Kuznetsova, M.L. Zheludkevich, M.G.S. Ferreira, Evaluation of self-healing ability in protective coatings modified with combinations of layered double hydroxides and cerium molibdate nanocontainers filled with corrosion inhibitors. Electrochim. Acta **60**, 31–40 (2012)
95. D.G. Shchukin, M. Zheludkevich, K. Yasakau, S. Lamaka, M.G.S. Ferreira, H. Möhwald, Layer-by-layer assembled nanocontainers for self-healing corrosion protection. Adv. Mater. **18**, 1672–1678 (2006)
96. D.G. Shchukin, S.V. Lamaka, K.A. Yasakau, M.L. Zheludkevich, M.G.S. Ferreira, H. Möhwald, Active anticorrosion coatings with halloysite nanocontainers. J. Phys. Chem. C **112**, 958–964 (2008)
97. D.G. Shchukin, H. Möhwald, Surface-engineered nanocontainers for entrapment of corrosion inhibitors. Adv. Funct. Mater. **17**, 1451–1458 (2007)
98. P. Zarras, N. Anderson, C. Webber, D.J. Irvin, J.A. Irvin, A. Guenthner, J.D. Stenger-Smith, Progress in using conductive polymers as corrosion-inhibiting coatings. Radiat. Phys. Chem. **68**, 387–394 (2003)
99. J.O. Iroh, R. Rajagopalan, Electrochemical polymerization of aniline on carbon fibers in aqueous toluene sulfonate solution. J. Appl. Polym. Sci. **76**, 1503–1509 (2000)
100. A.A. Syed, M.K. Dinesan, Review: polyaniline-a novel polymeric material. Talanta **38**, 815–837 (1991)
101. G. Paliwoda-Porebska, M. Rohwerder, M. Stratmann, U. Rammelt, L. Duc, W. Plieth, Release mechanism of electrodeposited polypyrrole doped with corrosion inhibitor anions. J. Solid State Electrochem. **10**, 730–736 (2006)
102. M. Rohwerder, A. Michalik, Conducting polymers for corrosion protection: What makes the difference between failure and success? Electrochim. Acta **53**, 1300–1313 (2007)
103. M. Rohwerder, L.M. Duc, A. Michalik, In situ investigation of corrosion localised at the buried interface between metal and conducting polymer based composite coatings. Electrochim. Acta **54**, 6075–6081 (2009)

104. M. Rohwerder, Conducting polymers for corrosion protection: a review. Int. J. Mater. Res. **100**, 1331–1342 (2009)
105. M. Rohwerder, S. Isik-Uppenkamp, C.A. Amarnath, Application of the Kelvin Probe method for screening the interfacial reactivity of conducting polymer based coatings for corrosion protection. Electrochim. Acta **56**, 1889–1893 (2011)
106. G. Williams, R.J. Holness, D.A. Worsley, H.N. McMurray, Inhibition of corrosion-driven organic coating delamination on zinc by polyaniline. Electrochem. Commun. **6**, 549–555 (2004)
107. A. Vimalanandan, L.-P. Lv, T.H. Tran, K. Landfester, D. Crespy, M. Rohwerder, Redox-responsive self-healing for corrosion protection. Adv. Mater. **25**, 6980–6984 (2013)

Chapter 3
Introduction to Depletion Interaction and Colloidal Phase Behaviour

Remco Tuinier

Abstract Efforts to explain physical properties of colloidal suspensions in terms of the forces that act between the colloidal particles go back to the beginning of the 20th century. In the second half of the last century theoretical progress clarified that the stability of colloidal particles is also affected by non-adsorbing polymers in solution, as first explained by Asakura and Oosawa in Japan using the excluded and free volume concepts. Here an introduction to the depletion interaction and resulting phase behaviour in colloidal suspensions is provided. The theory for the phase behaviour of colloidal dispersions is developed here starting from the Van der Waals theory for the as-liquid phase transition. Subsequently, the hard sphere fluid-solid phase transition is explained. Next, an attractive Yukawa hard-core model is used to outline the effects of varying the range of attraction on the phase behaviour of a colloidal suspension of attractive particles. Finally, the phase states that can be found in a colloidal hard sphere dispersion plus depletants are explained.

3.1 Introduction and Some History

3.1.1 The First Theory on Depletion Interaction

In the early 1950s the legendary Oosawa [1], at that time a young Associate Professor at Nagoya University in Japan, organized a winter symposium in Nagoya and invited a multidisciplinary group of Japanese scholars, mainly active in

R. Tuinier (✉)
Laboratory of Physical Chemistry, Department of Chemical Engineering and Chemistry and Institute for Complex Molecular Systems, Eindhoven University of Technology, 513, 5600 MB Eindhoven, The Netherlands
e-mail: r.tuinier@tue.nl

R. Tuinier
Van 't Hoff Laboratory for Physical and Colloid Chemistry, Department of Chemistry, Debye Institute, Utrecht University, Utrecht, The Netherlands

© Springer International Publishing Switzerland 2016
P.R. Lang and Y. Liu (eds.), *Soft Matter at Aqueous Interfaces*,
Lecture Notes in Physics 917, DOI 10.1007/978-3-319-24502-7_3

biology. Oosawa has a statistical mechanics background and he asked the group to present work on phenomena in biological systems where statistical physics could be helpful to understand certain mechanisms. During the meeting the 'aggregation' of particles under the influence of macromolecules was a re-occuring theme. It was observed in suspensions of red blood cells, bacterial cells, soil powder and gum latex particles. This inspired Oosawa to start work with Sho Asakura, then a graduate student, on the influence of polymers on the interaction between particles.

In 1953 P.J. Flory was invited by professor Yukawa to Tokyo and met Oosawa. Oosawa invited Flory to come to Nagoya University [1]. During Flory's visit Asakura and Oosawa explained their theoretical results on two particles immersed in a solution containing nonadsorbing polymer chains, showing the chains impose an *attractive* interaction between the particles. The very positive response of Flory, at that time Associate Editor of *J. Chem. Phys.* resulted in submission of this work, leading to the seminal paper in which Asakura and Oosawa [2] presented a statistical mechanical derivation of the interaction between two plates immersed in a solution of ideal nonadsorbing polymers. The theory of Asakura and Oosawa [2] is the first theoretical prediction of a depletion force. It will be explained in more detail later on. They showed that adding nonadsorbing polymer chains induce an effective attraction between particles with a hard core interaction. The attraction originates from repulsive interactions only, so it is a purely entropic effect.

This led to the discovery of the seminal Asakura-Oosawa depletion potential first published in their 1954 paper [2]. The term 'depletion' was probably introduced by Napper [3]. Oosawa indicated that the derivations and calculations were performed within a few weeks [4] and wished to note that he actually does not like the word 'depletion'.

3.1.2 Origin of the Depletion Force

The origin of the depletion effect is now first explained by regarding colloidal hard spheres in a solution of nonadsorbing polymer. The fact that polymers do not adsorb results in an effective depletion layer near the surface of the colloidal particles due to a loss of configurational entropy of a polymer chain in that region. In Fig. 3.1 a few colloidal spheres are depicted in a polymer solution. The depletion layers are indicated by the (dashed) circles around the spheres. When the depletion layers overlap the volume available for the polymer chains increases. Hence states in which the colloidal spheres are close together are more favourable. Therefore the polymers indirectly induce an effective attraction force between the spheres even though the direct colloid-colloid and colloid-polymer interactions are repulsive [5]. Vrij called this 'attraction through repulsion'.

At least in the limit of low depletant concentrations the attraction equals minus the product of the osmotic pressure and the overlap volume, indicated by the hatched region between the close spheres in Fig. 3.1. The picture sketched above became first clear in the 1950s through the work of Asakura and Oosawa [2, 6].

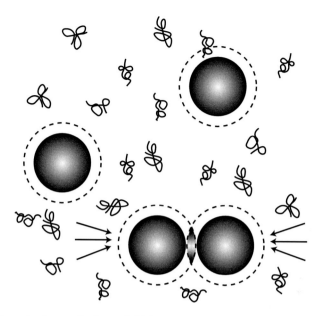

Fig. 3.1 Schematic picture of a few colloidal spheres in a polymer solution with nonadsorbing polymers. The depletion layers are indicated by the *short dashes*. When there is no overlap of depletion layers (*upper two spheres*) the osmotic pressure acting upon the spheres due to the polymers is isotropic. For overlapping depletion layers (*lower two spheres*) the osmotic pressure on the spheres is unbalanced; the excess pressure is indicated by the *arrows*

It was hardly noticed in the literature at first, seemed forgotten, but started to gain attention when Vincent et al. [7] and Vrij [5] started systematic experimental and theoretical work on colloid-polymer mixtures.

Below the standard expression used for the depletion interaction [2, 5, 6] is given. Consider two colloidal spheres each with volume $v_c = 4\pi R^3/3$ and diameter $2R$, surrounded by a depletion layer with thickness δ. In that case the depletion potential can be calculated from the product of $\Pi = n_b k_B T$, the (ideal) osmotic pressure of depletants with bulk number density n_b, times V_{ov}, the overlap volume of the depletion layers. Hence the Asakura-Oosawa-Vrij (AOV) depletion potential equals:

$$W(h) = \begin{cases} \infty & h < 0 \\ -\Pi\, V_{ov}(h) & 0 \le h \le 2\delta \\ 0 & h \ge 2\delta \end{cases} \tag{3.1}$$

with overlap volume $V_{ov}(h)$,

$$V_{ov}(h) = \frac{\pi}{6}(2\delta - h)^2(3R + 2\delta + h/2). \tag{3.2}$$

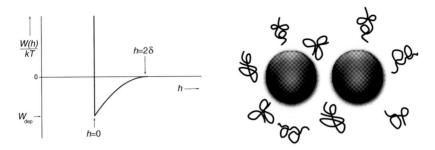

Fig. 3.2 Sketch of the depletion interaction between two hard spheres

In Fig. 3.2 the AOV interaction potential $W(h)$ is plotted. The minimum value of the potential W_{dep} is achieved when the particles touch ($h = 0$). Inspection of Eqs. (3.1) and (3.2) reveals that the *range* of the depletion attraction is determined by the size 2δ of the depletant, whereas the *strength* of the attraction increases with the osmotic pressure of the depletants, hence with the depletant concentration. Depletion effects offer the possibility to independently modify the range and the strength of attraction between colloidal particles. In dilute polymer solutions, the depletion thickness $\delta \approx 1.1R_g$ [8–10], so δ is close to the polymer's radius of gyration R_g.

It is noted that in the original paper of Asakura and Oosawa [6], where expression Eq. (3.1) was first derived, the polymers were regarded as (dilute) pure hard spheres. Vrij [5, 11] arrived at the same result by describing the polymer chains as penetrable hard spheres, see Sect. 3.2.

In a mixture of hard spheres and depletants (which can also be polymers, surfactant micelles, rodlike colloidal particles) a phase transition occurs upon exceeding a certain concentration of colloidal spheres and/or depletants. An important parameter that is used to describe the phase stability of colloid-polymer mixtures is the size ratio q,

$$q = \frac{R_g}{R}. \tag{3.3}$$

Phase diagrams of colloid-polymer mixtures are often described in terms of the volume fraction of colloids η and the *relative* polymer concentration:

$$\phi_p = \frac{n_b}{n_b^*} = \frac{\varphi}{\varphi^*}, \tag{3.4}$$

which is unity at the overlap concentration and can be regarded as the 'coil volume fraction' of polymer coils, exceeding unity in the semidilute concentration regime and beyond.[1] Commonly ϕ_p is used as the parameter for 'polymer concentration'. The overlap concentration in kg/m^3 or g/L equals

$$\frac{3M_p}{4\pi R_g^3 N_{av}} , \tag{3.5}$$

where M_p is the molar mass of the polymer and N_{av} is Avogadro's number. The osmotic pressure Π in Eq. (3.1) can, using Eq. (3.4), be rewritten as $\Pi v_p/k_BT = \phi_p$. This allows to write the depletion interaction for spheres that touch, $W_{dep}(h = 0) = W_{dep}$ as:

$$\frac{W_{dep}}{k_BT} = -\phi_p\left(1 + \frac{3}{2q}\right) , \tag{3.6}$$

which for small q boils down to $W_{dep} = -(3/2q)k_BT\phi_p$. This clarifies that at given ϕ_p the depletion force is very strong for small q.

Next a few examples are discussed where the effects of depletion were already noted long before Asakura and Oosawa rationalized the attractive interaction caused by depletants.

3.1.3 Early Observations

The aggregation of red blood cells (RBCs) in blood of human beings is found to be enhanced in case of for instance pregnancy or a wide range of illnesses, giving rather pronounced 'rouleaux'; clustered RBCs with their flat sides facing each other [12]. Rouleaux were described already more than 2 centuries ago. Enhanced RBC aggregation is observed for instance by measuring the sedimentation rate which can increase 100-fold in case of severe illnesses as compared to RBC sedimentation in healthy blood. The blood sedimentation test, based on monitoring aggregation of red blood cells, became a standard method for detecting illnesses. The relation between pathological condition, RBC aggregation and enhanced sedimentation rate was described for instance long time ago in [12–14]. It has been shown that adding macromolecules such as dextrans to blood also promotes rouleaux formation. Asakura and Oosawa [6] suggested that RBC aggregation is caused by depletion forces between the RBC's induced by serum proteins. The general picture is that

---

[1]The quantity n_b^* is the bulk polymer number density at which the polymer coils overlap. In terms of the volume fraction of polymer segments φ ($0 \le \varphi \le 1$), one then uses $\phi_p = \varphi/\varphi^*$, with φ^* the segment volume fraction at which the chains start to overlap: $\varphi^* = N_p v_s/v_p$, where N_p is the number of segments per polymer chain, v_s is the monomer (segment) volume, and $v_p = (4\pi/3)R_g^3$ the coil volume, so $\varphi^* \sim N_p/R_g^3$. The overlap number density n_b^* hence follows as $n_b^* = 3/(4\pi R_g^3)$.

red blood cells tend to cluster at elevated concentrations of the blood serum proteins, which act as depletants.

Large scale production of binder particles for paint production commenced about a century ago. In order to lower transport costs there was a significant interest in concentrating the polymeric latex. Centrifugation is highly energy consuming and thus expensive. Traube [15] showed that adding plant and seaweed polysaccharides led to a phase separation between a dilute and a concentrated phase with binder particles. Since the particles are lighter than the solvent the concentrated phase, with volume fraction $0.5 \leq \eta \leq 0.8$, floats on top. The lower phase is clear and hardly contains particles. Baker [16] and Vester [17] systematically investigated the mechanism that leads to what they called (enhanced) creaming. From the work of Baker [16] it can be concluded that the particles aggregate reversibly; upon dilution the latex particles can be resuspended. This suggests that bridging, which can also cause creaming [18], is not the driving force for enhanced creaming here.

3.1.4 Onset of Attention for Depletion After 1954

Not long after the publication of the work of Asakura and Oosawa, Sieglaff [19] demonstrated that a depletion-induced phase transition may occur upon adding polystyrene to a dispersion of microgel spheres in toluene. This demonstrated that the attractive depletion force is sufficiently strong to induce a phase separation. Sieglaff rationalized his findings in terms of the theory of Asakura and Oosawa. It took more than a decade before subsequent work was published.

Early systematic studies with respect to phase stability for colloid-polymer mixtures were performed in the 1970s by Vincent and co-workers [7, 20, 21]. They concentrated on mixtures of colloidal spheres (latex particles) plus nonadsorbing polymers such as polyethylene oxide (PEO). In the papers of Vincent et al. there is quite some variation in qualifying the demixing phenomena in colloid-polymer mixtures [20, 22–24]. These experiments were ahead of a full theoretical understanding of the phase behaviour of colloid-polymer mixtures.

Also in the 1970s Hachisu et al. [25] investigated aqueous dispersions of negatively charged polystyrene latex particles that undergo a colloidal fluid-to-solid phase transition upon lowering the salt concentration using dialysis or increasing the particle concentration. Under conditions where the latex dispersion is not ordered (fluid-like), Kose and Hachisu [26] added sodium polyacrylate to polystyrene latex particles (both components are negatively charged), and observed crystallization of the colloidal spheres. The authors suggested that the ordering is due to 'some attractive force'. When the polymer concentration is increased crystallization occurs faster. Since polymers and particles repel each other the crystallization process was probably induced by depletion interaction.

Theoretical work on depletion interactions and their effects on macroscopic properties such as phase stability started with work by Vrij [5] who considered the depletion interaction between hard spheres due to dilute nonadsorbing polymers described as penetrable hard spheres (see Sect. 3.2) and computed the second osmotic virial coefficient to explain the phase transitions observed by De Hek and Vrij [11]. By mixing aqueous hydroxyethylcellulose (HEC) with polymeric colloidal particles, Sperry [27, 28] and coworkers [29] observed phase separation and made a study on the effect of the structure of the colloid-rich phase as a function of the colloidal particle-free polymer size ratio $q = R_g/R$. Unstable systems at large q and not too high polymer concentrations are characterized by smooth interfaces, implying colloidal gas–liquid coexistence. For small q, demixed systems are characterized by irregular interfaces that indicate (colloidal) fluid-solid coexistence. This suggests that the width of the region where a colloidal liquid is found in colloid-polymer mixtures is limited.

The work of Sperry inspired Gast, Hall and Russel to develop a theory which might explain the experimental phenomena. Gast et al. [30] used thermodynamic perturbation theory (TPT) [31] to derive the free energy of a mixture of colloidal particles and polymers (described as phs; penetrable hard spheres), based on pair-wise additivity of the interactions between the colloids. They calculated the phase behaviour from the (perturbed) free energy which made it possible to assign the nature (i.e. colloidal gas, liquid or solid) of the coexisting phases as a function of size ratio q, the concentration (or formally activity within their approach) of the polymers, and the volume fraction of colloids. For small values of q, say, $q = R_g/R < 0.3$, increasing the polymer concentration broadens the hard sphere fluid-solid coexistence region; a (stable) colloidal fluid-solid coexistence is expected if the polymer chains are significantly smaller than the colloidal spheres (low q). Inside the unstable regions a (metastable) colloidal gas–liquid branch is located. For intermediate values of q, the gas–liquid coexistence curve crosses the fluid-solid curve and for large q-values mainly gas–liquid coexistence is found for $\eta < 0.49$, where η is the volume fraction of colloids. The results are in agreement with the findings of Sperry [27–29]. Experimentally, Gast et al. [32] later verified the predicted types of phase coexistence regions for a model colloid-polymer system. Colloid-polymer phase diagrams are commonly plotted in terms of the volume fraction of colloids η and the *relative* polymer concentration ϕ_p, defined in Eq. (3.4).

A semigrand canonical treatment for the phase behaviour of colloidal spheres plus nonadsorbing polymers was proposed by Lekkerkerker [33], who developed 'free volume theory' (also called 'osmotic equilibrium theory'), see Sect. 3.3.4. The main difference with TPT [30] is that free volume theory (FVT) accounts for polymer partitioning between the phases and corrects for multiple overlap of depletion layers, hence avoids the assumption of pair-wise additivity which becomes inaccurate for relatively thick depletion layers. These effects are incorporated through scaled particle theory (see for instance [34] and references therein). The resulting free volume theory (FVT) phase diagrams calculated by Lekkerkerker et al. [35] revealed that for $q < 0.3$ coexisting fluid-solid phases are predicted, whereas at low colloid volume fractions a gas–liquid coexistence is found for $q > 0.3$, as was predicted by TPT. A coexisting *three*-phase colloidal gas–

liquid-solid region, not present in TPT phase diagrams, was predicted by FVT for $q > 0.3$ and gained much attention. Experimental work [36, 37] demonstrated that this three-phase region indeed exists. Before the basics of phase behaviour of colloidal dispersions are discussed in more detail the focus is now first on the depletion interaction where more detailed derivations of the basic results are provided.

3.2 Depletion Interaction

In this section the depletion interaction between two flat plates and between two spherical colloidal particles is considered for penetrable hard spheres (phs). It is noted that besides polymers (for which PHS are a reasonable model when the polymers are small and dilute), small colloidal spheres, rods and plates can also act as depletants, see [38]. The penetrable hard sphere model, implicitly introduced by Asakura and Oosawa [2] and considered explicitly in detail by Vrij [5], is characterized by the fact that the spheres freely overlap each other but act as hard spheres with diameter 2δ when interacting with a wall or a colloidal particle.

3.2.1 Depletion Interaction Between Two Flat Plates

Consider Fig. 3.3 where two parallel flat plates in a polymer solution are sketched. The force per unit area, $K(h)$, between two parallel plates separated by a distance h,

Fig. 3.3 Schematic picture of two *parallel flat plates* in the presence of penetrable hard spheres (*dashed circles*)

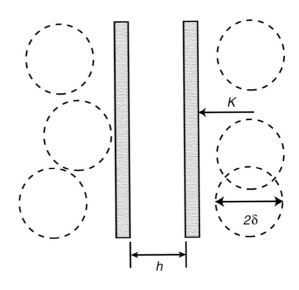

is the difference between the osmotic pressure Π_i inside the plates and the outside pressure Π_o

$$K = \Pi_i - \Pi_o. \tag{3.7}$$

Since the penetrable hard spheres behave thermodynamically ideally the osmotic pressure outside the plates is given by the Van 't Hoff law $\Pi_o = n_b k_B T$, where n_b is the bulk number density of the phs. When the plate separation h is equal to or larger than the diameter $\sigma(= 2\delta)$ of the penetrable hard spheres the osmotic pressure inside the plates is the same as outside, $\Pi_i = \Pi_o = n_b k_B T$. On the other hand, when the plate separation is less than the diameter of the penetrable hard spheres, no particles can enter the gap and $\Pi_i = 0$. This means that

$$K(h) = \begin{cases} -n_b k_B T & 0 \le h < 2\delta \\ 0 & h > 2\delta \end{cases}. \tag{3.8}$$

Since $K = -dW/dh$, integration yields the interaction potential $W(h)$ per unit area $W(h)$ between the plates

$$W(h) = \begin{cases} -n_b k_B T(2\delta - h) & 0 \le h < 2\delta \\ 0 & h \ge 2\delta \end{cases}. \tag{3.9}$$

3.2.2 Depletion Interaction Between Two Spheres

When the depletion zones with thickness δ around spherical colloidal particles with radius R start to overlap, i.e., when the distance $r(= h + 2R)$ between the centers of the colloidal particles is smaller than $2R + 2\delta$, a net force arises between the colloidal particles. It is useful to define an effective depletion radius R_d:

$$R_d = R + \delta. \tag{3.10}$$

The (attractive) force originates from an uncompensated (osmotic) pressure due to the depletion of penetrable hard spheres from the gap between the colloidal spheres. This is illustrated in Fig. 3.4 from which it can be deduced that the uncompensated osmotic pressure acts on the surface between $\theta = 0$ and $\theta_0 = \text{arc}\cos(r/2R_d)$. For obvious symmetry reasons only the component along the line connecting the centers of the colloidal spheres contributes to the total force. For the angle θ this component is $\Pi_o \cos\theta$ where the pressure is $\Pi_o = n_b k_B T$. The surface on which this force acts between θ and $\theta + d\theta$ equals $2\pi R_d^2 \sin\theta d\theta$. The total force between the colloidal spheres is obtained by integration:

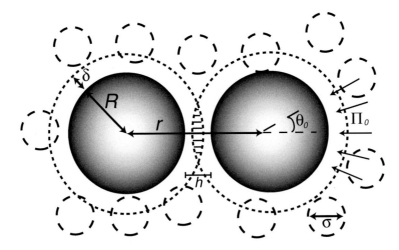

Fig. 3.4 Two hard spheres in the presence of penetrable hard spheres as depletants. The PHS impose an unbalanced osmotic pressure Π between the hard spheres resulting in an attractive force between them. The overlap volume of depletion layers between the hard spheres (*hatched*) has the shape of a lens with width $2\delta - h$ and height $2H = 2R_d \sin\theta_0$, where θ_0 is given by $\cos\theta_0 = r/2R_d$

$$K_s(r) = -2\pi n_b k_B T (R + \delta)^2 \int_0^{\theta_0} \sin\theta \cos\theta \, d\theta. \qquad (3.11)$$

Hence

$$\frac{K_s(r)}{n_b k_B T} = \begin{cases} \infty & r < 2R \\ -\pi R_d^2 [1 - (r/2R_d)^2] & 2R \leq r \leq 2R_d \\ 0 & r > 2R_d \end{cases} \qquad (3.12)$$

The minus sign in the right-hand side of Eq. (3.12) implies that the force is attractive. The depletion potential is now obtained by integration of the depletion force Eq. (3.12)

$$\begin{aligned} W_s(r) &= \int_r^{2R_d} K_s(r) dr \\ &= \infty & r < 2R \\ &= -n_b k_B T V_{\text{ov}}(r) & 2R \leq r \leq 2R_d, \\ &= 0 & r > 2R_d \end{aligned} \qquad (3.13)$$

with

$$V_{\text{ov}}(r) = \frac{4\pi}{3} R_d^3 \left[1 - \frac{3}{4} \frac{r}{R_d} + \frac{1}{16} \left(\frac{r}{R_d} \right)^3 \right] \qquad (3.14a)$$

$$V_{ov}(h) = \frac{\pi}{6}(2\delta - h)^2(3R + 2\delta + h/2) \tag{3.14b}$$

The result of Eq. (3.14a), in which r is the variable, was first obtained by Vrij [5] with PHS explicitly as depletants. In Eq. (3.14b) the variable is h and was already given (without explicit derivation) in Eq. (3.2). Both Eqs. (3.14a) and (3.14b) are frequently used in the literature.[2] The expression for $W_s(r)$ in Eq. (3.13) equals the osmotic pressure ($n_b k_B T$) times overlap volume V_{ov}. In the (Derjaguin) limit $\delta \ll R$ the attractive parts of the force (Eq. 3.11) and interaction potential (Eq. 3.13) between the spheres take on even simpler forms:

$$\frac{K_s(h)}{n_b k_B T} = -\pi R(2\delta - h) \tag{3.15}$$

and

$$\frac{W_s(h)}{n_b k_B T} = -2\pi R(\delta - h/2)^2. \tag{3.16}$$

For the contact potential $W_{dep} = W_s(h = 0)$ the result

$$\frac{W_{dep}}{k_B T} = -2n_b \pi R \delta^2 = -\frac{3}{2}\frac{\phi_d}{q} \tag{3.17}$$

is obtained, where ϕ_d is the relative concentration of penetrable hard spheres $4\pi\delta^3 n_b/3$, see also the definition of Eq. (3.4) for ϕ_p.

It is noted that in case of (small) hard spheres as depletants the results are identical to those above only in the dilute limit. For the minimum attraction between two spheres at contact the depletion force up to second order in hard sphere depletant volume fraction η_{ss} is [38, 39]:

$$\frac{W_{s,min}}{k_B T} = -\frac{3}{2}\frac{R}{\delta}\left(\eta_{ss} + \frac{1}{5}\eta_{ss}^2\right), \tag{3.18}$$

where δ now equals the radius of the small hard spheres. For small values of η_{ss} Eq. (3.17) is recovered for $\eta_{ss} = \phi_d$. At higher volume fractions the depletion attraction at contact between the big hard spheres by small hard spheres is however larger than the depletion due to penetrable hard spheres. Besides a depletion

---

[2]A smooth transition between these forms is:

$$V_{ov}(r) = \frac{2\pi}{3}(R_d - r/2)^2(2R_d + r/2).$$

attraction, depletion effects of small hard spheres also lead to a repulsive contribution to the interaction between two spheres. Near $h = 2\delta$ a positive maximum of the pair interaction is found with a maximum (up to second order in η_{ss})

$$\frac{W_{s,max}}{k_B T} = \frac{6R}{5\delta} \eta_{ss}^2.$$

3.3 Phase Behaviour of Colloidal Dispersions

Phase transitions are the result of physical properties of a collection of particles depending on the colligative properties. In Sect. 3.2 we focused on two-body interactions. Depletion effects are commonly not pair-wise additive [40–43]. Therefore, the prediction of phase transitions of particles with depletion interaction is not straightforward. As a starting point the Van der Waals model is recalled and applied to a collection of colloidal spheres with long-ranged attraction. Then a more advanced description for the thermodynamic properties of the pure colloidal dispersion are given. Subsequently, this is used to describe the phase behaviour of a collection of hard spheres plus additional Yukawa attraction, with a variable range of the interaction. Next, the basics of the free volume theory for the phase behaviour of colloids + depletants is explained. Only the simplest type of depletant, the penetrable hard sphere, is considered here. For experimental methods that enable measuring (depletion) interaction potentials between particles I refer to [44].

3.3.1 Phase Behaviour of a Van der Waals Fluid

The seminal equation of state of van der Waals [45] for the pressure P for N particles in a volume V reads:

$$P = \frac{N k_B T}{V - bN} - a \left(\frac{N}{V}\right)^2. \tag{3.19}$$

Here b is the excluded volume per particle, which is $4v_c$ for hard spheres. Using this equation van der Waals could implicitly demonstrate that a fluid can only phase separate when there is both excluded volume interaction (expressed via the contribution of the bN term) as well as attraction (the aN^2/V^2 term) between the molecules. It allows to describe the gas–liquid equilibria for a wide range of atomic and molecular substances and it revealed that the phase behaviour of low molecular systems is rather universal.

Let us now describe a colloidal dispersion of hard spheres plus an attraction in a very simple manner using the van der Waals model of Eq. (3.19). This can be done when considering the solvent in a colloidal dispersion as effective background. When performing computations on phase coexistence it is useful to use normalized quantities. Hence the normalized Helmholtz energy $\widetilde{F}$, the dimensionless chemical potential $\widetilde{\mu}$ and normalized pressure $\widetilde{\Pi}$ are introduced:

$$\widetilde{F} = \frac{Fv_c}{k_B TV}, \quad \widetilde{\mu} = \frac{\mu}{k_B T}$$

$$\eta = \frac{Nv_c}{V}, \quad \widetilde{\Pi} = \frac{\Pi v_c}{k_B T}. \tag{3.20}$$

For sake of completeness the definition of the volume fraction η is added. Note that instead of the pressure P of a molecular fluid here the normalized *osmotic* pressure Π is used because solvent is present.

The van der Waals equation can now be rewritten in dimensionless form:

$$\widetilde{\Pi} = \frac{\eta}{1 - 4\eta} - \gamma \eta^2, \tag{3.21}$$

with $\gamma = a/(k_B T v_c)$. In order to compute phase coexistence the osmotic pressure and the chemical potential for the van der Waals fluid are required. From thermodynamics $(P = -(\partial F/\partial V)_{N,T})$ it follows

$$\widetilde{\Pi} = -\left(\frac{\partial(\widetilde{F}/\eta)}{\partial(1/\eta)}\right)_{V,T} = \eta\left(\frac{\partial\widetilde{F}}{\partial\eta}\right)_{V,T} - \widetilde{F} = \eta\widetilde{\mu} - \widetilde{F}. \tag{3.22}$$

Hence the Helmholtz energy follows as:

$$\widetilde{F} = \eta \ln(\Lambda^3/v_c) + \eta \ln \eta - \eta - \eta \ln(1 - 4\eta) - \gamma \eta^2. \tag{3.23}$$

The last two terms are the result of the integration of Eq. (3.21) using Eq. (3.22). The other terms are the ideal contributions that follow from the ideal gas reference state [46]; Λ is the De Broglie wavelength.[3] Using

$$\widetilde{\mu} = \left(\frac{\partial\widetilde{F}}{\partial\eta}\right)_{N,T}, \tag{3.24}$$

the chemical potential follows as

$$\widetilde{\mu} = \widetilde{\mu}^+ + \ln\left(\frac{\eta}{1 - 4\eta}\right) + \frac{4\eta}{1 - 4\eta} - 2\gamma\eta, \tag{3.25}$$

---

[3] $\Lambda = h/\sqrt{2\pi m_c k_B T}$, with the colloid mass m_c and Planck's constant h.

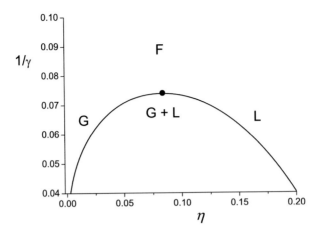

Fig. 3.5 Gas–liquid binodal following the van der Waals equation of state. The full *circle* is the critical point

with $\tilde{\mu}^+ = \ln \Lambda^3/v_c$. Now all ingredients to compute gas–liquid coexistence are available. The binodal gas–liquid coexistence curve follows from solving the coexistence conditions

$$\begin{aligned}
\tilde{\Pi}(\eta_G) &= \tilde{\Pi}(\eta_L) \\
\tilde{\mu}(\eta_G) &= \tilde{\mu}(\eta_L)
\end{aligned}, \qquad (3.26)$$

where η_G and η_L are the volume fractions of the particles in the gas and liquid phases, respectively. In Fig. 3.5 the binodal is plotted. The critical point (cp) follows analytically from Eq. 3.21 as $\gamma_{cp} = 27/2$ and $\eta_{cp} = 1/12$ and is indicated as full circle. The quantity γ^{-1} can be regarded as effective temperature. It is clear from Eq. (3.21) that the description of the fluid diverges at $\eta = 0.25$, which is in fact incorrect. Therefore the focus is first on more accurate expressions for the equations of state for a collection of pure hard spheres in the next section.

3.3.2 Phase Behaviour of a Hard Sphere Dispersion

The focus is first on the equation of state for the fluid phase of hard spheres interacting through the hard-sphere interaction

$$W(h) = \begin{cases} \infty & \text{for} \quad h < 0 \\ 0 & \text{otherwise} \end{cases}, \qquad (3.27)$$

Carnahan and Starling [48] found that the second and higher order virial coefficients for a collection of hard spheres can, to a good approximation, can be written as

Table 3.1 Values for the 2nd up to the 8th virial coefficient of hard spheres [47] in comparison with the Carnahan-Starling result Eq. (3.28). The numbers in the second and third column are B_i/v_c^{i-1} for $i = 2, 3,...0.8$

i	Exact/Numerical	CS Eq. (3.28)
2	4	4
3	10	10
4	18.36	18
5	28.22	28
6	39.82	40
7	53.34	54
8	68.53	70

$$B_{m+1} = (m^2 + 3m)v_c^m. \tag{3.28}$$

In Table 3.1 exact [47] virial coefficients are compared with the approximation given by Eq. (3.28). Inserting Eq. (3.28) into the general definition for the virial expansion of the (osmotic) pressure[4] [49],

$$\frac{\Pi_f^0 v_c}{k_B T} = \eta + \sum_{m=2} \frac{B_m}{v_c^{m-1}} \eta^m, \tag{3.29}$$

yields the Carnahan-Starling equation of state [48] for a fluid of (colloidal) hard spheres:

$$\tilde{\Pi}_f^0 = \frac{\Pi_f^0 v_c}{k_B T} = \frac{\eta + \eta^2 + \eta^3 - \eta^4}{(1 - \eta)^3}. \tag{3.30}$$

In Fig. 3.6 (left part) the (osmotic) pressure given by the Carnahan-Starling equation of state is compared to computer simulation data. Obviously, Eq. (3.30) is indeed very accurate.

From the Gibbs-Duhem relation $SdT - Vd\Pi + Nd\mu = 0$ the chemical potential can be calculated from the pressure. For constant T this relation may be written as

$$d\Pi = \frac{\eta}{v_c} d\mu. \tag{3.31}$$

Now μ follows as

$$\mu = k_B T \ln \frac{\Lambda^3}{v_c} + v_c \int_0^\eta \frac{1}{\eta'} \frac{d\Pi}{d\eta'} d\eta', \tag{3.32}$$

where $d\Pi/d\eta$ can be calculated from Eq. (3.30) for a fluid of hard spheres. The result for the chemical potential (normalized as $\tilde{\mu} = \mu/k_B T$) of a hard sphere in a fluid with volume fraction of hard spheres η follows now as

[4]the '0' refers to hard spheres and the subscript 'f' indicates a fluid phase.

$$\tilde{\mu}_f^0 = \ln\frac{\Lambda^3}{v_c} + \ln\eta + \frac{3-\eta}{(1-\eta)^3} - 3 \tag{3.33}$$

Using the standard thermodynamic result $\tilde{\Pi} = \eta\tilde{\mu} - \tilde{F}$, the resulting canonical free energy of the pure hard-sphere dispersion of a fluid is:

$$\tilde{F}_f^0 = \eta\left[\ln(\eta\Lambda^3/v_c) - 1\right] + \frac{4\eta^2 - 3\eta^3}{(1-\eta)^2}. \tag{3.34}$$

The first term on the right-hand side of Eq. (3.34) is the ideal contribution, while the second hard-sphere interaction term is the Carnahan-Starling equation of state [48].

To obtain the thermodynamic functions of the hard-sphere crystal one can use the cell model of Lennard-Jones and Devonshire [52]. For details I refer to our book [38]. The cell model result for the normalized Helmholtz energy of an fcc crystal is

$$\tilde{F}_s^0 = \eta\ln\left(\frac{27\Lambda^3}{8v_c}\right) - 3\eta\ln\left[\frac{\eta_{cp}}{\eta} - 1\right], \tag{3.35}$$

where $\eta_{cp} = \pi/3\sqrt{2} \simeq 0.74$ is the volume fraction at close packing. Using Eqs. (3.22) and (3.24) the dimensionless osmotic pressure and chemical potential become:

$$\tilde{\Pi}_s^0 = \frac{3\eta}{1 - \eta/\eta_{cp}}, \tag{3.36}$$

and

$$\tilde{\mu}_s^0 = \ln\frac{\Lambda^3}{v_c} + \ln\left[\frac{27}{8\eta_{cp}^3}\right] + 3\ln\left[\frac{\eta}{1-\eta/\eta_{cp}}\right] + \frac{3}{1-\eta/\eta_{cp}}. \tag{3.37}$$

The pressure given by Eq. (3.36) can be compared to computer simulation data and, as can be seen in Fig. 3.6 (right part), turns out to be highly accurate. The result for the chemical potential given by Eq. (3.37) is close to the computer simulation results. The constant on the right-hand side $\ln\left[27/8\eta_{cp}^3\right] = 2.1178$ is quite close to 2.1306, which can be abstracted from computer simulations [53]. The full free energy expression for the hard-sphere solid phase can now also be written as

$$\tilde{F} = 2.1178\eta + 3\eta\ln\left(\frac{\eta}{1-\eta/\eta_{cp}}\right) + \eta\ln\left(\Lambda^3/v_c\right). \tag{3.38}$$

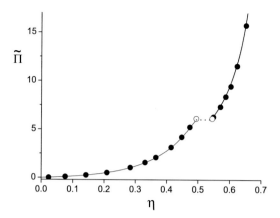

Fig. 3.6 The (*osmotic*) pressure of hard spheres. The *curves* are the Carnahan-Starling expression Eq. (3.30) for a fluid ($\eta \leq 0.494$) and the cell model result Eq. (3.36) for an fcc crystal (*solid curves*; $\eta \geq 0.545$). The full symbols are Monte Carlo computer simulation results [50]. The two *open symbols* correspond to the fluid-solid coexistence from simulation [51], the *dotted line* connects these binodal points

Solving the coexistence conditions

$$\begin{aligned}
\tilde{\Pi}_f^0(\eta_f) &= \tilde{\Pi}_s^0(\eta_s) \\
\tilde{\mu}_f^0(\eta_f) &= \tilde{\mu}_s^0(\eta_s)
\end{aligned} , \tag{3.39}$$

yields coexisting volume fractions $\eta_f = 0.491$, $\eta_s = 0.541$ and a coexistence pressure $\tilde{\Pi} = 6.01$. These values are indeed very close to computer simulation results, first accurately performed by Hoover and Ree [51], see the comparison in Fig. 3.6.

A collection of pure hard spheres is athermal; the thermodynamic properties are fully determined by entropy. At low densities the configurations of maximum entropy correspond to disordered arrangements. As the density increases crystalline arrangements lead to a more efficient packing and make more arrangements possible above some volume fraction, see Fig. 3.7. The fluid–crystal transition has been observed for instance in suspensions of sterically stabilized silica particles [54] and sterically stabilized PMMA particles [55] with low size dispersity. In addition to the fluid-crystal transition an amorphous glassy phase was observed above a volume fraction $\eta = 0.58$. For such high volume fractions the particles become so tightly trapped or caged that they do not crystallize but remain in long-lived metastable states, termed called colloidal glasses.

Accurate expressions for the equation of state of a fluid of hard spheres and for an fcc crystal of hard spheres have been presented above and will be used further on. Next it is interesting to account for additional attractions between the particles. As Boltzmann already pointed out [56] the attractive term of the van der Waals equation is only valid for long-ranged attractions. In order to investigate the effect

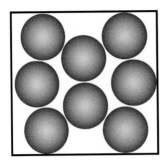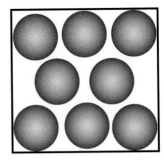

Fig. 3.7 Schematic pictures of a hard-sphere fluid (*left*) and hard spheres with 'crystalline' order (*right*); free volume entropy drives freezing

of the range of the attraction on the phase behaviour the focus is therefore now on particles with a hard-core plus Yukawa[5] attraction.

3.3.3 Phase Behaviour of a Dispersion of Hard Cores Plus Yukawa Attraction

Here a collection of hard-core spheres with hard-core diameter $2R$ plus a Yukawa attraction is considered. The hard-core Yukawa pair potential between two spheres can be written as[6]

$$W(h) = \begin{cases} \infty & \text{for } h < 0 \\ -\frac{\varepsilon}{1+h/2R}\exp(-\kappa h) & \text{otherwise} \end{cases}, \qquad (3.40)$$

where h equals $r - 2R$, with r the centre-to-centre distance. The strength of the attraction is the contact potential $W(0) = \varepsilon$ and the range is the screening length κ^{-1}. The relative range of attraction q_Y (with respect to the particle radius R) is defined as $1/\kappa R$.

For a collection of particles interacting through this pair interaction it is possible to derive an equation of state. Tang and Lu [57] solved the Ornstein-Zernicke (OZ) equation using the mean spherical approximation (MSA) closure in Fourier and Laplace space and found that each perturbation term in the contact potential can be solved analytically. For the first-order expansion, termed FMSA, this leads to relatively simple and rather accurate solutions for the thermodynamic properties [58, 59].

---

[5]The term Yukawa potential originally stems stems from the quantum mechanical theory of nuclear interactions. In a more general context, it is often used for potentials with a distance profile of the type $\exp\{-\kappa r\}/r$.

[6]Although here only Yukawa attractions are considered this description also holds for spheres interacting through a hard-core repulsive Yukawa interaction.

Tang et al. [59] derived an analytical expression for $\widetilde{F}$ (see Fig. 3.8 for an illustration of determining coexistence points from the free energy) that is accurate and which can be written in the Van der Waals-form, albeit the γ term is now dependent on η $(G \sim \gamma \eta^2)$. Tuinier and Fleer [60] found that this form can be simplified even more with almost no loss of accuracy. The final form is

$$\widetilde{F} = \widetilde{F}^0 - \beta \varepsilon G(\eta), \qquad (3.41)$$

where $\beta = 1/k_B T$ with a simple volume fraction-dependent function $G(\eta)$,

$$G(\eta) = \eta^2 \frac{a_0 + a_1 \eta}{b_0 + b_1 \eta + b_2 \eta^2}. \qquad (3.42)$$

The coefficients a_i and b_i depend only on the relative range of the Yukawa potential q_Y. They are expressed most easily in its inverse $k = 1/q_Y = \kappa R$:

$$a_0 = 4k^2 + 2k$$
$$a_1 = 2k^2 + 4k \qquad (3.43)$$

and

$$b_0 = \frac{2k^3}{3}$$
$$b_1 = \chi_1 - \frac{4k^3}{3} + 2k^2 - 1 \qquad (3.44)$$
$$b_2 = \chi_2 + \frac{2k^3}{3} - 2k^2 + 3k - 2,$$

where χ_1 and χ_2 are defined as:

$$\chi_1 = (2k + 1)\exp(-2k)$$
$$\chi_2 = (k + 2)\exp(-2k). \qquad (3.45)$$

Note that a volume fraction-independent γ survives in the low η-limit, for which G becomes $3q_Y(2 + q_Y)\eta^2$, so $\gamma = -3\beta \varepsilon q_Y(2 + q_Y)$.[7] However, in very concentrated systems G becomes proportional to η.

An analytical expression for the chemical potential is obtained from $\tilde{\mu} = \partial \widetilde{F}/\partial \eta$:

$$\tilde{\mu} = \tilde{\mu}^0 - \beta \varepsilon H(\eta), \qquad (3.46)$$

---

[7]This can be regarded as an explicit definition for γ within the van der Waals model when the attraction is described as a long-ranged Yukawa attraction.

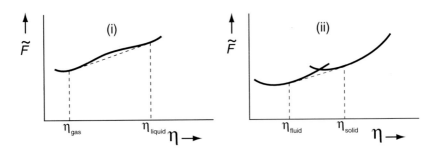

Fig. 3.8 The dimensionless Helmholtz energy $\widetilde{F} = \eta\tilde{\mu} - \widetilde{\Pi}$ as a function of volume fraction η. Schematic view of the common tangent construction (*straight lines*) to determine the phase coexistence in mixtures of colloidal hard spheres and phs. (i): gas–liquid coexistence, (ii): fluid-solid coexistence. The *dashed lines* represent the common tangent construction with intercept $-\widetilde{\Pi}$ and *slope* $\tilde{\mu}$

where $H(\eta)$ is given by:

$$H(\eta) = k\eta\frac{c_0 + c_1\eta + c_2\eta^2 + c_3\eta^3}{(b_0 + b_1\eta + b_2\eta^2)^2}, \tag{3.47}$$

with the following q-dependent coefficients:

$$
\begin{aligned}
c_0 &= \frac{8k^4}{3} + \frac{4k^3}{3} \\
c_1 &= (2k+1)\chi_1 + \frac{8k^4}{3} + \frac{14k^3}{3} + 3k^2 + 2k + 1 \\
c_2 &= (k+2)\chi_1 - \frac{4k^4}{3} - \frac{2k^3}{3} + 4k^2 + k + 1 \\
c_3 &= (2k+1)\chi_2 + \frac{4k^4}{3} - \frac{10k^3}{3} + 4k^2 - 3k - 2
\end{aligned}
\tag{3.48}
$$

The (osmotic) pressure follows from $\widetilde{\Pi} = \eta\tilde{\mu} - \widetilde{F}$:

$$\widetilde{\Pi} = \widetilde{\Pi}^0 + \beta\varepsilon J(\eta), \tag{3.49}$$

with $J(\eta)$ given by:

$$J(\eta) = G(\eta) - \eta H(\eta). \tag{3.50}$$

Now analytical expressions are available for both $\tilde{\mu}$ and $\widetilde{\Pi}$.

In principle this provides sufficient information to compute the binodal curves. In Fig. 3.9 gas–liquid binodals for attractive hard core Yukawa spheres are plotted for various ranges of the attraction κR. As an illustration computer simulation results are plotted as data points as well. It is clear that the unstable region shifts to

Fig. 3.9 Gas–liquid coexistences of a collection of hard-core attractive Yukawa spheres for three values of κR as indicated. Symbols are simulation results [61], the *solid curves* are the analytical expressions for the binodals. *Crosses* represent the theoretical critical points

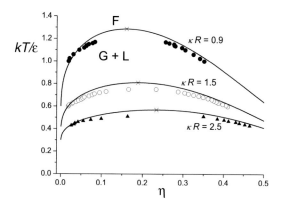

lower temperatures for shorter ranges of attraction; at identical values of $k_B T/\varepsilon = \beta\varepsilon$ there is similar attraction at close contact between the spheres but there is an additional attraction when the spheres are further apart in case of smaller κR.

In order to calculate the gas–liquid binodal ε is eliminated from Eqs. (3.46) and (3.49) to find the analytical coexistence relation:

$$\beta\varepsilon = \frac{\tilde{\mu}_l^0(\eta_l) - \tilde{\mu}_g^0(\eta_g)}{H(\eta_l) - H(\eta_g)} = \frac{\widetilde{\Pi}_l^0(\eta_l) - \widetilde{\Pi}_g^0(\eta_g)}{J(\eta_g) - J(\eta_l)}. \tag{3.51}$$

For both μ_g^0 and μ_l^0 Eq. (3.33) is used and for both Π_g^0 and Π_l^0 Eq. (3.30) is applied. For a given q the second and third parts of this equation relate the volume fractions at coexistence. For instance, a value for η_g is chosen and the corresponding η_l is solved using the second equality of Eq. (3.51). The first equality then tells to which Yukawa contact potential ε those binodal concentrations correspond. It is noted that the calculation of binodals in the analytical model involve nothing more than solving one equation in one unknown.

Fluid–solid coexistence is obtained analogously:

$$\varepsilon = \frac{\mu_s^0(\eta_s) - \mu_f^0(\eta_f)}{H(\eta_s) - H(\eta_f)} = \frac{\Pi_s^0(\eta_s)v - \Pi_f^0(\eta_f)v}{J(\eta_f) - J(\eta_s)}. \tag{3.52}$$

In this case μ_s^0 is used from Eq. (3.37) and Π_s^0 from Eq. (3.36). As in Eq. (3.51), the fluid parts are obtained from Eqs. (3.30) and (3.33).

In Fig. 3.10 a full phase diagram (gas–liquid and fluid–solid) is presented for $\kappa R = 2$ and compared to computer simulation results. It follows that there is a region where there is a stable fluid (low η, high T), a region where the fluid phase separates into a gas and a liquid below the critical point, a region where fluid and solid coexist (near $0.45 \lesssim \eta \lesssim 0.6$), a region where there is a solid phase (high η) and a gas–solid coexistence region (low T).

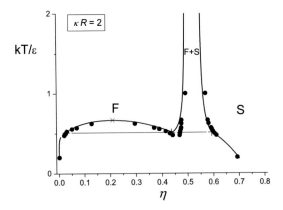

Fig. 3.10 Phase behavior of a dispersion of spherical hard-core attractive Yukawa particles with $\kappa R = 2$. Symbols are simulation results [62], the *solid curves* are the analytical results. *Cross* is the theoretical critical point, plusses identify the three coexisting phases of the triple points; the three plusses are connected through a *thin line*

Fig. 3.11 As Fig. 3.10 but for $\kappa R = 12.5$. The GL coexistence is now metastable, and there is no triple point

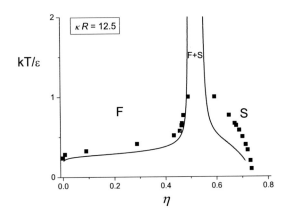

For large values of κR the gas–liquid region becomes metastable with respect to fluid-solid coexistence. See Fig. 3.11 where the fluid-solid binodals are plotted for $\kappa R = 12.5$. Stable liquid configurations require that the particles still attract one another when their interparticle distance fluctuates. Hence for short-ranged attractions the gas–liquid coexistence gets metastable.

3.4 Phase Behaviour of a Colloid-Polymer Mixture

Several theories have been developed that enable calculations of phase transitions in systems with depletion interactions. The first successful treatment accounting for the colligative thermodynamic properties mediated by depletion interactions

[24, 30] is thermodynamic perturbation theory [46, 49]. In this classical approach depletion effects can be treated as a perturbation to the hard-sphere free energy, as was done by Gast et al. [30]. Their work predicted that for a sufficient depletant concentration, the depletion interaction leads to a phase diagram with stable colloidal gas, liquid and solid phases for $\delta/R \geq 0.3$. For small depletants with $\delta/R \leq 0.3$ only colloidal fluid and solid phases are thermodynamically stable, and the gas–liquid transition is meta-stable. Although implementation of this theory is straightforward, it has the drawback that it does not account for depletant partitioning over the coexisting phases. Subsequent developments originate from liquid state approaches. Examples are density functional theory [63], PRISM [64] and the Gaussian core model [65].

3.4.1   Free Volume Theory

In the early nineties of the last century a theory that accounts for depletant partitioning over the coexisting phases was developed [35], which nowadays is commonly referred to as free volume theory (FVT) [66]. This theory is based upon considering the osmotic equilibrium between a (hypothetical) depletant and the colloid + depletant system. The depletants were simplified as penetrable hard spheres. See the sketch in Fig. 3.12.

This theory has the advantage that the depletant concentrations in the coexisting phases follow directly from the (semi)grand potential which describes the colloid plus depletant system. As illustrated in Fig. 3.13, the system can arrange itself such as to provide a larger free volume for the depletants by overlap of two depletion

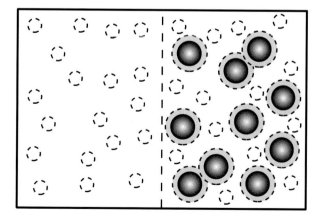

Fig. 3.12 A system (*right*) that contains colloids and penetrable hard spheres (*phs*) in osmotic equilibrium with a reservoir (*left*) only consisting of phs. A hypothetical membrane that allows permeation of solvent and phs but not of colloids is indicated by the *dashed line*. Solvent is considered as 'background'

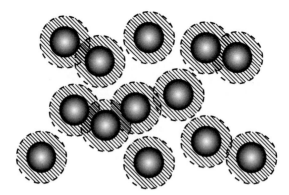

Fig. 3.13 Illustration of the free volume V_{free}: it is the unshaded volume not occupied by the colloids plus (partially overlapping) depletion layers

zones. This (entropic) physical origin of the phase transitions induced by depletion interactions is incorporated into the theory via the available volume for the depletants.

In FVT multiple overlap of depletion zones with thickness δ, see Fig. 3.14, is taken into account. Multiple overlap occurs for

$$\frac{\delta}{R} > \frac{2}{3}\sqrt{3} - 1 \simeq 0.15,$$

where three depletion zones start to overlap, see Fig. 3.14. Only for $\delta/R < 0.15$ is a colloid/depletant mixture pair-wise additive. For large δ/R a mixture of hard spheres plus penetrable hard spheres differs fundamentally from a mixture of hard-core spheres that directly attract one another [67]. This has a considerable influence on the

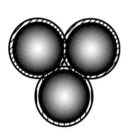 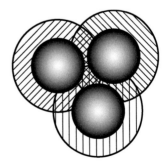

Fig. 3.14 Three hard spheres surrounded by depletion layers (*hatched areas*). When the depletion layers are thin (*left*) there is no multiple overlap of depletion layers; the system is pair-wise additive. For thicker depletion layers (*right*) multiple overlap of depletion layers occurs and depends on more than two-body contributions. The lowest value for δ/R where multiple overlap occurs follows from considering the triangle formed by the 3 particle centres; its edge is $2R + h$ at particle separation h. Multiple overlap starts when the *centre* of the triangle is a distance $R + \delta$ from the corners

topology of the phase diagram [68]. Multiple overlap of depletion layers widens the liquid window, which is the parameter range with phase transitions that include a stable liquid, in comparison with a pair-wise additive system [66].

The starting point of FVT is the calculation of the semigrand potential Ω describing the system of N_c colloidal spheres plus N_d depletants as depicted in Fig. 3.12.

$$\Omega(N_c, V, T, \mu_d) = F(N_c, N_d, V, T) - \mu_d N_d. \tag{3.53}$$

Using the thermodynamic relation

$$\left(\frac{\partial \Omega}{\partial \mu_d}\right)_{N_c, V, T} = -N_d, \tag{3.54}$$

one can write

$$\Omega(N_c, V, T, \mu_d) = F^0(N_c, V, T) - \int_{-\infty}^{\mu_d} N_d(\mu_d') d\mu_d'. \tag{3.55}$$

Here $F^0(N_c, V, T)$ is the (Helmholtz) free energy of the colloidal hard sphere suspension without added depletant as given by Eq. (3.34) (fluid) or Eq. (3.38) (solid). Note that Eq. (3.55) is still exact and can be used (with approximations) to compute the phase behaviour of hard spheres plus interacting depletants (small hard spheres, interacting polymers, hard rods) [38, 66, 69]. Below only the case of non-interacting depletants is treated.

The essential step within FVT is the calculation of the number of depletants in the system of hard spheres + depletants as a function of the chemical potential μ_d imposed by the depletants in the reservoir. In the calculations presented below the colloidal hard spheres have a radius R and the depletants are described as penetrable hard spheres with radius δ.

For the calculation of N_d the Widom insertion theorem [70] is used according to which the chemical potential of the depletants in the mixture of hard spheres and depletants can be written as

$$\mu_d^S = \text{const} + k_B T \ln \frac{N_d}{\langle V_{\text{free}} \rangle}. \tag{3.56}$$

Here $\langle V_{\text{free}} \rangle$ is the ensemble-averaged free volume for the depletants in the system 'S' of hard spheres, illustrated in Fig. 3.13. The chemical potential of the depletants in the reservoir is simply

$$\mu_d^R = \text{const} + k_B T \ln n_d^R, \tag{3.57}$$

where n_d^R is the number density of the depletants in the reservoir 'R'. By equating the depletant chemical potentials Eqs. (3.56) and (3.57) the result

$$N_d = n_d^R \langle V_{\text{free}} \rangle \tag{3.58}$$

is obtained. The average free volume obviously depends on the volume fraction of the hard spheres in the system but also on the chemical potential of the depletants. The activity of the depletants affects the average configuration of the hard spheres. Now the key approximation is made to replace $\langle V_{\text{free}} \rangle$ by the free volume in the pure hard sphere dispersion $\langle V_{\text{free}} \rangle_0$:

$$N_d = n_d^R \langle V_{\text{free}} \rangle_0. \tag{3.59}$$

This expression is correct in the limit of low depletant activity but is only an approximation for higher depletant concentrations. Substituting the approximate Eq. (3.59) into Eq. (3.55) and using the Gibbs-Duhem relation,

$$n_d^R d\mu_d = d\Pi^R, \tag{3.60}$$

gives

$$\Omega(N_c, V, T, \mu_d) = F^0(N_c, V, T) - \Pi^R \langle V_{\text{free}} \rangle_0, \tag{3.61}$$

where $\Pi^R = n_d k_B T$ is the (osmotic) pressure of the depletants in the reservoir.

As expressions are available for the free energy of the hard sphere system (both in the fluid and solid state, see Sect. 3.3.2) and for the pressure of the reservoir, the only remaining quantity to calculate is $\langle V_{\text{free}} \rangle_0$. According to the Widom insertion theorem expressed in Eq. (3.56):

$$\mu_d = \text{const} + k_B T \ln \frac{N_d}{\langle V_{\text{free}} \rangle_0}. \tag{3.62}$$

The chemical potential μ_d can however also be written in terms of the reversible work W required for inserting a depletant into the hard sphere dispersion:

$$\mu_d = \text{const} + k_B T \ln \frac{N_d}{V} + W. \tag{3.63}$$

The free volume fraction α now follows from combining Eqs. (3.62) and (3.63):

$$\alpha = \frac{\langle V_{\text{free}} \rangle_0}{V} = e^{-W/k_B T}. \tag{3.64}$$

3.4.2 Scaled Particle Theory

An expression for the work of insertion W can be obtained from scaled particle theory (SPT) [71]. The work W is calculated is by expanding (scaling) the size of the sphere to be inserted from zero to its final size: the radius of the scaled particle is $\lambda\delta$, with λ running from 0 to 1. In the limit $\lambda \to 0$, the inserted sphere approaches a point particle. In this limiting case it is very unlikely that the depletion layers overlap. The free volume fraction in this limit can therefore be written as

$$\alpha = 1 - \eta\left(1 + \frac{\lambda\delta}{R}\right)^3,$$ (3.65)

It then follows from Eq. (3.64) that

$$W = -k_B T \ln\left[1 - \eta\left(1 + \lambda\frac{\delta}{R}\right)^3\right] \quad \text{for } \lambda \ll 1.$$ (3.66)

In the opposite limit of a large inserted scaled particle $\lambda \gg 1$ the work of insertion W can be approximated as the volume work needed to create a cavity $\frac{4\pi}{3}(\lambda\delta)^3$ and is given by

$$W = \frac{4\pi}{3}(\lambda\delta)^3 \Pi^0 \quad \text{for } \lambda \gg 1,$$ (3.67)

where Π^0 is the (osmotic) pressure of the hard sphere dispersion. In SPT the above two limiting cases are connected by expanding W as a series in λ:

$$W(\lambda) = W(0) + \left(\frac{\partial W}{\partial \lambda}\right)_{\lambda=0} \lambda + \frac{1}{2}\left(\frac{\partial^2 W}{\partial \lambda^2}\right)_{\lambda=0} \lambda^2 + \frac{4\pi}{3}(\lambda\delta)^3 \Pi^0.$$ (3.68)

This yields

$$\frac{W(\lambda = 1)}{k_B T} = -\ln[1 - \eta] + \frac{3q\eta}{1 - \eta}$$
$$+ \frac{1}{2}\left[\frac{6q^2\eta}{1 - \eta} + \frac{9q^2\eta^2}{(1 - \eta)^2}\right],$$
$$+ \frac{\frac{4\pi}{3}q^3 R^3 \Pi^0}{k_B T}$$ (3.69)

where q is the size ratio between the depletant with radius δ and the hard sphere with radius R

$$q = \frac{\delta}{R}.$$ (3.70)

In Appendix, the result for the SPT osmotic pressure is derived:

$$\frac{\Pi^0 v_c}{k_B T} = \frac{\eta + \eta^2 + \eta^3}{(1 - \eta)^3}. \tag{3.71}$$

Inserting Eq. (3.71) into Eq. (3.69) and using Eq. (3.64) yields

$$\alpha = (1 - \eta)\exp[-Q(\eta)], \tag{3.72}$$

where

$$Q(\eta) = ay + by^2 + cy^3, \tag{3.73}$$

with $a = 3q + 3q^2 + q^3$, $b = \frac{9}{2}q^2 + 3q^3$ and $c = 3q^3$ and $y = \eta/(1 - \eta)$. In Fig. 3.15 the free volume fraction α predicted by SPT (Eq. 3.72) is compared to computer simulation results on hard spheres plus penetrable hard spheres for $q = 0.1$ as a function of η. As can be seen the agreement is very good, as for other q values [72]. Now all ingredients are available to compile the semigrand potential Ω given by Eq. (3.61).

From Ω the total pressure of the hard spheres + phs and the chemical potential of the hard spheres in the hard sphere + depletant system at given μ_d^0 are obtained:

$$\Pi_{tot} = -\left(\frac{\partial \Omega}{\partial V}\right)_{N_c, T, \mu_d} = \Pi^0 + \Pi^R\left(\alpha - n_c \frac{\partial \alpha}{\partial n_c}\right) \tag{3.74}$$

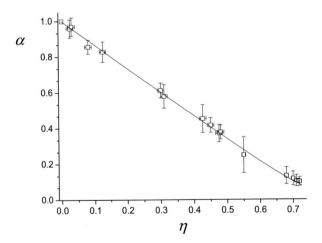

Fig. 3.15 Free volume fraction for penetrable hard spheres in a hard sphere dispersion for $q = \delta/R = 0.1$ as function of the hard sphere concentration. Data points are redrawn from Fortini et al. [50]. *Curve* is the SPT prediction of Eq. (3.72)

$$\mu_c = \left(\frac{\partial \Omega}{\partial N_c}\right)_{V,T,\mu_d} = \mu_c^+ - \Pi^R \frac{\partial \alpha}{\partial n_c}. \tag{3.75}$$

For non-interacting depletants Π^R is simply given by Van 't Hoff's law $\Pi^R = n_d^R k_B T$ or

$$\widetilde{\Pi}^R = \frac{\Pi^R v_c}{k_B T} = n_d^R v_d q^{-3} = \phi_d^R q^{-3}, \tag{3.76}$$

with ϕ_d^R the relative reservoir depletant concentration $n_d^R v_d$, where v_d is the volume of a depletant sphere. These penetrable hard spheres can, by definition, freely interpenetrate each other. It is useful to define the overlap condition of phs by $n^* v_d = 1$. At $n^* = 1/v_d$ the spheres fill the available space but (on average) do not yet interpenetrate; this happens only for $n > n^*$. Hence one may write $\widetilde{\Pi} = \phi_d q^{-3}$, where ϕ_d is the concentration of penetrable hard spheres relative to overlap.

3.4.3 Phase Diagrams

The phase behaviour of a system of hard spheres and depletants can now be calculated by solving the coexistence equations for a phase I in equilibrium with a phase II

$$\mu_c^I(n_c^I, \mu_d) = \mu_c^{II}(n_c^{II}, \mu_d), \tag{3.77}$$

$$\Pi^I(n_c^I, \mu_d) = \Pi^{II}(n_c^{II}, \mu_d). \tag{3.78}$$

For numerical computations of phase coexistence, it is convenient to work with dimensionless quantities. The dimensionless version of the free volume expression Eq. (3.61) for the (semi) grand potential is

$$\widetilde{\Omega} = \widetilde{F}^0 - \alpha \widetilde{\Pi}^R, \tag{3.79}$$

where $\widetilde{\Omega} = \Omega v_c / k_B T V$.

The sketch of Fig. 3.8 can also be drawn for the semigrand potential ($\widetilde{\Omega}$ instead of $\widetilde{F}$) as a function of the colloid volume fraction for given depletant reservoir concentration and size ratio q. A first criterion for two coexisting binodal compositions is equality of the slope because it corresponds to the chemical potential. The chemical potential of the colloids $\widetilde{\mu}_c$ can generally be expressed using the standard thermodynamic relation

$$\tilde{\mu}_c = \left(\frac{\partial \tilde{\Omega}}{\partial \eta}\right)_{\tilde{\Pi}^R, T, V}. \tag{3.80}$$

The (total) pressure is found from

$$\tilde{\Pi}_{tot} = \eta \tilde{\mu}_c - \tilde{\Omega}. \tag{3.81}$$

When two compositions can be connected through the common tangent (the thin straight lines in Fig. 3.8 connecting these compositions), binodal points are found; the intercepts of the extrapolated lines correspond to the total pressure $\tilde{\Pi}_{tot}$, see examples of scenarios for gas–liquid and fluid–solid coexistences in Fig. 3.8. For each depletant concentration the binodal compositions can be found in this manner; full phase diagrams can be constructed from such binodals.

For non-interacting depletants such as penetrable hard spheres the μ's and Π's in the phase coexistence Eqs. (3.77) and (3.78) can be written such that binodal colloid concentrations follow from solving one equation with a single unknown [66] as for the hard-core Yukawa spheres discussed earlier. Equations (3.74) and (3.75) can be rewritten as

$$\tilde{\mu} = \tilde{\mu}^0 + \tilde{\Pi}^R g(\eta) \tag{3.82}$$

$$\tilde{\Pi}_{tot} = \tilde{\Pi}^0 + \tilde{\Pi}^R h(\eta), \tag{3.83}$$

where $g = -\partial \alpha / \partial \eta$ and $h = \alpha + g\eta$, giving the following explicit expressions for g and h:

$$g(\eta) = e^{-Q(\eta)}\left\{1 + [1+y][a + 2by + 3cy^2]\right\}, \tag{3.84}$$

$$h(\eta) = e^{-Q(\eta)}\left\{1 + ay + 2by^2 + 3cy^3\right\}. \tag{3.85}$$

The gas–liquid binodal can be solved from the second and third parts of

$$\tilde{\Pi}^R = \frac{\tilde{\mu}_f^0(\eta_l) - \tilde{\mu}_f^0(\eta_g)}{g(\eta_g) - g(\eta_l)} = \frac{\tilde{\Pi}_f^0(\eta_l) - \tilde{\Pi}_f^0(\eta_g)}{h(\eta_g) - h(\eta_l)}, \tag{3.86}$$

where $\tilde{\mu}_f^0$ and $\tilde{\Pi}_f^0$ are only a function of η, see Eqs. (3.30) and (3.33). Hence, Eq. (3.86) gives a unique relation $\eta_l(\eta_g)$ at given q. For some value of η_g, within the region of η_g values where a colloidal gas coexists with a colloidal liquid, the corresponding value of η_l follows from the second equality of Eq. (3.86). The corresponding binodal depletant reservoir pressure $\tilde{\Pi}^R$ then follows from the first equality.

Similarly, the fluid-solid binodal can be obtained from

$$\widetilde{\Pi}^R = \frac{\tilde{\mu}_s^0(\eta_s) - \tilde{\mu}_f^0(\eta_f)}{g(\eta_f) - g(\eta_s)} = \frac{\widetilde{\Pi}_s^0(\eta_s) - \widetilde{\Pi}_f^0(\eta_f)}{h(\eta_f) - h(\eta_s)}, \tag{3.87}$$

where again $\tilde{\mu}_f^0$ is given by Eq. (3.33) and $\widetilde{\Pi}_f^0$ by Eq. (3.30); these are the fluid contributions. For colloidal dispersions in the solid state (fcc crystal) $\widetilde{\Pi}_s^0(\eta)$ and $\tilde{\mu}_s^0(\eta)$ are given by Eqs. (3.36) and (3.37), respectively.

Triple points have equal pressures and chemical potentials at colloidal gas, liquid *and* solid compositions. At the triple point expressions Eqs. (3.86) and (3.87) are connected through equal values for $\widetilde{\Pi}^R$ and, in principle, form a set of four equations from which the four coordinates of the triple point $(\eta_g, \eta_l, \eta_s, \widetilde{\Pi}^R)$ follow.

For large q ($q \geq 0.6$) the triple point can be approximated easily from Eqs. (3.82) and (3.83). The fluid-solid coexistence of the triple point occurs at nearly similar colloid concentrations as the pure hard sphere phase transition. For large q values, Eqs. (3.82) and (3.83) can be written as $\tilde{\mu}_f = \tilde{\mu}_f^0 = \tilde{\mu}_s^0$ and $\widetilde{\Pi}_f = \widetilde{\Pi}_f^0 = \widetilde{\Pi}_s^0$, because $g(\eta)$ and $h(\eta)$ vanish for large q. In the coexisting colloidal gas phase the colloid concentration is then extremely small so $\widetilde{\Pi}_g = \widetilde{\Pi}^R$, since $h(\eta) \to 1$,

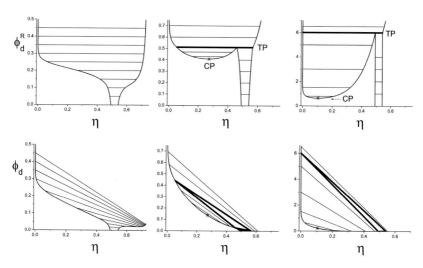

Fig. 3.16 Free volume theory predictions for the phase diagrams for hard spheres as depletants following Lekkerkerker et al. 1992 [35]. The *left* diagrams are for $q = 0.1$, *middle* $q = 0.4$, and *right* diagrams $q = 1.0$. *Upper* diagrams have depletant reservoir concentrations ϕ_d^R as ordinates, *lower* diagrams are in system depletant concentrations. *Triple lines* and *triangles* are indicated as *thick lines*. TP = triple point; CP = critical point (*asterisks* refer to the critical points). A few representative *tie-lines* are plotted as *thin lines*

implying $\widetilde{\Pi}^R = \widetilde{\Pi}_f^0 = \widetilde{\Pi}_s^0 = 6.01$ at the triple point. Hence, for large q the fluid-solid coexistence of the triple point occurs at nearly the same colloid concentrations as for the pure hard-sphere phase transition. The relative depletant concentration at the triple point now follows as $\phi_d^R \simeq \widetilde{\Pi}^R q^3 = 6.01 q^3$. As can be seen in Figs. 3.16 ($q = 1.0$) and 17 ($q = 0.6$) this is rather accurate.

The critical point can be found also as one equation in one unknown, for details, see [66]. The same applies to the *critical endpoint* (CEP), which corresponds to the q value where CP and TP coincide; it is the lowest q where a stable liquid is possible. See the extended discussions on liquid windows as related to the CEP in [66, 68].

In Fig. 3.16 phase diagrams are presented for $q = 0.1$, $q = 0.4$ and $q = 1.0$. As was already found by Gast et al. [30], for $q = 0.1$ there is only a fluid-crystal transition. For $\phi_d = 0$ the demixing gap is $0.491 < \eta < 0.541$ (see Sect. 3.3.2); with increasing depletant concentration this gap widens. For $q = 0.4$ there are a critical point (CP) and a triple point (TP) in the phase diagram, analogous to those found in simple atomic systems. At high depletant concentrations in the reservoir (above TP) a very dilute fluid (colloidal gas), coexists with a highly concentrated colloidal solid. Between TP and CP a colloidal gas (dilute fluid) coexists with a colloidal liquid (more concentrated fluid). At high volume fractions below the triple line, a colloidal liquid coexists with a colloidal solid phase. In the absence of depletant only the fluid-solid phase transition of a pure hard sphere dispersion remains. Increasing the depletant activity now plays a role similar to lowering the temperature in atomic systems. For larger q (see $q = 1.0$) the qualitative picture remains the same while the liquid window expands.

In the top diagrams of Fig. 3.16 the ordinate axis is the depletant concentration in the reservoir. The depletant concentrations in the system of coexisting phases can be obtained by using the relation

$$\phi_d = \alpha \phi_d^R.$$

Coexisting phases of course have the same μ_d and hence the same n_d^R but since the volume fractions of hard spheres and, hence, the free volume fractions α are different, n_d in the two (or three) phases are not the same, so the tie-lines are no longer horizontal. This is illustrated in the bottom diagrams of Fig. 3.16; now the ordinate axis gives the relative 'internal' or system concentrations ϕ_d. A few selected tie-lines are drawn to give an impression of depletant partitioning over the phases. Interestingly, the horizontal triple *line* in the presentation of the phase diagram at constant chemical potential μ_d (field-density representation) is now converted into a three-phase *triangle* system representation. The triple line connects three coexisting colloid concentrations at one fugacity (reservoir concentration).

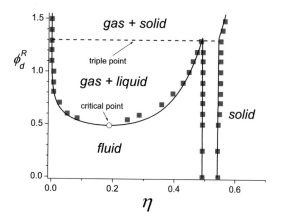

Fig. 3.17 Comparison of free volume theory (*curves*) with Monte Carlo computer simulations (data; [73]) for *q* = 0.6. *Open circle* = theoretical critical point

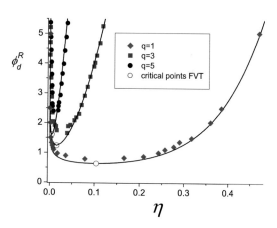

Fig. 3.18 Gas–liquid binodals for mixtures of HS + phs for large *q* values. *Curves* FVT; Data points: MC simulations by Dijkstra et al. [73] (*q* = 1) and Moncho-Jordá et al. [43] (*q* = 3 and *q* = 5)

In the system representation there are now three compositions (ϕ_d^1, η_g), (ϕ_d^2, η_l) and (ϕ_d^3, η_s). These three points in a (ϕ_d, η) plot form a triangle, within which there is three phase gas–liquid–solid coexistence.

As discussed in Sect. 3.3.2, the free volume theory is approximate in the sense that $\langle V_{free} \rangle$ is replaced by $\langle V_{free} \rangle_0$. To get an idea of the accuracy of the phase diagrams calculated with free volume theory the results for $q = 0.6$ are compared to computer simulations [73] in Fig. 3.17. The agreement is, given the fact that the free volume theory is approximate, very good. Also for $q = 0.1 - 1.0$ [73] and large q values [43] the agreement with simulations is striking. As a final illustration of the accuracy of FVT, colloidal gas–liquid binodals are plotted for $q = 1$, 3 and 5 in Fig. 3.18 and are compared to Monte Carlo computer simulation results.

Acknowledgments This text is highly inspired by several parts of the book I wrote with Henk Lekkerkerker and I thank him for the wonderful collaborations. I also acknowledge Agienus Vrij, Alvaro Gonzalez Garcia and Maartje S. Feenstra for useful discussions.

Appendix

As was the original objective of SPT [71], the pressure Π^0 of the hard sphere system can be obtained from the reversible work of inserting an identical sphere $(q = 1)$

$$\frac{W}{k_B T} = -\ln[1 - \eta] + \frac{6\eta}{1 - \eta} + \frac{9\eta^2}{2(1 - \eta)^2} + \frac{4\pi R^3 \Pi^0}{3k_B T}, \qquad (3.88)$$

to obtain the chemical potential of the hard spheres

$$\mu_c^0 = \text{const} + k_B T \ln \frac{N_c}{V} + W. \qquad (3.89)$$

Applying the Gibbs-Duhem relation

$$\frac{\partial \Pi^0}{\partial n_c} = n_c \frac{\partial \mu_c^0}{\partial n_c}$$

one obtains

$$\frac{\Pi^0 v_c}{k_B T} = \frac{\eta + \eta^2 + \eta^3}{(1 - \eta)^3}, \qquad (3.90)$$

the SPT expression for the pressure of a hard sphere fluid [71], which preceded the slightly more accurate Carnahan-Starling equation Eq. (3.30), which contains an additional η^4-term.

References

1. F. Oosawa. *Hyo-Hyo Rakugaku*. Autobiography, Nagoya, 2005
2. S. Asakura, F. Oosawa, J. Chem. Phys. **22**, 1255 (1954)
3. D.H. Napper, *Polymeric Stabilization of Colloidal Dispersions* (Academix Press, Oxford, 1983)
4. As explained by Oosawa during the "Nagoya Symposium on Depletion Forces: Celebrating the 60th anniversary of the Asakura-Oosawa theory" held on March 14 and Japan 15 in 2014 in Nagoya
5. A. Vrij, Pure Appl. Chem. **48**, 471 (1976)
6. S. Asakura, F. Oosawa, J. Pol. Sci. **33**, 183 (1958)
7. R. Li-In-On, B. Vincent, F.A. Waite, ACS Symp. Ser. **9**, 165 (1975)
8. E. Eisenriegler, J. Chem. Phys. **79**, 1052 (1983)
9. A. Hanke, E. Eisenriegler, S. Dietrich, Phys. Rev. E **59**, 6853 (1999)
10. G.J. Fleer, A.M. Skvortsov, R. Tuinier, Macromolecules **36**, 7857 (2003)
11. H. De Hek, A. Vrij, J. Colloid Interface Sci. **84**, 409 (1981)
12. R. Fåhraeus, Physiol. Rev. **9**, 241 (1929)

13. R. Fåhraeus, Acta. Med. Scand. **55**, 1 (1921)
14. J.E. Thysegen, Acta Med. Scand. Suppl. **134**, 1 (1942)
15. J. Traube, Gummi Zeitung **39**, 434 (1925)
16. H.C. Baker, Inst. Rubber Ind. **13**, 70 (1937)
17. C.F. Vester, Kolloid Z. **84**, 63 (1938)
18. E. Dickinson, Food Hydrocolloids **17**, 25 (2003)
19. C. Sieglaff, J. Polym. Sci. **41**, 319 (1959)
20. C. Cowell, R. Li-In-On, B. Vincent, F.A. Waite, J. Chem. Soc. Faraday Trans. **74**, 337 (1978)
21. B. Vincent, P.F. Luckham, F.A. Waite, J. Colloid Interface Sci. **73**, 508 (1980)
22. B. Vincent, J. Edwards, S. Emmett, A. Jones, Colloids Surf. **17**, 261 (1986)
23. B. Vincent, Colloids Surf. **24**, 269 (1987)
24. B. Vincent, J. Edwards, S. Emmett, R. Croot, Colloids Surf. **31**, 267 (1988)
25. S. Hachisu, A. Kose, Y. Kobayashi, J. Colloid Interface Sci. **55**, 499 (1976)
26. A. Kose, S. Hachisu, J. Colloid Interface Sci. **55**, 487 (1976)
27. P.R. Sperry, H.B. Hopfenberg, N.L. Thomas, J. Colloid Interface Sci. **82**, 62 (1981)
28. P.R. Sperry, J. Colloid Interface Sci. **87**, 375 (1982)
29. P.R. Sperry, J. Colloid Interface Sci. **99**, 97 (1984)
30. A.P. Gast, C.K. Hall, W.B. Russel, J. Colloid Interface Sci. **96**, 251 (1983)
31. J.A. Barker, D. Henderson, Rev. Mod. Phys. **48**, 587 (1976)
32. A.P. Gast, W.B. Russel, C.K. Hall, J. Colloid Interface Sci. **109**, 161 (1986)
33. H.N.W. Lekkerkerker, Colloids Surf. **51**, 419 (1990)
34. H. Reiss, J. Phys. Chem. **96**, 4736 (1992)
35. H.N.W. Lekkerkerker, W.C.K. Poon, P.N. Pusey, A. Stroobants, P.B. Warren, Europhys. Lett. **20**, 559 (1992)
36. S.M. Ilett, A. Orrock, W.C.K. Poon, P.N. Pusey, Phys. Rev. E **51**, 1344 (1995)
37. F. Leal-Calderon, J. Bibette, J. Biais. Europhys. Lett. **23**, 653 (1993)
38. H.N.W. Lekkerkerker, R. Tuinier, *Colloids and the Depletion Interaction* (Springer, Berlin, 2011)
39. Y. Mao, M.E. Cates, H.N.W. Lekkerkerker, Phys. A **222**, 10 (1995)
40. E.J. Meijer, D. Frenkel, J. Chem. Phys. **100**, 6873 (1994)
41. M. Dijkstra, J.M. Brader, R. Evans, J. Phys. Condens. Matter **11**, 10079 (1999)
42. P.G. Bolhuis, A.A. Louis, Macromolecules **35**, 1860 (2002)
43. A. Moncho-Jordá, A.A. Louis, P.G. Bolhuis, R. Roth, J. Phys. Condens. Matter **15**, S3429 (2003)
44. D. Kleshchanok, R. Tuinier, P.R. Lang, *J. Phys Condens. Matt.* **20**, 073101 (2008)
45. J.D. van der Waals, Doctoral thesis. A.W. Sijthoff, Leiden, 1873
46. D.A. McQuarrie, *Statistical Mechanics* (University Science Books, Sausalito, 2000)
47. A. Malijevský, J. Kolafa. *Introduction to the Thermodynamics of Hard Spheres and Related Systems, in: 'Theory and Simulation of Hard-Sphere Fluids and Related Systems',* Lecture Notes in Physics, vol 753 (Springer, Berlin, 2008)
48. N.F. Carnahan, K.E. Starling, J. Chem. Phys. **51**, 635 (1969)
49. J.-P. Hansen, I.R. McDonald, *Theory of Simple Liquids* (Academic Press, San Diego, 1986)
50. A. Fortini, M. Dijkstra, R. Tuinier, *J. Phys. Condens. Matt.* **17**, 7783–7803 (2005)
51. W.G. Hoover, F.H. Ree, J. Chem. Phys. **49**, 3609 (1968)
52. J.E. Lennard-Jones, A.F. Devonshire, Proc. Roy. Soc. **163A**, 53 (1937)
53. D. Frenkel, A.J.C. Ladd, J. Chem. Phys. **81**, 3188 (1984)
54. C.G. de Kruif, P.W. Rouw, J.W. Jansen, A. Vrij, J. de Phys. **46**, C3–295 (1985)
55. P.N. Pusey, W. Van Megen, Nature **320**, 340 (1986)
56. J. Levelt Sengers, *How Fluids Unmix* (Edita KNAW, Amsterdam, 2002)
57. Y. Tang, B.C.-Y. Lu, J. Chem. Phys. **99**, 9828 (1993)
58. Y. Tang, J. Chem. Phys. **118**, 4140 (2003)
59. Y. Tang, Y.-Z. Lin, Y.-G. Li, J. Chem. Phys. **122**, 184505 (2005)
60. R. Tuinier, G.J. Fleer, J. Phys. Chem. B **110**, 2045 (2006)
61. K.P. Shukla, J. Chem. Phys. **112**, 10358 (2000)

62. M. Dijkstra, Phys. Rev. E **66**, 021402 (2002)
63. J.M. Brader, R. Evans, M. Schmidt, Mol. Phys. **101**, 3349 (2003)
64. M. Fuchs, K.S. Schweizer, J. Phys. Condens. Matter **14**, R239 (2002)
65. P.G. Bolhuis, A.A. Louis, J.P. Hansen, Phys. Rev. Lett. **89**, 128302 (2002)
66. G.J. Fleer, R. Tuinier, Adv. Colloid Interface Sci. **143**, 1–47 (2008)
67. R. Tuinier, M.S Feenstra. *Langmuir* (2014). doi:10.1021/la5023856
68. G.J. Fleer, R. Tuinier, Phys. A **379**, 52 (2007)
69. D.G.A.L. Aarts, R. Tuinier, H.N.W. Lekkerkerker, J. Phys. Condens. Matter **14**, 7551 (2002)
70. B. Widom, J. Chem. Phys. **39**, 2808 (1963)
71. H. Reiss, H.L. Frisch, J.L. Lebowitz, J. Chem. Phys. **31**, 369 (1959)
72. E.J. Meijer, Computer Simulation of Molecular Solids and Colloidal Dispersions, PhD thesis. Utrecht University, Utrecht, 1993
73. M. Dijkstra, R. van Roij, R. Roth, A. Fortini, Phys. Rev. E **73**, 041409 (2006)

Part II
General Physics of Aqueous Interfaces

Chapter 4
Thermodynamics of Interfaces in Soft-Matter Systems

Gerhard H. Findenegg

Abstract Thermodynamics of interfaces provides a framework to relate measurable quantities to other important yet not directly accessible equilibrium properties of interfacial systems. For liquid/gas and liquid/liquid interfaces (fluid interfaces) the interfacial tension and its dependence on temperature and composition can be measured, while the adsorbed amounts of the components are not accessible. Conversely, for solid/fluid interfaces the adsorbed amount can be measured but the interfacial tension (free energy) is not accessible. For both cases the Gibbs equation represents a bridge between the two kinds of quantities. In this chapter we explain the application of the Gibbs equation with a focus on soft matter systems. We also discuss the meaning of surface excess amounts and their relation to (absolute) surface concentrations which appear in adsorbate equations of state. Finally we briefly touch the additional features of charged interfaces and of ionic equilibria at interfaces.

4.1 Thermodynamic Quantities and Relations

4.1.1 Introduction

Interfaces represent thin regions between macroscopic phases in which the properties gradually change from those of one adjacent phase to those of the other. Because of this inhomogeneous nature of interfaces, a thermodynamic treatment of their properties has been a challenge to generations of scientists. Different formalisms have been developed to cope with this problem. One intuitively appealing way of treating a surface is to consider it as a distinct phase of finite thickness and volume, so that adsorption and related phenomena can be treated analogous to

G.H. Findenegg (✉)
Stranski Laboratory of Physical and Theoretical Chemistry,
Technical University Berlin, 10623 Berlin, Germany
e-mail: findenegg@chem.tu-berlin.de

© Springer International Publishing Switzerland 2016
P.R. Lang and Y. Liu (eds.), *Soft Matter at Aqueous Interfaces*,
Lecture Notes in Physics 917, DOI 10.1007/978-3-319-24502-7_4

phase equilibria. This approach was pursued, among others, by Bakker [1] and Guggenheim [2]. However, the clarity of the surface phase formalism is deceptive, as there is no way to define unambiguously the thermodynamic properties of this surface phase. An alternative formalism for defining the thermodynamic properties was proposed by Gibbs in 1878 [3]. In this treatment the surface is regarded as a mathematical dividing plane between the two macroscopic phases, and the properties of the interface are defined as a *surface excess* relative to a hypothetical reference system in which all properties remain uniform up to the dividing plane. Surface excess quantities often have no intuitively simple interpretation, but their importance lies in the fact that they represent measurable quantities.

In this short review we focus on adsorption phenomena from liquid phases at different types of interfaces (liquid/gas, liquid/liquid and liquid/solid). The treatment is limited to interfaces at which specific curvature effects can be neglected. We show how surface excess quantities suitably defined for these types of interfaces can be determined experimentally and interpreted in terms of physical models. The examples chosen are related to soft matter at aqueous interfaces.

4.1.2 Surface Tension

The liquid/vapour interface of fluids represents an inhomogeneous region in which the local density changes from high to low values at a length scale of a few molecular diameters. We consider a flat interface in the plane xy as sketched in Fig. 4.1. When neglecting gravity, a force balance on an infinitesimal cube centred at point x, y, z in the system gives [4]

$$\frac{\partial p_{xx}(x,y,z)}{\partial x} = \frac{\partial p_{yy}(x,y,z)}{\partial y} = \frac{\partial p_{zz}(x,y,z)}{\partial z} = 0 \qquad (4.1)$$

where p_{xx}, p_{yy}, and p_{zz}, are the pressures exerted on surfaces normal to the x, y, and z axis, respectively. Since the local density depends only on z, the coordinate normal to the interface, but not on the position in the xy plane, the pressure components p_{xx}, p_{yy}, and p_{zz} can be only functions of z. Also, because of the condition of isotropy in the xy plane, $p_{xx} = p_{yy}$. Equation 4.1 can therefore be written as

$$\frac{\partial p_{zz}(z)}{\partial z} = 0 \text{ or } p_{zz} = \text{const} \equiv p_N$$
$$p_{xx} = p_{yy} = p_T(z) \qquad (4.2)$$

Hence in a two-phase system with planar interface the conditions for hydrostatic equilibrium are: (i) the normal pressure p_N is constant and equal to the pressure p of the coexistent bulk phases; (ii) sufficiently far from the interface, the transverse pressure p_T is also equal to p, but in the interfacial region $p_T(z) \neq p_N$. It can be shown that the interfacial tension γ is given by [1]

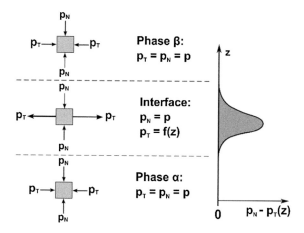

Fig. 4.1 Components of the pressure tensor across the interface of phases α and β: The normal component p_N is independent of the position, the transverse component p_T is a function of z and assumes negative values in the interfacial region. Accordingly, the pressure difference $p_N - p_T$ has positive values in this region but is zero elsewhere

$$\gamma = \int_{-\infty}^{\infty} [p - p_T(z)]dz \tag{4.3}$$

The integration can be taken from $-\infty$ to $+\infty$ because $p_T(z)$ differs from p only near the interface. Equation 4.1 can be taken as a mechanical definition of the interfacial tension. An idea of the magnitude of the stress acting within the interface is obtained by the following elementary calculation. The width of the surface of simple liquids at their normal boiling point is about 1 nm and a typical value of the surface tension is 30 mN m^{-1}. Since $p = 1$ bar this means that the average value of $p_T(z)$ in the interface is about -300 bar. This large value makes it plausible that the surface tension can dominate phenomena at the mesoscopic and even at a macroscopic level. The expression for γ in Eq. (4.3) is obtained by considering the isothermal reversible work δW for increasing the surface area of the interface by an increment δA at constant volume, i.e., $\gamma = \delta W/\delta A$. Hence in the language of thermodynamics the surface tension represents the change in free energy F of the two-phase system with an infinitesimal change of the surface area at constant temperature T and volume V, viz., $\gamma = (\partial F/\partial A)_{T,V}$, expressed in SI units in J m^{-2}.

4.1.3   Adsorption as a Surface Excess

Adsorption generally stands for the enrichment of substances at an interface, but different situations prevail at different types of interface. For example, gas adsorption leads to a higher *density* of the gas near the surface. At liquid/solid interfaces, on the other hand, enrichment of one component of a mixture goes at the expense of the other component(s), causing changes in *composition* near the surface, and adsorption may be viewed as a displacement of solvent by the solute in the surface layer of the liquid.

The solid phase often represents a more or less inert external medium. This is different in the case of fluid interfaces (liquid/gas or liquid/liquid), where all components may be present at significant concentrations in both phases. In this latter situation it is conceptually difficult to define the adsorbed amount of a component. This problem was solved by Gibbs by introducing the concept of *surface excess quantities* and *relative adsorption*. To rationalize this concept consider the concentration profiles $c_k(z)$ of the components of a binary mixture ($k = 1, 2$) across a liquid/vapour interface, where the local concentration $c_k(z)$ changes from c_k^l to c_k^g in a monotonic or nonmonotonic manner (Fig. 4.2). The surface excess amount of component k is now defined as the difference between the known total amount n_k and the amount in a hypothetical reference system in which the two phases would extend up to a mathematical dividing plane located at some position z_0. The surface excess amount of component k is then given by

$$n_k^\sigma = n_k - (c_k^l V^l + c_k^g V^g) \tag{4.4}$$

For given values of n_k and total volume $V = V^l + V^g$, and known concentrations c_k^l and c_k^g, the value of the surface excess n_k^σ is not yet defined in a singular way but depends on precisely how the volume V is divided into the volumes V^l and V^g. For a geometric interpretation of the surface excess n_k^σ consider a cylindrical volume with the concentration profile $c_k(z)$ along the cylinder axis. The overall amount n_k of component k in this cylinder is obtained by integration of $c_k dV = c_k(z) A dz$, where A represents the basal area of the cylinder (and thus the area of the interface).

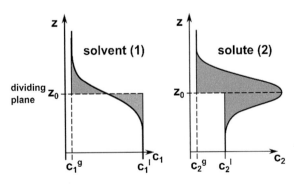

Fig. 4.2 Sketch of the concentration profiles $c(z)$ of solvent (*left*) and solute (*right*) across the liquid/vapour interface. In the Gibbs convention the dividing plane (position z_0) is chosen such that for the solvent (component 1) the two integrals on the *right-hand side* of Eq. (4.5) become equal in magnitude with opposite sign, so that $\Gamma_1^\sigma = 0$. With this choice of z_0, Eq. (4.5) yields a well-defined surface excess Γ_2^σ for the solute, called the relative adsorption $\Gamma_2^{(1)}$

For the surface excess amount per unit area, i.e., the surface excess concentration $\Gamma_k^\sigma = n_k^\sigma/A$ (units: mol/m^2) of a binary mixture we thus obtain:

$$\Gamma_k^\sigma = \int_{-\infty}^{z_0} \left[c_k(z) - c_k^l\right]dz + \int_{z_0}^{\infty} \left[c_k(z) - c_k^g\right]dz \qquad (4.5)$$

The superscript σ on n_k^σ and Γ_k^σ indicates 'surface excess' but, as explained above, these quantities need to be specified by choosing a suitable location of the dividing plane (z_0). The following specifications are important for the different kinds of interfaces:

Relative adsorption (Gibbs prescription): This is mostly adopted to quantify the adsorption at the liquid/gas interface of *solutions* (solute components k) in a solvent (component 1): In a geometric way the relative adsorption of the solutes can be rationalized by placing the dividing surface at a position z_0 such that $\Gamma_1^\sigma = 0$ ('equimolar' dividing surface for the solvent). Relative surface excess concentrations are denoted as $\Gamma_k^{(1)}$, where the superscript (1) indicates "relative to the solvent". Depending on the concentration profiles of solvent and solute the relative surface excess of a solute can be positive or negative. Experimentally, the relative adsorption $\Gamma_k^{(1)}$ of solutes adsorbed at the liquid/air interface can be obtained by surface tension measurements as a function of solute concentration (see Eq. (4.17)).

Reduced adsorption: This is used to quantify adsorption from mixtures in which no component is distinguished as the solvent. In a geometric way the reduced adsorption of the components can be imagined by placing the dividing surface at a position z_0 such that $\sum_{k=1}^{n} \Gamma_k^\sigma = 0$. Reduced surface excess concentrations are denoted as $\Gamma_k^{(n)}$, where the superscript (n) indicates that the sum of the surface excess amounts of all n components is zero. Hence for a binary system, $\Gamma_2^{(n)} = -\Gamma_1^{(n)}$. The reduced surface excess $\Gamma_k^{(n)}$ and similar specifications are mostly used to characterize adsorption at solid/liquid interfaces, where $\Gamma_k^{(n)}$ can be measured from the change in concentration before and after equilibration with the adsorbent (see Eq. 4.42). Relative adsorption and reduced adsorption are interrelated by $\Gamma_2^{(1)} = \Gamma_2^{(n)}/(1-x_2^l)$, where x_2^l is the mole fraction of component 2 in the liquid phase.

Two-solvent relative adsorption: Adsorption of solutes at liquid/liquid interfaces is usually defined relative to the solvents of both phases. Denoting the two solvents as components 1 and 2, then according to this prescription $\Gamma_1^\sigma = 0$ and $\Gamma_2^\sigma = 0$. The surface excess concentration of solutes ($k = 3, 4, \ldots$) relative to the two solvents is denoted as $\Gamma_k^{(1,2)}$. This definition, which goes beyond the original Gibbs formalism, implies that we are placing two 'equimolar' dividing surfaces, one for each solvent. Hence the volume of the system is no longer the sum of the two bulk phases α and β, but now given by $V = V^\alpha + V^\beta + V^\sigma$, where the excess volume V^σ may be positive or negative. The surface excess concentrations $\Gamma_k^{(1,2)}$ can be obtained from measurements of the interfacial tension as a function of concentration (see Eq. 4.18).

As mentioned above, the surface excess concentrations $\Gamma_k^{(1)}$, $\Gamma_k^{(n)}$, and $\Gamma_k^{(1,2)}$ are measurable quantities based on clear operational definitions. Working with these quantities has the disadvantage, however, that they lack a simple physical interpretation. To overcome this problem, physical models of the interfacial layer have been introduced. Usually it is assumed that the surface layer has a uniform composition with concentrations Γ_k^s of the individual components k. Such a surface phase model will be introduced in Sect. 4.2.3, where it will be shown how the surface concentrations can be calculated from the experimentally accessible surface excess concentrations.

4.1.4 Gibbs Adsorption Equation

The Gibbs formalism of surface excess quantities outlined above can be applied to all extensive thermodynamic quantities of the system (internal energy U, enthalpy H, entropy S, Helmholtz free energy F, Gibbs free energy G, etc.) except the volume V. The surface excess X^σ of a quantity X is defined as [5]

$$X^\sigma = X - X^\alpha - X^\beta \tag{4.6}$$

where X represents the value for the entire two-phase system, while X^α and X^β relate to the homogeneous phases α and β, when their volumes extend up to the dividing surface. $V^\sigma = 0$ is implicit in the Gibbs formalism, because the bulk phases extend up to the dividing surface. Remarkably, thermodynamic relations between the excess quantities can be formulated just as if it was a separate phase. The most important relation between the surface excess quantities is the Gibbs equation, which has the general form

$$d\gamma = -s^\sigma dT - \sum_{k=1}^{n} \Gamma_k^\sigma d\mu_k \tag{4.7}$$

Here, $s^\sigma = S^\sigma/A$ is the surface excess entropy per unit area, Γ_k^σ is the surface excess concentration of component k defined relative to the same convention for the Gibbs dividing surface as s^σ, and μ_k is the chemical potential of component k at the given temperature T and composition of the system. According to Eq. (4.7) the Gibbs equation relates changes in surface tension to changes in temperature and the chemical potential of the solutes.

The chemical potential of component k in the surface can be defined by [5]

$$\mu_k^s = \left(\partial F/\partial n_k^\sigma\right)_{T,V,A,n_j^\sigma} \tag{4.8}$$

Here F is the Helmholtz free energy $(F = U - TS)$ of the whole system and A is the area of the interface. The condition of equilibrium with respect to diffusion of component k to the interface from the adjacent phases α and β can be shown to be

$$\mu_k^\alpha = \mu_k^s = \mu_k^\beta = \mu_k \tag{4.9}$$

where μ_k^α and μ_k^β are the chemical potentials in the adjacent bulk phases. Hence μ_k in Eq. (4.7) is the common value of the chemical potential throughout the system. Alternatively, the chemical potential can be defined by

$$\mu_k^a = \left(\partial F/\partial n_k^\sigma\right)_{T,V,\gamma,n_j^\sigma} \tag{4.10}$$

Here the superscript a is used to differentiate this chemical potential from that defined by Eq. (4.8). Equilibrium between a liquid phase and the interface is then shown to exist when

$$\mu_k^l = \mu_k^s = \mu_k^a - \gamma \underline{a}_k \tag{4.11}$$

where $\underline{a}_k = \left(\partial A/\partial n_k^\sigma\right)_{T,V,\gamma,n_j^\sigma}$ is the partial molar area of component k in the surface. Equation 4.11 expresses the fact that when the derivative $\partial F/\partial n_k^\sigma$ is taken at constant interfacial tension γ rather than constant surface area A, the dependence on γ must be added by the term $\gamma \underline{a}_k$.

4.2   Fluid Interfaces

4.2.1   Surface of Pure Liquids and Liquid/Liquid Interface of Partially Miscible Binary Systems

For the liquid/vapour interface of a pure fluid, when choosing the 'equimolar' dividing surface ($\Gamma^\sigma = 0$), the Gibbs equation (Eq. 4.7) reduces to $s^\sigma = -(d\gamma/dT)$. For simple liquids both γ and $-(d\gamma/dT)$ decrease monotonically with increasing temperature from the triple point to the critical point. Water is not a simple liquid. It has a high surface tension (72.0 mN m^{-1} at 25 °C) and $-(d\gamma/dT)$ exhibits a maximum near 200 °C. When the critical temperature T_c of a fluid is approached along the vapour/liquid coexistence line, the densities of the liquid and vapour phase become equal and the interface vanishes at the critical point. Close to T_c the vanishing of the surface tension follows a universal law [6]

$$\gamma(T) = \gamma_0(T_c - T)^m \tag{4.12}$$

where γ_0 is a material constant and the index m has a universal value ($m \approx 1.25$). From Eqs. (4.7) and (4.12) the surface excess entropy as a function of temperature then becomes

$$s^\sigma(T) = m\gamma_0(T_c - T)^{m-1} \qquad (4.13)$$

This relation tells us that pure liquids have a positive excess entropy that decreases progressively as the critical point is approached. Physically, $s^\sigma > 0$ means that molecules in the topmost layers of the liquid phase have a higher free volume and thus a higher translational entropy than those in the interior of the liquid.

From a thermodynamic point of view, the liquid/liquid interface of binary systems with a lower miscibility gap behaves similar to the liquid/vapour interface of a pure fluid. The interfacial tension $\gamma(T)$ again follows Eq. (4.12) as the critical temperature (consolute temperature) T_c is approached, and with $s^{(1,2)} = -(\partial\gamma/\partial T)_p$ we find a positive surface excess entropy which falls off steeply near T_c, as shown in the graphs at the left-hand side of Fig. 4.3. Systems with an upper miscibility gap exhibit a different behaviour, as shown on the right-hand side of Fig. 4.3. In this case phase separation starts at a lower critical point and the interfacial tension increases with temperature. The temperature dependence of γ and $s^{(1,2)}$ can again be described by Eqs. (4.12) and (4.13) when replacing $(T_c - T)^m$ by $(T - T_c)^m$ and introducing a minus sign on the r.h.s. of Eq. (4.13). Hence the liquid/liquid interface of systems with an upper miscibility gap exhibits a negative surface excess entropy $s^{(1,2)}$. Examples of systems with an upper miscibility gap are aqueous systems of proton acceptors (e.g., ethers or polyethers). Such systems typically have negative values of the excess enthalpy and excess entropy of mixing in the bulk liquid state ($H^E < 0, S^E < 0$) [7].

Does the generic behaviour of liquid surfaces also apply to complex liquids? In the past decades interesting model systems have been studied, in which colloidal particles dispersed in a solvent replace the molecules of a simple fluid, and a non-adsorbed polymer is added to tune the interaction between the colloid particles (see Chap. 3 by R. Tuinier). In a certain range of polymer concentrations phase separation into a colloidal liquid (rich in colloid and poor in polymer) and a colloidal gas (poor in colloid and rich in polymer) occurs (Fig. 4.4) [8, 9]. Experiments and theoretical work have shown that in such systems, at states well away from the critical point, the interfacial tension scales as the thermal energy ($k_B T$) divided by the square of the particle diameter d, i.e., [10]

$$\gamma \sim \frac{k_B T}{d^2} \qquad (4.14)$$

For particles of diameter 25 nm this factor is of order 1 μN/m, i.e., about 4 orders of magnitude lower than the surface tension of molecular liquids. Theoretical and experimental studies also indicate a rapid decrease of the interfacial tension with decreasing difference in the particle number density in the two phases,

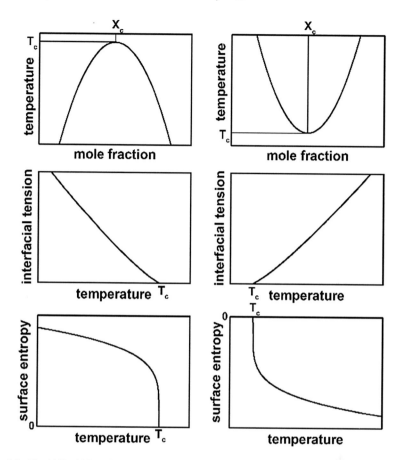

Fig. 4.3 Liquid/liquid interface in binary systems of partial miscibility in the liquid state: Systems with an upper critical solution temperature (*left*) or a lower critical solution temperature (*right*). The graphs show the liquid/liquid coexistence curve (*top*), the temperature dependence of the interfacial tension γ (*middle*), and the temperature dependence of the surface excess entropy $s^{(1,2)}$ (*bottom*) in the vicinity of the critical temperature T_c

$\gamma \sim \left(\rho^l - \rho^g \right)^4$, as the critical point is approached. This again is in agreement with the behaviour of simple liquids.

4.2.2 Adsorption at Fluid Interfaces

Adsorption at the liquid/vapour interface of a solvent (component 1) is expressed commonly by the relative surface excess concentrations $\Gamma_k^{(1)}$ of the solutes ($k = 2$, 3...). Hence at constant temperature the Gibbs adsorption equation takes the form

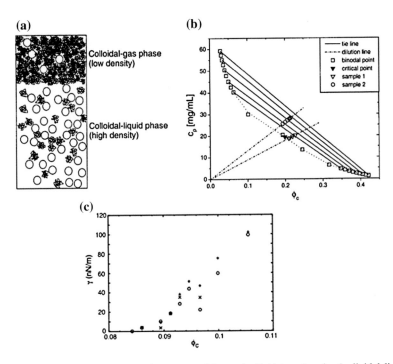

Fig. 4.4 A colloid–polymer suspension separated into a 'colloidal gas' and a 'colloidal liquid' phase: **a** sketch of the two-phase system; **b** phase diagram (polymer concentration c_P versus colloid volume fraction φ_C) showing coexistent colloid-rich (*liquid*-like) and polymer-rich (*gas*-like) phases and the critical point of the binodal curve (Reproduced from Ref. [8] with permission. Copyright 1999, American Chemical Society); **c** interfacial tension γ versus colloid volume fraction φ_C near the critical point showing analogy with the behaviour of simple liquids near their liquid/vapour critical point (Reproduced from Ref. [9] with permission. Copyright 2004, American Association for the Advancement of Science)

$$dy = -\sum_{k=2}^{n} \Gamma_k^{(1)} d\mu_k \qquad (4.15)$$

The chemical potential of solutes in the bulk solution is given by

$$\mu_k^l = \mu_k^o + RT \ln a_k \qquad (4.16)$$

The standard chemical potential μ_k^o refers to the hypothetical state of an ideal dilute solution. The activity a_k can be expressed either as $a_k = c_k f_k$ or $a_k = x_k f_k$, where c_k is the concentration, x_k the mole fraction and f_k the appropriate activity coefficient of component k in the unsymmetrical (Henry) convention ($f_k \rightarrow 1$ as $x_k \rightarrow 0$). Using the differential form of Eq. (4.16) at constant temperature, $d\mu_k = RT d\ln a_k$, the Gibbs equation for a single nonionic solute (2) may be written

$$\Gamma_2^{(1)} = -\frac{1}{RT}\left(\frac{\partial \gamma}{\partial \ln a_2}\right)_T \tag{4.17}$$

According to this relation the relative surface excess $\Gamma_2^{(1)}$ can be determined from the experimentally well accessible dependence of the surface tension on the activity of the solute. Because of the logarithmic activity scale, the choice of the concentration units (mol/L or mol/kg, etc.) is irrelevant, as long as the corresponding activity coefficient is used. The necessity of using activities in the Gibbs equation has been discussed in the literature [11]. For qualitative considerations the activity may be replaced by concentration. Equation 4.17 then indicates that solutes which lower the surface tension $(d\gamma/dc_2 < 0)$ are positively adsorbed $(\Gamma_2^{(1)} > 0)$, while solutes causing an increase in surface tension $(d\gamma/dc_2 > 0)$ are negatively adsorbed $(\Gamma_2^{(1)} < 0)$. At the surface of water, hydrophilic and well-hydrated solutes (inorganic salts, but also glycerine, glycine, etc.) are negatively adsorbed, while hydrophobic solutes (hexane, benzene) and amphiphilic substances (surfactants, etc.) are positively adsorbed.

For the adsorption of solutes at liquid/liquid interfaces it is convenient to express the Gibbs equation in terms of surface excess concentrations relative to the two solvents, $\Gamma_k^{(1,2)}$ (see Sect. 4.1.3). For a single solute (component 3) this gives by analogy with Eq. (4.17)

$$\Gamma_3^{(1,2)} = -\frac{1}{RT}\left(\frac{\partial \gamma}{\partial \ln a_3}\right)_{T,p} \tag{4.18}$$

According to this relation it is the relative adsorption $\Gamma_k^{(1,2)}$ that is directly accessible from measurements of the interfacial tension as a function of the activity a_k. Note that at equilibrium the concentration of a solute k in the two coexistent liquid phases a and β can be grossly different, but the thermodynamic activity is equal, i.e., $a_k = c_k^\alpha f_k^\alpha = c_k^\beta f_k^\beta$.

Adsorption of surfactants from aqueous solutions: As a case study we consider the determination of the adsorption of surfactants at the air/water interface [12–15]. Nonionic surfactants of alkyl chain length C_{12} or greater are strongly adsorbed at the air/water and oil/water interface. The surface tension derivative $(d\gamma/d\ln c_2)$ reaches a high negative limiting value, indicating a high limiting adsorption at bulk concentrations well below the critical micelle concentration (cmc). Above the cmc no further decrease in surface tension occurs $(d\gamma/d\ln c_2 = 0)$. This can be explained by the formation of micellar aggregates, so that the concentration of monomeric surfactant—and hence its activity a_2—remains constant above the cmc [16].

For an ionic surfactant, R^-Na^+, the general Gibbs equation (Eq. 4.7) takes the form ($T = $ const) [12]

$$-d\gamma = \Gamma^\sigma_{Na^+} d\mu_{Na^+} + \Gamma^\sigma_{R^-} d\mu_{R^-} + \Gamma^\sigma_{H^+} d\mu_{H^+} + \Gamma^\sigma_{OH^-} d\mu_{OH^-} + \Gamma^\sigma_{H_2O} d\mu_{H_2O} \quad (4.19)$$

where we consider all ionic species and the solvent. In the bulk solution the chemical potentials are interrelated by the Gibbs-Duhem relation

$$n_{Na^+} d\mu_{Na^+} + n_{R^-} d\mu_{R^-} + n_{H^+} d\mu_{H^+} + n_{OH^-} d\mu_{OH^-} + n_{H_2O} d\mu_{H_2O} = 0 \quad (4.20)$$

Combining these two relations and neglecting the terms in H^+ and OH^- against the concentration of the surfactant ions leads to

$$-d\gamma = \left(\Gamma^\sigma_{Na^+} - \Gamma^\sigma_{H_2O} \frac{n_{Na^+}}{n_{H_2O}} \right) d\mu_{Na^+} + \left(\Gamma^\sigma_{R^-} - \Gamma^\sigma_{H_2O} \frac{n_{R^-}}{n_{H_2O}} \right) d\mu_{R^-} \quad (4.21)$$

For electrical neutrality in the solution and the surface we also have $n_{Na^+} = n_{R^-} = n_{NaR}$ and $\Gamma^\sigma_{Na^+} = \Gamma^\sigma_{R^-} = \Gamma^\sigma_{NaR}$, so that

$$-d\gamma/RT = \left(\Gamma^\sigma_{NaR} - \Gamma^\sigma_{H_2O} \frac{n_{NaR}}{n_{H_2O}} \right) d\ln(a_{Na^+} a_{R^-}) \quad (4.22)$$

Introducing the mean activity $a_\pm = (a_{Na^+} a_{R^-})^{1/2}$ and noting that the expression in brackets is the surface excess concentration of the surfactant relative to water, $\Gamma^{(H_2O)}_{NaR} = \Gamma^{(1)}_{NaR}$, we find

$$-d\gamma/RT = 2\Gamma^{(1)}_{NaR} d\ln(a_\pm) \quad (4.23)$$

The factor 2 in this relation arises because both the surfactant ion R^- and counterion Na^+ must be adsorbed to maintain electroneutrality. Accordingly, $d\gamma/d\ln a_\pm$ is twice as large as for a nonionic surfactant. The mean ion activity coefficient needed to evaluate the Gibbs relation can be taken from the extended Debye-Hückel equation in the form [15]

$$\log f_\pm = \frac{-0.5115|z_+ z_-|\sqrt{I}}{1 + 1.316\sqrt{I}} + 0.055I \quad (4.24)$$

where z_+ and z_- are the charge numbers of cation and anion, I is the ionic strength of the solution expressed in molar units, and the numerical constants apply for a temperature of 25 °C.

Is the adsorption of the ionic surfactant at the air/water interface affected by the addition of an inert electrolyte? If a non-adsorbed electrolyte (say, NaCl) is present in large excess, an increase in the concentration of R^-Na^+ causes a negligible increase of the Na^+ concentration, so that $d\mu_{Na^+}$ is negligible. Consideration of the Gibbs-Duhem equation then shows that $d\mu_{Cl^-}$ is also negligible, and thus

$$-d\gamma/RT = \Gamma_{R^-}^{(1)} dln(a_{R^-}) \tag{4.25}$$

The activity coefficient f_{R^-} depends on the ionic strength which is determined by the excess of NaCl and is therefore constant, so that

$$-d\gamma/RT = \Gamma_{R^-}^{(1)} dln(c_{NaR}) \tag{4.26}$$

where $\Gamma_{R^-}^{(1)}$ is again the surface excess of the surfactant relative to water and c_{NaR} is its concentration in solution. Hence the factor 2 in the Gibbs equation has disappeared and the ionic surfactant in excess electrolyte is adsorbed as if it was a nonionic surfactant.

4.2.3 Surface Phase Model

A concept adopted explicitly or implicitly in many treatments of adsorption from solution is that of a distinct *surface phase*, i.e., a layer of finite thickness located between the two bulk phases and affected by the interfacial tension [17, 18]. The surface phase is usually supposed to be of uniform composition. On the basis of such a model the measured surface excess concentrations can be expressed by the concentration or mole fraction difference between the surface phase (superscript s) and bulk liquid phase (superscript l). Specifically, if the surface phase consists of an amount n^s of solvent plus solute, with a mole fraction x_2^s of the solute, then the reduced surface excess $\Gamma_2^{(n)}$ and the relative surface excess $\Gamma_2^{(1)}$ of the solute can be expressed as

$$\Gamma_2^{(n)} = \left(1 - x_2^l\right)\Gamma_2^{(1)} = \left(x_2^s - x_2^l\right)n^s/A \tag{4.27}$$

where A is the surface area and x_2^l is the mole fraction of solute in the bulk solution. If n^s is known, Eq. (4.27) may be used to calculate x_2^s from the measured surface excess concentration. However, the value of n^s depends on what assumptions are made about the nature of the surface phase. In practice, meaningful results can be obtained by this approach only if there is independent evidence that the surface phase consists of a single monolayer of molecules. For adsorption from binary mixtures the condition that this surface layer is completely covered is then that

$$\Gamma_1^s a_{1,0} + \Gamma_2^s a_{2,0} = 1 \tag{4.28}$$

Here, Γ_1^s and Γ_2^s represent the (absolute) surface concentrations (amount per unit area), given by $\Gamma_k^s = n_k^s/A = x_k^s n^s/A$, and the quantities $a_{k,0}$ denote the partial molar areas of solvent and solute in the surface phase. These cross-sectional areas may be estimated from molecular models. By combining Eqs. (4.27) and (4.28) we obtain an explicit expression to convert surface excess concentrations $\Gamma_2^{(n)}$ to absolute surface concentrations Γ_2^s of the solute in the monolayer surface phase:

$$\Gamma_2^s = \frac{x_2^l + \underline{a}_{1,0}\Gamma_2^{(n)}}{1 - (\underline{a}_{2,0} - \underline{a}_{1,0})\Gamma_2^{(n)}} \frac{n^s}{A} \tag{4.29}$$

In the particular case when the two components have the same size $(\underline{a}_{1,0} = \underline{a}_{2,0} = \underline{a}_0)$, then $n^s = A/\underline{a}_0$ and Eq. (4.29) leads to

$$\underline{a}_0 \Gamma_2^s = x_2^l + \underline{a}_0 \Gamma_2^{(n)} \tag{4.30}$$

This relation shows that in the case of strong adsorption from dilute solutions, when $\underline{a}_0 \Gamma_2^{(n)} \gg x_2^l$, the surface excess concentration becomes nearly equal to the absolute concentration of the solute in the surface phase, i.e., $\Gamma_2^{(n)} \cong \Gamma_2^{(1)} \cong \Gamma_2^s$. Hence in such cases it is justified to treat surface excess concentrations as true concentrations of the solute in the topmost layer of the liquid phase. If the condition $\underline{a}_0 \Gamma_2^{(n)} \gg x_2^l$ does not apply, Eq. (4.29) may be used to calculate Γ_2^s from the measured surface excess concentration.

4.2.4 Surface Equation of State and Adsorption Isotherm

Monolayers of strongly adsorbed substances have some resemblance with insoluble monolayers of lipids at the water surface [17]. The decrease in surface tension from the value of pure water, γ_0, to a value γ corresponding to a given surface concentration Γ^s, can be interpreted as a lateral pressure $\Pi = \gamma_0 - \gamma$ exerted by the monolayer film. In the case of water-insoluble monolayers (so-called *Langmuir films*) a certain amount of lipid, commonly expressed by the number of molecules N) is placed on a well-defined surface area A, and the surface concentration $\Gamma^s = N/A$ of the lipid can be varied by increasing or decreasing A using suitable barriers to keep all the lipid molecules within the area A. The dependence of the film pressure Π on the area A at constant temperature and constant number of lipid molecules ($\Pi - A$ isotherm) has a formal analogy with the pressure-volume diagram of a given amount of gas at constant temperature ($P - V$ isotherm). In the case of water-soluble substances (e.g., surfactants) the surface concentration of the substance is independent of the surface area but can be controlled via the adsorption isotherm $\Gamma^s = \Gamma^s(c)$, i.e., by changing its concentration c in the subphase. Surface films of this kind are called *Gibbs films*. In this section we explore the functional dependence $\Pi = \Pi(\underline{a})$, where $\underline{a} = A/N = 1/\Gamma^s$ is the mean area per adsorbed molecule in the Gibbs film at the given concentration c in the subphase. The relation

$\Pi = \Pi(a, T)$ is called the *monolayer equation of state* or two-dimensional (2D) equation of state of the adsorbed substance [19], by analogy with the equation of state $p = p(V, T)$ of a fluid in the bulk (3D) state.

Experimentally, the monolayer equation of state is obtained by the following sequence of steps:

(i) Determination of the film pressure isotherm $\Pi = \Pi(c, T)$ by surface tension measurements as a function of the concentration c.
(ii) Calculation of the appropriate surface excess concentration isotherm, e.g., $\Gamma^{(1)} = \Gamma^{(1)}(c, T)$ from the surface tension data by application of the Gibbs equation.
(iii) Choice of a surface phase model and conversion of the surface excess concentrations to the model-based surface concentrations $\Gamma^s(c, T)$.
(iv) Determination of the monolayer equation of state $\Pi = \Pi(a, T)$ by correlating film pressure $\Pi(c, T)$ and surface concentration $\Gamma^s(c, T)$ data corresponding to the same bulk concentration c, noting that $a = 1/\Gamma^s$.

Model equations of state and adsorption isotherms: Drawing on the analogy between 'two-dimensional' (2D) monolayers and three-dimensional (3D) bulk fluids, monolayer equations of state of increasing complexity have been proposed:

$$\begin{array}{ll} \Pi a = k_B T & \text{2D perfect gas} \\ \Pi(a - a_0) = k_B T & \text{2D Volmer} \\ \left(\Pi + \frac{\alpha}{a^2}\right)(a - a_0) = k_B T & \text{2D van–der–Waals} \end{array} \qquad (4.31)$$

In these relations k_B is the Boltzmann constant, a_0 represents the cross-sectional area of a molecule in the monolayer (analogous to the 'co-volume' in the 3D equation of state), and α is a measure of attractive lateral interaction between adsorbed molecules. Each of these equations of state can be converted to a corresponding adsorption isotherm with the Gibbs adsorption equation. For ideal solution behaviour of the bulk phase all model adsorption isotherm equations can be expressed in the form $Kc = f(\Gamma^s) = \tilde{f}(a)$, where K is a constant which depends on the units in which the equilibrium concentration c of the solute in the bulk phase is expressed (molar concentration, mole fraction, etc.). Specifically, at low surface concentration, when the surface film behaves as a *2D perfect gas,* the adsorption isotherm becomes

$$Kc = \Gamma^s = \frac{1}{a} \qquad (4.32)$$

Low surface concentration means that $1/\Gamma^s = a \gg a_0$. If this condition is no longer met, deviations from linear adsorption isotherms occur. In this regime it is convenient to express the adsorption isotherm and monolayer equation of state in terms of $\theta = a_0/a$, the fraction of surface occupied by the adsorbed molecules.

For mobile monolayer films without long-range lateral interactions (2D Volmer) the adsorption isotherm and equation of state are

$$K_V c = \frac{\theta}{1-\theta} \exp \frac{\theta}{1-\theta}$$

$$\Pi \underline{a}_0 = k_B T \frac{\theta}{1-\theta}$$

(4.33)

For monolayer films with long-range interactions (2D van-der-Waals or Hill-deBoer) these relations are modified to

$$K_V c = \frac{\theta}{1-\theta} \exp \left[\frac{\theta}{1-\theta} - \frac{2\alpha}{\underline{a}_0 k_B T} \theta \right]$$

$$\Pi \underline{a}_0 = k_B T \frac{\theta}{1-\theta} - \frac{\alpha}{\underline{a}_0} \theta^2$$

(4.34)

The following important conclusions emerge from these model isotherms:

(1) An adsorbate conforming to the 2D perfect gas exhibits a linear adsorption isotherm. This is a generic behaviour at very low surface concentrations ($\underline{a} \gg \underline{a}_0$). Accordingly, all adsorption systems must exhibit a linear adsorption isotherm at sufficiently low concentrations c. If the surface concentration Γ^s and bulk concentration c are both expressed on a molar basis, then the adsorption constant K has the dimension of a length. The adsorption constant K_V appearing in Eqs. (4.33) and (4.34) can be converted to K by $K = K_V / \underline{a}_0$.

(2) At higher surface concentrations the interaction between the adsorbed molecules comes into play. The Volmer equation applies when the adsorbed molecules interact only by short-range repulsive forces. This can be tested experimentally by writing the Volmer equation of state in the form, $k_B T / \Pi = \underline{a} - \underline{a}_0$. Hence, a graph of $k_B T / \Pi$ versus $\underline{a}$ should be linear down to the lowest values of $\underline{a}$ and the extrapolation to $k_B T / \Pi = 0$ gives the co-area $\underline{a}_0$ of the adsorbed molecules.

(3) In the absence of attractive lateral interactions the film pressure at a given surface coverage θ is inversely proportional to the size of the adsorbed molecules (co-area $\underline{a}_0$):

$$\Pi = \frac{k_B T}{\underline{a}_0} \frac{\theta}{1-\theta}$$

(4.35)

Figure 4.5 shows the film pressure as a function of surface coverage for two different values of the co-area: $\underline{a}_0 = 0.2$ nm^2 (typical of amphiphiles with small head groups, e.g. alkanols), and $\underline{a}_0 = 20$ nm^2 (typical for globular proteins) [20]. It can be seen that at half-coverage of the surface ($\theta = 0.5$) the film pressure of the small amphiphile is already high (20 mN m^{-1}), while for the protein it is still very low (0.2 mN m^{-1}), and a marked increase of Π occurs only at very high surface coverage. This example shows that for proteins and other large adsorbate molecules

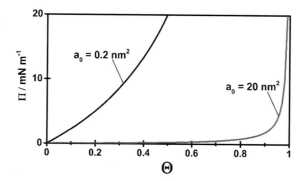

Fig. 4.5 Surface pressure Π as a function of surface coverage θ for adsorbed films of molecules with small or large cross-sectional area a_0. When a_0 is small (0.2 nm^2), Π rises steeply with θ; when a_0 is large (20 nm^2), Π stays very low up to high surface coverage θ (results for the Volmer model, Eq. (4.35), for 20 °C)

it can be dangerous to draw conclusions about the adsorbed amount from film pressure measurements.

(4) The 2D van-der-Waals equation can be used to represent systems in which attractive lateral interactions between adsorbed molecules cannot be neglected. Figure 4.6 shows results for the surface pressure $\Pi(c)$, the surface equation of state $\Pi(a)$, and the adsorption isotherm $\theta(c)$ for the 2D vdW model (Eq. 4.34), computed for several values of the reduced interaction parameter $\alpha^* = \alpha/a_0 k_B T$. The isotherms for $\alpha^* = 0$ represent the 2D Volmer model. Positive values of α^* correspond to attractive lateral interactions between the molecules in the film. It can be seen that increasing α^* causes higher values of the surface pressure Π and adsorption (surface coverage θ) at given concentration $c^* = c * K_V$. On the other hand, at a given mean area per adsorbed molecule (a), increasing lateral interaction between the molecules (increasing α^*) causes a decrease in the surface pressure Π, as can be seen in the graphs showing the surface equation of state $\Pi(a)$. Attractive lateral interactions are commonly observed for amphiphilic substances (e.g., fatty alcohols) adsorbed at the air/water interface. These molecules are adsorbed even more strongly at oil/water interfaces, but in that case the adsorbed film can be represented by the Volmer equation [21]. This remarkable behaviour is attributed to the fact that the attractive interactions between the alkanol chains are screened when they are surrounded by hydrocarbon molecules of the oil phase.

Models of localized monolayer adsorption: In many cases adsorption involves some sort of *binding* to specific adsorption sites. Hence adsorbed molecules are no longer free to move ("mobile") but "localized". The Langmuir equation is the prototype of isotherms for localized monolayer adsorption. It assumes that the surface constitutes M equivalent adsorption sites, each of which can accommodate one adsorbed molecule. If N^s molecules are adsorbed, the surface coverage is $\theta = N^s/M$. The Langmuir model assumes that no lateral interactions between adsorbed moleculesexist. For this reason it is also called *2D ideal lattice gas* model.

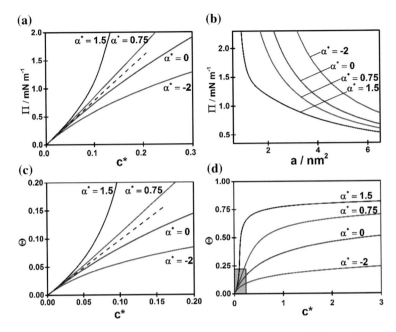

Fig. 4.6 Behaviour of adsorbed surface films according to the van der Waals model (Eq. 4.34): surface pressure $\Pi(c)$ (*upper left*), surface equation of state $\Pi(\underline{a})$ (*upper right*), and adsorption isotherm $\theta(c)$ (*lower left* and *right*; where the *graph* at the *left* shows the low concentrations region enlarged). All results refer to 20 °C and a molecular co-area $\underline{a}_0 = 0.5$ nm^2. The values of the interaction parameter α are given in reduced units ($\alpha^* = \alpha/\underline{a}_0 k_B T$), the bulk concentration is expressed in dimensionless units ($c^* = K_V c$)

The adsorption isotherm and monolayer equation of state of the Langmuir model are [14, 19]

$$K_L c = \frac{\theta}{1-\theta}$$
$$\Pi\underline{a}_0 = -k_B T ln(1-\theta) \tag{4.36}$$

When lateral interactions are introduced in this model on a mean-field basis, the Frumkin-Fowler-Guggenheim (FFG) isotherm is obtained:

$$K_L c = \frac{\theta}{1-\theta} \exp\left(-\frac{2\alpha}{k_B T}\theta\right)$$
$$\Pi\underline{a}_0 = -k_B T \left[ln(1-\theta) + \frac{\alpha}{k_B T}\theta^2\right] \tag{4.37}$$

As in the van-der-Waals equation (Eq. 4.34), positive values of the interaction parameter α correspond to attractive lateral interactions, negative α to repulsive lateral interactions. Repulsive interactions can be significant in the adsorption of polyions (proteins, etc.). Attractive lateral interactions can play a role, for example, in the adsorption of ionic surfactants to an oppositely charged surface, when the hydrophobic tails tend to aggregate due to the hydrophobic effect. Like the van-der-Waals isotherm, the Frumkin isotherm indicates that a phase separation into a dilute and a dense 2D phase occurs at low temperatures, when $|\alpha/k_BT|$ exceeds some critical value. This phase separation causes a step-wise increase of the surface coverage.

4.2.5 Standard Free Energy of Adsorption

Standard free energies of adsorption are used to characterize the strength of adsorption to interfaces. Of particular interest are values for very dilute systems, when interactions between adsorbed solute molecules are absent. From the Gibbs equation we obtain for 1 mol of adsorbate

$$-d\gamma = d\Pi = \Gamma_2^{(1)}d\mu_2 \cong \frac{1}{N_A\underline{a}}d\mu_2 \qquad (4.38)$$

where the last relation applies only to the regime of the 2D perfect gas (N_A is the Avogadro constant). Inserting the equation of state, $N_A\underline{a} = RT/\Pi$, gives after integration

$$\mu_2^a = \mu_2^{o,a} + RTln\Pi \qquad (4.39)$$

At equilibrium μ_2^a is equal to the chemical potential of the component in the bulk solution (Eq. 4.16). For highly dilute solutions the activity can be replaced by the concentration and we obtain the following expression for the standard free energy of adsorption

$$\Delta_a G^o \equiv \mu_2^{o,a} - \mu_2^{o,l} = -RTln(\Pi/c_2)_{c_2\to 0} \qquad (4.40)$$
$$(\Pi/c_2)_{c_2\to 0} = \exp(-\Delta_a G^o/RT)$$

where $\mu_2^{o,a}$ and $\mu_2^{o,l}$ represent the standard chemical potential of the solute in the surface and bulk solution. The standard enthalpy and standard entropy of adsorption can be obtained from the temperature dependence of $\Delta_a G^o$, viz.

$$\Delta_a H^o = \frac{d(\Delta_a G^o/T)}{d(1/T)} \qquad (4.41)$$
$$T\Delta_a S^o = \Delta_a H^o - \Delta_a G^o$$

Experimentally this is achieved by measuring the initial slope of film pressure isotherms $(\Pi/c_2)_{c_2\to 0}$ at a number of temperatures T and using Eq. (4.40) to

Table 4.1 Standard free energies $\Delta_a G^o$, enthalpies $\Delta_a H^o$ (in units of J mol^{-1}), and entropies $T\Delta_a S^o$(in units of J K^{-1} mol^{-1}) of adsorption of n-carboxylic acids at the air/water interface (20 °C); n is the number of carbon atoms in the tail group (from Ref. [12])

acid	n	$-\Delta_a G^0$	$\Delta\Delta_a G^0$	$-\Delta_a H^0$	$T\Delta_a S$
propionic	2	6.8		24.4	−17.6
butyric	3	10.1	3.3	17.8	−7.7
pentanoic	4	13.4	3.3	12.1	1.3
hexanoic	5	15.6	2.1	7.5	8.1
heptanoic	6	19.0	3.4	10.1	8.9
octanoic	7	22.7	3.7		
nonanoic	8	25.6	2.9		
decanoic	9	29.7	4.1		

determine $\Delta_a G^o(T)$. Care must be taken to ascertain that all film pressure data correspond to the initial linear regime of the film pressure isotherm.

Table 4.1 shows results for the adsorption of carboxylic acids at the free water surface [12]. It can be seen that $-\Delta_a G^o$ increases nearly linearly with the chain length (n) of the hydrophobic tail, with an average increment $\Delta\Delta_a G^o$ of ca. -3 kJ mol^{-1} per CH$_2$ group. $\Delta_a H^o$ is negative, indicating that adsorption of a carboxylic acid molecule at the water surface is an exothermic process, but the values become less exothermic with increasing chain length. $T\Delta_a S^o$ shows a most interesting dependence on the chain length: It is negative for short chains, indicating a loss of degrees of freedom upon adsorption. With increasing chain length $T\Delta_a S^o$ becomes less negative and assumes high positive values for hexanoic and heptanoic acid. This is attributed to the *hydrophobic effect*: Highly oriented water molecules forming a hydrogen-bonded 'cage' around the hydrophobic tails of the solute molecule in solution are released when the hydrocarbon tail leaves the aqueous medium on adsorption. The average increment in entropy $\Delta\Delta_a S^o$ is about 23 JK^{-1}mol^{-1} per CH$_2$ group. If two water molecules are oriented at each CH$_2$ group, the entropy of orientation per water molecule ($\Delta_{or}S/R \cong 1.3$) is about half the entropy of melting of ice ($\Delta_{sl}S/R \cong 2.6$). The results of Table 4.1 suggest that for longer chain lengths n the entropy term $T\Delta_a S^o$ becomes larger in magnitude than the enthalpy term $\Delta_a H^o$. Hence, there is a predominantly entropic driving force for adsorption of the higher alkanoic acids at the water surface, due to the release of $2n$ oriented water molecules on adsorption.

4.3 Liquid/Solid Interfaces

Adsorption of surfactants and polymers to solid/liquid interfaces is a broad field with a diversity of applications, from controlling the wettability of macroscopic surfaces to the stabilization of colloidal dispersions. Adsorption of biomolecules

onto micron- or nano-sized particles is used to immobilize biomarkers and drugs
and many recent studies deal with methods to control the release of adsorbed drugs
for applications in the pharmaceutical field. Traditionally, solid surfaces have been
classified into hydrophilic and hydrophobic, or 'high-energy' (inorganic) and
'low-energy' (organic), but many surfaces are heterogeneous and combine hydrophilic and hydrophobic behaviour. For example, carbons or organic polymer surfaces may contain ionizable surface groups (like –COOH) which at higher pH will
be ionized and form a hydrophilic site. In this short article only a few aspects of
adsorption of *soft matter* from aqueous solutions to solid surfaces can be touched.

4.3.1 Measurement of Adsorption

A wide variety of experimental methods is available to study adsorption at
solid/liquid interfaces [22]. Adsorption onto flat macroscopic surfaces can be
measured by ellipsometry and optical reflectometry (also by neutron or X-ray
reflectometry in favourable cases, see chapter D.12 by J. Daillat), surface spectroscopic methods (see chapter D.15 by M. Hoffmann et al.) and quartz
microbalance techniques. Adsorption onto particulate solids (powders and colloids)
with high specific surface area can be determined directly from the
adsorption-induced change in composition of the liquid phase. The reduced surface
excess concentration $\Gamma_k^{(n)}$ of a component k (defined in Sect. 4.1.3) is directly
related to the change in composition and given in terms of the mole fraction before
and after equilibration, x_k^0 and x_k^l, by [18]

$$\Gamma_k^{(n)} = \frac{n^l(x_k^0 - x_k^l)}{m_s a_s} \tag{4.42}$$

where n^l is the amount of solution, m_s is the mass and a_s the specific surface area of
the adsorbent. For solutions of polymers and other large molecules it is more
convenient to express adsorption by the *volume-reduced* surface excess concentration $\Gamma_k^{(v)}$ which is defined operationally by

$$\Gamma_k^{(v)} = \frac{V^l(c_k^0 - c_k^l)}{m_s a_s} \tag{4.43}$$

where V^l is the volume of a given amount of solution, c_k^0 and c_k^l are the concentrations of component k before and after equilibration with the adsorbent. With
some simplification (additivity of the volumes of the components on mixing, no
adsorption-induced volume changes), $\Gamma_k^{(v)}$ is related to the volume fraction profile
$\varphi_k(z)$ of the component in the boundary layer

$$\Gamma_k^{(v)} = \frac{1}{V_k^*} \int_0^\infty [\varphi_k(z) - \varphi_k^l] dz \qquad (4.44)$$

where V_k^* is the molar volume of component k. For a two-component system of solvent (1) and solute (2) this implies that $V_1^* \Gamma_1^{(v)} = -V_2^* \Gamma_2^{(v)}$, i.e., the ratio of the volume-reduced surface excess concentrations of solvent and solute is inversely proportional to the ratio of their molar volumes. This conforms to the intuitive picture of adsorption as a displacement of solvent molecules by the solute. For example, a protein molecule of a volume 1000 times the volume of water molecules will displace 1000 water molecules from the surface region, and $\Gamma_{water}^{(v)} = -1000 \Gamma_{protein}^{(v)}$.

4.3.2 Thermodynamic Relations

We have seen that for fluid interfaces the interfacial tension γ and its dependence on temperature and concentration of the components represents the primary experimental source of information on the interface. The interfacial tension of a liquid phase against a solid, which in the following will also be denoted by γ, is experimentally not accessible. However, the Gibbs equation forms a basis to determine γ from the measured adsorption. For a binary mixture at constant temperature we have

$$-d\gamma = \Gamma_1^{(n)} d\mu_1 + \Gamma_2^{(n)} d\mu_2 = \Gamma_2^{(n)} (d\mu_2 - d\mu_1) \qquad (4.45)$$

because $\Gamma_1^{(n)} = -\Gamma_2^{(n)}$. With the Gibbs-Duhem relation $x_1^l d\mu_1 + x_2^l d\mu_2 = 0$ this yields

$$-d\gamma = \Gamma_2^{(n)} d\mu_2 (1 + x_2^l/x_1^l) = \frac{\Gamma_2^{(n)}}{1 - x_2^l} d\mu_2 \qquad (4.46)$$

Integration of this relation over the composition range from pure solvent (component 1) to a solution of mole fraction x_2^l then yields

$$\gamma_1^* - \gamma(x_2^l) = RT \int_0^{x_2^l} \frac{\Gamma_2^{(n)}}{1 - x_2^l} d\ln(x_2^l f_2^l) \qquad (4.47)$$

In this relation, γ_1^* is the interfacial tension of the solid against pure liquid 1 and $\gamma(x_2^l)$ the tension against the solution of composition x_2^l. It can be shown [18] that $\gamma_1^* - \gamma(x_2^l)$ is equivalent to the difference in Gibbs free energies of wetting of the solid by pure solvent and a solution of composition x_2^l. This difference, in turn, corresponds to the free energy change of displacement of pure solvent by the solution that causes the adsorption $\Gamma_2^{(n)}$. Accordingly, the left-hand side of Eq. (4.47) is called the *Gibbs free energy of displacement* and is denoted by $\Delta_{12}G$ (J m^{-2}). An equivalent relation for $\Delta_{12}G$ can be derived when the adsorption is expressed by the volume-reduced surface excess. In the limit of ideal dilute solutions (when $x_2^l \ll 1$) this relation simplifies to

$$\Delta_{12}G(c_2) = -RT \int_0^{c_2} \frac{\Gamma_2^{(v)}}{c_2^l} dc_2^l \tag{4.48}$$

This relation can be used to determine Gibbs free energies of displacement from measured surface excess isotherms. When such isotherm measurements have been performed for several temperatures, the enthalpy and entropy of displacement, $\Delta_{12}H$ and $\Delta_{12}S$, can be determined by relations analogous to Eq. (4.41). However, for solid/liquid interfaces the enthalpies of wetting and enthalpies of displacement can also be determined directly by isothermal titration calorimetry (ITC) or isothermal flow calorimetry (IFC). By combining adsorption measurement with calorimetric studies the thermodynamics of the adsorption system can be fully characterized.

As an example, Fig. 4.7 shows the thermodynamic functions $\Delta_{12}G$, $\Delta_{12}H$ and $T\Delta_{12}S$ for the displacement of water by a short-chain nonionic surfactant (C$_8$E$_4$) at a hydrophilic glass surface [23]. The Gibbs free energy $\Delta_{12}G$ decreases with increasing concentration (i.e., with increasing adsorption) of the surfactant.

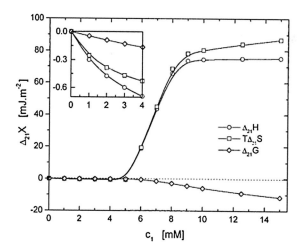

Fig. 4.7 Thermodynamic characterization of the adsorption of the surfactant C$_8$E$_4$ from aqueous solutions onto CPG silica: enthalpy ($\Delta_{12}H$), entropy ($T\Delta_{12}S$), and Gibbs free energy ($\Delta_{12}G$) as functions of the displacement of water (2) by surfactant (1) at 25 °C. The *inset* shows the behaviour at low concentrations c_1 on an enlarged scale (Reproduced from Ref. [23] with permission. Copyright 1997, American Chemical Society)

However, the overall change in $\Delta_{12}G$ is rather small, due to enthalpy/entropy compensation. In the low-concentration region (shown by the inset in Fig. 4.7) the displacement of water by the surfactant is dominated by the exothermic enthalpy of displacement, which is attributed to a direct contact of the head groups with the hydrophilic surface. This initial adsorption step is connected with a decrease in entropy. At higher concentrations the enthalpy and entropy both change sign and the displacement of water by surfactant becomes entropy-controlled. This is a signature of the hydrophobic aggregation of the surfactant tails at the surface [23].

4.3.3   Electrical Nature of Solid/Aqueous Solution Interfaces

Electrostatic interactions between surface charges and oppositely charged ionic groups of solute molecules are often determinant for the adsorption from aqueous media. For instance, in ionic solids (e.g., silver halides) ions of one charge dissolve preferentially, leaving behind a surface of opposite charge. Alternatively, one type of ions of the solution may be adsorbed preferentially, again causing a charge separation at the surface. Many inorganic oxide surfaces (Al_2O_3, SiO_2, TiO_2, etc.), exhibit a pH dependent surface charge according to the scheme

$$MOH_2^+ \overset{H^+}{\leftrightarrow} MOH \overset{H^+}{\leftrightarrow} MO^-$$

In all cases the charge on the solid surface (characterized by a charge density σ^0) must be neutralized by oppositely charged counterions in the nearby solution, thus creating an *electric double layer*. The structure of this layer is sketched in Fig. 4.8. According to the classical Stern model [24, 25] the solution side of the double layer is subdivided somewhat artificially into two parts: the inner part (Stern layer) and the outer part (Gouy layer or diffuse layer). The Stern layer, in the words of J. Lyklema [25], is 'where all the complications regarding finite ion size, specific adsorption, discrete charge, surface heterogeneity etc. reside', while the diffuse layer is by definition ideal, obeying Poisson-Boltzmann statistics. The border line between the Stern layer (thickness d) and the diffuse layer is called the outer Helmholtz (oH) plane. The net charge per unit area of this diffuse layer is σ^d. In modern treatments the Stern layer is further subdivided into an inner and an outer region. The centres of specifically adsorbed ions (i.e., ions adsorbed by non-electrostatic interactions) are located in the inner Helmholtz plane (iH), with a charge density σ^i. Ions which are not specifically adsorbed and remain hydrated can approach the surface no closer than the outer Helmholtz plane. In some cases, *super-equivalent* specific adsorption can lead to a change in sign of the potential ψ^i at the iH plane, connected with a charge reversal of the diffuse ion layer (see Fig. 4.8c). This can occur, for instance, in the adsorption of highly charged

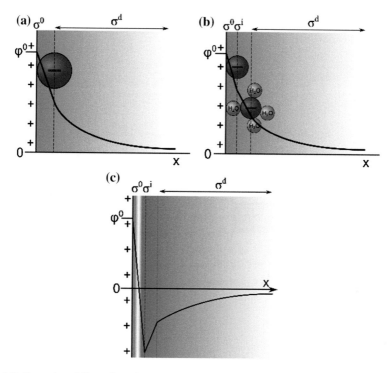

Fig. 4.8 Examples of Gouy-Stern layers: **a** only finite counterion size (*upper left*); **b** ion size and specific adsorption (*upper right*); **c** ion size and super-equivalent specific adsorption. All double layers have the same surface potential ψ^0, while the surface charge σ increases from (**a**) to (**c**) (after Ref. [25])

polymers to an oppositely charged surface. In any case, however, charge neutrality requires that

$$\sigma^0 + \sigma^i + \sigma^d = 0 \qquad (4.49)$$

The situation when $\sigma^0 = 0$ is called the *point of zero charge* (pzc), and when the solid surface plus specifically adsorbed ions has zero net charge (i.e., when $|\sigma^0| = |\sigma^i|$) is called *isoelectric point* (iep). At the isoelectric point the potential ψ^d (Fig. 4.8) is zero. The potential at the slip plane (zeta-potential ζ) is very similar to ψ^d and also zero at the iep.

A central part of the Stern theory is to determine the specifically adsorbed charge σ^i as a function of the surface charge σ^0. For the adsorption of small ions i, the Langmuir adsorption equation can be adopted for this purpose, i.e., $K_i c_i = \theta_i/(1 - \theta_i)$, where $\theta_i = N_i/N_s$ is the fraction of adsorption sites occupied by the ions and c_i is the concentration in the solution. For a charged adsorbate the electrostatic contribution to the free energy of adsorption, $z_i F \psi^i$, has to be introduced, so that instead of K_i we have $K_i \exp(-z_i F \psi^i / RT)$. Writing the Langmuir equation explicit in θ_i and introducing $\sigma^i = z_i e N_i$, we obtain the Stern equation

$$\sigma^i = z_i e N_s \frac{c_i K_i \exp(-z_i y^i)}{1 + c_i K_i \exp(-z_i y^i)} \tag{4.50}$$

where e is the elementary charge, F is the Faraday constant, and $y^i = F\psi^i / RT$. Hence it is possible to calculate σ^i once the surface potential ψ^i is known. This, however, generally requires some model assumptions [25], and is beyond the scope of this article.

As an example of super-equivalent specific adsorption, Fig. 4.9 shows results for the binding of the basic protein lysozyme to silica nanoparticles [26]. The silica surface is nearly uncharged below pH 4, but becomes negatively charged at higher pH due to the deprotonation of silanol groups. Lysozyme has a positive net charge up to its isoelectric point at pH 11. Figure 4.9a shows that binding of the protein starts when the silica surface becomes negatively charged, and it leads to an over-charging of the surface, as indicated by the positive zeta-potential of the silica

Fig. 4.9 Binding of lysozyme to silica nanoparticles for a fixed overall amount of protein (corresponding to 50 molecules per particle), and the effect on the zeta potential of the particles, as a function of pH: **a** adsorbed amount expressed as protein mass per unit area and number of protein molecules per particle; **b** zeta potential of the particles in the absence and presence of lysozyme (Reproduced from Ref. [26] with permission. Copyright 2011, American Chemical Society)

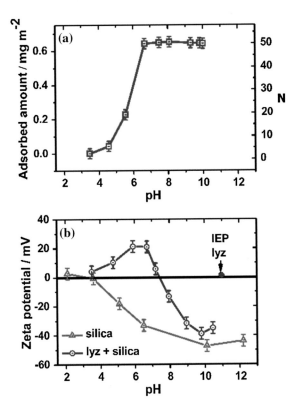

particles in the presence of the protein, although the silica particles without protein have a negative zeta potential. At higher pH the zeta potential decreases and becomes negative. This can be attributed to the increasing negative charge density of the silica surface and the decreasing positive net charge of the adsorbed protein as its isoelectric point is approached [26].

4.3.4 Adsorption as Ion Exchange

Ion exchange represents an important mechanism for adsorption to charged surfaces. This process can be dominated by electrostatic attraction of the ionic group of the adsorbate with an oppositely charged surface site. In this case adsorption is expected to be accompanied by a high exothermic adsorption enthalpy. However, in the case of protein or polyelectrolyte adsorption onto oxide surfaces, in many cases only weakly exothermic or even endothermic enthalpies of adsorption are observed, indicating that the driving force must include an entropic contribution that outweighs the enthalpic contribution. This can be rationalized from the fact that adsorption of the ionic group at the oppositely charged surface site involves the formation of an ion pair and the release of two counterions. In addition, it may also involve the release of water molecules hydrating the counterions. Models for this process indicate that the entropy gain amounts to ca. k_BT for each counterion or water molecule released [27].

The adsorption of a charged species on a charged site can be represented by an ion equilibrium reactions. For example, for a negative protein group P^- adsorbing to a positive site $-R^+$

$$-R^+Cl^- + P^-Na^+_{aq} = -R^+P^- + Na^+_{aq} + Cl^-_{aq} \qquad (4.51)$$

Here, the species are depicted with the counterions associated with the charges. The equilibrium constant of this reaction expressed in concentration units is [28]

$$K_{ads} = \frac{RP[NaCl]^2_{aq}}{R[P^-Na^+]_{aq}} \qquad (4.52)$$

where RP represents the fraction of sites occupied by the protein and R is the fraction of vacant sites. In the absence of specific attractive interactions between P^- and $-R^+$ the adsorption is driven by the release of the counterions Na^+ and Cl^-. According to Eq. (4.52) the fraction of occupied sites should decrease when the salt concentration is increased, as it is indeed often observed.

Acknowledgments I thank J. Meissner for his support in the preparation of this manuscript. Financial support by the German Research Foundation (DFG) in the framework of IRTG 1524 is also gratefully acknowledged.

References

1. G. Bakker, *Kapillarität und Oberflächenspannung*, vol. 6 of Handbuch der Experimentalphysik (ed. W. Wien, F. Harms and H. Lenz), Chap. 10, Akad. Verlagsges., Leipzig (1928)
2. E.A. Guggenheim, *Thermodynamics*, 5th edn. (North Holland, Amsterdam, 1967) p. 45
3. J.W. Gibbs, *The collected works of J. Willard Gibbs*, 2 volumes (Longmans, Green, New York, 1928)
4. H.T. Davis, *Statistical Mechanics of Phases, Interfaces, and Thin Films*, Chap. 7 (VCH Publishers, New York, 1996)
5. R. Defay, I. Prigogine, A. Bellemans, translated by D.H. Everett, *Surface Tension and Adsorption*, Chaps. 5–7 (Longmans, Green, London, 1966)
6. J.S. Rowlinson, B. Widom, *Molecular Theory of Capillarity*, Chap. 9 (Clarendon Press, Oxford, 1982)
7. J.S. Rowlinson, F.L. Swinton, *Liquids and Liquid Mixtures*, Chap. 5 (Butterworths, London, 1982)
8. E.H.A. de Hoog, H.N.W. Lekkerkerker, J. Schulz, G.H. Findenegg, J. Phys. Chem. B **103**, 10657–10660 (1999)
9. D.G.A.L. Aarts, M. Schmidt, H.N.W. Lekkerkerker, Science **304**, 847–850 (2004)
10. D.G.A.L. Aarts, J.H. van der Wiel, H.N.W. Lekkerkerker, J. Phys. Condens. Matter **15**, S245–S250 (2003)
11. R. Strey, Y. Viisanan, M. Aratono, J.P. Kratohvil, Q. Yin, S.E. Friberg, J. Phys. Chem. B **103**, 9112–9116 (1999)
12. A. Couper, in *Surfactants*, ed. by Th.F. Tadros, Chap. 2 (Academic Press, London, 1984)
13. M.J. Rosen, J.T. Kunjappu, *Surfactants and Interfacial Phenomena*, 4th edn. (John Wiley & Sons, 2012)
14. V.B. Fainerman, D. Möbius, R. Miller (eds.), *Surfactants: Chemistry, Interfacial Properties, Applications*. In: D. Möbius, R. Miller (eds.), Studies in Interface Science, vol 13 (Elsevier, Amsterdam, 2001)
15. V.B. Fainerman, E.V. Aksenenko, N. Mucic, A. Javadi, R. Miller, Soft Matter **10**, 6873–6887 (2014)
16. S. Lehmann, G. Busse, M. Kahlweit, R. Stolle, F. Simon, G. Marowsky, Langmuir **11**, 1174–1177 (1995)
17. R. Defay, I. Prigogine, A. Bellemans, translated by D.H. Everett, *Surface Tension and Adsorption*, Chaps. 12–14 (Longmans, Green, London, 1966)
18. D.H. Everett, *Colloid Science. A Specialist Periodical Report*, vol 1, Chap. 2 (The Chemical Society, London, 1973)
19. A.I. Rusanov, J. Chem. Phys. **120**, 10736–10747 (2004)
20. C.J. Beverung, C.J. Radke, H.W. Blanch, Biophys. Chem. **81**, 59–80 (1999)
21. R. Aveyard, B.J. Briscoe, Trans. Faraday Soc. **66**, 2911–2916 (1970)
22. B.P. Binks (ed.), *Modern Characterization Methods for Surfactant Systems*. Surfactant Science Series, vol 83 (Marcel Dekker, New York, 1999)
23. Z. Király, R.H.K. Börner, G.H. Findenegg, Langmuir **13**, 3308–3315 (1997)
24. O. Stern, Z. Elektrochem. **30**, 508 (1924)
25. J. Lyklema, *Fundamentals of Interface and Colloid Science*, vol 2, Chap. 3 (Academic Press, London, 1995)
26. B. Bharti, J. Meissner, G.H. Findenegg, Langmuir **27**, 9823–9833 (2011)
27. J.B. Schlenoff, H.H. Rmaile, C.B. Bucur, J. Am. Chem. Soc. **130**, 13589–13597 (2008)
28. J.B. Schlenoff, Langmuir **30**, 9625–9636 (2014)

Chapter 5
Dynamics of Surfactants and Polymers at Liquid Interfaces

Benoît Loppinet and Cécile Monteux

Abstract In the first part, we provide the background describing the adsorption dynamics in surfactant solutions, including the cases of kinetically and diffusion controlled dynamics. The second part deals with the dynamics of polymer molecules at interfaces. We provide a review of the mechanisms involved in the adsorption dynamics of polymer molecules at liquid interfaces. We show that there are energy barriers, of steric or electrostatic nature which tend to slow down the adsorption of polymer molecules in comparaison to surfactant systems. We also review results concerning the surface shear and compression properties of adsorbed and spread polymer monolayers at liquid interfaces. In particular, we describe existing models describing the mushroom to brush transition, which occurs in polymer layers as the surface concentration increases. At last, we review recent experimental results concerning the diffusion of polymer molecules at interfaces.

5.1 Introduction

The dynamics of amphiphilic molecules at liquid interfaces, including the adsorption dynamics as well as the response of adsorbed layers to external stresses and deformation plays a crucial role in several processes, where liquid interfaces are

B. Loppinet (✉)
IESL-FORTH, Heraklion, Greece
e-mail: benoit@iesl.forth.gr

C. Monteux
École Supérieure de Physique et de Chimie Industrielles de la Ville de Paris (ESPCI),
ParisTech, PSL Research University, Sciences et Ingénierie de la Matière Molle,
CNRS UMR 7615, 10 rue Vauquelin, 75231 Paris Cedex 05, France
e-mail: cecile.monteux@espci.fr

C. Monteux
Sorbonne-Universités, UPMC Univ Paris 06, SIMM, 10 rue Vauquelin,
75231 Paris Cedex 05, France

© Springer International Publishing Switzerland 2016
P.R. Lang and Y. Liu (eds.), *Soft Matter at Aqueous Interfaces*,
Lecture Notes in Physics 917, DOI 10.1007/978-3-319-24502-7_5

rapidly created or deformed. For example, industrial processes in the food or cosmetic industry often involve foaming or emulsifying processes, where bubbles and drops are rapidly created. To stabilize such fresh air-water or oil-water interfaces, amphiphilic molecules, such as surfactants or polymers, are used and their adsorption dynamics need to be faster than the rate of interface creation. Moreover, emulsions or foams are often forced to flow in confined geometries, leading to stretching and shearing of the liquid interfaces. As a result, the rheological behaviour of a foam or an emulsions strongly depends on the ability of the thin-liquid films and bubble interfaces to deform, hence on the response of the amphiphilic layer to stretching and shearing of the interface.

In this chapter, we review existing theories and experimental studies concerning the adsorption dynamics, diffusion and rheological properties of interfacial layers of surfactants and polymers at liquid interfaces. In the first part, we present derivation for the adsorption dynamics of soluble surfactants at interfaces. In the second part, we overview the case of diffusion dynamics of polymer chains near surfaces.

5.2 Adsorption Dynamics of Surfactants

When a clean and fresh interface is created the surfactant molecules which are close to the surface adsorb at the surface, creating a depletion zone, named the 'subsurface' (see Fig. 5.1). This concentration gradient between the subsurface and the bulk concentration induces a diffusion flux of the molecules to the interface. In most practical cases, the adsorption dynamics is a diffusion-controlled process, meaning that the adsorption/desorption transfer from the subsurface to the surface is very fast. However in some cases, such as for concentrated solutions or charged surfactants, where the diffusion is fast, the adsorption/desorption process of the surfactant at the interface controls the adsorption dynamics. This case is known as the

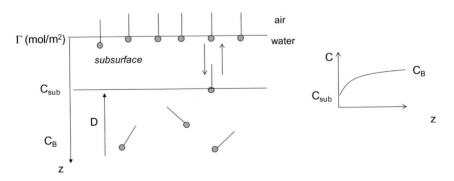

Fig. 5.1 Schematic drawing of the adsorption dynamics of surfactants at an interface, illustrating the diffusion process from the bulk phase to the subsurface as well as the adsorption/desorption transfer from the subsurface to the interface

kinetically-limited process. Below we describe the diffusion-limited and the kinetically limited adsorption models. The detailed derivation can be found in the book from Joos [1].

5.2.1 Diffusion Controlled Adsorption

Using dimensional analysis, one finds that the typical diffusion time needed to obtain a saturation surface excess, Γ_∞, in mol/m^2 scales as

$$\tau_{diff} \approx \frac{1}{D}\left[\frac{\Gamma_\infty}{C_B}\right]^2 \tag{5.1}$$

with C_B the bulk concentration, in mol/m^3 and D the diffusion coefficient of the surfactant, in m^2/s. Equation 5.1 shows that the diffusion transfer is faster for large concentrations and diffusion coefficients.

To obtain the time-variation of the surface excess, $\Gamma(t)$, one has to solve the standard diffusion equation

$$\frac{\partial C}{\partial t} = D\frac{\partial^2 C}{\partial z^2} \tag{5.2}$$

with the following boundary condition for the conservation of mass at the interface

$$\frac{\partial \Gamma}{\partial t} = D\left[\frac{\partial C}{\partial z}\right]_{z=0} \tag{5.3}$$

where z is the coordinate normal to the interface. Far from the interface, the following boundary condition applies $C = C_B$ at $z \to \infty$.

We obtain the Ward and Tordai (WT) equation below

$$\Gamma(t) = 2\sqrt{\frac{D}{\pi}}\left[C_B\sqrt{t} - \int_0^{\sqrt{t}} C_{sub}(t-\tau)d\sqrt{\tau}\right] \tag{5.4}$$

The first term of the rhs represents the diffusion to the interface while the second term on the rhs represents the back diffusion.

This equation contains two unknown functions, C_{sub} and Γ. If the adsorption process is diffusion-limited, there is a local equilibrium between the surface and the subsurface and the Langmuir isotherm gives the relation between $\Gamma(t)$ and C_{sub}.

The short time and long time limits of the WT equation can be easily derived.

For short times, $t \to 0$, the back diffusion term of the WT equation can be neglected and one can use the Henry isotherm,

$$\pi \approx k_B T \Gamma \qquad (5.5)$$

where $k_B T$ is the thermal energy and π is the surface pressure, defined as $\pi = \gamma - \gamma_0$, where γ_0 is the surface tension of the pure liquid and γ is the surface tension of the surfactant laden interface.

Combining Eqs. 5.4 and 5.5 we find that

$$\gamma(t) \approx t^{1/2} \qquad (5.6)$$

where $\gamma(t)$ is the surface tension of the surfactant solution as a function of time. For long times, $t \rightarrow \infty$, $C_{sub} \sim C_B$ and the Ward and Tordai equations rewrites

$$\gamma - \gamma_{eq} = \frac{\Gamma^2}{C_B} \left(\frac{\pi}{Dt} \right)^{1/2} \qquad (5.7)$$

5.2.2 Kinetically-Limited Adsorption

When the diffusion flux to the interface is very fast, for example in concentrated surfactants solutions, the transfer from the subsurface to the interface is the limiting step to the adsorption process. The adsorption flux writes

$$\frac{d\Gamma}{dt} = k_{ads} C_B \left[1 - \frac{\Gamma(t)}{\Gamma_\infty} \right] - k_{des} \left[\frac{\Gamma(t)}{\Gamma_\infty} \right] \qquad (5.8)$$

where k_{ads} is the adsorption constant in m/s and k_{des}, in mol s^{-1}m^{-2}, is the desorption constant.

The first term on the rhs represents the adsorption flux which varies linearly with $\left[1 - \frac{\Gamma(t)}{\Gamma_\infty} \right]$, representing the amount of free adsorption sites available for the surfactants at the interface. The second term is the desorption flux and scales with $\Gamma(t)/\Gamma_\infty$ the amount of surfactants already adsorbed.

The adsorption/desorption constants can be expressed as a function of an adsorption/desorption energy using an Arrhenius-type law,

$$k_{ads} \approx k_{ads}^0 e^{-E_{ads}/k_B T} \qquad (5.9)$$

with E_{ads} is the adsorption energy. The desorption constant writes

$$k_{des} \approx e^{-E_{des}/k_B T} \qquad (5.10)$$

with E_{des}, the desorption constant.

Equation 5.10 shows that the probability for surfactant molecules to adsorb or desorb, hence the adsorption or desorption fluxes, increase when the adsorption or desorption energy decreases.

The general solution of Eq. 5.10 writes

$$\Delta\gamma(t) \approx \Delta\Gamma(t) \approx \exp\{-kt\} \tag{5.11}$$

with

$$k = \frac{k_{ads}C_B + k_{des}}{\Gamma_\infty} \tag{5.12}$$

At equilibrium, $d\Gamma/dt = 0$ and we obtain

$$\Gamma = \Gamma_\infty \left(\frac{k_{ads}C_B}{k_{des} + k_{ads}C_B}\right) \tag{5.13}$$

Equation 5.13 shows that the surface excess increases for a decreasing k_{des} hence for increasing desorption energy, meaning that more hydrophobic surfactants tend to adsorb in denser layers at interfaces.

In the case of ionic surfactants, adsorbed surfactants create an electrostatic barrier, which repels other surfactants from the bulk and slows down their adsorption [2]. In this case, the adsorption constant depends on the surface excess and salt concentration, as salt screens this electrostatic repulsion.

5.3 Dynamics of Adsorbed and Spread Polymer Layers at Liquid Interfaces

5.3.1 Amphiphilic Soluble Copolymers

Similarly to surfactants, which contain a hydrophilic and a hydrophobic part, making them surface active, there exist several types of amphiphilic polymers, which contain both hydrophilic and hydrophobic moieties (Fig. 5.2). The hydrophobic moieties can either be part of the backbone or grafted as a pendant group. Block copolymers are composed of alternate hydrophilic and hydrophobic blocks while for random copolymers the hydrophobic moieties are randomly distributed on the backbone. In the case of grafted moieties, let us mention three specific cases: the comb polymers, where the size of the pendant group is of the order of the length of the backbone and the polysoaps, for which each monomer is grafted with a pendant group. Lastly for telechelic polymers the hydrophobic moieties are situated at both chain ends.

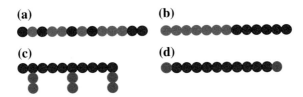

Fig. 5.2 Types of amphiphilic polymers, **a** random copolymer, **b** block copolymer, **c** hydrophobicaly-modified polymer with grafted anchors, **d** telechelic polymer

5.3.2 Pancake and Quasi-brush Regimes

In a series of papers [3–6] Daoud described theoretically, surface transitions for layers of **block copolymers**, containing N sequences of Z_A hydrophobic monomers and Z_B hydrophilic monomers, as a function of surface coverage, Γ.

In the dilute regime, the chains are far from each other, the hydrophilic blocs form 3D coils swollen in the liquid phase while the hydrophobic blocs form 2D pancakes at the interface (Fig. 5.2a). As seen in chapter C.9 by J.-U. Sommer the radius of the 2D pancakes scales as

$$r_{A\parallel} \sim a Z_A^{v2D=3/4} \tag{5.14}$$

with a the length of a monomer and $v^{2D} = 3/4$ the Flory exponent for a chain in good solvent in 2 dimensions.

The thickness of the pancakes is of the order of the monomer size,

$$r_{A\perp} \sim a \tag{5.15}$$

The size of the 3D hydrophilic coils scales as

$$r_{B\perp} \sim r_{B\parallel} \sim N^{v3D} \sim a Z_B^{3/5} \tag{5.16}$$

The 2D radius of the chains writes

$$R_{\parallel} \sim r_{A\parallel} N^{3/4} \sim a N^{3/4} Z_A^{3/4} \tag{5.17}$$

For higher surface coverage, as the chains start overlapping—but not the blocks-, a first 2D semi-dilute regime is expected. For higher surface concentration, the interface enters a semi-dilute regime of the block sequences, first on the side of the interface with the larger sequences and later on both sides of the interface. The overlap concentrations and regimes encountered depend on the ratio $\alpha = Z_A/Z_B$,

On the gas side, for increasing surface coverage, we encounter successively a 2D semi-dilute regime of the hydrophobic pancakes, then a quasi-melt when the pancakes are expulsed into the gas phase and lastly a quasi-brush. The thickness of the quasi-brush is larger than the size of the hydrophobic sequences, which stretch into the gas phase. On the liquid side, when the hydrophilic blocks overlap, the 3D hydrophilic coils stretch in the direction perpendicular to the interface and form a quasi-brush in the solution.

5.3.3 Surface Pressure and Surface Compressibility of the Layers

Below, we present a summary of the derivation from Daoud for the scaling laws of surface pressure with surface excess $\pi \sim \Gamma^y$ in the dilute, 2D semi-dilute and brush regimes.

In the dilute regime, the surface pressure writes like a 2D perfect gas, $\pi = RT\Gamma$.

When the surface coverage increases, the molecules start overlapping at the interface. The overlap surface excess Γ_{pol}^* is obtained when the area occupied by the $N(Z_A + Z_B)$ monomers of a chain is of the order of $R_{\parallel}^2$. We obtain

$$\Gamma_{pol}^* \sim \frac{N(Z_A + Z_B)}{R_{\parallel}^2} \sim a^{-2} Z_B^{-1/5} N^{-1/2} (1 + \alpha) \qquad (5.18)$$

The correlation length, ξ, has to be independent of N and writes

$$\xi \sim R_{\parallel} \left(\frac{\Gamma}{\Gamma_{pol}^*} \right)^m \sim N^0 \qquad (5.19)$$

From Eqs. 5.19 and 5.18, we find that $m = -3/2$ hence we obtain

$$\xi \sim a^{-2} (1 + \alpha)^{3/2} Z_B^{3/10} \Gamma^{-3/2} \qquad (5.20)$$

Then the surface pressure writes,

$$\pi \sim \frac{k_B T}{\xi^2} \sim \Gamma^3 \qquad (5.21)$$

For the other semi-dilutes regimes, the full calculation also leads to $y = 3$.

In the case of the quasi-brushes, the overlap surface excess is expressed as below

$$\Gamma_{block}^* \sim \frac{Z_B + Z_A}{r_{B\parallel}^2} \sim a^{-2} Z_B^{-1/5} (1 + \alpha) \qquad (5.22)$$

The correlation length scales like

$$\xi_B \sim Z_B^{1/2} \Gamma^{-1/2} \qquad (5.23)$$

and one deduces that the surface pressure scales as

$$\pi \sim \frac{k_B T}{\xi_B^2} \sim k_B T a \Gamma \tag{5.24}$$

These scaling laws can be tested experimentally by two different methods. The first one consists in spreading the polymer layer, compressing it using a Langmuir trough and measuring the surface pressure as a function of surface coverage using a Wilhelmy plate. Aghie-Beguin and Daoud proposed a second method using the pendant drop experiment. A pendant drop of polymer solution is formed at the tip of the syringe. While the polymer is adsorbing at the interface and the surface pressure is rising, the area of the drop, A, is oscillated. From the oscillations, one can obtain the surface elasticity

$$\varepsilon = \frac{d\pi}{d \ln A} = \frac{-d\pi}{d \ln \Gamma} \tag{5.25}$$

Over the course of time, as the polymer layer is adsorbing, the surface pressure rises; one can therefore obtain ε as a function of π. Moreover from Eq. 5.25, y writes

$$y = \frac{d \ln \pi}{d \ln \Gamma} = \frac{\varepsilon}{\pi} \tag{5.26}$$

Therefore the exponent y corresponds to the slope of the $\varepsilon(\pi)$curves.

Daoud and Aghié-Beghin investigated experimentally several **triblock copolymers**, either PEO-PPO-PEO or PPO-PEO-PPO with varying degrees of hydrophobicity, PEO being Poly(ethyleneoxide) and PPO Poly(propyleneoxide). For all the eight copolymers, they found the first 2D semi-dilute regime with y = 3. At larger surface pressures, the elasticity either decreased or remained constant and then increased again for the polymers with moderate hydrophobicity, results which were not satisfactorily described by the model.

Other experimental studies concerning PEO-PPO-PEO copolymers were published by the Miller [7, 8] and Rubio [7, 9] groups. These authors reported qualitatively similar results—a maximum of the elasticity with the surface pressure. For larger surface pressures they observe a second maximum of the compression elasticity. The authors attribute the first decrease of the elasticity to the desorption of the PEO monomers into the liquid to form a brush and the second decrease to the desorption of PPO blocks into the water along with the PEO segments, instead of a desorption into the air phase as described by Daoud.

In the literature, there are other systems which present a maximum of the compression elasticity. Barentin [10, 11] observed a plateau in the surface pressure isotherm, hence a maximum of the elasticity, for Langmuir monolayers of **telechelic PEO** spread at the air-water interface. The authors also attributed this behaviour to a mushroom to brush transition. At low surface concentrations, the chains lie flat at the interface and the surface pressure is the same as the one predicted by scaling laws for a 2D semi-dilute regime, $\pi = \Gamma^3$. At larger compressions, the PEO monomers desorb from the interface and form a 'brush', anchored by the two hydrophobic moieties. A second maximum of the elasticity can be observed at larger surface coverage, when the hydrophobic anchors desorb from the interface (Fig. 5.3). The desorption time writes

$$\tau_{des} \sim \exp\left\{\frac{n\delta}{k_B T}\right\} \tag{5.27}$$

where n is the number of carbon atoms of the hydrophobic chain end and δ is the desorption energy of a -CH$_2$- unit.

After desorption of the hydrophobic moiety from the interface it is expulsed from the brush because of the gradient of chemical potential between the brush and the solution. The expulsion velocity results from a balance between osmotic pressure in the brush and viscous friction during the expulsion process. According to Wittmer [12], it writes

$$k_B T d^{-1} \sim v_{exp\,ulsion} N \zeta \tag{5.28}$$

where d is the distance between the polymer chains in the brush and ζ a friction coefficient. The typical expulsion time scales like

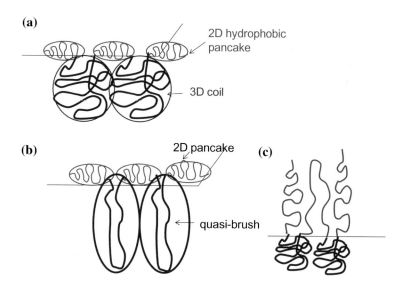

Fig. 5.3 Transitions upon compression for amphiphilic block copolymer adsorbed onto a liquid/air interface. **a** Structure of a single chain adsorbed onto the interface, **b** brush regime on the liquid side when the hydrophilic blocks are overlapping, **c** brush regime on the gas side, when the hydrophobic blocks overlap

$$\tau_{exp} \sim \frac{h}{v_{exp\,ulsion}} \sim N^2 \tag{5.29}$$

with $h \sim \Gamma^{-1/3}$ the thickness of the layer.

In the case of the telechelic polymers, Barentin et al. find that $\tau_{exp} \sim 10^{-5}s \ll \tau_{des}$. Therefore the relaxation dynamics of the layer is controlled by the desorption dynamics. The authors confirmed experimentally that the relaxation of the surface pressure after a strong compression depends on the number of carbon atoms of the hydrophobic moiety.

5.3.4 Adsorption Kinetics

The adsorption kinetics of soluble polymers spontaneously adsorbing from a bulk solution is usually very slow because of energy barriers of various origin. Below we present the main mechanisms which slow down the adsorption process of polymers at interfaces.

5.3.5 Adsorption Barrier Due to Exchange Between Micelles and Unimers in the Bulk Solution

Theodoly et al. [13] investigated experimentally the adsorption kinetics of three types of block copolymers where one block is hydrophilic, PAA, Poly (acrylic acid), and the second block has an increasing hydrophobicity, namely PAA-PDEGA (PDGEBA stands for poly(diethylene glycol ethyl ether acrylate) PAA-PBA (PBA stands for Poly (buthyl acrylate)) and PAA-PS, (PS stands for polystyrene). Using the pendant drop method, they found that for the most hydrophobic core, PS, the surface activity is very weak. Indeed, the frozen micelles do not adsorb at the interface and there are no free polymer molecules in the solution. In the case of PAA-PBA, the adsorption time scales like $\tau_{ads} \sim C_B^{-1}$, which rules out diffusion limited kinetics. In that case, the adsorption kinetics is controlled by the unimer extraction from the micelles. Finally for the most hydrophilic copolymer, PAA-PDEGA, the adsorption is controlled by the diffusion of the unimers in the solution and scales as $\tau_{ads} \sim C_B^{-2}$.

5.3.6 Adsorption Barrier Due to Diffusion-Reptation of Chains Through the Layer

Ligoure and Leibler [14] and Johner and Joanny [15] described theoretically the kinetics of adsorption of amphiphilic polymers: a telechelic polymer dissolved in a solution in ref. 14, and a diblock copolymer in ref 15 and show that there are two successive regimes. At short times, of the order of one second, the adsorption is limited by the diffusion of chains to the interface. The typical time scales as

$$\tau_{\substack{diff \\ saturation}} \approx \frac{1}{D}\left(\frac{1/a^2}{N_A C_B}\right)^2 = 0.3s \tag{5.30}$$

where $N_A = 6.02 \times 10^{23}$ mol^{-1} is the Avogadro number. Afterwards, the chains start to overlap at the interface and to stretch strongly in the direction perpendicular to the interface, forming a brush. The additional chains approaching the interface have to stretch and diffuse by reptation through the brush already adsorbed (Fig. 5.4).

The adsorption flux scales as

$$J_{in} \sim -K(C_B - C_{surface}) \tag{5.31}$$

with

$$K \sim e^{-N(N_A a^2 \Gamma)^\alpha} \tag{5.32}$$

Where $\alpha = 5/6$ or $2/3$ respectively in references 14 and 15. K is a kinetic constant which represents the energy barrier due to stretching of the chain.

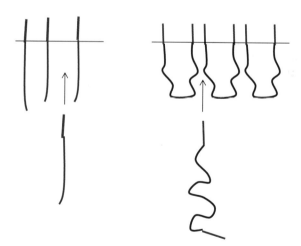

Fig. 5.4 Energy barrier due to the stretching of a chain which is required to adsorb at an interface where polymer molecules are already adsorbed and form a quasi-brush. On the *left*, case of a block copolymer, the situation described in Johner and Joanny's work. On the *right*, case of a telechelic polymer, situation described in Ligoure and Leibler's article

The desorption flux simply scales like

$$J_{out} \approx \Gamma e^{-\varepsilon} \tag{5.33}$$

with ε the attraction energy in units of $k_B T$ between the chains and the interface, determined by the hydrophobic anchors.

In this second regime the adsorption of the chains is much slower than the first one and the surface density increases logarithmically with time.

This logarithmic law was verified experimentally by Millet et al. [16, 17] using polyacrylic acid polymers statistically grafted with hydrophobic anchors. They found that the adsorption dynamics did not depend on the polymer concentration, nor the grafting density nor the salt concentration but only on the molecular mass, as predicted by Ligoure and Leibler.

5.3.7 Case of Polyelectrolytes: Adsorption Barrier Due to Electrostatics

In the case of charged molecules, the first chains which adsorb at the interface create a negative potential which slows down the adsorption of additional chains. Theodoly et al. [18] observed that electrostatic effects also control the adsorption of random PS/PSS, polystyrene/polysteren sulfonate copolymers at the air-water interface. The authors found that the surface tension reaches a minimum with the degree of sulfonation. The surface tension of the PS/PSS with the highest degree of sulfonation remains equal to that of pure water. Highly charged polymers are not hydrophobic enough to be surface active. However for sufficiently large polymer concentrations or salt concentrations, the surface tension decreases. For intermediate degrees of sulfonation, the polyelectrolyte is more surface active and the surface tension is lower than that of pure water. Unexpectedly, for the lowest degree of sulfonation, the surface tension is close to that of pure water although the molecules are hydrophobic. In the bulk solution the chains form hydrophobic globules which are surrounded by a cloud of counter-ions which is repelled by the interface. The addition of salt enables to screen the electrostatic interaction and to decrease the surface tension.

To probe the desorption dynamics of the PS/PSS layers, the same authors washed out the bulk solution either with pure water or with a salt solution. In these experiments, the adsorbed amount remains constant, however a lower surface pressure is reached when the salt solution is used for the rinsing process. This result illustrates that the surface pressure originates from the electrostatic repulsions between the charged monomers at the interface and it can be screened by the salt.

5.4 Dynamics of Adsorbed Chains

Polymer chains are dynamical objects, with very large number of internal degrees of freedom. Many of the characteristic properties of polymers find their origin in the specifics of long chain dynamics, from the single coil Brownian motion to the rubber visco-elasticity. The dynamics of polymer melts and solutions have been studied for many years. The current understanding of the mechanisms for macro-molecular motions permits a good microscopic description of dynamics of chains in solutions or in melts [19].

The case of confined polymer chains and in particular the effect of the presence of a surface or an interface on the polymer conformation and also on the polymer dynamics has long attracted scientific attention [20, 21]. Such dynamics are important not the least because they determine the mechanical properties of the interfaces.

A large effort has been directed towards the understanding the adsorption mechanisms and the characterization of structure and conformation of the polymer chains, i.e. the static average properties, chain conformation, extend and so on. In many cases, the effects of the confinement are satisfactorily established. The dynamical properties often turned out to be more difficult to access, especially in experiments. The limited amount of available experimental data is mostly due to experimental difficulties, often related to low experimental signal related to the small numbers of involved monomers, but also to the difficulties to control the sample and its preparation to the level required for the clear and clean data interpretation.

Numerical simulations have been extensively used in order to provide support and validation for theories. Their results also can be compared to experiments. The comparison is not always straightforward, as the details of the real systems are often difficult to take into account in the models [22].

In this part, we briefly overview some cases of dynamics of polymer chains next or at an interface, mostly from an experimentalist's viewpoint.

5.4.1 Single Chains on a Solid Substrate

The case of adsorption of a single polymer chain onto a (slightly attractive) surface has long attracted experimental attention. The conformation of adsorbed hains is predicted to consist of loops trains and tails (e.g. chapter A. 1 by I. Voets), designing the difference sequences between adsorbed monomers. The (attractive) interaction between monomer and the substrate is an important parameter, which can be experimentally varied depending on the specifics of the polymer/solvent/substrate systems. The conformation of adsorbed chains is often difficult to check experimentally, (though it can be achieved in some cases through neutron scattering for example). Computer simulations have been used extensively to address this question. Regarding dynamics, a question, connected to the one of the conformation of the adsorbed chain, is the motion of such chains, and in particular the centre of mass diffusion. This has been addressed through simulation and scaling laws been observed and found to depend on exact conditions [23]. In particular, a clear scaling of the diffusion coefficient with the molecular mass $D \sim M^{-1}$ was identified. It was attributed to Rouse like dynamics, where the friction is dependent of the number of contacts with the surface, i.e. proportional to the molecular mass.

It turned out to be difficult to establish the single chain diffusion experimentally and it might not yet be fully achieved. One of the difficulties was to be able to control the type of confinement. Researcher from S. Granick's group attempted to study diffusion of polymer chains adsorbed on a solids, using fluorescence based methods [24–27]. Fluorescence correlation spectroscopy was used to measure diffusion of single particles. Fluorescence recovery after photo bleaching was also used in some cases. After long efforts, the situation recently seemed to clarify. Strong adsorption was expectedly found to lead to immobilization, where the chain centre of mass does not move at all. But weaker adsorption was observed to lead to an adsorption/desorption mechanism to be the dominant motion. This was only recently realized in the case of polymers near to solid surface. For an intermediate regime it seems to be possible to observe chains crawling on the surface. This was established for polymer/solid cases in a recent study of polystyrene adsorbed onto quartz surfaces [27, 28]. The authors were able to control the surface roughness of the solid substrate, through thermal treatment, and they could identify two regimes. In the case of a smooth surface, the chains were observed to diffuse, with a diffusion coefficient that was slowing down with increasing molecular mass as $D \sim M^{-1}$. That was identified as the expected Rouse like motion, where the friction is proportional to the number of monomers. For rougher surfaces, the same authors observed slower diffusion with a stronger slow down with molecular mass, $D \sim M^{-3/2}$ which was attributed to a motion along the chain path length (reputation like). They also noticed that for large M, the scaling seems to depart and go back the M^{-1} behaviour.

This was hypothesized to be due to the onset of the adsorption/desorption mechanism, that also had long been predicted, and that has been observed in various type of systems and described in term of Levy flight. Note that such adsorption desorption mechanism can lead to an apparent 2D diffusion coefficient, which is somewhat faster than what would be expected in the case of adsorbed chains and which is of a different nature.

The case of polyethylene oxide adsorbed onto silica surfaces has also been largely studied experimentally by the Granick's group. A similar picture also emerged for this system but now with three different regimes. For this system variation of the diffusion with the surface coverage has been reported [26]. The results showed some strong concentration dependence of D, at relatively low surface coverage an unanticipated speed-up of the diffusivity was observed. This was rationalized by the formation of a multilayer as observed in simulations. At larger surface coverage, the diffusivity was observed to drop significantly reaching very

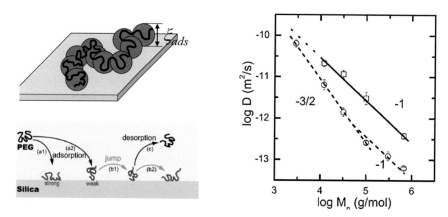

Fig. 5.5 Schematic of an adsorbed chain onto a solid surface. Diffusion coefficient of single polystyrene chains on polished and annealed quartz for different molecular masses and corresponding scaling laws [from Ref. 24] schematic of adsorption desorption hoping for PEO on silica [from Ref. 27]

slow diffusion. That was attributed to interchain interactions and possibly entanglement between chains (Fig. 5.5).

The issue of polymer chains in contact with solid surfaces has received a large attention, particularly connected to the case of mixtures of polymer and solid particles, i.e. composites, and in particular to better understand their mechanical properties. The effect of confinement on the polymer chains dynamics and especially on the fast segmental dynamics is a much debated issue at present, with different studies apparently leading to conflicting conclusions. Many studies have been devoted to the possible change of glass transition temperature and changes in segmental dynamics for chains in the immediate vicinity of a solid substrate or of the surface. It is expected to have large impact on the mechanical properties of the systems. This remains as a very active research field [29–32]. Polymer thin films and possible change of the glass transition temperature, and possible changes of the segmental dynamics have been much studied.

5.4.2 Polymer Chains at Water Surface

Another case of interest is the one of a fluid substrate, and more particularly the case of polymer chains deposited at air-water interfaces.

There has been a long interest into the possibility of forming Langmuir polymer monolayers on top of a water surface. Water insoluble chains can be spread on the water-air interface using a dilute solution of the polymer in a volatile good solvent A Langmuir through with moving barriers offers an easy way to vary the surface concentration and the interfacial tension can be monitored with a Whilhelmy plate. This has been widely used to prepare polymer Langmuir monolayers. The measured changes of interfacial tension interpreted in terms of surface pressure are in good agreement with the expected scaling laws for polymers in two dimensions. The evolution of the pressure with increasing concentration confirms the existence of dilute, semi-dilute, concentrated regimes. A characteristic coil size R is deduced from the onset of the semi-dilute regime. In particular the solvent quality is deduced from the molecular mass dependence of the coil size ($R \sim N^{3/4}$ good solvents and $R \sim N^{1/2}$ for theta solvents). Note that the conformation of individual chains in such quasi two dimensional systems is difficult to assess directly. The overall homogeneity of the layer over a large area is often questionable and can be checked by various techniques like for example Brewster Angle Microscopy.

The mechanical properties of such layers have been studied and eventually one would like to relate them to a microscopic theory of polymer dynamics, as it is done in 3D polymer solutions and melts. This has not been achieved yet.

Somewhat surprisingly, the diffusion of dilute chains at the air water interface has not been much studied experimentally. One could expect the diffusion to be similar to the case of adsorbed chains onto solid substrate with Rouse like dynamics where $D \sim M^{-1}$. Such scaling has indeed been observed for large DNA chains adsorbed onto fluid phospholipids layers [33]. However, in polymer chains on top of water one may expect hydrodynamic interaction between monomers of the same chain through the water phase to have an effect on the overall diffusion. In standard three dimensional polymer solutions, the hydrodynamic interactions are important as they lead to the coil diffusion to scale with the coil size as $D \sim R^{-1}$. Hydrodynamic interactions are incorporated in the Zimm model which leads to the correct scaling, equivalent to the one of a rigid object. The case of dilute chains with hydrodynamic interaction has also been treated via computer simulation [34]. The effect of hydrodynamic interaction on the diffusion and the dynamical scaling. One may expect a weak (logarithmic) dependence of the diffusion coefficient with the molecular mass.

At larger concentrations, in the semi-dilute regime, the chains start to interact. At even larger concentration the layer may resemble a two dimensional polymer melt. Surface light scattering and interfacial rheology have been used to study the non-dilute regimes, mainly with the objective to characterize the mechanical properties of the monolayers and try to relate them to the underlying structure, and to a lesser extend to attempt to characterize the dynamics of the layers.

Surface light scattering techniques provide a way to assess the local segmental dynamics of such systems, measuring the spectrum of capillary waves and their modification by the presence of polymer monolayer [35–37]. The analysis is often put in terms of the dilational modulus. The capillary surface waves are rather complicated waves, and their propagation implies several types of deformation of the visco-elastic surface. As the frequency domain accessible through this type of measurement is in a rather high frequency (kHz) range, the deduced modulus will correspond to the high frequency limit of the chains dynamics, known as segmental dynamics. It is sensitive to local fluctuations, where the polymeric nature of the layer is expected to have limited effect.

Interfacial rheology has undergone important experimental developments in the last decade. Various approaches are used to measure the interfacial shear modulus, and its elastic and viscous parts. Experiments are reaching a state that can allow systematic studies and a fine understanding of the mechanical properties of the interface in relation with its microscopic origin. Interfacial rheology has been applied to polymer monolayers at air-water interface and it has been recently reviewed in a book by Jan Vermant, which should appear around the same time as these lecture notes. The interpretation of mechanical behaviours in terms of molecular motion (2d molecular rheology, as an analogy of the 3d polymer molecular rheology) remains limited in those systems. The number of polymers studied remain rather limited. Moreover, some experimental issues regarding the exact preparation conditions of the layer are important and makes the reproducibility and comparison of the experiments difficult to reach.

The most systematic study concerns a broad family of acrylate polymers. The change of chemistry leads to change of the polymer glass transition T_g and also change the 2d water solvent quality spanning good and theta conditions. Detailed studies were done using various interfacial rheology techniques for a range of molecular masses, in an effort to determine the variation of the mechanical properties. Scaling approach were used to compare exponent derived from expected scaling laws to experimental results.

Several questions remained not fully answered, concerning the underlying structure and its relation to the mechanical properties. In particular the question of whether a coils can overlap in a Langmuir monolayer and to what extent, and how it does depend on the specific chemistry is not fully established. Does it resemble more non-interpenetrating coils in a colloid like behaviour or do monolayers allow some coils overlap and entanglement, forming through small loops in the direction normal to the interface, as depicted in Fig. 5.6 (right) ?

Note that true two dimensional polymer melts have been studied through computer simulation [34, 38]. They are different from 3D melts in the fact that chains can not overlap, and therefore they should not present entanglement. The dynamics would then be expected to be of Rouse type for all given molecular masses. Hydrodynamic interaction could become relevant.

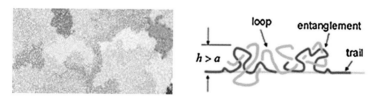

Fig. 5.6 Schematic of two dimensional melt realistic view of semi-dilute polymer chains in semi-dilute regime at air water interface (adapted from 37, 38)

5.4.3 Dynamics of Polymer Brushes

Non adsorbing polymer chains that are anchored by one end to the surface represent another type of systems. In the case of large grafting densities, they form what is known as polymer brushes where the polymer chains are extended. Polymer brushes have been and still are very actively researched because of their practical relevance. Most recent studies of the studies focused on the responsiveness of such systems. Alexander and De Gennes proposed some simple scaling laws for the thickness as a function of the grafting density based on a scaling blob model, (See also chapter of C.9 by J.-U. Sommer) which was mostly confirmed.

But comparatively little has been done on the experimental part to uncover the dynamics of such grafted chains. Polystyrene grafted chains have been studied using evanescent wave dynamic light scattering [39]. The results were in agreement with the blob picture developed for polymer brushes [40, 41]. The diffusion length scale was found to be small and fairly similar to the average distance between grafted points. An observed fast diffusive process was attributed to a cooperative mode, in analogy with semi-dilute polymer solutions in good solvent. This corresponds to the polymer-solvent motion within one blob. A hydrodynamic blob size was deduced from the diffusion coefficient using the Stokes-Einstein relation. The measured values were found to be in good agreement with the average distance between grafting points. This supports the Alexander-De Gennes blob models where the blob size is imposed by the distance between grafting points.

The dynamics of the same brush in a theta solvent revealed more complex behaviour, with a slower, stronger relaxation developing as the solvent quality was decreased (Figure 5.7).

Another approach to probe dynamics of those thicker systems is to follow the diffusion of tracers [42, 43]. The tracer particles' diffusion inside the layer is expected to relate to some of the dynamic features of the layer. However, the relation between the diffusion of the tracers and the polymer dynamics is often not straightforward due to interaction. One still needs to establish the relation between the motion of the tracer and the dynamics of the polymer chains.

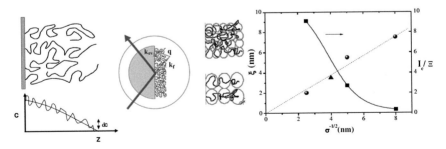

Fig. 5.7 Concentration fluctuation and blobs in polymer brushes schematics of polymer brush and associated concentration profile, indicating fluctuation dc. Relation between blob size and grafting density. Hydrodynamic blob size ξ as a function of grafting density (adapted from 40, 41)

5.5 Conclusion

Understanding the adsorption dynamics of polymer chains at interfaces as well as the dynamics of the chains at the interface remains a challenge. In particular, the microscopic understanding of the dynamical properties is still in an early state.

Acknowledgments The authors acknowledge financial support from ANR JCJC INTERPOL.

References

1. P. Joos, V.B. Fainerman, *Dynamic surface phenomena* (VSP, Zeist, 1999)
2. A. Bonfillon, F. Sicoli, D. Langevin, Dynamic surface tension of ionic surfactant solutions. J. Colloid Interface Sci. **168**, 497–504 (1994)
3. V. Aguié-Béghin, E. Leclerc, M. Daoud, R. Douillard, Asymmetric multiblock copolymers at the gas-liquid interface: phase diagram and surface pressure. J. Colloid Interface Sci. **214**, 143–155 (1999)
4. R. Douillard, M. Daoud, V. Aguié-Béghin, Polymer thermodynamics of adsorbed protein layers. Curr. Opin. Colloid Interface Sci. **8**, 380–386 (2003)
5. R. Douillard et al., State equation of β-Casein at the air/water interface. J. Colloid Interface Sci. **163**, 277–288 (1994)
6. A. Hambardzumyan, V. Aguié-Béghin, M. Daoud, R. Douillard, β-Casein and symmetrical triblock copolymer (PEO−PPO−PEO and PPO−PEO−PPO) surface properties at the Air−Water Interface. Langmuir **20**, 756–763 (2004)
7. B.A. Noskov, S.-Y. Lin, G. Loglio, R.G. Rubio, R. Miller, Dilational viscoelasticity of PEO−PPO−PEO triblock copolymer films at the air–water interface in the range of high surface pressures. Langmuir **22**, 2647–2652 (2006)
8. A.M. Díez-Pascual et al., Adsorption of water-soluble polymers with surfactant character. Dilational viscoelasticity. Langmuir **23**, 3802–3808 (2007)
9. M.G. Muñoz, F. Monroy, F. Ortega, R.G. Rubio, D. Langevin, Monolayers of symmetric triblock copolymers at the air–water interface. 2. Adsorption kinetics. Langmuir **16**, 1094–1101 (2000)
10. C. Barentin, P. Muller, J.F. Joanny, Polymer brushes formed by end-capped poly(ethylene oxide) (PEO) at the air–water interface. Macromolecules **31**, 2198–2211 (1998)
11. C. Barentin, J.F. Joanny, Surface pressure of adsorbed polymer layers. Effect of sticking chain ends. Langmuir **15**, 1802–1811 (1999)
12. J. Wittmer, A. Johner, J.F. Joanny, K. Binder, Chain desorption from a semidilute polymer brush: a Monte Carlo simulation. J. Chem. Phys. **101**, 4379 (1994)
13. O. Théodoly, M. Jacquin, P. Muller, S. Chhun, Adsorption kinetics of amphiphilic diblock copolymers: from kinetically frozen colloids to macrosurfactants. Langmuir **25**, 781–793 (2009)
14. C. Ligoure, L. Leibler, Thermodynamics and kinetics of grafting end-functionalized polymers to an interface. J. Phys. **51**, 1313–1328 (1990)
15. A. Johner, J.F. Joanny, Block copolymer adsorption in a selective solvent: a kinetic study. Macromolecules **23**, 5299–5311 (1990)
16. F. Millet, P. Perrin, M. Merlange, J.-J. Benattar, Logarithmic adsorption of charged polymeric surfactants at the air–water interface. Langmuir **18**, 8824–8828 (2002)
17. F. Millet et al., Adsorption of hydrophobically modified poly(acrylic acid) sodium salt at the air/water interface by combined surface tension and X-ray reflectivity measurements. Langmuir **15**, 2112–2119 (1999)

18. O. Théodoly, R. Ober, C.E. Williams, Adsorption of hydrophobic polyelectrolytes at the air/water interface: conformational effect and history dependence. Eur. Phys. J. E **5**, 51–58 (2001)
19. M. Rubinstein, R.H. Colby, *Polymer Physics* (Oxford University Press, Oxford, 2003)
20. G.J. Fleer, M.A. Cohen Stuart, J.M.H.M. Scheutjens, T. Cosgrove, B. Vincent (eds.) *Polymers at Interfaces* (Chapman and Hall, London, 1993)
21. R.A.L Jones, R.W. Richards, *Polymer at Surfaces and Interfaces* (Cambridge University Press, Cambridge, 1999)
22. A. Milchev, Single-polymer dynamics under constraints: scaling theory and computer experiment, J. Phys. Condens. Matter **23** 103101 (2011)
23. D. Mukherji, G. Bartels, M.H. Müser, Scaling laws of single polymer dynamics near attractive surfaces. Phys. Rev. Lett. **100**, 068301 (2008)
24. J. Zhao, S. Granick, How polymer surface diffusion depends on surface coverage. Macromolecules **40**, 1243 (2007)
25. S. Granick, S.C. Bae, Molecular motion at soft and hard interfaces: from Phospholipid bilayers to polymers and lubricants. Annu. Rev. Phys. Chem. **58**, 353 (2007)
26. J.S. Wong, L. Hong, S.C. Bae, S. Granick, Polymer surface diffusion in the dilute limit. Macromolecules **44**, 3073 (2011)
27. C. Yu, J. Guan, K. Chen, S.C. Bae, S. Granick, Single-molecule observation of long jumps in polymer adsorption. ACSNano **7**, 9735 (2013)
28. M.J. Skaug, J.N. Mabry, D.K. Schwartz, Single-molecule tracking of polymer surface diffusion. J. Am. Chem. Soc. **136**(4), 1327–1332 (2014)
29. M.D. Ediger, J.A. Forrest, Dynamics near free surfaces and the glass transition in thin polymer films: a view to the future. Macromolecules **47**, 471 (2014)
30. A. Papon, H. Montes, M. Hanafi, F. Lequeux, L. Guy, K. Saalwächter, Glass-transition temperature gradient in nanocomposites: evidence from nuclear magnetic resonance and differential scanning calorimetry. Phys. Rev. Lett. **108**, 065702 (2012)
31. M. Krutyeva, A. Wischnewski, M. Monkenbusch, L. Willner, J. Maiz, C. Mijangos, A. Arbe, J. Colmenero, A. Radulescu, O. Holderer, M. Ohl, D. Richter, Effect of nanoconfinement on polymer dynamics: surface layers and interphases. Phys. Rev. Lett. **110**, 108303 (2013)
32. M. Tress, E.U. Mapesa, W. Kossack, W.K. Kipnusu, M. Reiche, F. Kremer, Glassy dynamics in condensed isolated polymer chains. Science **341**(6152), 1371–1374 (2013)
33. B. Maier, J.O. Radler, DNA on fluid membranes: a model polymer in two dimensions. Macromolecules **33**, 7185–7194 (2000)
34. B.J. Sung, A. Yethiraj, Dynamics of two-dimensional and quasi-two-dimensional polymers. J. Chem. Phys. **138**(23), 234904 (2013)
35. P. Cicuta, I. Hopkinson, Recent developments of surface light scattering as a tool for optical-rheology of polymer monolayers. Colloids Surf. A **233**, 97–107 (2004)
36. A.R. Esker, C. Kim, H. Yu, Polymer monolayer dynamics. Adv. Polym. Sci. **209**(1), 59–110 (2007)
37. D. Langevin, F. Monroy, Interfacial rheology of polyelectrolytes and polymer monolayers at the air–water interface. Curr. Opin. Colloid Interface Sci. **15**(4), 283–293 (2010)
38. A.N. Semenov, H. Meyer, Anomalous diffusion in polymer monolayers. Soft Matter **9**(16), 4249–4272 (2013)
39. R. Sigel, Light scattering near and from interfaces using evanescent wave and ellipsometric light scattering. Curr. Opin. Colloid Interface Sci. **14**(6), 426–437 (2009)
40. V.N. Michailidou, B. Loppinet, O. Prucker, J. Ruhe, G. Fytas, Cooperative diffusion of end-grafted polymer brushes in good solvents. Macromolecules **38**(21) 8960–8962 (2005)
41. V.N. Michailidou, B. Loppinet, D.C. Vo, O. Prucker, J. Ruhe, G. Fytas, Dynamics of end-grafted polystyrene brushes in theta solvents. J. Polym. Sci. Part B Polym. Phys. **44**(24), 3590–3597 (2006)

42. S. Wang, Y. Zhu, Molecular diffusion on surface tethered polymer layers: coupling of molecular thermal fluctuation and polymer chain dynamics. Soft Matter **6**, 4661–4665 (2010)
43. A. Vagias, P. Košovan, K. Koynov, C. Holm, H.-J. Butt, G. Fytas, Dynamics in stimuli-responsive poly(N-isopropylacrylamide) hydrogel layers as revealed by fluorescence correlation spectroscopy. Macromolecules **47**(15), 5303–5312 (2014)

Chapter 6
Water–Water Interfaces

R. Hans Tromp

Abstract The theory and experimental properties of interfaces between phase separated aqueous polymer solutions are discussed. These interfaces are characterized by a very low interfacial tension (0.01–1 µN/m), accumulation of solvent and in some cases a spontaneous curvature and an electric potential (Donnan potential). The procedure of calculating interfacial tension of both flat and curved interfaces is presented, and an overview of the methods to measure this interfacial tension and interface potential is given.

Keywords Water–water interface · Interfacial tension · Interfacial curvature · Interfacial potential

6.1 General Aspects of Water–Water Interfaces

Water–water interfaces are interfaces between aqueous volumes which differ in composition. These interfaces are found between volumes co-existing in thermodynamic equilibrium or between unmixed aqueous solutions in direct contact which would lose free energy by mixing but are kept in a separated meta-stable state by slow kinetics. In this treatise, only the former case will be considered, i.e. water–water interfaces between co-existing liquid solutions in thermodynamic equilibrium, called phases. These phases arise from phase separation. Later on, the definition of water–water interfaces will be widened slightly, including curved interfaces of meta-stable droplets of one phase, surrounded by the other phase.

R. Hans Tromp (✉)
NIZO Food Research, Kernhemseweg 2, 6718 ZB Ede, The Netherlands
e-mail: Hans.Tromp@NIZO.com

R. Hans Tromp
Van 't Hoff Laboratory for Physical and Colloid Chemistry, Debye Institute for Nanomaterials Science, University of Utrecht, Padualaan 8, 3584 CH Utrecht, The Netherlands

© Springer International Publishing Switzerland 2016
P.R. Lang and Y. Liu (eds.), *Soft Matter at Aqueous Interfaces*,
Lecture Notes in Physics 917, DOI 10.1007/978-3-319-24502-7_6

Many properties of water–water interfaces are the same as those of other solvent–solvent interfaces, separating phases with other solvents than water as solvent.
However, in one aspect the case of water as a solvent differs from the case of other
solvents. Due to the high solubility of salts and polyelectrolytes in water, water–
water interfaces may carry an electric potential, which is in general absent or very
weak in systems with non-aqueous solvents.

Water–water interfaces exist in any phase separated aqueous system. The simplest, and most common system contains water soluble non-gelling polymers 1 and
2, which separates at sufficiently high concentration of each polymer and sufficiently low temperature into two volumes (phases), each rich in either one of the
polymers 1 and 2. The water content of the phases may or may not be equal. The
process may be further modified or complicated by the presence of salt and by
conformational changes of the polymer chains during the separation process.

An important class is formed by mixtures of polysaccharides and proteins, which
phase separate typically at concentrations larger than 3–5 % of each. The phase
separation of aqueous food proteins and food polysaccharides solutions, often
accompanied by gelation of one or both polymers [9, 18], is an important mechanism of structure formation in food matrices [28]. Such systems may offer a
low-calorie alternative to structuring by fat or oil-containing emulsions. Most of the
examples in this discussion of water–water interfaces will be taken from experiments carried out with aqueous mixtures of non-gelling gelatin with dextran or
pullulan. Such mixtures are ideal model systems because they form clear solutions
unless when phase separating, and show (close to) Newtonian rheological behavior
at all relevant concentrations.

Figure 6.1 shows a cartoon of a solvent–solvent interface, meant to show that the
interface only 'exists' for large molecules, not for solvent, salt or fractions of

Fig. 6.1 Schematic impression of a water–water interface, formed by the phase separation of two
aqueous solutions containing two different polymers. The interface is permeable for water, salt and
the low molar mass fraction of the polydisperse polymers

smaller polymers in the case of polydisperse polymers. This permeability the interface is further supported by comparing the molar mass distribution of a polydisperse polymer, in this case dextran, in the two coexisting phases (Fig. 6.2). In the phase poor in dextran, the molar mass of dextran is an order of magnitude smaller than that in the phase rich in dextran [13]. In fact, the small molar mass polymers are not as fully involved in the phase separation as the polymers with larger molar masses.

Another demonstration of the permeability of the interface, in this case for water and low molar mass salt is given in Fig. 6.3. Here, density matching is observed at the same molarities of salts which differ by about a factor of three in molar mass. This indicates that water is redistributed when salt is added, possibly accompanied by a change in the polymer composition of the phases. Figure 6.3 also highlights the role of water in the phase diagram. The continuous water matrix cannot be considered as an inert medium, and as a consequence, the mixtures cannot be treated as polymer blends.

This chapter will focus on the fundamental aspects of water–water interfaces, as they are found in non-gelling ternary systems (polymer 1, polymer 2 and water) in which the phase separation gives rise to a phase A and a phase B, each of which is rich in one of the two polymers (segregative phase separation). The other important case, in which one of the phases is rich in each of the two polymers, and the other rich in water (complex coacervation or aggregative phase separation) [5] will not be considered.

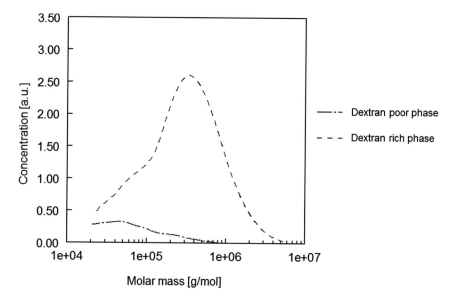

Fig. 6.2 Mass distribution of dextran in the two coexisting phases of a phase separated mixture of non-gelling gelatin and dextran

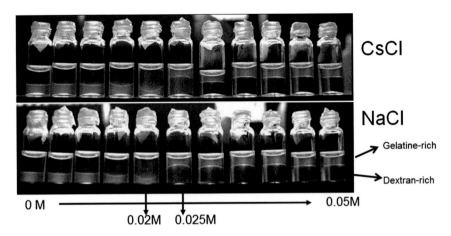

Fig. 6.3 Water redistribution as a result of adding salt. For two types of inert, monovalent salt, significantly differing in molar mass, the density matching due to redistribution of water when increasing the salt molarity takes place at the same salt molarity (polymer composition 5 % gelatin/5 %dextran 200 kDa)

6.2 Phase Diagrams

The conditions for phase separation are determined by the behavior of the free energy density of the mixture, as a function of the composition and temperature. For three components, the simplest case giving rise to water–water or solvent–solvent inter-faces, this free energy density is given by the mean-field Flory-Huggins equation [5, 14, 27]:

$$
\frac{f}{k_B T} = \frac{\varphi_1}{V_1} \log \varphi_1 + \frac{\varphi_2}{V_2} \log \varphi_2 + \frac{\varphi_3}{V_3} \log \varphi_3 + \frac{1}{2}\chi_{12}\varphi_1\varphi_2 \left(\frac{1}{V_1} + \frac{1}{V_2} \right)
$$
$$
+ \frac{1}{2}\chi_{13}\varphi_1\varphi_3 \left(\frac{1}{V_1} + \frac{1}{V_3} \right) + \frac{1}{2}\chi_{23}\varphi_2\varphi_3 \left(\frac{1}{V_2} + \frac{1}{V_3} \right) \tag{6.1}
$$

in which the first three terms are the entropy density of mixing, and the last three the free energy density of interaction of mixing. φ_i is the volume fraction of component i, V_i is the molecular volume of component i, and $\chi_{i,j}$ the interaction energy between molecules of type i and j, in units of $k_B T$ (k_B is Boltzmann's constant, and T the absolute temperature). $\chi_{i,j}$ is proportional to T^{-1} for purely enthalpic interactions, but may also contain a temperature independent part, due to an entropic contri-bution, e.g. from molecular ordering in the hydration sphere of the polymer [15].

Only two of the three variables can be independently varied because

$$
\varphi_1 + \varphi_2 + \varphi_3 \equiv 1 \tag{6.2}
$$

From now on, φ_1 and φ_2 will be the volume fractions of the two polymers, and φ_3 that of water

Examples of the energy landscape described by (6.1) are shown in Fig. 6.4.

Phase equilibrium exists if no energy or pressure changes occur when thermal fluctuations cause the exchange between the phases of small quantities of material, which cause small changes in the volume fractions. Therefore, the exchange chemical potential μ_1 and μ_2 of polymers 1 and 2 (in units of $k_B T/V_1$, with $\alpha_2 = V_1/V_2$ and $\alpha_3 = V_1/V_3$)

$$\mu_1 = \frac{V_1}{k_B T}\frac{\partial f}{\partial \varphi_1}\bigg|_{\varphi_2} = \log \varphi_1 + 1 - \alpha_3 \log(1 - \varphi_1 - \varphi_2) - \alpha_3$$
$$+ \frac{1}{2}(\chi_{12}[1 + \alpha_2] - \chi_{13}[1 + \alpha_3] - \chi_{23}[\alpha_2 + \alpha_3])\varphi_2 + \frac{1}{2}\chi_{13}(1 + \alpha_3)(1 - 2\varphi_1)$$

$$(6.3)$$

and

$$\mu_2 = \frac{V_1}{k_B T}\frac{\partial f}{\partial \varphi_2}\bigg|_{\varphi_1} = \alpha_2 \log \varphi_2 + \alpha_2 - \alpha_3 \log(1 - \varphi_1 - \varphi_2) - \alpha_3$$
$$+ \frac{1}{2}(\chi_{12}[1 + \alpha_2] - \chi_{13}[1 + \alpha_3] - \chi_{23}[\alpha_2 + \alpha_3])\varphi_1 + \frac{1}{2}\chi_{23}(\alpha_2 + \alpha_3)(1 - 2\varphi_2)$$

$$(6.4)$$

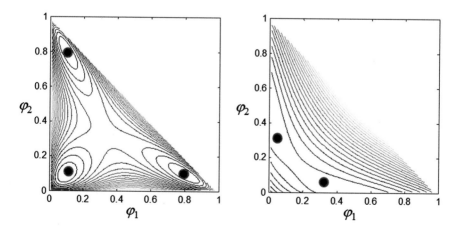

Fig. 6.4 Free energy landscapes in the case of coexistence of three phases, each rich in one of the components 1, 2 and 3 (*left*) and coexistence of two phases, each rich in one of the components 1 and 2 (*right*). The solvent concentration decreases from the *bottom left corner*. The *lines* are connecting locations of equal energy. Coexisting compositions, which are in 'valleys' in the energy landscape, are indicated

as well as the pressure

$$p = \frac{k_B T}{V_1}(\mu_1 \varphi_1 + \mu_2 \varphi_2) - f \qquad (6.5)$$

in the co-existing phases should be equal. Because of Eq. (6.2), only two of the three μ_i are needed for fixing the composition of coexisting phases. The composition of the phases at three phase co-existence can be obtained from finding sets of three points in the energy landscape $f(\varphi_1, \varphi_2)$ where the gradient vectors have the same size and direction, and lie in the same plane. This is expressed by the following set of equations:

$$\mu_{1,A}(\varphi_{1,A}, \varphi_{2,A}) = \mu_{1,B}(\varphi_{1,B}, \varphi_{2,B}) = \mu_{1,C}(\varphi_{1,C}, \varphi_{2,C})$$
$$\mu_{2,A}(\varphi_{1,A}, \varphi_{2,A}) = \mu_{2,B}(\varphi_{1,B}, \varphi_{2,B}) = \mu_{2,C}(\varphi_{1,C}, \varphi_{2,C}) \qquad (6.6)$$
$$p_A(\varphi_{1,A}, \varphi_{2,A}) = p_B(\varphi_{1,B}, \varphi_{2,B}) = p_C(\varphi_{1,C}, \varphi_{2,C})$$

The phases are labeled A, B and C. These equations form a set of six, with six unknowns, and can therefore be (numerically) solved.

Each of the three phases is richest in one of the three components. Normally, though, there are only two phases, say A and B, each rich in one of the two polymers, with the solvent-rich phase missing. The solvent-rich phase is missing because there is no (strong) energetically unfavorable interaction between the solvent and the dissolved polymers, so a solvent-rich phase is not stable. In this case there are only three equations, with six unknowns. The extra information is provided by putting in values for any three of these four quantities: the total volume and the total number of molecules of type 1, 2 or 3. If the latter three are taken, the total volume is

$$V = V_1 n_1 + V_2 n_2 + V_3 n_3 \qquad (6.7)$$

and additional relations between the volume fractions can be written down (with V_A and V_B the volumes of phases A and B, respectively, and $n_{i,A}$ and $n_{i,B}$ the number of molecules of type i in these phases):

$$\varphi_{1,A} = \frac{V_1 n_{1,A}}{V_A}; \varphi_{1,B} = \frac{V_1(n_1 - n_{1,A})}{V - V_A}$$
$$\varphi_{2,A} = \frac{V_2 n_{2,A}}{V_A}; \varphi_{2,B} = \frac{V_2(n_2 - n_{2,A})}{V - V_A} \qquad (6.8)$$

This reduces the number of unknowns to three, i.e. $n_{1,A}$, $n_{2,A}$ and V_A.

In most practical cases, phase separation induced by changing the water content is more relevant than temperature induced phase separation. Therefore, compositional phase diagrams of mixing will be treated in detail. Representative examples of phase diagrams are shown in Figs. 6.5 and 6.6. For equal degrees of

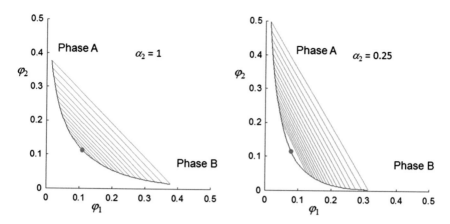

Fig. 6.5 Calculated phase diagrams of mixing, for equal ($\alpha_2 = 1$) and unequal ($\alpha_2 = 0.25$) degrees of polymerization with equal water affinities ($\chi_{13} = \chi_{23} = 0$). $\chi_{12} = 9$, $N_1 = 1000$ and $N_3 = 1$. *Tie lines* and the *critical points (full dots)* are indicated

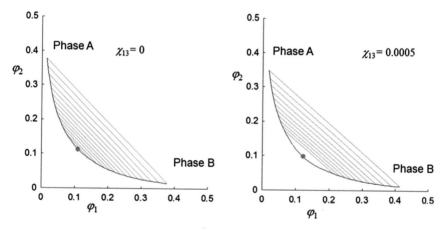

Fig. 6.6 Calculated phase diagram of mixing for equal ($\chi_{13} = 0$) and unequal ($\chi_{13} = 0.0005$) water affinities and equal degrees of polymerization ($N_1 = N_2 = 1000$, so $\alpha_2 = 1$). $\chi_{12} = 9$, $\chi_{23} = 0$, $N_3 = 1$. *Tie lines* and the *critical points (full dots)* are indicated

polymerization of the two polymers ($N_1 = N_2 = 1000$), the diagram is precisely symmetric relative to the $\varphi_1 = \varphi_2$ line, whereas a difference in degrees of polymerization ($N_1 = 1000$, $N_2 = 4000$, so $\alpha_2 = 0.25$) causes coexistence between phases of unequal polymer concentrations. The asymmetry corresponds to a stronger separation of the large polymer than for the small polymer. The asymmetry due to difference in water affinity is more subtle. The binodals overlap in the symmetric case, but coexisting total compositions $\varphi_1 + \varphi_2$ are not equal, in contrast with the case of identical water affinities. The inequality corresponds to a stronger separation for the polymer with the most unfavorable interaction with the solvent. The

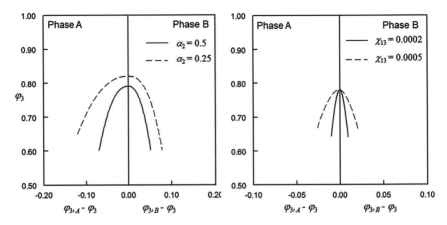

Fig. 6.7 Calculated water content of coexisting phase phases A and B *versus* the overall water fraction φ_3 in the case of different degrees of polymerization and equal water affinity ($\chi_{13} = \chi_{23} = 0$, *left*), and in the case of equal degrees of polymerization ($\alpha_2 = 1$, *right*). $\chi_{12} = 9$, $N_3 = 1$, $N_1 = 1000$. Note the difference in scale on the *horizontal axis*

asymmetry also corresponds to a difference in the water content of the phases as is shown in Fig. 6.7. In general, for practically relevant values of α_1, α_2, α_3, χ_{12} and difference in water affinity ($\chi_{13} \neq \chi_{23}$), the effect of 'water separation' is much smaller if due to a difference in water affinity, than if due to a difference in degree of polymerization. The deviation of the water content in the phases from the overall water content is not symmetric, corresponding to unequal phase volumes, even in the case of equal volume fractions of the two polymers.

The critical point of mixing, here defined as the composition where the length of the tie line approaches zero, moves away from the $\varphi_1 = \varphi_2$ line in the case of asymmetries in the polymer size or water affinity. For unequal degrees of polymerization the critical point shifts to higher or lower water contents, dependent on whether the average molar mass decreases or increases.

The experimental phase diagram of mixing (Fig. 6.8) of gelatin and dextran [12] with molar weights of 170 and 282 kDa, respectively shows a symmetrically located critical point in spite of the apparently large difference between the molar weights of the polymers. However, the difference in degree of polymerization is much smaller, considering the difference in monomer mass. With monomer molar masses 100 and 160 g/mol of gelatin and dextran, respectively, the value of α is estimated to be about 1. This estimate is made less accurate by the polydispersity of both polymers (see Fig. 6.2). The location of the critical point, close to 3.5 %/3.5 % suggest a value for χ_{12} of $2/(1 - \varphi) \cong 30$ [27].

An interesting issue is the phase volume ratio as a function of the overall water content. This issue is encountered when diluting a system of coexisting phases. Figure 6.9 shows a representative example of what is observed when diluting a phase separated mixture. Depending on whether the starting concentrated system has a phase volume ratio above or below a certain value, as determined by the

Fig. 6.8 Experimental phase diagram of mixtures of gelatin (approximately 170 kDa) and dextran (app. 282 kDa). The estimated location of the *critical point* of mixing is indicated

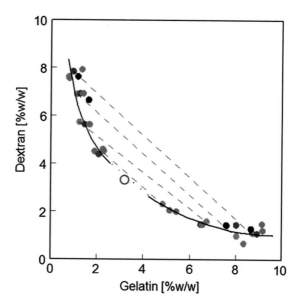

asymmetry in size and water affinity, the phase volume ratio diverges up or down with progressive dilution. This is also seen in the calculated phase volume ratios (Fig. 6.10). On dilution towards the critical water content, the phase volume fraction of phase A (rich in polymer 2) tends towards 1 in the case of a larger water affinity of polymer 2, or towards zero in the case of a larger polymer size in phase A. In the former case water, which is added prefers the phase in which there is the least of the polymer with an unfavorable interaction with water, which is phase A. In the latter case additional water is taken up by the phase which contains the most of the larger polymer, which experiences the strongest driving force for dilution.

6.3   The Interface

6.3.1   The Blob Model

The free energy of mixing Eq. (6.1) predicts the composition of coexisting phases (within the mean field approximation). At or near the interface the calculated phase diagram of mixing cannot be fully correct, because the demixing polymers are 'forced' to have contact. This 'forced contact' leads to an interfacial energy, often called an interfacial tension when expressed per unit area. An infinitely sharp transition in composition would amount to an infinitely low local entropy, and is therefore thermodynamically unacceptable. Therefore, the systems adapts by forming a compositional gradient (see also Chap. 4 by G. H. Findenegg). This gradient, however, contains energy which has to be included in the free energy of

R. Hans Tromp

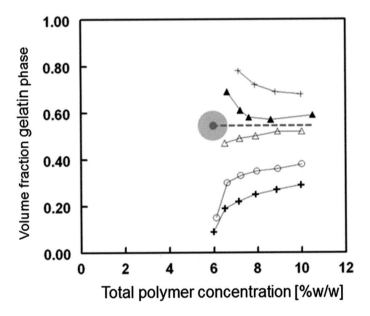

Fig. 6.9 Phase volume fraction of the gelatin-rich phase as a function of overall water content. *Top* series of phase separated gelatin (non-gelling)/dextran (180 kDa) systems; *bottom* the phase volume fraction on dilution for different starting ratios

mixing close to the interface. In order to apply Eq. (6.1) in the border region between the coexisting phases, it is extended by two gradient terms [6, 8].

Before we use the expression for the free energy of mixing at the interface, we first introduce a simplification provided by the so called 'blob' model [16], which is valid in the semi-dilute concentration regime of overlapping chains, and convenient as it disposes of the solvent as a separate component (for details of the blob model see also Chap. 10 by J.-U. Sommer). It is assumed that the solvent is 'good', i.e. the chain is fully hydrated and repels therefore itself. In other words, the solvent causes the chain to swell. In that case, it may be realized that for the polymer chains the solvent is only over short distance different from simply 'space'. Over these short distances, usually called the correlation length, ξ, the polymer chains takes an

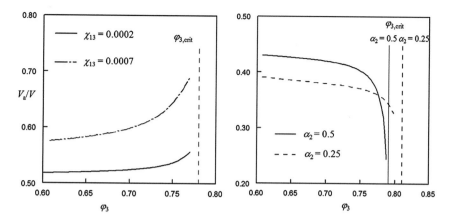

Fig. 6.10 Phase volume fraction V_A/V as a function of water content for different water affinities (*left*), and different degrees of polymerization (*right*) of the two polymers. $\chi_{12} = 9$; $\chi_{23} = 0$; $N_3 = 1$; $N_1 = 1000$

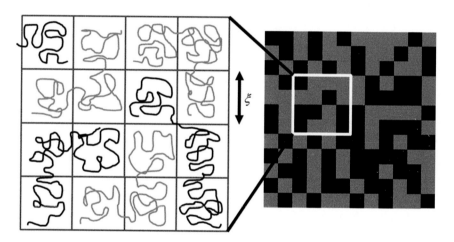

Fig. 6.11 Illustration of the blob model. The *black* and *gray squares* of size ξ in the *right-hand panel* represent blobs and form random walks in three dimensions. Inside a blob, a subsection of the chain (a blob) forms a self-avoiding structure due to the energetically favorable interaction with the solvent

extended configuration which it would not have, if there was no solvent. Beyond the distance ξ, the correlation between the chain elements is the same as in a melt of chain elements of size ξ and, therefore, the configuration of the blobs is that of a random walk. In Fig. 6.11 the blob model is further illustrated. The renormalization of the polymer chain by dividing it in elements of size ξ makes it possible to write the free energy of mixing without an explicit role of the solvent [6]:

$$\frac{F_{blob}}{Vk_{B}T} = \frac{1}{\xi(c)^3}\left[\frac{\varphi}{N_b(c)}\log\varphi + \frac{1-\varphi}{N_b(c)}\log(1-\varphi) + u(c)\varphi(1-\varphi) + K\right] \quad (6.9)$$

where, for simplicity it is assumed that the two polymers have the same degree of polymerization. $K = 0.024$ [6, 7] and accounts for the free energy of mixing of the monomers inside a blob with solvent. Because the solvent is good, K is set by the properties of a self-avoiding walk and is therefore a constant in the concentration. The solvent only appears via the concentration dependence of ξ. It can be shown that [6]:

$$\xi(c) = 0.43R_g\left(\frac{c}{c*}\right)^{\frac{v}{1-3v}} \quad (6.10)$$

and the number of blobs per chain

$$N_b(c) = \frac{N}{\xi^3 c} \quad (6.11)$$

and

$$u(c) = u_{crit}\left(\frac{c}{c_{crit}}\right)^{\frac{\chi}{3v-1}} = \frac{2}{N_{b,crit}}\left(\frac{c}{c_{crit}}\right)^{\frac{\chi}{3v-1}} \quad (6.12)$$

where c is the polymer concentration, N the number of monomer per chain (see below), R_g the radius of gyration at concentration below the overlap concentration $c*$, c_{crit} the critical concentration of mixing, $N_{b,crit}$ the number of blobs per chain at c_{crit}, u_{crit} the interaction energy at c_{crit} and $\chi \cong 0.22$. For a good solvent, the polymer radius of gyration

$$R_g \cong bN^v \quad (6.13)$$

in which b is the Kuhn length and $v = 3/5$. $c*$ is the overlap concentration, above which hydrated coils start to overlap. It is calculated by setting the overall monomer concentration equal to the internal monomer concentration of the chain:

$$c* \approx \frac{3}{4\pi}\frac{N}{R_g^3} \quad (6.14)$$

The specific effects of polymer-solvent interactions are in the R_g, which together with $c*$ and the monomer concentration c determines the blob size. Equation 6.12 provides experimental access to the theory through N, R_g and c_{crit}, which can be measured.

Equation 6.9 can be used to calculate phase diagrams of mixing, just as Eq. (6.1). For asymmetric cases (unequal blob sizes of polymers 1 and 2 due to different degrees of polymerization or unequal Kuhn lengths) it is convenient to introduce an effective blob size (taking into account that the relative contribution of small blobs should be higher because they have per monomer more contact points with the other blobs and therefore a higher interaction energy per monomer)

$$\frac{1}{\zeta_{eff}} = \frac{1}{2}\left(\frac{1}{\zeta_1} + \frac{1}{\zeta_2}\right) \tag{6.15}$$

So Eq. (6.9) becomes

$$\frac{F}{Vk_BT} = \frac{\varphi}{N_{b,1}\zeta_1^3}\log\varphi + \frac{1-\varphi}{N_{b,2}\zeta_2^3}\log(1-\varphi) + \frac{\omega}{\zeta_{eff}^3}\varphi(1-\varphi) \tag{6.16}$$

where φ is the volume fraction of blobs of polymer 1. Figure 6.12 shows some examples of the phase diagram when the interaction strength ω is varied. The interaction strength ω is a renormalized form of u, which is only dependent on the distance to the critical point c_{crit} and the asymmetry in blob size, not on the degree of polymerization:

$$\omega = \frac{2u}{\zeta_{eff}^3}\left(\frac{1}{N_{b,A}\zeta_A^3} + \frac{1}{N_{b,B}\zeta_B^3}\right)^{-1} = \frac{(1+\sqrt{\alpha})^2}{(1+\alpha)}\left(\frac{c}{c_{crit}}\right)^{\frac{-\chi-1}{1-3v}} \tag{6.17}$$

where $\alpha = N_A/N_B = N_{b,A}\zeta_A^3/N_{b,B}\zeta_B^3$, the ratio of the degrees of polymerization.

In Fig. 6.12 it is clearly seen, just as in Fig. 6.5, that the difference in size of phase separated polymers results in coexisting concentrations which are more extreme for the larger polymer than for the smaller polymer.

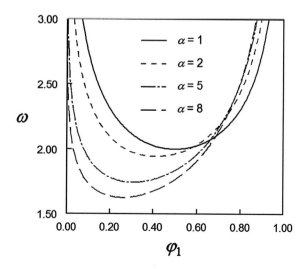

Fig. 6.12 The temperature-composition phase diagram calculated using Eq. (6.16) for some values of the ratio of degrees of polymerization. ω represents the interaction strength between the polymers and changes with temperature via c_{crit}

6.3.2 The Flat Interface

In order to calculate the interface profile by using Eq. (6.9), it has to be extended with gradient energies. Because the system contains solvent, there is, apart from the composition gradient $\vec{\nabla}\varphi(z)$, which also exists in coexisting blends, also a concentration gradient $\vec{\nabla}c(z)$, where z is the distance from the interface. The local free energy density for equal blob sizes in the two phases becomes [6]

$$\frac{f[\varphi(z), c(z)]}{k_B T} = \frac{\varphi}{N_b \xi^3}\log\varphi + \frac{1-\varphi}{N_b \xi^3}\log(1-\varphi) + \frac{u}{\xi^3}\varphi(1-\varphi) + \frac{K}{\xi^3}$$
$$+ \frac{\left|\vec{\nabla}\varphi\right|^2}{24\xi\varphi(1-\varphi)} + \frac{\left|\vec{\nabla}c\right|^2}{24\xi c^2} \qquad (6.18)$$

where the z-dependence of φ, c and N_b has been omitted by for the sake of clarity. In the case of unequal blob sizes, an additional term representing the gradient energy of the cross-correlation of composition and concentration would have to be included. In order to calculate the interface tension it is convenient to use the grand potential $\Omega(T, \mu_\varphi, \mu_c) = F - V\mu_\varphi\varphi - V\mu_c c = \Omega_{ex} - pV$ as the representation for the free energy, because the interface is able to exchange both thermal energy as well as mass with the bulk (defined here as the part of the volume without gradients). The interfacial tension is the excess grand potential per unit area due to the presence of a gradient, integrated over all distances to the interface:

$$\frac{\Omega_{ex}}{A k_B T} = \int_V dz \left[\frac{1+\beta}{2N_b\xi^3}\log\frac{1+\beta}{2} + \frac{1-\beta}{2N_b\xi^3}\log\frac{1-\beta}{2} + \frac{u}{4\xi^3}(1-\beta^2) + \frac{K}{\xi^3} + \frac{\left|\vec{\nabla}\beta\right|^2}{24\xi(1-\beta^2)} + \frac{\bar{u}^2\left|\vec{\nabla}\varepsilon\right|^2}{24\xi(1-\bar{u}\varepsilon)^2} - \mu_\eta\beta - \mu_\varepsilon\varepsilon + \bar{p} \right]$$
$$(6.19)$$

with the composition variable

$$\beta(z) = 2\varphi(z) - 1 \qquad (6.20)$$

and the concentration variable

$$\varepsilon(z) \equiv \frac{c(z) - \bar{c}}{\bar{u}\bar{c}} \qquad (6.21)$$

and $\bar{p}$ the pressure.

Symbols with a bar are bulk values, far from the interface. μ_β is the coexistence value of exchange chemical potential of the polymers, which is zero for equal degrees of polymerization and equal random chain statistics for both polymers, μ_ε is the value of the chemical potential of the solvent when the phases are coexisting. In order to get the equilibrium interfacial tension, Ω_{ex} has to be minimized with respect to the profiles $\beta(z)$ and $\varepsilon(z)$, using the Euler-Lagrange equation. Because the two polymers have an unfavorable interaction energy, the concentration of polymer will we suppressed at the interface, in favor of an increased solvent concentration. In other words, the interface is expected to be enriched in solvent. The minimization of Eq. (6.19) can be done relatively easily, by introducing the assumption that this enrichment of solvent at the interface is small:

$$|\bar{u}\varepsilon \ll 1| \tag{6.22}$$

The resulting expression for the interfacial tension

$$\sigma = \frac{k_B T}{\bar{\xi}^2}\left(\frac{\bar{u}}{6}\right)^{1/2}(1 - \Delta_1 - \bar{u}\Delta_2) \tag{6.23}$$

contains three terms. The first represents the interfacial tension for the limiting case of infinite degree of polymerization. The only reason the interfacial tension is not infinite in that case (and the interfacial profile not infinitely sharp) is the fact that the value of $\bar{u}$ is finite, which allows some interpenetration of the phases. The second term Δ_1 represents the fact that the degree of polymerization is finite, which increases the miscibility and lowers the interfacial energy. The third term, Δ_2 represents the effect of accumulation of solvent at the interface, which suppresses the interpenetration of the two polymers, and therefore lowers the interfacial tension.

The details of the procedure can be found in Broseta et al. [6] and Tromp and Blokhuis [30]. Some results will be given here. Figure 6.13 shows the interfacial

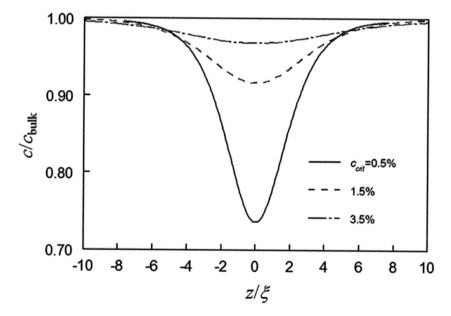

Fig. 6.13 Interfacial profile of the total polymer concentration for three degrees of incompatibility (expressed in terms of the critical concentration of mixing). $R_g = 18$ nm

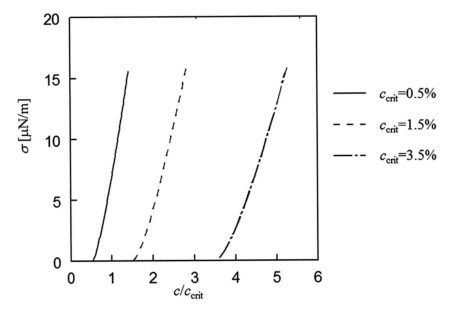

Fig. 6.14 Interfacial tension calculated using Eq. (6.23) for three different degrees of incompatibility. $R_g = 18$ nm

profiles for some experimentally relevant cases of incompatibility. The incompatibility is set by c_{crit} (the smaller c_{crit}, the stronger the incompatibility), for which values between 0.5 and 3.5 % are experimentally relevant. It is seen that the reduction at the interface of the total polymer concentrations is 25 % for the strongest incompatibility. The width of the interface is calculated to be about 4 blob sizes, which is typically a few nm. Figure 6.14 shows the interfacial tension as a function of c_{crit}. The typical values calculated are in the range 1–20 μNm^{-1}, which is in agreement with the results of experiments, as shown below. In particular, close to the critical point, the effect of accumulation of solvent at the interface on the interfacial tension is significant, as seen in Fig. 6.15.

6.3.3 The Curved Interface

Figure 6.16 shows a representative example of the microscopic structure of a phase separated solution (10 % gelatin and 10 % dextran) in a meta-stable stage, developing into a macroscopically separated system by droplet coalescence. Often, this meta-stable droplet stage is quite stable, during many hours, or even days. In a single second, water molecules are able to diffuse over distances of the order of 100 μm, i.e. many times the characteristic size of the droplets during the meta-stable life time of the structure. This diffusion of water is not hindered by the presence of the

Fig. 6.15 Effect of solvent
accumulation at the interface
on the interfacial tension,
calculated using Eq. (6.23), as
a function of concentration,
for three experimentally
relevant values of the
incompatibility

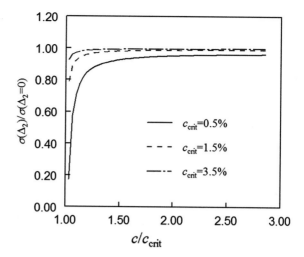

Fig. 6.16 Microscope image
of a water-in-water emulsion,
obtained after stirring a phase
separated mixture of aqueous
non-gelling fish gelatin and
dextran solution. The
gelatin-rich phase is the
droplet phase

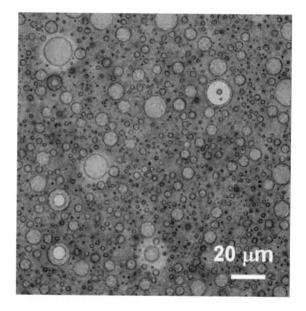

interfaces, which are permeable for small molecules. Therefore, on the time scale of
the coarsening of the structure, the solvent distribution can be considered in
equilibrium with the state of curved interfaces. The curvature of the interfaces
suggests a pressure difference between in the inside and outside of the droplets
(with a droplet radius of 10 μm and an interfacial tension of 1 μNm^{-1} typically of
the order of 0.2 Pa). This pressure difference corresponds to a water distribution
across the curved interface which is slightly different from that across a flat interface. As a consequence, it is expected that also the polymer composition of the

droplets and the continuous phase are slightly different from that of phases with a flat separating interface, and, as a consequence, the interfacial tension is slightly different as well. It was considered interesting to investigate if the curvature of the interface affects stability of the droplets via an effect of the curvature on the interfacial tension. This might be the case in particular for different degrees of polymerization or different blob sizes at the two sides of the interface, because, the gradient energy is determined by the distance scale in the structure of the liquid, i.e. the blob size (see Eq. (6.19)).

The (meta)stability of the system in Fig. 6.16 suggests a value for the chemical potential which is higher than μ_{coex}, the value for flat interfaces in actual thermodynamic equilibrium. Therefore, the interfacial profile has to be calculated, using Eq. (6.16), for a value of μ larger than μ_{coex}. For simplicity, the asymmetry is assumed to be due to a difference in the degree of polymerization (and not to a difference in blob size) and the solvent gradient is ignored, so the concentration dependence of ξ and u vanishes and the free energy density can be expressed as (with Eqs. (6.15)–(6.17)) and returning to φ as the composition variable:

$$f(\varphi) = \frac{uk_BT}{\xi^3} \left[\frac{2\varphi}{(1+\alpha)\omega} \log\varphi + \frac{2\alpha(1-\varphi)}{(1+\alpha)\omega} \log(1-\varphi) + \varphi - \varphi^2 \right] \quad (6.24)$$

The excess grand potential of the curved interface, assumed to be a droplet, is

$$\Omega_{ex}[\varphi] = \int_V d\vec{r} \left[f(\varphi) - \mu\varphi + m(\varphi)\left|\vec{\nabla}\varphi\right|^2 \right] \quad (6.25)$$

with the gradient energy

$$m(\varphi) = \frac{k_BT\varphi}{12\xi(1-\varphi)} \quad (6.26)$$

The origin of the spherical coordinates is at the center of the droplet. In order to minimize Ω_{ex} with respect to the composition profile $\varphi(r)$ the compositions far from the interface have to be known. These are obtained by calculating the phase diagram for two phases according to Eq. (6.6), but now for $\mu > \mu_{coex}$. The calculation of the phase diagram results in the compositions far from the interface inside φ_{in} and outside φ_{out} the droplet (it is assumed that $\varphi_{in} > \varphi_{out}$). With

$$f(\varphi_{in}) - f(\varphi_{out}) - V\mu(\varphi_{in} - \varphi_{out}) = -p_{in} + p_{out} = -\Delta p \quad (6.27)$$

the pressure difference is obtained, which corresponds to the chosen value of $\mu > \mu_{coex}$. Using φ_{in} and outside φ_{out}, the composition profile $\varphi(r)$ minimizing Ω_{ex} and the corresponding minimum value of Ω_{ex} are obtained from Eq. (6.25) by applying the Euler-Lagrange equation. The curvature R^{-1} corresponding to the chosen non-coexistence value of μ is obtained from the interfacial profile assuming that the radial distance R where interface is located is such that

$$4\pi \int_0^\infty [\varphi(r) - \varphi_{out}] r^2 dr = \frac{4\pi}{3} R^3 (\varphi_{in} - \varphi_{out}) \tag{6.28}$$

or, in other words, deviations of φ from φ_{in} inside and φ_{out} outside the interface cancel.

With R and Δp known, the interfacial tension can be obtained from the excess grand potential per unit area of a droplet:

$$\Omega_{ex} \equiv \frac{\Omega + p_{out} V}{4\pi R^2} = -\frac{\Delta p R}{3} + \sigma(R) \tag{6.29}$$

The details of these calculations can be found in [30].

A much less cumbersome way to calculate the effect of asymmetry in the polymer properties on the interfacial tension is based on the approximation that the interfacial tension for droplets with small curvatures R^{-1} ($R >>$ interfacial width, which is certainly the case for the system shown in Fig. 6.16) can be written as an expansion in the curvature:

$$\sigma(R) = \sigma_\infty - \frac{4k\sigma_\infty}{R_s} \frac{1}{R} + \frac{2k + k_G}{R^2} + \cdots \tag{6.30}$$

where σ_∞ is the interfacial tension of a flat interface, R_s the spontaneous curvature, k the bending rigidity and k_G the Gaussian rigidity of the interface. The curvature R^{-1} is taken positive for an interfacial curvature centered on the droplet. Expressions have been derived which give the spontaneous curvature and the interfacial rigidities using the interfacial profile of the flat interface [30].

The essential result of the analysis of the interfacial tension as a function curvature is shown in Fig. 6.17. Here, the excess free energy per unit volume of a droplet is plotted against the curvature of the droplet. The reference point of the excess is the flat interface of the same area as that of the droplet. This excess free energy is plotted for three cases: equal degrees of polymerization ($\alpha = 1$), and larger polymer inside or outside the droplet. In all cases, the energy of a droplet increases when growing from very small sizes because the interfacial area increases. Beyond a certain preferential size, further growth leads to a decreasing energy, because the decrease in internal pressure outweighs the increase in interface area. The preferential size, as well as the height of the energy barrier varies with the asymmetry of the polymer sizes. A droplet containing larger polymers than the continuous phase has to cross a higher barrier at a smaller size and will therefore be more stable (or grow more slowly) than a droplet containing smaller polymers than the continuous phase.

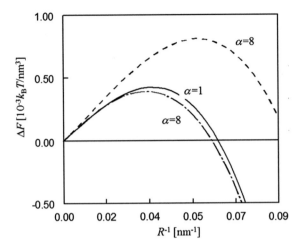

Fig. 6.17 Excess interfacial energy of a collection of monodisperse phase droplets relative to a flat interface of the same size, as a function of droplet curvature, with asymmetric ($\alpha = 8$) and symmetric interfaces ($\alpha = 1$). The total volume of the droplets is fixed. $u_{crit} = 0.0061$, $\xi_{crit} = 2.6$ nm, $c/c_{crit} = 1.8$. *Dashed line* droplets rich in large polymers in a continuous phase rich in small polymer; *Dash-dotted line* vice versa

It turns out that for experimentally relevant systems the droplet sizes for which the curvature affects the interfacial tension are fairly small, in the order of tens of nm. Therefore, in the system shown in Fig. 6.16 the curvature is probably irrelevant. However, the size distribution, which develops in a stage of much smaller droplets than can be observed in a light microscope, may be influenced by a curvature dependence of the interfacial tension [19]. Also the profile of the spinodal ring [8], as observed in small angle light scattering and reflecting very early stage phase separation could be affected.

6.4   Measuring the Interfacial Tension

The interfacial tension of water–water interfaces is in the order of 0.01–10 µN/m. Therefore, the classical way to measure interfacial tension by mechanically deforming the interface on a macroscopic length scale and measuring the applied force (or vice versa) is not suitable for water–water interfaces, because the intended deformation is too easily obscured by convective flows, which also deform the interface. Another issue would be the extremely small forces to be measured. Instead, the best way to measuring these very low interfacial tensions is under static conditions or from observation of microscopic deformation.

Under static conditions the interfacial tension can be obtained from the shape of the interface near a wall when one of the phases is wetting this wall [1]. An example

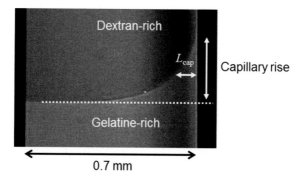

Fig. 6.18 Confocal microscopy image of the interface between phase separated non-gelling gelatin and dextran solutions (3.6 % gelatin/2.4 % dextran) in a flat capillary. The gelatin solution is fully wetting the glass wall

of such a situation is shown in Fig. 6.18. The characteristic decay length of the height of the interface when moving away from the wall is the capillary length, which is a function of the interfacial tension and the difference between the densities of the phases ($\Delta\rho$):

$$L_{cap} = \sqrt{\frac{2\sigma}{g\Delta\rho}} \tag{6.31}$$

with g the acceleration of gravity. $\Delta\rho$ is usually quite small (0.1–1 kg m^{-3}) and is obtained by collecting the two phases and the measuring the density of each by e.g. a density meter with a oscillating U-tube sensor. L_{cap} is obtained from fitting the shape of the interface to the appropriate minimal surface expression [4]. For the case shown in Fig. 6.18 L_{cap} was found to be 80 μm. With $\Delta\rho$ 0.25 kg m^{-3}, this gives an interfacial tension of about 0.01 μNm^{-1}.

An often used method for measuring low interfacial tensions is the spinning drop method, which uses the deformation from a spherical to an extended shape of a droplet of one phase surrounded by the other phase inside a cylindrical container, while the container is rotating around its axis (see Fig. 6.19). The deformation of the enclosed droplet at a certain spinning rate ω is determined by the interfacial tension and the density difference. At sufficiently large elongation, the interfacial tension is given by [22, 33].

$$\sigma = \frac{\Delta\rho\omega^2 R^3}{4} \tag{6.32}$$

where R is the radius of the cap of the extended droplet. Results for a water–water phase separated systems can be found in e.g. Ryden and Albertsson [25] and Scholten et al. [26].

Fig. 6.19 Schematic
representation of the setup of
the spinning drop technique

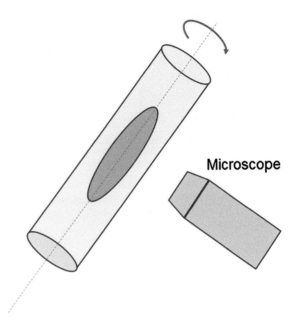

Microscope

Another common method to measure the interfacial tension of water–water interfaces is to follow the relaxation of shear induced microscopic structure, after the shearing is stopped. Usually, the structure consists of droplets extended to ellipsoids, cylinders or even band-like or thread-like structures. In the most common case, the retraction or relaxation of an ensemble of droplets is measured, either by light scattering [2], or by direct microscopic observation [3, 10, 17]. An example is given in Fig. 6.20.

The deformation is defined as the Taylor ratio of the 2-dimensional gray value correlation at level 0.5:

$$D = \frac{A_{flow} - A_{vort}}{A_{flow} + A_{vort}} \tag{6.33}$$

where A_{flow} and A_{vort} are the sizes of the contour at level 0.5 of the correlation pattern along central horizontal and vertical lines, respectively. After stopping the shear, the deformation decays exponentially, assuming that no coalescence takes place, which is usually justified at early times:

$$D = D_0 e^{-t/\tau} \tag{6.34}$$

For a droplet of radius R, and a viscosity inside the droplet, η_{in}, and outside, η_{out}, τ is given by [23]

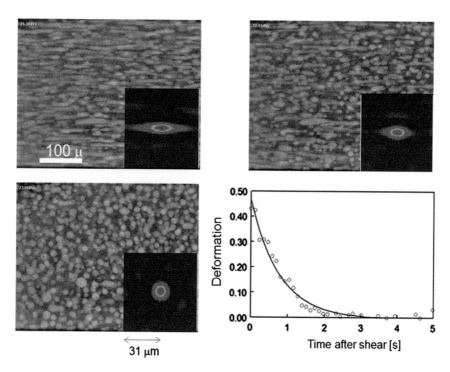

Fig. 6.20 Stills from a microscope movie of the relaxation in the flow/vorticity plane of the droplet shape after shearing (6 s^{-1}, in *horizontal direction*), and the quantitative average deformation. The *full line* is a fit of Eq. (6.34) to the data. The result is $\tau = 0.47$ s. The system contains 5 % non-gelling gelatin and 10 % pullulan. The *insets* are the *gray value* correlation charts

$$\tau = \frac{\eta_{out} R}{\sigma} \frac{(19\lambda + 16)(2\lambda + 3)}{40(\lambda + 1)} \tag{6.35}$$

where $\lambda = \eta_{in}/\eta_{out}$. Assuming a droplet size of 5 μm, and with the viscosities of the pullulan phase, 3 Pa s, and of the gelatin phase, 0.03 Pa s, the example in Fig. 6.20 results in $\sigma = 23$ μNm^{-1}.

A faster and easier method is to interpret the 2-dimensional light scattering pattern of the sheared droplet suspension, which should be in a thin layer, to avoid multiple scattering. The setup is in that case a glass plate and a rotating glass disk, with a laser beam crossing the sample layer in perpendicular direction. The scattering pattern is projected onto a screen and recorded. The disadvantage of this method is the fact that the droplet size is unknown. Also other shear induced structures, such as shear bands are not easily recognized in the scattering pattern. The relaxation of such patterns after shearing may therefore be erroneously interpreted in terms of droplet retraction. This method may therefore only be used for comparing qualitatively differences in the interfacial tension between different samples of the same kind.

Other shear-induced structures than extended and retracting droplets can be used to measure interfacial tension. When exposed to sufficient shear, phase separated systems with sufficient difference in viscosity between the phases develop cylindrical shear bands, extended in the flow direction, [31], the nature of which is not yet fully understood. After stopping the shear, these bands become unstable and develop Rayleigh instabilities. The rate by which they develop is set by the radius, inner and outer viscosities and the interfacial tension. The shape of the band is, at early times after stopping the shear, when the amplitude of the growing instability is still much smaller than the initial band width, given by

$$R(x,t) = R(0) + A(t)\cos(xk + \varphi) \tag{6.36}$$

where $k = 2\pi/l$, with l the wave length of the instability, and φ a phase factor. $A(t)$ is the amplitude of the instability, whose time dependence is exponential:

$$A(t) = A(0)e^{qt} \tag{6.37}$$

The growth rate q is obtained from a microscopic observation of the time dependence of the band shape (see Fig. 6.21) and used in an expression which relates it to the interfacial tension [29]:

$$\sigma = \frac{2q\eta_{out}R(0)}{\Omega(\lambda)} \tag{6.38}$$

The function $\Omega(\lambda)$ is tabulated in Tomotika [29]. In the example shown the result is $0.9\ \mu Nm^{-1}$.

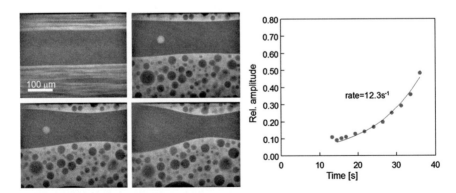

Fig. 6.21 Stills from a microscope movie of a shear-induced cylindrical band after stopping the shear (*left*), and the development of the amplitude of the instability in units of the initial band width (*right*). The system is a mixture of 5 % non-gelling gelatin (200 kDa) and 5 % dextran 282 kDa

6.5   The Interfacial Potential

A special feature of water–water interfaces is the possibility of electric potentials across the interface (Donnan potentials [11]). Because of the high dielectric constant of water, charges can be separated relative easily by thermal energies at room temperature. When one of the phase separating polymers, say α, is a polyelectrolyte, counter ions on one side of the interface, which is permeable for small solutes, can spread across the interface to the other side. The spreading is suppressed by the buildup of an electric potential. The result is a balance between small ion entropy and electric potential

$$\psi_D = -\frac{N_{av}k_BT}{F}\log\frac{c_\alpha^+}{c_\beta^+} = \frac{N_{av}k_BT}{F}\log\frac{c_\alpha^-}{c_\beta^-} \tag{6.39}$$

where $c_i^\pm$ are concentrations of small ions, and the index β refers to the phase containing the neutral polymer. F is Faraday's constant and N_{av} is Avogadro's number. Far from the interface charge neutrality is maintained:

$$c_\alpha^+ + zc = c_\alpha^- \tag{6.40}$$

and

$$c_\beta^+ = c_\beta^- = c_s \tag{6.41}$$

in which c is the concentration of polyelectrolyte, c_s the low molar mass salt concentration, and z is the number of positive charges on a polymer chain. For total absence of polyelectrolyte in the phase β, the interfacial or Donnan potential can be expressed by

$$\psi_D = \frac{N_{av}k_BT}{F}\operatorname{arcsinh}\frac{zc}{2c_s} \approx \frac{N_{av}k_BT}{F}\frac{zc}{2c_s} \tag{6.42}$$

It can be shown, that for the case in which there is also polyelectrolyte in the phase rich in non-polyelectrolyte, the Donnan potential is expressed by

$$\psi_D \approx \frac{N_{av}k_BT}{F}\frac{z\Delta c}{2c_s} \tag{6.43}$$

where Δc is the difference in polyelectrolyte concentration between the phases. It can be seen that by adding salt, the Donann potential decreases and becomes zero when the salt concentration is much higher than the concentration of counterions of the polyelectrolyte.

6.6   Measuring the Interfacial Potential

The electric or Donnan potential between the phases can be measured using electrochemical reference electrodes, such as Ag/AgCl electrodes [20, 21, 24] (For the measurement of interfacial potentials see also Chap. 2 by C.D. Fenández-Solis et al.). In this way the chemical potential difference of chloride ions between the phases is measured, which is a measure for the electric potential difference. This is described in detail in Vis et al. [32]. Care should be taken to avoid artefacts from streaming potentials.

Figure 6.22 shows some representative results. With increasing separation of gelatin the absolute value of the interface potential increases. This dependence is proportional as predicted by Eq. (6.43). The typical number of charges per gelatin chain, which can be calculated from the slopes in Fig. 6.22 is plotted in Fig. 6.23 and turned out to be in range of 0–5. This range is expected for a polyelectrolyte chain which consists largely of non-ionizable amino acids. It turns out that the sign of the electrical potential changes when the pH crosses the isoelectric pH of gelatin, which is about 7–8. This is a strong indication that the potential measured is indeed an interfacial potential.

The interfacial potential discussed here differs from the regular membrane potential in the sense that this potential arises spontaneously across an interface without quenched degrees of freedom or actively maintained gradients, such as in a living cell.

It remains to be studied what is the effect of an electric interfacial potential on the interfacial profile and the interfacial tension. In may be expected that the interfacial tensions decreases due the separated charges on either side, because the

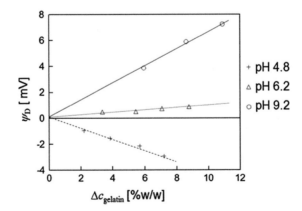

Fig. 6.22 Measurements of the interfacial electric potential at low (but ubiquitous) salt (~5 −10 mM) as a function of the difference in the mass fraction of gelatin in the two phases for some pH values. Below the isoelectric point, a positive potential is observed, whereas above the isoelectric point, a negative potential is found. The *lines* are linear fits through the origin. Adapted with permission from Vis et al. [32]. Copyright 2014 American Chemical Society

Fig. 6.23 Charge per chain (in units of elementary charges) calculated from the slopes in Fig. 6.22, at $\Delta c_{gelatin} = 5.8(\pm 0.1)\%$

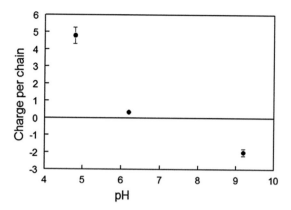

interfacial width may increase. However, the question is how to compare charged and non-charged interfaces in such a way that the comparison is meaningful, i.e. at the same distance from the critical point. For this purpose, the phase diagram separately including positive and negative ions has to be calculated.

Another effect of charges on the interface may be the mutual repulsion of droplets in water–water emulsions. This would be an interesting aspect in the study of the stability of water–water emulsions.

References

1. D.G.A.L. Aarts, J.H. van der Wiel, H.N.W. Lekkerkerker, J. Phys.: Condens. Matter **15**, S245–S250 (2003)
2. Y.A. Antonov, P. van Puyvelde, P. Moldenaers, Int. J. Biol. Macromol. **34**, 29 (2004)
3. S. Assighaou, L. Benyahia, Phys. Rev. E **77**, 036305 (2008)
4. G.K. Batchelor, *An Introduction to Fluid Dynamics* (Cambridge University Press, Cambridge, 1967)
5. K. Bergfeldt, L. Piculell, P. Linse, J. Phys. Chem. **100**, 3680–3687 (1996)
6. D. Broseta, L. Leibler, J.-F. Joanny, Macromolecules **20**, 1935 (1987)
7. D. Broseta, L. Leibler, A. Lapp, Europhys. Lett. **2**, 733 (1986)
8. J.W. Cahn, J.E. Hilliard, J. Chem. Phys. **28**, 258 (1958)
9. I. Capron, T. Nicolai, D. Durand, Food Hydrocolloids **13**, 1–5 (1999)
10. P. Ding, B. Wolf, W.J. Frith, A.H. Clark, I.T. Norton, A.W. Pacek, J. Colloid Interface Sci. **253**, 367–376 (2002)
11. F.G. Donnan, Z Elektrochem. Angew. Phys. Chem. **17**, 572 (1911)
12. M.W. Edelman, E. de Hoog, R.H. Tromp, Biomacromolecules **2**, 1148–1154 (2001)
13. M.W. Edelman, R.H. Tromp, E. van der Linden, Phys. Rev. E **67**, 021404 (2003)
14. P.F. Flory, *Principles of Polymer Chemistry* (Cornell University Press, Ithaca, 1953)
15. J.R. Fried, *Polymer Science and Technology,* 2nd edn. (Prentice Hall, New York, 2003), Chap. 3
16. P.G. de Gennes, *Scaling Concepts in Polymer Physics* (Cornell University Press, Ithaca, 1979)
17. S. Guido, M. Simeone, A. Alfani, Carbohydr. Polym. **48**, 143–152 (2002)

18. N. Lorén, A.-M. Hermansson, M.A.K. Williams, L. Lundin, T.J. Foster, C.D. Hubbard, A.H. Clark, I.T. Norton, E.T. Bergström, D.M. Goodall, Macromolecules **34**, 289–297 (2001)
19. M.P. Moody, P. Attard, Phys. Rev. Lett. **91**, 056104 (2003)
20. J.T.G. Overbeek, J. Colloid Sci. **8**, 593 (1953)
21. J.T.G. Overbeek, Prog. Biophys. Biophys. Chem. **6**, 58 (1956)
22. H.M. Princen, I.Y.Z. Zia, S.G. Mason, J. Colloid Interface Sci. **23**, 99–107 (1967)
23. J.M. Rallison, Ann. Rev. Fluid Mech. **16**, 45–66 (1984)
24. M. Rasa, B.H. Erné, B. Zoetekouw, R. van Roij, A.P. Philipse, J. Phys.: Condens. Matter **17**, 2293 (2005)
25. J. Ryden, P.-A. Albertsson, J. Colloid Interface Sci. **37**, 219 (1971)
26. E. Scholten, R. Tuinier, R.H. Tromp, H.N.W. Lekkerkerker, Langmuir **18**, 2234–2238 (2002)
27. R.L. Scott, J.Chem.Phys. **17**, 279–284 (1949)
28. V. Tolstoguzov, Biotech. Adv. **24**, 626 (2006)
29. S. Tomotika, Proc. Roy. Soc. London. Series A **150**, 322–337 (1935)
30. R.H. Tromp, E.M. Blokhuis, Macromolecules **46**, 3639−3647 (2013)
31. R.H. Tromp, E.H.A. de Hoog, Phys. Rev. E **77**, 031503 (2008)
32. M. Vis, V.F.D. Peters, R.H. Tromp, B.H. Erne, Langmuir **30**, 5755-5762 (2014)
33. B. Vonnegut, Rev. Sci. Instr. **13**, 6–9 (1942)

Part III
Theoretical Aspects

Chapter 7
Basics of Statistical Physics

W.J. Briels and J.K.G. Dhont

Abstract Statistical physics is the theory that relates macroscopic properties of materials to their microscopic constitution. Obviously a detailed study of systems containing on the order of 10^{23} particles is out of the question, actually even useless, so we must resort to statistical methods. There are many ways to introduce statistical mechanics, of which we will present two. Besides this, there are several different 'models', called ensembles, to arrive at a relation between microscopic and macroscopic properties of materials. Although these ensembles in principle are different, for sufficiently large systems they provide identical results. While developing statistical physics we will automatically construct thermodynamics, a macroscopic theory of invaluable help in describing for example phase transitions. In principle no knowledge of thermodynamics is assumed in order to understand the present notes, although obviously it will be of great help to have some acquaintance with it.

7.1 Introduction

Dear reader, first read carefully the Abstract.

Now you know that we are going to introduce several ways to calculate average properties of macroscopic systems. The various ways to perform averages depend on the way the system is prepared, *i.e.* on the way the system interacts with its surroundings. Once we have chosen a particular way to perform the calculations, we are said to use a particular ensemble.

W.J. Briels (✉)
University of Twente, Enschede, The Netherlands
e-mail: w.j.briels@utwente.nl

W.J. Briels · J.K.G. Dhont
Julich, Germany
e-mail: j.k.g.dhont@fz-juelich.de

© Springer International Publishing Switzerland 2016
P.R. Lang and Y. Liu (eds.), *Soft Matter at Aqueous Interfaces*,
Lecture Notes in Physics 917, DOI 10.1007/978-3-319-24502-7_7

Calculating averages for macroscopic variables implies that these variables fluctuate around their averages. In this situation it is natural to ask for the mean square deviations of these fluctuations from the corresponding averages. While the averages are the same in all ensembles, the fluctuations will be different. For sufficiently large systems the fluctuations will be negligible, and one macroscopic theory emerges, called Thermodynamics.

It will be clear that we cannot treat all subjects of statistical physics that might interest one or another of this class. We have made our own choice and will feel glad in case you enjoy the selection we made. In case you identify subjects that we left out and that you think would interest you even more than the selection we made, it means we made you see beyond the present notes and we will be even more satisfied.

7.2   The Micro Canonical Ensemble

In this section we will introduce some concepts of statistical physics in an almost natural way. We will do this in the so-called microscopic ensemble. It will take a while before we will explain this name. For now just follow our reasoning.

7.2.1   Entropy

Let us start by asking a question. How can physicists have great confidence in their knowledge of the chemical constitution of the outer layer of the sun. After every step of a multistage synthesis, chemists want to know what was the result of their last step. How do they quickly get information about the chemical structure that resulted from their last synthetic step? The answer in both cases is the same, by spectroscopic measurements. Physicists measure emission and absorption lines of light that comes from the sun and thereby get information about the possible energies available for the atoms that exist in the sun's outer layer. Parts of these energy spectra will be recognized as the fingerprints of particular atoms. Similarly chemists use infrared spectra as fingerprints of particular atomic groups in the molecules they have synthesized. The moral of this story is that the distribution of possible energies acts as a fingerprint to tell one molecule from another.

Now let's consider a macroscopic system. How can we tell one system from another? Well, by measuring its distribution of possible energies you will say. That is exactly what we are going to do. In order to do so we start with the ground state, call it state number one and write down its energy. Next we go to the state with the lowest energy but one, call it state number two and write down its energy. In case there are states with the same energy we number them in whichever order before going to the next energy level. The number of states in one level we call the degeneracy Ω of that level. Once we are done we have enumerated all states and written down their energies in increasing order. This is what we call the energy spectrum, which is the thing that is characteristic for the contents of the system.

We are now in possession of the energy spectrum of our system, meaning that we have enumerated all possible states of the system and given their energies. In mathematical terms we have produced a map from the natural numbers to the real numbers

$$n \rightarrow E_n.$$

The enumeration has been done such that the energies satisfy $E_n \leq E_{n+1}$ for all n.

We want to describe this information in a somewhat different way. To this end, we define $\Phi(U)$ to be the number of states with energy less than U. $\Phi(U)$ is zero for sufficiently low energies, then jumps to one when U becomes larger than the ground state energy E_1, and next increases by a value $\Omega(E_n)$ every time when U passes a possible energy E_n. Notice that Φ is a kind of continuation of the discrete index n to the real numbers. When all energies are non-degenerate, *i.e.* when all Ω are equal to one, this is clear. With Ω's different from zero the map

$$\Phi \rightarrow U_\Phi$$

does not discriminate between degenerate states. In a few cases it is possible to calculate $\Phi(U)$ exactly.

Consider a crystal composed of N atoms, which each vibrate around a given equilibrium position. This model is called Einstein crystal. It is a simple matter to calculate $\Omega(U)$ and next $\Phi(U)$ for this model. Later you may do this as an exercise, here we just give the final result

$$\Phi(U) = \left[\left(\frac{e}{3\hbar\omega} \frac{U}{N} \right)^3 \right]^N \frac{1}{\sqrt{6\pi N}}.$$

Here ω is the vibration frequency of each of the atoms and e is the base of the natural logarithm; $\hbar$ is Planck's constant divided by 2π.

A second model for which we may easily calculate $\Phi(U)$ is an ideal gas consisting of N atoms in a volume V, obtaining

$$\Phi(U) = \left[\left(\frac{me}{3\pi\hbar^2} \frac{U}{N} \right)^{3/2} \frac{V}{N} e \right]^N \frac{1}{\sqrt{6\pi N}}.$$

Obviously, in this case there is a third parameter besides N and U on which Φ depends, which is the volume V.

The two functions above are clearly very different and are characteristic for the system they correspond to. So Φ can be used to tell one system from another. Actually any monotonous function of Φ can be used to characterize the system. It will turn out to be very useful to define

$$S = k_B \ln \Phi.$$

Here k_B is called Boltzmann constant, which we will specify further down these notes. S is called entropy and will play an important role in all further considerations.

An important property of Φ is the fact that it can be written as

$$\Phi(U, V, N) = \left[\varphi \left(\frac{U}{N}, \frac{V}{N} \right) \right]^N aN^b,$$

with φ being some simple function of U/N and V/N, and with a and b some constants that are independent of U, V or N. This is characteristic for all systems. Taking the logarithm we find

$$S(U, V, N) = k_B N \ln \left[\varphi \left(\frac{U}{N}, \frac{V}{N} \right) \right],$$

where we have neglected contributions in the right hand side that are negligible compared to the one proportional to N. From this it immediately follows that

$$S(xU, xV, xN) = xS(U, V, N).$$

So if we take two systems, one of which has energy, volume and number of particles twice as big as the other, then its entropy is also twice as big as that of the other. We say that S is an extensive quantity. It is not difficult to prove that only the logarithm can turn Φ into an extensive quantity.

It will be of great help to be a bit flexible with the definition of entropy. To this end, notice that Φ can be written as

$$\Phi(U) = \sum_n \Theta(U - E_n).$$

Here n runs through all states of the system, and $\Theta(x)$ is the Heaviside function, which equals zero for $x \leq 0$ and unity for x larger than zero. We may rewrite this as

$$\Phi(U) = \sum_{E_m \leq U} \Omega(E_m),$$

where the sum now runs over all possible energies (levels, not states) less than U. It may be shown that for a macroscopic system the overwhelming contribution to this sum comes from the last energy level. In that case S can equally well be defined by

$$S = k_B \ln \Omega.$$

We suggest that you are happy with this and neglect the explanation given below.

Let us now give a quick argument why Φ can be replaced by Ω in the definition of entropy. Roughly speaking Ω is equal to $\partial\Phi/\partial U$, more exactly

$$\frac{\partial\Phi}{\partial U} = \Omega(U)\rho_L(U),$$

where ρ_L is the number of energy levels per energy increase dU. With the particular property of Φ mentioned above

$$\frac{\partial\Phi}{\partial U} = \Phi\frac{\varphi'}{\varphi},$$

where

$$\varphi' = \frac{\partial\varphi}{\partial u},$$

with $u = U/N$. As a result

$$\Phi = \Omega\rho_L\frac{\varphi}{\varphi'}.$$

After taking the logarithm, only the $\ln\Omega$ term is proportional to N. Notice that we assume that the energy level density is approaching a constant with increasing energy, or at least is not growing like a power of N.

7.2.2 Thermodynamics

Now consider an isolated macroscopic system in the sense that no particles can leave or enter the system, that no external force can perform work on the system, nor can any other form of energy be transmitted to the system. You may think of a system in a rigid Styrofoam box. For this box, the energy U, the volume V and the number of particles N are well defined. Therefore also the entropy $S = S(U, V, N)$ is well defined. Our next discussion will be simpler if we concentrate on the energy U as a function of S, V and N, instead of on the entropy S as a function of U, V and N. We now ask, how will the energy change if we change entropy, volume and number of particles? From a mathematical point of view the answer is simple:

$$dU = \left(\frac{\partial U}{\partial S}\right)_{V,N} dS + \left(\frac{\partial U}{\partial V}\right)_{S,N} dV + \left(\frac{\partial U}{\partial N}\right)_{S,V} dN.$$

where the subscripts indicate the variables which are kept constant during the individual steps. What does it mean in physical terms? The second term in the right hand side has a simple interpretation; the others are a bit more difficult. So let's start with the second term.

Consider a change of volume of a closed system (i.e. no particles can leave or enter the system, but energy can be exchanged with the surroundings) at constant entropy. Recall that constant entropy means that the number of states with energy below the actual energy is constant. This means that the number of states below the new energy $U + dU$ is equal to the number of states below the original energy U. If we change volume all energies of the system will change as shown in Fig. 7.1. Keeping the number of states with energy less than U constant simply means that the change of energy of the system dU is equal to the change of energy of the state with energy $E_n = U$, i.e. the actual state of the system. In order to make sure that the system does not jump to a new state and only follows the actual energy in its dependence on volume, we must perform the process very slowly; processes like these are called reversible adiabatic. The change of energy is by definition the work w performed on the system in order to change its volume. This we know from high-school is given by $w = -p^{ext}dV$, with p^{ext} the applied pressure on the piston. For adiabatic processes $p^{ext} = p$, with p being that particular pressure that must be applied to a freely sliding piston to keep its position constant, and so

$$w = -pdV.$$

p is called the pressure of the system. As a result, we have

$$\left(\frac{\partial U}{\partial V}\right)_{S,N} = -p.$$

In order to interpret the differential quotient of entropy and energy we will make a little detour. To this end we write

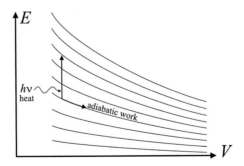

Fig. 7.1 Change of energies with changing volume. In order to keep the number of states with energy below the actual energy the process must be performed adiabatically

$$dS = \left(\frac{\partial S}{\partial U}\right)_{V,N} dU + \left(\frac{\partial S}{\partial V}\right)_{U,N} dV + \left(\frac{\partial S}{\partial N}\right)_{U,V} dN.$$

We now concentrate on changes with $dS = dN = 0$ and divide by dV, obtaining

$$0 = \left(\frac{\partial S}{\partial U}\right)_{V,N} \left(\frac{\partial U}{\partial V}\right)_{S,N} + \left(\frac{\partial S}{\partial V}\right)_{U,N},$$

and finally

$$p = -\left(\frac{\partial U}{\partial V}\right)_{S,N} = \frac{\left(\frac{\partial S}{\partial V}\right)_{U,N}}{\left(\frac{\partial S}{\partial U}\right)_{V,N}} = \left(\frac{\partial S}{\partial V}\right)_{U,N} \left(\frac{\partial U}{\partial S}\right)_{V,N}.$$

For an ideal gas, knowing S as a function of U, V and N, it can be shown that

$$p = \frac{Nk_B}{V} \left(\frac{\partial U}{\partial S}\right)_{V,N}.$$

We deliberately have not evaluated the last factor in this equation. From experiment we know that

$$p = \frac{nRT}{V},$$

where $n = N/N_{Av}$, with N_{Av} being Avogadro's number, R is the gas constant and T is absolute temperature. This result strongly advises us to choose the till now unspecified Boltzmann constant according to

$$k_B = \frac{R}{N_{Av}}.$$

We then find that for an ideal gas

$$\left(\frac{\partial U}{\partial S}\right)_{V,N} = T.$$

This is not yet very spectacular, but *alas* at least we have produced something.

Now let us turn this into something more spectacular. From high school you may remember that in the old days temperature was measured by putting the system in thermal contact with a small body of ideal gas until equilibrium was obtained, after which the temperature of the system was declared to be equal to the temperature of the ideal gas thermometer. Later we will see that at thermal equilibrium between systems A and B we have

$$\left(\frac{\partial U}{\partial S}\right)^{(A)}_{V,N} = \left(\frac{\partial U}{\partial S}\right)^{(B)}_{V,N}.$$

This implies that we may identify $(\partial U/\partial S)_{V,N}$ and the ideal gas temperature T. For the time being we will define

$$\left(\frac{\partial U}{\partial N}\right)_{S,V} = \mu.$$

This new quantity μ is called the chemical potential and it plays a similar role in equilibria as pressure and temperature. To be continued.

With our findings so far we have

$$dU = TdS - pdV + \mu dN.$$

This is the fundamental equation of thermodynamics. An alternative way to write the fundamental equation of thermodynamics is

$$dS = \frac{1}{T}dU + \frac{p}{T}dV - \frac{\mu}{T}dN.$$

We have seen that U, S, V and N are extensive quantities; they grow proportional to the size of the system. By their very definition, T, p and μ are intensive variables; they do not change when we combine two identical systems (actually they tell us if systems are identical or not, apart from their size). There is one interesting thing that we should mention about these intensive variables. Because of extensivity we have

$$U(xS, xV, xN) = xU(S, V, N).$$

Differentiating this equation with respect to x we find

$$TS - pV + \mu N = U.$$

This implies

$$dU = TdS - pdV + \mu dN + SdT - Vdp + Nd\mu.$$

Comparing this with the fundamental equation of thermodynamics we find

$$SdT - Vdp + Nd\mu = 0.$$

This is called the Gibbs-Duhem equation. It says that out of the three intensive variables T, p and μ only two can be varied independently. Fixing two intensive variables fixes the other. This means that for example the chemical potential is a function of temperature and pressure. Giving the values of two intensive variables only fixes the internal state of the system. In order to know the size of the system at least one extensive quantity must be known.

Exercise:

Prove the following equation:

$$N! = \int_0^\infty dx \, x^N e^{-x} = \int_0^\infty dx \, e^{N \ln x - x}.$$

Next develop $n \ln x - x$ in a Taylor series up to second order around its maximum and calculate the integral, obtaining

$$N! = \left(\frac{N}{e}\right)^N \sqrt{2\pi N}.$$

This approximation for $N!$ is called Stirling equation. You will need

$$\int_{-\infty}^\infty dx \, e^{-cx^2} = \sqrt{\frac{\pi}{c}}.$$

Exercise:

The states of an Einstein crystal are indexed by $3N$ quantum numbers $n_1, n_2, \ldots, n_{3N-1}, n_{3N}$, which each take integer values from 0 to ∞. The corresponding energies read

$$E_{n_1, \ldots n_{3N}} = \hbar\omega(n_1 + 1/2) + \cdots + \hbar\omega(n_{3N} + 1/2) = \hbar\omega\left(\frac{3N}{2} + M\right)$$

$$M = n_1 + \cdots + n_{3N}$$

In order to calculate the number of states $\Omega(U)$ corresponding to energy U, you must distribute

$$M = \frac{U}{\hbar\omega} - \frac{3}{2}N$$

quanta of energy $\hbar\omega$ over $3N$ oscillators. To this end you write down M crosses. Next you decide how many quanta you give to the first oscillator, *i.e.* you decide about the value of n_1. If it is zero, you put a bar to the left of the very first cross, if it is larger than zero you put the bar between the crosses indexed n_1 and $n_1 + 1$ respectively. Next you decide about the number of quanta you give to the second oscillator, *i.e.* you decide about the value of n_2. If it is zero you put a bar right after the previous bar, if it is non-zero you put the bar after the next n_2 crosses, *i.e.* between the crosses indexed $n_1 + n_2$ and $n_1 + n_2 + 1$ respectively. Next you repeat this procedure for all remaining oscillators.

Show that

$$\Omega(U) = \frac{(M + 3N - 1)!}{M!(3N - 1)!},$$

and if you want

$$\Phi(M) = \frac{(M + 3N)!}{M!(3N)!}.$$

From this calculate $T = T(U)$ and invert this to find $U = U(T)$. Next calculate the specific heat $C_V = \partial U / \partial T$.

Note:

The Φ for the Einstein crystal given in the main text is obtained from the one given in this exercise assuming that M is much larger than $3N$. It will therefore not give the correct behavior of the specific heat at low temperatures.

7.2.3 Irreversible Processes and Equilibria

Now consider what happens to the entropy in case we let a spontaneous process occur. How do we do this? As an example, we start with an isolated system which is constrained in the sense that energy cannot be freely distributed all over the system, like in Fig. 7.2, where two parts of the system are separated by an insulating wall. The system consists of a part A with energy U_A and a part B with energy U_B. A spontaneous process may be induced by suddenly turning the insulating wall into a heat conducting wall. As a result of removing the constraint, suddenly many more states become available. Calling Ω^c the number of states of the constrained system and Ω^u the number of states of the unconstrained system, we have $\Omega^u \geq \Omega^c$ and hence with $\Delta S = S^u - S^c$:

$$\Delta S \geq 0.$$

This is the second law of thermodynamics. It says that spontaneous processes are accompanied by an increase of entropy.

We now want to translate this inequality into something more useful. To this end, let's analyze the process in more detail. First notice

$$\Omega^c = \Omega^{(A)}(U_A) \cdot \Omega^{(B)}(U_B)$$

$$\Omega^u = \sum_n \Omega^{(A)}(U_A + E_n) \cdot \Omega^{(B)}(U_B - E_n).$$

Fig. 7.2 Two systems isolated from the outside world. Initially (*left panel*) the wall between both systems is rigid and thermally insulating. When this wall is made thermally conducting (*right panel*) energy can be exchanged between both systems

Here the sum runs over all possible values of E_n such that $U_A + E_n$ is a possible energy for system A and at the same time $U_B - E_n$ is a possible energy for system B. The first factor in each term in the sum for Ω^u is steeply increasing with increasing values of E_n while the second factor is steeply decreasing with increasing E_n. As a result, the product of the two factors will be maximal for some $E_n = E_{\max}$. The product may then be approximated as

$$\Omega^{(A)}(U_A + E_n) \cdot \Omega^{(B)}(U_B - E_n) = \Omega^{(A)}(U_A + E_{\max}) \cdot \Omega^{(B)}(U_B - E_{\max})$$
$$\cdot \exp\left[-\frac{(E_n - E_{\max})^2}{2\sigma_{AB}^2}\right].$$

Here σ_{AB} determines the width of the product in the left hand side as a function of E_n. The unconstrained number of states within this approximation reads

$$\Omega^u = \Omega^{(A)}(U_A + E_{\max}) \cdot \Omega^{(B)}(U_B - E_{\max}) \cdot \sum_n \exp\left[-\frac{(E_n - E_{\max})^2}{2\sigma_{AB}^2}\right].$$

The last factor, *i.e.* the sum over all possible E_n may be approximated by

$$\sum_n \exp\left[-\frac{(E_n - E_{\max})^2}{2\sigma_{AB}^2}\right] \approx \rho_L \int dE \exp\left[-\frac{(E - E_{\max})^2}{2\sigma_{AB}^2}\right] \approx \rho_L \sqrt{2\pi\sigma_{AB}^2}$$

with ρ_L the density of energy levels. With these results we obtain

$$S^c = k_B \ln \Omega^{(A)}(U_A) + k_B \ln \Omega^{(B)}(U_B)$$

$$S^u = k_B \ln \Omega^{(A)}(U_A + E_{\max}) + k_B \ln \Omega^{(B)}(U_B - E_{\max})$$

where in the second equation we have ignored non-extensive terms (Fig. 7.3).

The final result of the analysis given above is that an amount of energy $E_{\max}$ has been transferred from system B to system A such that

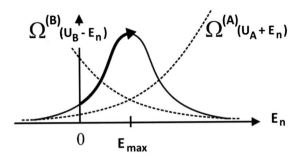

Fig. 7.3 Second law of thermodynamics. Initially E_n is constrained to be zero. When the wall separating the two systems is changed to be thermally conducting, energy can be exchanged and new states become available. The total entropy is equal to the entropy with E_n constrained to E_{max}

$$S^u(U_A + U_B) = S^{(A)}(U_A + E_{max}) + S^{(B)}(U_B - E_{max})$$

is maximal. With this redistribution of energy both systems are in equilibrium and no further processes take place. An additional exchange of energy will lead to no increase of the entropy, so at equilibrium

$$dS = dS^{(A)} + dS^{(B)} = 0.$$

Applying thermodynamics we get

$$\frac{1}{T^{(A)}} dU^{(A)} + \frac{1}{T^{(B)}} dU^{(B)} = 0.$$

Since $dU^{(A)} = -dU^{(B)}$ we get as our criterion for equilibrium

$$T^{(A)} = T^{(B)}.$$

Very similar arguments can be used more generally. Suppose we have a liquid and a vapor in contact with each other such that besides energy also volume and particles can be exchanged. The entropy will be maximal in case further exchange of energy, volume and/or particles does not change the entropy. With $dU^{(A)} = -dU^{(B)}$, $dV^{(A)} = -dV^{(B)}$ and $dN^{(A)} = -dN^{(B)}$ the criterion for equilibrium reads

$$dS = \left(\frac{1}{T^{(A)}} - \frac{1}{T^{(B)}}\right) dU^{(A)} + \left(\frac{p^{(A)}}{T^{(A)}} - \frac{p^{(B)}}{T^{(B)}}\right) dV^{(A)} - \left(\frac{\mu^{(A)}}{T^{(A)}} - \frac{\mu^{(B)}}{T^{(B)}}\right) dN^{(A)} = 0.$$

So, finally

$$T^{(A)} = T^{(B)}$$
$$p^{(A)} = p^{(B)}$$
$$\mu^{(A)} = \mu^{(B)}$$

Generalizations will pose no problems. The general concept is that redistributing extensive quantities over parts of a system in equilibrium does not lead to a change of entropy. To be clear, 'redistributing' implies that the total values of the extensive quantities that are redistributed do not change.

A useful alternative criterion for irreversible processes applies for systems with constant temperature. To find this criterion, put the system in a huge thermostat of temperature T and allow for possible exchange of energy between the thermostat and the system. Since the thermostat is assumed to be infinitely large, its temperature does not change as a result of possible exchange of energy between system and thermostat. Remove existing constraints in the system and let the spontaneous process run. Before and after the process the temperature of the system is the same as that of the thermostat. The criterion for irreversibility says

$$\Delta S + \Delta S^{Th} \geq 0.$$

Here ΔS^{Th} is the change of entropy of the thermostat as a result of the energy exchange with the system while the spontaneous process was running in the system. Because the thermostat is huge, we may use the fundamental equation of thermodynamics to get $\Delta S^{Th} = \Delta U^{Th}/T$. Now, since the energy ΔU^{Th} gained by the thermostat must be lost by the system, we have $\Delta U^{Th} = -\Delta U$, where ΔU is the increase of energy of the system as a result of the spontaneous process in the system. The irreversibility criterion therefore reads

$$\Delta S - \frac{\Delta U}{T} \geq 0.$$

Introducing the Helmholtz free energy $F = U - TS$, we finally get

$$\Delta F \leq 0.$$

Therefore the change of Helmholtz free energy during a spontaneous process is negative, and the Helmholtz free energy attains a minimum at equilibrium.

7.2.4 Probabilities

It is time to give meaning to the adjective 'statistical' in statistical physics.

Most of you probably have heard about the principle of 'equal a priori probabilities'. By this we mean that if we interrogate the state of a system we will find one out of the Ω possible states and that we will find all of these equally often if we repeat the experiment a huge number of times. As a working hypothesis (ergodic hypothesis) we further assume that when we perform a measurement, the system runs through a great many of its possible states and that the average that is measured is equal to the average over all possible states. So, if we measure a quantity A with value A_n in state n, the result will be

$$\langle A \rangle = \frac{1}{\Omega} \sum_n A_n = \sum_n \frac{1}{\Omega} A_n = \sum_n P_n A_n.$$

In the last step we have introduced the probability P_n to find state n. 'Equal a priori probabilities' means that all P_n are equal; since $\sum_n P_n = 1$, we must have $P_n = 1/\Omega$.

Now, in case we are interested in the probability P_A to find a particular property A that is shared by Ω_A states, we may write

$$P_A = \frac{\Omega_A}{\Omega} = e^{(S_A - S)/k_B}.$$

Here S_A is the entropy of a system constrained to have property A. The above equation was used extensively by Einstein. We will use it in the next section on the Canonical Ensemble.

Let us end this section by explaining the word 'ensemble'. One way of introducing probabilistic arguments is by imagining a huge collection, ensemble in French, of macroscopically similar systems, which only differ in their microscopic realization. The number of times that a particular microscopic state occurs in this collection is taken to be proportional to its probability. Averaging a quantity A with the proper weight is then equivalent to taking a non-weighted average over the so created ensemble.

7.3   The Canonical Ensemble

Most of you will remember from high school the barometric height equation

$$p(z) = p(0) \exp\left(-\frac{mgz}{k_B T}\right).$$

This equation gives the partial pressure of molecules with mass per molecule equal to m as a function of height z above the earth's surface. Here k_B is the Boltzmann constant and g is the gravitational acceleration. It is a very simple exercise to derive this equation. Now, using the equation of state for ideal gases, we may transform the above equation into

$$\rho(h) = \rho(0) \exp\left(-\frac{mgh}{k_BT}\right),$$

with $\rho(h)$ the density at height h. We interpret this result as telling that the probability density to find a molecule of mass m in the earth's gravitational field at height z above the earth's surface is proportional to $\exp(-mgz/(k_BT))$. The probability density $P(z)$ is related to the probability $\Pr(z \leq Z \leq z + dz)$ to find Z between z and $z + dz$ by $\Pr(z \leq Z \leq z + dz) = P(z)dz$.

A very similar result was obtained by Maxwell who derived that the probability density to find the momenta of a molecule to be at (p_x, p_y, p_z) is given by

$$P(p_xp_y, p_z) \propto \exp\left(-\frac{p_x^2 + p_y^2 + p_z^2}{2mk_BT}\right).$$

Here $p_x = v_x/m$ with v_x the velocity in x-direction and m the mass of the molecule, and similarly for the y- and z-directions. Also this equation can be easily derived.

Investigating all components of the position vector $\vec{R}$ and the momentum vector $\vec{P}$ we may write

$$\Pr(x \leq R_x \leq x + dx, \ldots, p_z \leq P_z \leq p_z + dp_z) = P(\vec{r}, \vec{p})d^3rd^3p,$$

with $\vec{r} = (x, y, z)$ and $\vec{p} = (p_x, p_y, p_z)$, $d^3r = dxdydz$ and $d^3p = dp_xdp_ydp_z$. As an intermezzo, let us tell you that in classical mechanics the concept of momentum is generalized such that, among other things, d^3rd^3p is invariant. Another good reason for using momenta instead of velocities is that MS Word produces ugly v's while p's are great. Now, combining both results above we have

$$P(\vec{r}, \vec{p}) \propto \exp\left(-\frac{mgz}{k_BT} - \frac{p_x^2 + p_y^2 + p_z^2}{2mk_BT}\right).$$

Introducing the total energy per molecule $\varepsilon = mgz + (p_x^2 + p_y^2 + p_z^2)/2m$ we have

$$P(\vec{r}, \vec{p}) \propto \exp\left(-\frac{\varepsilon(\vec{r}, \vec{p})}{k_BT}\right).$$

This says that the relevant probability density is proportional to natural base to the power minus energy divided by k_BT, which is called Boltzmann factor.

In the following we will generalize this result to systems containing N particles in a volume V. We now define the probability distribution $P(r^{3N}, p^{3N})$ by

$$\Pr(\vec{r}_1 \le \vec{R}_1 \le \vec{r}_1 + d\vec{r}_1, \ldots, \vec{p}_N \le \vec{P}_N \le \vec{p}_N + d\vec{p}_N) = P(r^{3N}, p^{3N}) \frac{d^{3N}r d^{3N}p}{h^{3N}},$$

with $d^{3N}r = d^3r_1 \ldots d^3r_N$ and $d^{3N}p = d^3p_1 \ldots d^3p_N$; the factor h^{-3N}, with h being Planck's constant, has been introduced to make the probability distribution dimensionless. We will find

$$P(r^{3N}, p^{3N}) \propto \exp\left(-\beta H(r^{3N}, p^{3N})\right).$$

Here $\beta = 1/(k_B T)$ and H is the Hamiltonian with value $H(r^{3N}, p^{3N}) = E$.

The collection of points (r^{3N}, p^{3N}) is called phase space. In the following we will divide phase space into little cubes of size $d^{3N}r d^{3N}p$. This turns phase space into a countable collection of states. We next introduce the probability P_n to find the system in cube n:

$$P_n = P(r^{3N}, p^{3N}) \frac{d^{3N}r d^{3N}p}{h^{3N}}.$$

The expectation value of a variable $A(r^{3N}, p^{3N})$ may then be written as:

$$\langle A \rangle = \frac{1}{h^{3N}} \int d^{3N}r \int d^{3N}p P(r^{3N}, p^{3N}) A(r^{3N}, p^{3N}) = \sum_n P_n A_n.$$

Here A_n is the value of $A(r^{3N}, p^{3N})$ in cube n. The good thing is that we can now use the same notation in the classical and the quantum mechanical case. That's it, we can start.

7.3.1 Thermodynamics

We start by putting a system in a huge thermostat. By definition system and thermostat have the same temperature T and may exchange energy. Although the system's energy may fluctuate, we will assume that a well-defined average energy U exists. The total of system and thermostat will be isolated, so we can use the micro canonical ensemble to calculate the thermodynamic properties of the total. Letting U^{tot} be the energy of system plus thermostat, we write the number of states going with this energy as

$$\Omega^{tot}(U^{tot}) = \sum_n \Omega^{Th}(U^{tot} - E_n).$$

Here n runs through all states of the system, while E_n is the energy of state n. When the energy of the system is E_n, that of the thermostat is $U^{tot} - E_n$. The number of states the thermostat can assume with this particular energy is

$\Omega^{Th}(U^{tot} - E_n)$. Most of the time E_n will be near U. Therefore we will write $\Omega^{Th}(U^{tot} - E_n) = \Omega^{Th}(U^{Th} - \Delta E_n)$, where by definition $\Delta E_n = E_n - U$ and $U^{Th} = U^{tot} - U$, *i.e.* the total energy minus the system's energy. Now we use our main result from the microscopic ensemble to relate the number of states of the thermostat to its entropy: $\Omega^{Th}(U^{Th} - \Delta E_n) = \exp(S^{Th}(U^{Th} - \Delta E_n)/k_B)$. Since the thermostat is huge, ΔE_n will be small compared to U^{Th}, so we write

$$S^{Th}(U^{Th} - \Delta E_n) = S^{Th}(U^{Th}) - \frac{1}{T}\Delta E_n$$

to first order in ΔE_n. Combining everything we get

$$\Omega^{tot} = e^{S^{Th}/k_B}\sum_n e^{-\Delta E_n/k_B T} = e^{S^{Th}/k_B}e^{U/k_B T}Q,$$

where

$$Q = \sum_n e^{-E_n/k_B T}.$$

Finally we calculate the total entropy

$$S^{tot} = S^{Th}(U^{Th}) + \left[\frac{U}{T} + k_B \ln Q\right].$$

We interpret this equation by letting $S^{Th}(U^{Th})$ be the entropy of the thermostat and by letting $S^{tot} = S^{Th} + S$, with S the entropy of the system. For the latter we then have

$$S = \frac{U}{T} + k_B \ln Q.$$

This result is equivalent to

$$F = -k_B T \ln Q,$$

where, as before, $F = U - TS$ is the Helmholtz free energy.

Notice that by definition $Q = Q(T, V, N)$, and that therefore we obtain the free energy as a function of T, V and N. With $dF = dU - TdS - SdT$ and the fundamental equation of thermodynamics we have

$$dF = -SdT - pdV + \mu dN.$$

We are therefore able to calculate entropy, pressure and chemical potential as function of temperature, volume and number of particles, and from that all other thermodynamic properties.

7.3.2 Probabilities

As we have seen, a system at constant temperature can run through a great many of
states with variable energies. In this section we will obtain the probability with
which each of these states occurs. Calculations are simple.

With every state n of the system, the thermostat can occur in one of $\Omega^{Th}(U^{Th} - \Delta E_n)$ states, all of which are equally probable. The probability to find state
n therefore is

$$P_n \propto \Omega^{Th}(U^{Th} - \Delta E_n) \propto e^{-E_n/k_B T}.$$

In the last step we removed all factors not depending on the state n. Using the
fact that the sum of all probabilities must be equal to one, we may calculate the
constant of proportionality, obtaining

$$P_n = \frac{e^{-E_n/k_B T}}{Q}.$$

Here Q is our friend from the previous section.

In this derivation we have used the results already obtained when treating the
micro canonical ensemble. It is possible to obtain the above result without invoking
the micro canonical ensemble. Of course similar ingredients must go into the
derivation, but it is still instructive to see the derivation, in particular if you didn't
really like the Taylor expansion used in the previous section. So, here we go. The
probability to find state n of the system is

$$P_n = \frac{\Omega^{Th}(U^{Th} - \Delta E_n)}{\Omega^{tot}}.$$

Now, $\Omega^{Th}(U) = C \cdot U^{cN^{Th}}$ with c being some small constant and the constant of
proportionality C not dependent on U. Therefore

$$\Omega^{Th}(U^{Th} - \Delta E_n) = \Omega^{Th}(U^{Th}) \cdot \left(1 - \frac{\Delta E_n}{U^{Th}}\right)^{cN^{Th}}.$$

For convenience, let us write $cN^{Th} = M$ and reshuffle things a bit, obtaining

$$\Omega^{Th}(U^{Th} - \Delta E_n) = \Omega^{Th}(U^{Th}) \cdot \left(1 - \frac{1}{U^{Th}/M} \frac{\Delta E_n}{M}\right)^{M}.$$

Now, if the thermostat is huge compared to the system, we may assume that U^{Th}/M becomes constant with increasing size of the thermostat. Calling this constant $1/\beta$
we obtain

$$\Omega^{Th}(U^{Th} - \Delta E_n) = \Omega^{Th}(U^{Th}) \cdot \left(1 - \frac{\beta \Delta E_n}{M}\right)^M = \Omega^{Th}(U^{Th}) \cdot e^{-\beta \Delta E_n}.$$

In the last step we have used the mathematical identity

$$\lim_{M \to \infty} \left(1 - \frac{x}{M}\right)^M = e^{-x}.$$

Putting this expression for $\Omega^{Th}(U^{Th} - \Delta E_n)$ into the expression for P_n above, we get

$$P_n = \frac{e^{-\beta E_n}}{\sum_m e^{-\beta E_m}}.$$

This is our old friend if we manage to prove that $\beta = 1/(k_B T)$.

Assuming we already know about thermodynamics, we can follow a quick route to obtain the relation between our present microscopic treatment and thermodynamics. First notice that, with the actual energy fluctuating, it is clear that the thermodynamic energy U should be the average energy:

$$U = \sum_n E_n P_n = \frac{1}{Q} \sum_n E_n e^{-\beta E_n}.$$

From this we easily obtain

$$U = -\frac{\partial \ln Q}{\partial \beta}.$$

For an ideal gas one finds with a quantum mechanical description

$$Q = \left(\frac{2\pi m}{\beta h^2}\right)^{3N/2} \frac{V^N}{N!},$$

from which it follows that

$$U = \frac{3N}{2\beta}.$$

Experiment then tells us that indeed $\beta = 1/(k_B T)$. Since β is a property of the thermostat it is irrelevant that we used the ideal gas to obtain this result.

Next, assuming we know already about thermodynamics, we may write

$$U = F + TS = F - T\left(\frac{\partial F}{\partial T}\right)_{N,V} = \left(\frac{\partial F/T}{\partial 1/T}\right)_{N,V}.$$

We then have

$$-k_B \frac{\partial \ln Q}{\partial 1/T} = \frac{\partial F/T}{\partial 1/T}.$$

From this we conclude

$$F = -k_B \ln Q + C_{N,V}.$$

If we use quantum mechanical energies to calculate the partition function Q, we may put $C_{N,V} = 1$, to obtain results in agreement with experiments. If we use classical mechanics to calculate the partition function Q, experiment forces us to use $C_{N,V} = k_B \ln N!$. From now on we assume that in the classical case a factor of $1/N!$ is included in the definition of the partition function, i.e.

$$Q = \frac{1}{h^{3N} N!} \int d^{3N} r \int d^{3N} p \exp\left(-H(r^{3N}, p^{3N})/k_B T\right).$$

If you are happy with this, be happy and go to the next section.

For the diehards we present an alternative route to arrive at the previous results, but now without assuming you know anything about thermodynamics. To this end let us see how the average energy changes when we perturb the system? Well

$$dU = d\left(\sum_n E_n P_n\right) = \sum_n E_n dP_n + \sum_n P_n dE_n.$$

In the first term we replace E_n by $-(\ln P_n + \ln Q)/\beta$ and use $\sum_n dP_n = 0$ to obtain

$$dU = -\beta^{-1} d\left(\sum_n P_n \ln P_n\right) + \sum_n P_n dE_n.$$

Suppose we adiabatically change the volume of the system. Adiabatically means that we do not allow the system to change state. Adopting the picture of an ensemble, when no system changes its state during the adiabatic volume change means that the probabilities P_n do not change, and so that the first term in dU equals zero. The second term on the right hand side is just the average work to be performed on the system to achieve this change of volume, so

$$dU_{adiabatic} = \sum_n P_n dE_n = -p dV.$$

If we moreover introduce the entropy

$$S = -k_B \sum_n P_n \ln P_n,$$

we obtain

$$dU = (k_B \beta)^{-1} dS - p dV.$$

Don't be disturbed by the fact that this entropy looks different from what you have seen so far; remember that we pretend to know nothing about thermodynamics. In any case, we will prove in the next section that this is essentially the same entropy as the one we met in the micro canonical ensemble.

Notice that the last equation implies

$$(k_B \beta)^{-1} = \left(\frac{\partial U}{\partial S} \right)_{V,N}.$$

As in the micro canonical ensemble, we will use the ideal gas law to introduce temperature. In the present case we have

$$p = \left(\frac{\partial S}{\partial V} \right)_{U,N} \left(\frac{\partial U}{\partial S} \right)_{V,N} = \left(\frac{\partial S}{\partial V} \right)_{U,N} \frac{1}{k_B \beta}.$$

To calculate the first factor in the right hand side we must know $S = S(U, V, N)$. In the present case, *i.e.* the canonical ensemble, we have

$$U = -\frac{\partial \ln Q}{\partial \beta}$$

$$S = -k_B \beta \frac{\partial \ln Q}{\partial \beta} + k_B \ln Q$$

The second of these expressions follows from the definition of entropy using $\ln P_n = -\beta E_n - \ln Q$. So, with the partition function Q for the ideal gas given above, we find

$$U = \frac{3N}{2\beta}$$

$$S = N k_B \ln \left[\left(\frac{2\pi m}{\beta h^2} \right)^{3/2} e^{5/2} \frac{V}{N} \right]$$

Eliminating β in favor of U we get

$$S = N k_B \ln \left[\left(\frac{4\pi m}{3 h^2} \right)^{3/2} e^{5/2} \left(\frac{U}{N} \right)^{3/2} \frac{V}{N} \right].$$

Putting everything together yields

$$p = \left(\frac{\partial S}{\partial V} \right)_{U,N} \frac{1}{k_B \beta} = \frac{N}{\beta V}.$$

So, we conclude

$$\beta = \frac{N_{Av}}{RT}.$$

Since β is a property of the thermostat, this relation holds fully generally, independent of the fact that it was obtained on the basis of an application to the ideal gas. Again we choose $k_B = R/N_{Av}$ and find

$$dU = TdS - pdV + \mu dN.$$

We have added a contribution from changes of the number of particles. Since we will find that the newly defined entropy is basically the same as before, it is clear that the chemical potential μ is the same as before. More about this later.

We end this section by introducing the Helmholtz free energy as before

$$F = U - TS = \sum_n E_n P_n + k_B T \sum_n P_n \ln P_n.$$

Introducing again $E_n = -k_B T (\ln P_n + \ln Q)$ we obtain

$$F = -k_B T \ln Q.$$

So, from statistical physics we get Q as function of T, V and N, and therefore also F as function of T, V and N. This allows us to calculate all thermodynamic properties of the system.

Exercise:

The energies of a system of N non-interacting particles in a volume V are indexed by $3N$ quantum numbers $n_1, n_2, \ldots, n_{3N-1}, n_{3N}$, which each take values from 1 to ∞. The energies are given by

$$E_{n_1,\ldots,n_{3N}} = \frac{h^2}{8mL^2} \left(n_1^2 + \cdots + n_{3N}^2 \right)$$

Sequences of quantum numbers which differ by a mere permutation refer to one and the same state and should only be counted once when calculating the partition function. Show that the partition function is given by

$$Q = \left(\frac{2\pi m k_B T}{h^2}\right)^{3N/2} \frac{V^N}{N!}.$$

Next calculate the partition function using classical mechanics, *i.e.* according to

$$Q = \frac{1}{h^{3N} N!} \int d^{3N} r \int d^{3N} p \exp\left(-\frac{p_{1,x}^2 + \cdots + p_{3N,z}^2}{2mk_B T}\right).$$

You will need

$$\int_{-\infty}^{\infty} dx e^{-cx^2} = \sqrt{\frac{\pi}{c}}.$$

Without the 'unexpected factor' $1/N!$ the results would have been different, so now you know why it has been added.

7.3.3 Fluctuations

We now have two different ways to calculate thermodynamic properties, and it is not immediately clear that they both give the same result. In the canonical ensemble the energy of the system can fluctuate, while in the micro canonical ensemble it is fixed. Both ensembles can only be equivalent when fluctuations of energy in the canonical ensemble are negligible.

Let us denote by $P(E_n)$ the probability to find the system with energy E_n. This means that the system exists in one of the $\Omega(E_n)$ states having energy E_n. Therefore

$$P(E_n) = \Omega(E_n)\frac{e^{-\beta E_n}}{Q}.$$

Now, $\Omega(E_n)$ is steeply increasing with increasing values of E_n, while $e^{-\beta E_n}/Q$ is a decreasing function of E_n. Since exponentials finally kill any algebraically increasing function, for high enough values of E_n the probability to find energy E_n will be zero. Consequently the product of the two factors will have a maximum at some value of E_n. Assuming that $P(E_n)$ will be symmetric around this maximum, the maximum occurs at the average energy U. The width of the distribution of energies may be characterized by the standard deviation

$$\sigma_E = \sqrt{\left\langle (E - \langle E \rangle)^2 \right\rangle}.$$

Here the pointy brackets denote an average

$$\langle E \rangle = \sum_n E_n P_n$$

$$\langle E^2 \rangle = \sum_n E_n^2 P_n$$

etc. The standard deviation therefore is the square root of the average of the square of the deviation of the energy from its average.

It is an easy task to calculate σ_E for the canonical ensemble. To this end we first note

$$\sigma_E^2 = \left\langle E^2 - 2E\langle E \rangle + \langle E \rangle^2 \right\rangle = \langle E^2 \rangle - \langle E \rangle^2.$$

The latter expression may easily be calculated according to

$$\frac{\partial^2 \ln Q}{\partial \beta^2} = \frac{\partial}{\partial \beta} \frac{1}{Q} \frac{\partial Q}{\partial \beta} = \frac{1}{Q} \frac{\partial^2 Q}{\partial \beta^2} - \left[\frac{1}{Q} \frac{\partial Q}{\partial \beta} \right]^2 = \sigma_E^2$$

Exercise:

Check the last identity.

End of exercise.

This is a beautiful result, since we now have

$$\sigma_E^2 = \frac{\partial}{\partial \beta} \frac{\partial \ln Q}{\partial \beta} = -\frac{\partial U}{\partial \beta} = k_B T^2 \frac{\partial U}{\partial T}.$$

What we are really interested in, is the standard deviation relative to the energy itself. This tells us at which decimal the energy is fluctuating. In general the energy U is proportional to $N k_B T$ with a proportionality constant α depending weakly on density. Therefore

$$\frac{\sigma_E}{U} = \frac{1}{\sqrt{N\alpha}}.$$

So, with increasing number of particles the relative uncertainty in the energy quickly goes to zero.

We end this section discussing the equivalence of ensembles. According to the arguments in the last paragraph

$$\Omega(E_n)e^{-E_n/k_B T} = \Omega(U)e^{-U/k_B T} \exp\left[-\frac{(E_n - U)^2}{2\sigma_E^2}\right]$$

The partition function then reads

$$Q = \sum_n \Omega(E_n)e^{-E_n/k_B T} = \Omega(U)e^{-U/k_B T} \sum_n \exp\left[-\frac{(E_n - U)^2}{2\sigma_E^2}\right]$$
$$\approx \Omega(U)e^{-U/k_B T}\rho_L\sqrt{2\pi\sigma_E^2}.$$

The sums in these expressions are over all possible energies, *i.e.* over all energy levels, not over states. The free energy then reads (with $F = -k_B T \ln Q$):

$$F = U - k_B T \ln \Omega(U),$$

where we have ignored terms not proportional to N. From this we conclude

$$S(U) = k_B \ln \Omega(U).$$

So, within the approximation of ignoring terms not proportional to N, the micro canonical ensemble and the canonical ensemble will give identical results. The good thing about this is that calculating Q is much easier than calculating Ω.

7.4 Interacting Systems, Very Simple Application

The most interesting application of statistical physics of course is to interacting systems. In the previous section we have seen that the major task with statistical physics is to calculate the partition function. In the (semi) classical approximation this reads

$$Q = \frac{1}{h^{3N}N!}\int d^{3N}r \int d^{3N}p \exp\left[-\frac{p_{1,x}^2 + \cdots + p_{N,z}^2}{2mk_B T} - \frac{\Phi_N(r^{3N})}{k_B T}\right].$$

Here $\Phi_N(r^{3N})$ is the potential energy of the system. Performing the integrals over the momenta we get

$$Q = \frac{1}{\Lambda^{3N} N!} Z_N$$

with

$$\Lambda = \sqrt{\frac{h^2}{2\pi m k_B T}},$$

and

$$Z_N = \int d^{3N} r \exp\left[-\frac{\Phi_N(r^{3N})}{k_B T}\right].$$

Λ is called the thermal de Broglie length and Z_N is the configuration integral. So, our main problem is to calculate the configuration integral.

It is convenient at this point to notice that averages of functions $f(r^{3N})$ which only depend on the configuration r^{3N}, may be obtained without any reference to the momenta:

$$\langle f \rangle = \frac{1}{Z_N} \int d^{3N} r f(r^{3N}) \exp\left[-\frac{\Phi_N(r^{3N})}{k_B T}\right].$$

You should have no problem to prove this by integrating over the momenta once in the numerator and once in the denominator.

We will treat one simple example to illustrate the complexity of calculating configuration integrals. Consider a system of roughly spherical molecules. The potential energy then reads

$$\Phi_N(r^{3N}) = \sum_{i=1}^{N-1} \sum_{j=i+1}^{N} \varphi(r_{i,j}),$$

with r_{ij} being the distance between molecules i and j. In many cases the interaction between two molecules can be described by the Lennard-Jones potential:

$$\phi(r) = 4\epsilon\left[\left(\frac{\sigma}{r}\right)^{12} - \left(\frac{\sigma}{r}\right)^{6}\right],$$

with ϵ being the depth of the potential and σ the diameter of the molecule.

Before going to our example, let us first give you one more bit of formal development, which allows us to understand the concept of chemical potential a bit better.

7.4.1 Chemical Potential

It is instructive to have a look at the statistical expression for the chemical potential. Recall that

$$\mu = \left(\frac{\partial F}{\partial N}\right)_{T,V},$$

from which

$$\mu = k_B T \ln \Lambda^3 + k_B T \ln N - k_B T \frac{\partial \ln Z_N}{\partial N}.$$

Approximating

$$\frac{\partial \ln Z_N}{\partial N} = \ln Z_{N+1} - \ln Z_N = \ln \frac{Z_{N+1}}{Z_N}$$

we get

$$\mu = k_B T \ln \left(\frac{N}{V}\Lambda^3\right) - k_B T \ln \left(\frac{1}{V}\frac{Z_{N+1}}{Z_N}\right).$$

The last term in this equation has an interesting meaning. Let us write

$$\Phi_{N+1}(r^{3(N+1)}) = u(r^{3N}, \vec{r}_{N+1}) + \Phi_N(r^{3N}),$$

where $u(r^{3N}, \vec{r}_{N+1})$ is the interaction between particle $N + 1$ and the remaining first N particles and $\Phi_N(r^{3N})$ represents the sum of the interactions between the first N particles. We then write

$$\frac{1}{V}\frac{Z_{N+1}}{Z_N} = \frac{1}{V}\int d^3 r_{N+1} \frac{\int d^{3N} r e^{-\beta u(r^{3N}, \vec{r}_{N+1})} e^{-\beta \Phi_N(r^{3N})}}{\int d^{3(N-1)} r e^{-\beta \Phi_{N-1}(r^{3(N-1)})}} = \langle e^{-\beta u}\rangle.$$

This means that we take the average of $\exp(-\beta u(r^{3N}, \vec{r}_{N+1}))$ on all configurations of the first N particles. Next the operator $(1/V)\int d^3 r_{N+1}$ averages this over all positions of the $(N + 1)$th particle. So the result is the average of the exponent of minus the work to bring an additional particle into a system of N particles divided by $k_B T$, hence the notation with the pointy brackets. Next we take minus the logarithm and multiply by $k_B T$:

$$-k_B T \ln \left(\frac{1}{V}\frac{Z_{N+1}}{Z_N}\right) = -k_B T \ln \left\langle e^{-u/k_B T}\right\rangle$$

Roughly stated, we have calculated the average of the work to be performed to bring the particle from vacuum into the system. The particular way of averaging, *i.e.* by first exponentiation and next taking the logarithm, is to get rid of infinities. Without exponentiation, particles put into the system such as to overlap with one of the original particles would contribute infinitely much, while now they contribute zero to the average.

Combining our results, we obtain

$$\mu = -k_B T \ln\left(\frac{V/N}{\Lambda^3}\right) - k_B T \ln\left\langle e^{-u/k_B T}\right\rangle.$$

The chemical potential is more negative the larger the volume per particle and the lower the energy per particle.

7.4.2   Van der Waals' Equation of State

In order to calculate the configuration integral we divide the system's volume into small cubes of size $\Delta = \sigma^3$. The integral may then be replaced by a sum over all configurations obtained by distributing all molecules over the various little cubes:

$$Z_N = \sum_{conf} e^{-\beta\Phi_N(conf)}\Delta^N.$$

Configurations with cubes containing more than one molecule have energies equal to infinity and will not contribute to the sum. So the sum must run over all configurations with no cube containing more than one molecule. Next, we assume that the range of the attractive well in the potential energy contribution of a pair of molecules is less than σ. So, only pairs of molecules which are nearest neighbors will contribute to the total potential energy. This model may strike you as extremely simple, yet its configuration sum cannot be calculated exactly except for two dimensional systems. The mathematics to perform the sum in the two dimensional case is extremely involved and there is no hope that it will be ever extended to three dimensional systems.

Let's go for the next approximation. We will assume that all configurations have roughly the same energy $\bar{\Phi}_N$. The configuration integral then becomes

$$Z_N = \Omega e^{-\beta\bar{\Phi}_N}\Delta^N,$$

with Ω being the number of configurations not having any cube being occupied more than once. Now the problem becomes simple:

$$\Omega = \frac{M!}{(M-N)!}$$

$$\bar{\Phi}_N = \frac{1}{2}Nz\frac{N}{M}(-\varepsilon).$$

Here $-\varepsilon$ is the contribution to the energy of one nearest neighbor pair of molecules, and z is the coordination number of the lattice, or the number of nearest neighbors with each cube. Moreover M is the total number of cubes, i.e. $M = V/\Delta$. With these results we calculate the free energy of the system as

$$F = -Nk_BT\ln\Lambda^3 - k_BT\ln\left(\frac{M!}{N!(M-N)!}\right) - \frac{1}{2}Nz\frac{N}{M}\varepsilon - Nk_BT\ln\Delta.$$

The pressure then becomes

$$p = -\frac{k_BT}{\Delta}\ln\left(1-\frac{N\Delta}{V}\right) - \frac{1}{2}z\varepsilon\Delta\frac{N^2}{V^2}.$$

Finally we make use of an approximate equation for the logarithm

$$\ln(1+x) = \frac{1}{1+\frac{x}{2+\frac{x}{3+\frac{4x}{4+\cdots}}}} \approx \frac{1}{1+x/2}.$$

The result then is

$$p = \frac{Nk_BT}{V-Nb} - a\frac{N^2}{V^2}$$
$$a = z\varepsilon\Delta/2$$
$$b = \Delta/2.$$

This is van der Waals' equation of state. Clearly the term proportional to a accounts for attractions between molecules and the one depending on b accounts for repulsions between molecules.

7.4.3 Carnahan and Starling

Exercise:

In this exercise you will obtain the virial equation of state (to first order). To start the derivation, define the activity a by writing the chemical potential as

$$\mu = k_B T \ln \Lambda^3 + k_B T \ln \alpha.$$

From the results in Sect. 7.4.1. A we have

$$\frac{1}{\alpha} = \frac{1}{N} \int d^3 r_{N+1} \frac{\int d^{3N} r e^{-\beta u(r^{3N}, \vec{r}_{N+1})} e^{-\beta \Phi_N(r^{3N})}}{\int d^{3N} r e^{-\beta \Phi_N(r^{3N})}}.$$

The explicit equation for $u(r^{3N}, \vec{r}_{N+1})$ reads

$$u(r^{3N}, \vec{r}_{N+1}) = \sum_{i=1}^{N} \varphi(r_{i,N+1}).$$

We define

$$e^{-\beta \varphi(r)} = 1 + f(r),$$

so

$$e^{-\beta u(r^{3N}, \vec{r}_{N+1})} = (1 + f(r_{1,N+1}))(1 + f(r_{2,N+1}))...(1 + f(r_{N,N+1})).$$

We must evaluate this product and put it into the integral above. In this exercise we restrict ourselves to terms of first order in $f(r_{i,N})$, so

$$e^{-\beta u(r^{3N}, \vec{r}_{N+1})} = 1 + f(r_{1,N+1}) + f(r_{2,N+1}) \cdots + f(r_{N,N+1}).$$

Argue that this leads to

$$\frac{1}{\alpha} = \frac{1}{\rho} + \int d^3 r_{N+1} \frac{\int d^{3N} r f(r_{N,N+1}) e^{-\beta \Phi_N(r^{3N})}}{\int d^{3N} r e^{-\beta \Phi_N(r^{3N})}}.$$

Here $\rho = N/V$, i.e. the number density of the system. Now comes the trick. You must prove that this may be written as

$$\frac{1}{\alpha} = \frac{1}{\rho} - 2B_2,$$

Where B_2 is given by

$$B_2 = -\frac{1}{2} \int d^3 r f(r) = \frac{1}{2} \int d^3 r \left(1 - e^{-\beta \varphi(r)} \right).$$

Conclude that to first order

$$\alpha = \rho + 2B_2\rho^2$$
$$\rho = \alpha - 2B_2 z^2.$$

Now comes another trick:

$$p = \int_0^\alpha d\alpha \left(\frac{\partial p}{\partial \alpha}\right)_T = \int_0^\alpha d\alpha \left(\frac{\partial p}{\partial \mu}\right)_T \left(\frac{\partial \mu}{\partial \alpha}\right)_T = \int_0^\alpha d\alpha \rho \frac{k_B T}{\alpha}.$$

We have used the Gibbs-Duhem relation $(\partial p/\partial \mu)_T = \rho$. Try to find this relation in these notes. Next plug in the equation for the density that we just have found and perform the integrations to get

$$\frac{p}{k_B T} = \alpha - B_2 \alpha^2.$$

Finally plug in the equation for α and conclude

$$\frac{p}{\rho k_B T} = 1 + B_2 \rho.$$

This is the virial equation of state to first order in the density.
Next we consider a so-called hard sphere gas, with interactions

$$\varphi(r) = \infty \quad r \leq \sigma$$
$$= 0 \quad r \geq \sigma$$

Calculate B_2 for this gas.
In the seventies of the last century, theoretical physicists have extended the analysis to higher powers of the density and found

$$\frac{p}{\rho k_B T} = 1 + 4\eta + 10\eta^2 + 18.365\eta^3 + 28.24\eta^4 + 39.5\eta^5 + 56.6\eta^6 + \cdots$$

Here η is the volume fraction of hard spheres:

$$\eta = \frac{\pi}{6}\sigma^3 \frac{N}{V}.$$

Our friends, the theoreticians, even went a bit further but calculating coefficients of large powers of η became increasingly difficult. Then came two engineers who found that the coefficients of η^n can be well approximated by $n^2 + 3n$. Assuming this to be correct to all orders and summing the series to infinite order they found

$$\frac{p}{\rho k_B T} = 1 + \sum_{n=1}^{\infty} (n^2 + 3n)\eta^n = \frac{1 + \eta + \eta^2 - \eta^3}{(1 - \eta)^3}.$$

Can you do the summation? By the way, the engineers were called Carnahan and Starling.

Finally, let us give the free energy for the hard sphere system:

$$A = -Nk_B T \ln\left(\frac{e}{\Lambda^3} \frac{V}{N}\right) + Nk_B T \frac{\eta(4 - 3\eta)}{(1 - \eta)^2}.$$

Show that this indeed gives rise to the equation of state given above.

Chapter 8
Interfaces of Binary Mixtures

Reinhard Sigel

Abstract Methods to derive an interface concentration profile in a two component system are discussed on the basis of squared gradient theories. Starting point is the description of soft matter systems, where the correlation length of fluctuations becomes the relevant length scale. A phase diagram which contains bulk and interface phase transitions is used as a road map to the involved phenomena. The LANDAU theory and the FLORY-HUGGINS theory as a typical representative of soft matter mean field theories are outlined as motivations for the squared gradient approach. A brief discussion of bulk properties forms the basis for the discussion of the interface profile of a two component system and the wetting behavior of this system at a substrate.

8.1   Introduction: Soft Matter at Interfaces

This contribution discusses the application of concepts of interface science to a representative soft matter system. While interface science is based on a statistical mechanics language, it usually does not specify a palpable formula for the free energy of a specific system. The concepts thus remain on an abstract and formal level. With an atomistic length scale in mind with detailed and complicated interactions, it might be difficult to write down explicitly a system free energy which is simple enough for further calculations. For soft matter systems, on the other hand, effective statistical mechanical models of suitable simplicity do exist, and thus it is possible to apply the interface formulas. It is not the aim of the soft matter models to describe the atomic length scale, since a description of polymer chains, liquid crystals, or colloids based on first principles would be even more complicated than the description of ensembles based on atoms or low molecular weight molecules. Instead, only the most relevant properties are included in a soft matter model, while other molecular details are summarized as effective parameters. An example are thermotropic nematic liquid crystals, where anisotropic molecules in a liquid state might either self organize with a preferential orientation in a nematic phase, or with random orientation in

R. Sigel (✉)
German University in Cairo (GUC), New Cairo 11835, Egypt
e-mail: reinhard.sigel@guc.edu.eg

© Springer International Publishing Switzerland 2016
P.R. Lang and Y. Liu (eds.), *Soft Matter at Aqueous Interfaces*,
Lecture Notes in Physics 917, DOI 10.1007/978-3-319-24502-7_8

the isotropic phase. The typical soft matter approach assumes these molecules as rod-like entities with only excluded volume interactions. While such a description simplifies the detailed molecular interactions between the constituent molecules to a maximum extend, it offers a good description of the nematic to isotropic phase transition. The correlation length of fluctuations ξ becomes the relevant length scale, and a description even of microscopic averages and fluctuations down to this length scale is possible and successful. A second example concerns polymer chains. The most relevant property is here the connectivity of the molecules in a polymer chain, while the molecular interactions are otherwise strongly simplified to excluded volume interactions. The only local parameters of a polymer chain which enter the models are a persistence length l_{ps} and a molecular friction coefficient. The persistence length of a polymer coil is small for flexible polymer chains, while it is large for stiff polymer chains. The molecular friction coefficient is required to access dynamic phenomena like the relaxation dynamics. On this simplified basis, advanced approaches like the reptation model are able to describe phenomena which are as complicated as the visco-elasticity of polymer melts in rheological experiments.

The application of interface science concepts to soft matter system offers traceable models which describe interfaces on the soft matter length scale ξ. Beside elucidating the interface concepts on specific examples, another important topic of soft matter enters the description of interfaces. For soft matter, suitable degrees of freedom show weak and "soft" restoring forces which are driven by the thermal energy $k_B T$ only. Here, k_B is the BOLTZMANN constant and T the absolute temperature. An example are again thermotropic liquid crystals. The wide technical use of these systems in liquid crystalline displays relies on the possibility to change the direction of preferred alignment in a nematic phase easily by an electric field of moderate strength. The orientation of this direction which is called liquid crystalline director is a soft degree of freedom.

A simple second example is a rubber band, i.e. a weakly cross-linked polymer melt. A rubber band is easily stretched by moderate mechanical forces. Restoring forces to bring the rubber band back to its original length are caused by entropy elasticity, not by molecular interactions. The stretching of the rubber band is also a soft degree of freedom. It can be stated that the technical important properties of soft matter are mostly due to the soft degrees of freedom in these materials. On this background, a specific question of the **SOMATAI** initial training network is to identify soft degrees of freedom of these materials at the interface. Those degrees of freedom will react to moderate external forces with a strong response, which might become the starting point of new technological applications.

Which degrees of freedom at interfaces could be soft? So, for which degrees of freedom a deviation from the equilibrium position experiences only weak restoring forces? One can think about a polymer brush, or the two dimensional analog systems of a nematic phase or a polymer melt. And how could we investigate experimentally the softness of these systems? For an answer, we refer to another consequence of softness. Since the restoring forces are weak, the thermal energy $k_B T$ leads to a sizable activation of the soft degrees of freedom. Thus, thermal fluctuations have a significant magnitude in soft matter system. Such fluctuations can be detected by a scattering experiment, in case the fluctuations are connected to the scattering contrast (see Chap. 11 by A. C. Völker et al. and Chap. 12 by J. Daillant). Static scattering experiments yield the root mean squared thermal amplitude of a fluctuation, while dynamic scattering experiments (e.g. dynamic light scattering) provide the relaxation dynamics of a fluctuation, which in many cases is an exponential decay with a relaxation time τ. The scattering experiment sets up the scattering vector $\mathbf{q}$, which defines the wave length $2\pi/|\mathbf{q}|$ and the geometry of the fluctuations. At an interface, it is of importance to distinguish the wavelength $2\pi/q_{\parallel}$ of a fluctuations within the interface plane, described by the parallel component $q_{\parallel}$ of $\mathbf{q}$, and the fluctuation extend perpendicular to the interface. Interface-bound fluctuations which penetrate the bulk only to a limited extend are usually not well characterized by the perpendicular component $q_{\perp}$ of $\mathbf{q}$, even if the complex nature of this parameter in interface sensitive scattering experiments is taken into account [1]. Still, $q_{\perp}$ determines the weighting with which fluctuations with different profiles contribute to a scattering experiment.

Most interfacial degrees of freedom are not soft and are only weakly excited by the thermal energy. An example are capillary waves at a liquid-fluid interface. Since a wave enhances the interface area, capillary waves are suppressed by the interface tension γ, especially at short wavelengths (large $q_{\parallel}$). For large wavelengths, it is the density difference between the two phases which suppresses capillary waves and yields a flat, horizontally oriented interface. For extremely low interface tension, also capillary waves become soft and reach a sizable thermal amplitude [2]. In liquid crystalline displays, it is the non-soft interface anchoring of the preferred nematic orientation at the interfaces of a display device which provides the restoring force for the bulk orientation.

A suitable soft matter system to elucidate and study concepts of interfacial science and interfacial fluctuations are mixtures of polymers A and B, described by the FLORY-HUGGINS theory. Depending on the interaction strength between the two polymers which form the mixture, there is either a one-phase, mixed system, or a phase separation into two phases. The theory is usually applied for the description of bulk phases, or phase de-mixing kinetics in the bulk. Here, we study interface effects of the theory. The mixture is brought in contact with an interface of a third material, and the wetting transitions between partial and complete wetting will be investigated. For partial wetting, a non-zero contact angle is formed, while complete wetting is characterized by contact angle zero. Furthermore, there are internal interfaces between the two phases. Depending on the width of the interface region, one distinguishes a weak segregation limit (WSL) with a smooth transition, a strong segregation limit (SSL) with a more sharp transition and even a super strong segregation limit (SSSL) with a step-like concentration change. These cases are also important in the description of block-copolymer systems, where A and B are not separate polymer coils, but linked blocks of a common polymer coil.

For a theoretical description of soft matter systems, the application of mean-field theories offers a first approach. These theories are often based on intuitive arguments. The implemented physical mechanisms which determine the behavior of a theory can thus be followed directly. The essential simplification in a mean field theory is the introduction of the mean field, which covers the multi-particle interactions of the constituents on a simplified basis. These interactions are pre-averaged, and form the mean field. The theoretical description of a complicated many particle system is reduced in this way to the description of the single particle behavior in the presence of the mean field. The single particle behavior is dependent on the strength of the mean field. On the other hand, the strength of the mean field is a suitable average over the single particle behavior. The combination of these two dependencies leads to an equation, where the mean field is calculated as an average which depends on the mean field. This closure of the mean field theory is the essential step, and its solution is called a self-consistent solution. We can take advantage of the common background of mean field theories for different soft matter classes and will apply concepts which were developed for liquid crystals to polymer mixtures described by the FLORY-HUGGINS theory.

The mean field concept fails when the description by averaged interactions is not appropriate. This situation happens when there are large fluctuations within a system, especially very close to a second order phase transition, which is also called a critical point. Another system known to have large fluctuations is a semi-dilute polymer solution. Due to the large fluctuations, the concept of an averaged local mean field is no longer appropriate and no longer successful. Very close to a critical point, there is interesting physics which can be described by different theoretical concepts, e.g. scaling arguments. By inverting the failure argument of mean field theories into a positive criterion it can be stated, that apart from conditions very close to critical points and other possible situations with strong fluctuations, mean field theories describe a soft matter system usually rather well. This finding might be traced back to a coincidence: for technical applications, we are mainly interested in systems with soft matter properties around room temperature. The temperature of the critical point T_{cp} is usually also in this range, let's say around $T_{cp} \sim 300$ K. When we perform measurements with a temperature deviation ΔT from the critical point, the relative deviation $\Delta T/T_{cp}$ typically remains small. A deviation of $\Delta T \sim 60$ K which would be an impressing range for a nematic phase might appear large at first sight, however, the correlation length of fluctuations is still enhanced compared to the molecular length scale by the proximity of the critical point $\Delta T/T_{cp} \sim 0.2$. Thus, a theoretical description on the length scale ξ based on suitable averages of molecular interactions is sufficient and successful. On this basis, it is expected that soft matter interfaces can also be described successfully down to a length scale ξ by mean field theories, as long as one does not approach the critical point too closely. Equivalent to mean field theories is the LANDAU theory, or the squared gradient theory introduced by VAN DER WAALS.

The outline of this contribution is as follows: as a start, Sect. 8.2 provides an introduction to wetting transitions, with a general phase diagram of bulk and interface phase transitions. As a second step, the LANDAU theory and the FLORY-HUGGINS theory are briefly introduced in Sect. 8.3, as examples of squared gradient theories. The FLORY-HUGGINS theory describes a similar phase diagram, where, however, the temperature is replaced by the interaction parameter χ. For the equivalent CAHN HILLIARD theory, only few literature hints will be provided. Since interfaces are boundaries of a bulk phase, a brief outline of bulk properties in Sect. 8.4 forms the basis of a discussion of interfaces, which will be treated in Sect. 8.5. A general approach for the calculation of interface properties within the squared gradient theories is illustrated for the LANDAU theory.

8.2 Wetting Transitions

The phase diagram of a binary mixture might serve as a roadmap to interfacial phenomena. Such a mixture is composed of two constituents A and B. Volume additivity is assumed and the volume V contains the volume fraction ϕ of B and $1 - \phi$ of A. Attractive interactions are usually stronger between molecules of the same kind [3], and thus the internal energy contribution U to the free energy $F = U - TS$ favors a de-mixing of the mixture. The entropic contribution S, on the other hand, is higher in a mixed state. Since the entropy is weighted with the temperature T, it wins at high T and there is generally a mixed state at high T. At low T, interactions might lead to a de-mixing. We follow the discussion of Bonn and Ross for the description of wetting phenomena [4], and combine it with physical arguments provided by Strobl [5]. The shape of the phase diagram is depicted in Fig. 8.1.

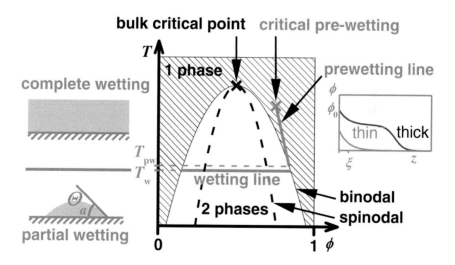

Fig. 8.1 Phase diagram of a binary mixture. The bulk behavior is illustrated by *black lines*, while the interface behavior is drawn in *grey*. The *left two insets* illustrate the cases of complete wetting for a temperature above the wetting temperature T_w and partial wetting for $T < T_w$. The *right inset* displays the difference between a *thick* and a *thin* interface layer, which occur at the *left* and the *right side* of the pre-wetting line, respectively

8.2.1   Bulk Behavior of the Mixture

We first focus on the black lines and labels in Fig. 8.1 which indicate the bulk behavior of the mixture. In the shaded one phase region, the A-B blend is in the mixed state. The border to the de-mixed state is indicated by a line which is called binodal $T_{bin}(\phi)$. A system prepared at a point (ϕ, T) which falls into the two phase region is not thermodynamically stable. Phase separation sets in and two phases are created. One phase is rich in A and the other rich in B. Entropy effects are still present and no pure phases are created in general. The resulting compositions of the two phases are found on the two points of the binodal curve which are at the temperature set by the experiment. So, for a system prepared at point (ϕ, T) in the two phase region, one finds the compositions of the two resulting phases by moving at the same T to the left and to the right until the binodal is hit. In order to find the volumes occupied by the A-rich phase and the B-rich phase, one can use the known compositions of the de-mixed phases and determine the volumes of the A-rich phase and the B-rich phase in a way so the total content of A and B match the original preparation conditions (ϕ, T).

The spinodal line $T_{spin}(\phi)$ in the two phase range separates two regions with different mechanisms of phase separation. In the region between the binodal and the spinodal, the mixture is meta-stable. It might remain mixed for a while, although the mixed state is not the thermodynamic equilibrium. Phase separation occurs here via nucleation and growth. A thermal fluctuation might lead to a large enough volume element with surplus of one component, let's say A. This fluctuation acts as a nucleation point which can initiate a phase separation of the whole system. The formation of such a nucleus involves the creation of an interface between the A-rich fluctuation and the remaining system. A small nucleus has an unfavorable interface to volume ratio and so the fluctuation rather decays than initiates the phase separation. Only if the fluctuations produce a large enough nucleus it survives and initiates macroscopic phase separation. Note, that this mechanism of meta-stability involves interface effects between the nucleus and the surrounding. The meta-stability indicates, that the de-mixing is a first order phase transition.

In the region below the spinodal, a mixture is unstable and decays immediately. Concentration fluctuations of all wavelengths no longer have restoring forces, but are even enhanced by thermodynamics which favors de-mixing for these unstable situations. The interplay of diffusion which requires more time to set up a concentration fluctuation of large wavelength and thermodynamic driving force which increases with the wavelength of a fluctuation results in a de-mixing, where the wavelength of the fastest growing concentration fluctuation sets the length scale where de-mixing starts. In later stages of de-mixing, there is a coarse graining and building up of better defined interfaces between A-rich and B-rich regions.

The binodal and the spinodal meet in the phase diagram in the bulk critical point. In this point, the de-mixing becomes a second order phase transition. It is here that strong concentration fluctuations occur, since restoring forces are weak. In the region very close to the critical point, mean-field theories are no longer suitable and scaling arguments can predict the right behavior. When approaching the critical point, the difference in the A-rich phase and the B-rich phase gets smaller and finally vanish (recall that the de-mixing process leads to an A-rich phase and a B-rich phase which are the two points of the binodal at the same temperature as the original instable mixed state; at the critical point, the compositions of these phases become identical). For a temperature above the critical phase transition temperature, a distinction between the two phases is no longer possible. The situation is in analogy to the critical point of a single phase system. When the pressure is varied, the first order boiling transition becomes second order at the critical pressure, and for higher pressure it is no longer possible to distinguish liquid and vapor. The bulk critical point is of special interest for experimentalists. At this value of ϕ, a T jump starting from the one phase region to the two phase region directly reaches a composition where spinodal de-mixing can be observed, without the necessity to cross a region of nucleation and growth in the T-jump.

8.2.2 Interface Behavior of the Mixture

When the binary mixture gets into contact with a substrate—either the container, a test sample, a colloidal particle, or a gas phase—either A or B molecules are preferentially absorbed at the interface. The interface behavior of the mixture is included to Fig. 8.1 in grey color [6], where we assume that the interactions result in a preferable adsorption of A to the interface. The two phase region is split by the horizontal wetting line, which separates regions of partial and complete wetting. For complete wetting, a film of macroscopic thickness of one phase covers the interface—for our assumption of preferential adsorption of A molecules it is the A-rich phase. For partial wetting, the interface might be covered by a microscopic thin film of the component with preferential adsorption, however, no homogeneous film of macroscopic thickness evolves. Instead, drops of the preferentially adsorbed phase are observed at the interface. The 'CAHN argument' established by Cahn in 1977 indicates, that close to the bulk critical point there is always complete wetting [6]. The transition of partial to complete wetting is an interface phase transition. Usually, it is of first order, which implies a sudden, non-steady change in the thickness of the absorbed layer from microscopic to macroscopic thickness. A further consequence of the first order nature of this interface phase transition is the existence of meta-stable states, so a meta-stable microscopic thin film and partially wetted interface for $T > T_w$ or a meta-stable macroscopically thick film for $T < T_w$ (as a bulk analogy consider the first order boiling transition of water, where superheating of water and supercooling of vapor in the absence of nucleation sites are possible meta-stable states). Also a thermodynamic contribution (beside pinning) for the difference in advancing and receding contact angles of a liquid drop on a solid substrate can be discussed on the basis of meta-stability. The CAHN argument was discussed by Bonn and Ross, who also reviewed scientific papers describing a continuous wetting transition which can be achieved under certain, suitable conditions.

The preferential adsorption of one component to the substrate persists into the one phase region. The typical size of thermal fluctuations of the local composition defines the correlation length ξ. This length is small and compares to molecular dimensions for a location (ϕ, T) far away from the bulk critical point, however, it increases and diverges when approaching the bulk critical point. The preferential adsorption of one component at a substrate can be compared to a bulk fluctuation, and so the thickness of a resulting interface layer is equal to ξ. More exactly, an exponential concentration profile is found at the interface with a decay length equal to ξ. Close to the bulk critical point where fluctuations become enhanced, the preferential interface adsorption of one component can thermodynamically stabilize a fluctuation, and an interface layer with thickness larger than ξ evolves. For our assumption of preferential A adsorption to the substrate, such thick layers are found on the B-rich side of the phase diagram in Fig. 8.1 in case the temperature is larger than the pre-wetting temperature T_{pw} and only slightly above $T_{bin}(\phi)$. The interface phase transition within the bulk one-phase region between a thin film with thickness ξ and a thick film is called pre-wetting transition. It can be either a first or second order transition, or a supercritical change. When changing ϕ, the first order pre-wetting transition occurs for different temperatures. The trace of these temperatures of first order pre-wetting transitions is captured by the pre-wetting line. The pre-wetting line ends at the critical pre-wetting point, where the pre-wetting transition becomes a second order phase transition. For compositions ϕ closer to the bulk critical point and higher temperatures, there is a super-critical continuous change from a thin to a thick layer. In Fig. 8.1, the starting temperature T_{pw} of the pre-wetting line is distinguished from T_w. Usually, these two temperatures coincide, and in this case the wetting transition is of first order. It turns out, however, that for the LANDAU theory which is discussed as main example in Sect. 8.5.2 a different behavior with $T_w < T_{pw}$ is encountered. For this case, there is no jump in the contact angle like in a first order wetting transition, but a continuous change from zero to finite values.

A broad overview over the interface behavior of polymers also from the experimental side is provided by the book of Jones and Richards [7]. Examples of experimental and theoretical results of wetting on colloidal particles are found in references [8, 9].

8.2.3 The Contact Angle

For completeness, the connection between contact angle and interface tension is briefly repeated. The interface tension γ between two phases can be expressed either as specific interface energy with the dimension energy per interface area, or as force per width. The second option, which indicates the required force to increase the size of an interface of fixed width is used to motivate the YOUNG DUPRE equation

$$\gamma_{bS} = \gamma_{aS} + \gamma_{ab}\cos(\Theta_a). \tag{8.1}$$

Fig. 8.2 Balance of the force per length of an A-rich phase and a B-rich phase at a solid substrate S

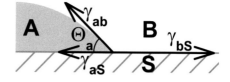

Here, γ_{ab}, γ_{aS}, and γ_{bS} are the interface tensions between the A-rich phase and the B-rich phase, the A-rich phase and the substrate, and the B-rich phase and the substrate, respectively. For the usage of indices, we generally use capital letters A and B for the pure A and B phase only, while the A-rich and the B-rich phases are indicated by the lower case letters a and b. The contact angle inside the A-rich droplet is denoted as Θ_a. Figure 8.2 illustrates the acting forces: the force per width to the left composed of γ_{aS} and the component parallel to the interface $\gamma_{ab} \cos(\Theta)$ is counter balanced by the force per area γ_{bS} to the right. Only for $\gamma_{bS} < \gamma_{aS} + \gamma_{ab}$, (8.1) can be used for the calculation of the contact angle, while for $\gamma_{bS} > \gamma_{aS} + \gamma_{ab}$ complete wetting of the A-rich phase occurs. This relation is the basis for the spreading coefficient of the A-rich phase

$$S_a = \gamma_{bS} - \gamma_{aS} - \gamma_{ab}. \tag{8.2}$$

For a positive value of S_a, a drop of A-rich phase will tend to cover the interface to the substrate. For any contact of the B-rich phase with the substrate, the interface energy can be reduced by replacing the interface of the substrate to B-rich phase by two interfaces, one of them substrate to A-rich phase and the other one A-rich phase to B-rich phase. In analogy to (8.2), a spreading coefficient S_b for the B-rich phase can be defined in order to decide, if there is complete or partial wetting of the substrate by the B-rich phase. Here, we always assume that the A-rich phase preferentially adsorbs to the substrate. Bonn and Ross argue that the spreading coefficient in equilibrium cannot attain positive values, since any contact of the B-rich phase will have been eliminated after complete wetting [4]. Such a concept does not fit to the definition (8.2), but rather an alternate quantity defined as $S_a' = \min(S_a, 0)$. We stick to (8.2) as definition, interpret the spreading coefficient as thermodynamic tendency to wet an interface, and tolerate positive values of S_a. Meta-stable partial wetting states and pinning can lead to situations with positive S_a without complete wetting.

8.3 Squared Gradient Theories

There are several approaches to justify the phenomenological description of Sect. 8.2 and the roadmap of Fig. 8.1. On a mean-field level and for short ranged interactions, they typically lead to an increment of a thermodynamic potential of the form:

$$\Delta\Omega[\phi] = \int_V \left\{ \Delta\omega\left(\phi(\mathbf{r}), T\right) + \tfrac{1}{2}\kappa\left[\nabla\phi(\mathbf{r})\right]^2 \right\} d^3r. \tag{8.3}$$

Here, $\Delta\Omega$ is an increment of the grand canonical potential and $\Delta\omega$ is its density, so the grand canonical potential increment of a volume element divided by the size of this element. The free energy increment ΔF can be expressed by the free energy density Δf and the squared gradient of ϕ similar to (8.3). In the description based on increments $\Delta\Omega$ or ΔF it is not required, to consider other contributions which are independent of ϕ. Any constant background terms in addition to these increments do not affect the subsequent calculations which are based on derivatives of $\Delta\Omega$ or ΔF. The result of (8.3) depends on the composition profile $\phi(\mathbf{r})$, as indicated by the squared brackets on the left side. A system optimizes $\phi(\mathbf{r})$ for a minimal value of $\Delta\Omega$ or ΔF. A brief introduction to the mathematical tools to perform this optimization is provided in the appendix, Sect. 8.8.

Interactions between neighboring volume elements are considered by the second term in the integral (8.3), which involves a phenomenological elastic constant κ. In bulk, a system in thermodynamic equilibrium is in a homogeneous state, i.e. a state with a constant value of ϕ for all locations $\mathbf{r}$ and thus vanishing gradient $\nabla\phi$ everywhere. Deviations from a homogeneous ϕ result in a higher value of $\Delta\Omega$. The increase in $\Delta\Omega$ for inhomogeneous states is described by the square of $\nabla\phi$ in (8.3). The square can be considered as the second term in a TAYLOR expansion, where the zero and first order terms vanish in order to match the condition of minimum bulk value of $\Delta\Omega$ for a homogeneous state. Since only short range interactions are assumed in the theory, higher terms in the TAYLOR series can be neglected. In different fields, (8.3) is addressed either as LANDAU-GINZBURG functional [5], or squared gradient theory [4], which is traced back to VAN DER WAALS. This section provides an overview over two famous approaches which lead to (8.3). A third approach with very similar arguments is the CAHN HILLIARD theory [10–12]. A historical overview and an application to simulations is described by Lamorgese et al. [13].

8.3.1 LANDAU *Theory*

LANDAU **Free Energy Density** The canonical approach to statistical mechanics (see also Chap. 7 by W.J. Briels and J.K.G. Dhont) starts with the calculation of the partition function Z, and the free energy F results as

$$F = -k_{\mathrm{B}} T \ln(Z). \tag{8.4}$$

Here, k_{B} is the BOLTZMANN constant. When the system is divided to cells, and E_i denotes the energy contained in cell i, Z reads [14]

$$Z = \sum_i \exp\left\{-\frac{E_i(\phi)}{k_{\mathrm{B}} T}\right\}. \tag{8.5}$$

Instead of the sum over the cells, one can integrate over the distribution ρ of ϕ values in the system

$$Z = \int d\phi \, \rho(\phi) \exp\left\{-\frac{E_i(\phi)}{k_B T}\right\}. \tag{8.6}$$

An improvement of (8.6) takes interactions between the cells into account. Due to the assumed short range nature of the interactions, it is sufficient to consider interactions between neighbouring cells. The interaction between two neighboring cells scales with the difference of their ϕ values divided by the cell size. For neighboring cells with identical ϕ, the interaction is zero, while a deviation in ϕ leads to a positive contribution. In the limit of infinitesimal small cell size, this parameter reduces to the derivative $\nabla\phi$ of $\phi(\mathbf{r})$. The weighted sum over different profiles $\phi(\mathbf{r})$ is achieved by a functional integration $\mathbf{D}\phi(\mathbf{r})$ over different profiles $\phi(\mathbf{r})$, which leads to

$$Z[\phi] = \int \mathbf{D}\phi(\mathbf{r}) \exp\left\{-\frac{1}{k_B T} \int_V d^3r \left[f(\phi(\mathbf{r}), T) + E'(\phi, \nabla\phi, T)\right]\right\}. \tag{8.7}$$

The function

$$f(\phi, T) = E(\phi) - k_B T \ln(\rho(\phi)), \tag{8.8}$$

which is called LANDAU free energy density, takes energetic contributions $E(\phi)$ and entropic contributions $k_B T \ln(\rho(\phi))$ into account. Inhomogeneities are considered in (8.7) by the gradient term $E'(\phi, \nabla\phi, T)$. The latter can be expressed as a TAYLOR series in $\nabla\phi$. The zero order term of this series vanishes, as the interaction between neighboring cells with the same ϕ is zero. The first order term as well as higher odd order terms can be excluded by a symmetry argument. If they would have a non-zero value, it could be reverted by mirroring the coordinates. Since the thermodynamic properties do not depend on the choice of the coordinates, such contributions can be excluded. For weak gradients, $E'(\phi, \nabla\phi, T)$ is usually reduced to the second order term $\frac{1}{2}\kappa(\nabla\phi)^2$, which leads to the squared gradient term in (8.3). The elastic constant κ might depend on ϕ and T.

It is now required to show that F in the conventional equation (8.4) and the integration over f have the same information content. With this equivalence, it is possible to consider f as a free energy density and to discuss a system based on f instead of F. The exponential function in (8.7) is at maximum for the value of ϕ which yields the minimum value of f. The maximum is very sharp, since the integration in the exponential function is over V, which can be made very large in the thermodynamic limit. Any deviation from the minimum of f is multiplied by a large volume factor and weighted exponentially. For this reason, the sharp maximum of the exponential function and thus the minimum of f dominate the integration in (8.7) and thus determine the thermodynamic behavior of the system.

As a quantitative example, we consider the homogeneous mixed bulk state described by the equilibrium value ϕ_{eq}. It is not required to consider the squared gradient term in the discussion of the bulk equilibrium, since this term increases f for inhomogeneous states, which are thus away from the minimum of f. Due to this simplification, the functional integration in (8.7) is only over constant paths and thus reduces to a conventional integration. An expansion of f close to the minimum reads

$$f(\phi) \approx f(\phi_{eq}) + \frac{1}{2}\frac{d^2 f}{d\phi^2}\left(\phi - \phi_{eq}\right)^2. \tag{8.9}$$

Since ϕ_{eq} is the equilibrium value, there is no linear term in the expansion (8.9). Inserting (8.9) into (8.7) results in

$$Z = \int d\phi \, \exp\left[-\frac{V}{k_B T}\left(f(\phi_{eq}) + \frac{1}{2}\frac{d^2 f}{d\phi^2}\left(\phi - \phi_{eq}\right)^2\right)\right]. \tag{8.10}$$

The integration in (8.10) contains a Gauss function with maximum at ϕ_{eq} which can be made arbitrarily narrow in the thermodynamic limit $V \to \infty$. The integration leads to a constant multiple $cf(\phi_{eq})$ of $f(\phi_{eq})$, and a combination of (8.4) and (8.10) results in $F = Vf(\phi_{eq}) + \ln(c)$. The effect of c is just a shift of the reference point of F which has no influence on thermodynamic properties, and thus, F and f have the same information content.

LANDAU **Assumption** Within the LANDAU approach, the transition between an ordered, interaction dominated state at low T (here, a state within the two phase region) and a disordered, entropy dominated state at high T (here, a state in the one phase region) is considered as an order to disorder transition. In many applications of the LANDAU approach, the disordered state has a higher symmetry, since symmetry operations like rotations, translations, or reflections leave the homogeneous, high temperature, disordered phase unchanged. The ordered state, on the other hand, might be not invariant under one of these symmetry operations. An example are liquid crystals, where the isotropic phase has complete rotational symmetry, while the nematic phase has only cylindrical symmetry. The theoretical description of the phase transition is usually based on an order parameter, which vanishes in the disordered state and is non-zero in the ordered state. In case of different symmetries of the two phases, it is possible to construct order parameters based on the symmetry operations. With the LANDAU assumption, $f(\phi)$ is expressed as a power series in the order parameter, and predictions for the phase transition and correlation length can be derived by simple analysis of the behavior of $f(\phi)$. We will discuss an example in Sect. 8.5.2.

8.3.2 FLORY-HUGGINS *Theory*

The theoretical description of polymer mixtures (also called polymer blends) known today as FLORY-HUGGINS theory was developed independently by Huggins and Flory

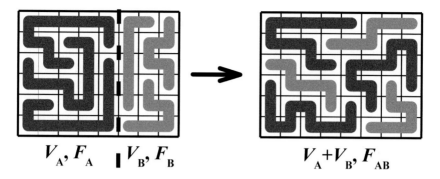

Fig. 8.3 Illustration of the FLORY-HUGGINS theory, in which polymers on a lattice are described. The free energy of mixing is the free energy difference between a configuration where polymers can mix (*right side*) and a situation where the two polymers are in separate volumes (*left side*)

[15, 16]. In its original form, it describes polymers on a lattice, as illustrated in Fig. 8.3. Strobl presents this theory in the form of a mean field theory with only limited connections to the lattice background and then transforms it to the form of a LANDAU theory [5]. We briefly summarize his approach to provide a second path to squared gradient theories.

FLORY-HUGGINS as Mean Field Theory The polymer blend is composed of n_A polymer chains of A monomers with chain length N_A and n_B polymer chains of B monomers with chain length N_B. As a remainder of the original lattice theory, a cell volume v_c and the number of nearest neighbors z_{eff} enter the theory. So, N_A and N_B are not necessarily the number of chemical identical repeat units in a polymer chain. Instead, they are defined as the overall volume of a polymer chain divided by v_c. Since there are no restrictions for the choice of v_c, it is possible to set it equal to the volume of a chemical A monomer v_A. With this choice, N_A becomes the number of chemical repeat units in an A chain. With the volume of a chemical B monomer v_B and the number of chemical repeat units in a B chain N_B', the adapted number of B monomers for the FLORY-HUGGINS description results as $N_B = N_B' v_B / v_A$. Based on the assumption of volume additivity in the A B blend, the volumes $V_A = n_A N_A v_c$ and $V_B = n_B N_B v_c$ occupied by A and B chains, respectively, and the total volume $V = V_A + V_B$ are defined. The volume fraction results as $1 - \phi = V_A / V$, with $\phi = V_B / V$. The parameters are not independent, but fulfill the relations

$$n_A = \frac{V(1 - \phi)}{v_c N_A} \quad \text{and} \quad n_B = \frac{V\phi}{v_c N_B}. \tag{8.11}$$

A prediction of the mixing behavior of the blend is based on the free energy of mixing.[1]

---

[1] Strobl uses in his book the term free enthalpy instead of free energy, which is employed here. In order to describe experiments theoretically, the free enthalpy would be more appropriate, since experiments are in most cases performed at constant pressure, not at constant volume. However, the

$$\Delta F_{\mathrm{mix}} = F_{\mathrm{AB}} - (F_{\mathrm{A}} + F_{\mathrm{B}}). \tag{8.12}$$

Here, F_{AB} is the free energy when the A and B chains are in a common volume $V_{\mathrm{A}} + V_{\mathrm{B}}$, while F_{A} and F_{B} are the free energies for the states when the A chains are located in volume V_{A}, separated from the B chains which occupy the volume V_{B}. The two situations are depicted in Fig. 8.3.

A first energetic contribution to ΔF_{mix} arises from A-B contacts, which are formed in the mixing process. The attraction between unlike monomers is usually weaker than between the same monomers [3], so the total cohesive energy is weaker in the mixed state. Since the cohesive binding energy enters the free energy with a negative sign, the difference of (8.12) results in a positive energy contributions E_{AB} for each A-B contact. From the point of the lattice theory, there are $(1 - \phi)V/v_{\mathrm{c}}$ cells which contain an A monomer. Each of them has z_{eff} neighboring cells, and on average ϕz_{eff} neighboring cells which contain a B monomer. Similarly, one can start from the total number of cells with B monomers $\phi V/v_{\mathrm{c}}$ and determine the average number of neighbors $(1 - \phi)z_{\mathrm{eff}}$ of each B cell which contain an A monomer. When both contributions are added, the number of A-B contacts is counted twice, so we have to divide by a factor 2. As a result, the total energy contribution from A-B contacts becomes

$$\Delta U = \frac{V}{v_{\mathrm{c}}}\phi(1 - \phi)z_{\mathrm{eff}}E_{\mathrm{AB}} = \frac{V}{v_{\mathrm{c}}}\phi(1 - \phi)\chi k_{\mathrm{B}}T. \tag{8.13}$$

The second form of (8.13) introduces the FLORY-HUGGINS interaction parameter χ, which expresses the total interaction energy of a cell $z_{\mathrm{eff}}E_{\mathrm{AB}}$ in units of the thermal energy $k_{\mathrm{B}}T$. The calculation of the average number of A-B contacts in the derivation of (8.13) is a typical mean field argument. The presence of A and B monomers is considered in an averaged way. In case of strong fluctuations, the A monomers occupy correlated regions in space, as also the B monomers do. For such a case, (8.13) over estimates the number of A-B contacts, and the mean field description fails.

A second contribution to ΔF_{mix} stems from the increase of translational entropy of the chains due to the larger total volume $V_{\mathrm{A}} + V_{\mathrm{B}}$ in the mixed state, instead of only V_{A} for the A chains and V_{B} for the B chains before mixing. It is calculated in the same way as the translational entropy of an ideal gas and reads

$$T\Delta S = k_{\mathrm{B}}T\left(n_{\mathrm{A}}\ln\frac{V}{V_{\mathrm{A}}} + n_{\mathrm{B}}\ln\frac{V}{V_{\mathrm{B}}}\right). \tag{8.14}$$

(Footnote 1 continued)

FLORY-HUGGINS theory contains the volume V as explicit parameter, not the pressure P. Therefore, it is formally a free energy, not a free enthalpy. Since the volume change of polymer blends upon mixing or heating is usually negligible, there is basically no difference between the free energy and the free enthalpy.

With $V_A/V = 1 - \phi$, $V_B/V = \phi$, and (8.11), it is possible to eliminate n_A, n_B, V_A, and V_B for ϕ, N_A and N_B. The total free energy of mixing $\Delta F_{mix} = \Delta U - T\Delta S$ results as combination of (8.13) and (8.14)

$$\Delta F_{mix} = k_B T \frac{V}{v_c} \left[\frac{\phi}{N_B} \ln(\phi) + \frac{1 - \phi}{N_A} \ln(1 - \phi) + \chi\phi(1 - \phi) \right]. \tag{8.15}$$

In addition to translational entropy, polymer chains have also configurational entropy. In the FLORY-HUGGINS theory it is assumed, that the configurations for the polymer chains are not affected by the surrounding chains and remain the same for the un-mixed and the mixed state. As a consequence, the configurational entropy cancels in the difference of (8.12). In experiments on real polymer systems, in contrast, the mixing might have an effect of the configurational entropy of polymers. It is considered as an entropic contribution to χ. In (8.13), χ was introduced as a purely energetic contribution $\chi = z_{eff} E_{AB}/(k_B T)$. A completely energetically determined χ thus has a T^{-1} temperature dependence, while a deviation from this temperature dependence hints to contributions due to configurational entropy. More details can be found in [5].

The translational entropy contribution of A and B chains in (8.15) is divided by the degrees of polymerization N_A and N_B, respectively. Each chain has only 3 translational degrees of freedom, no matter how many monomers it contains; most entropy is assigned to conformational entropy of the chains, which, however, does not enter the free energy increment (8.15). As a consequence, the tendency of trans-lational entropy to induce mixing is reduced, and energetic interactions dominate in cases of long chains, leading to a de-mixed blend. Only for polymers with very similar monomers or specific interactions between A and B like H bridges, mixing might be favored. A phase diagram similar to Fig. 8.1 can be drawn for polymer mixtures, with χN_A on the y axis instead of T. For $\chi \sim T^{-1}$, the phase diagram is inverted, with a mixed state at low values of χN_A and a two phase region at high values. For symmetric polymer mixtures ($N_A = N_B$), the critical point is at $\phi = 0.5$ and $\chi N_A = 2$.

Transformation to a Squared Gradient Theory The resulting mean field formula (8.15) allows predictions of the behavior of the homogeneous bulk phase and the transition from a mixed to a de-mixed state. There is, however, no information con-tained about the spatial behavior. So, it is not possible to calculate from (8.15) the amplitude of bulk fluctuations with finite wavelength, or the wetting behavior at an interface. In both cases there are concentration gradients involved. For such kind of calculations of local properties, it is required to move from the total free energy (8.15) to a free energy density

$$\Delta f_{mix} = \frac{\Delta F_{mix}}{V} = \frac{k_B T}{v_c} \left[\frac{\phi}{N_B} \ln(\phi) + \frac{1 - \phi}{N_A} \ln(1 - \phi) + \chi\phi(1 - \phi) \right]. \tag{8.16}$$

Fig. 8.4 Illustration of
interactions between
different volume elements,
where each element is
described by the
FLORY-HUGGINS free
energy density

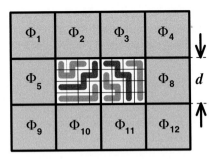

This density is not yet sufficient for a description, since also interactions of neighboring volume elements need to be considered. The situation is sketched in Fig. 8.4 with an argument very similar as for the justification of (8.7). The values of the free energy in cubic volume elements of side length d is calculated on the basis of (8.16), and interactions between neighboring volume elements i and j are considered by an additional term $d\kappa/2(\phi_i - \phi_j)^2$. This term vanishes if elements i and j have identical compositions ($\phi_i = \phi_j$), while for deviations and thus inhomogeneous concentration profiles it provides a positive free energy penalty. The lattice model free energy calculation reads

$$\Delta F = \sum_i \left\{ \Delta f_{mix}(\phi_i, T) \, d^3 + \sum_{\text{neighbors } j} \frac{d}{2}\kappa(\phi_i - \phi_j)^2 \right\}. \qquad (8.17)$$

In the limit $d \to 0$, the interaction term becomes a squared gradient, and the free energy can be written as

$$\Delta F = \int_V \left\{ \Delta f_{mix}(\phi(\mathbf{r}), T) + \tfrac{1}{2}\kappa\,[\nabla\phi(\mathbf{r})]^2 \right\} d^3 r. \qquad (8.18)$$

Equation (8.18) shows the structure of the squared gradient equation (8.3), with the explicit equation[2] (8.16) for $\Delta f_{mix}(\phi, T)$.

8.4 Bulk Behavior

A brief discussion of bulk properties provides the foundation for the description of interfaces. An excellent extended presentation of bulk properties, e.g. phase separation mechanisms is found in the book of Strobl [5].

---

[2]The formal difference between the total free energy density in (8.3) and the free energy increment to a constant background is of minor importance, as any calculation is based on derivatives of the free energy, where any constant shift of the energy scale cancels out.

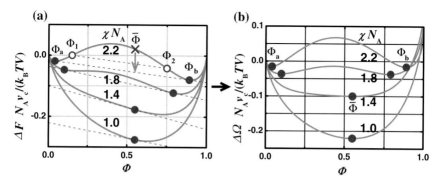

Fig. 8.5 Free energy (**a**) and grand canonical potential (**b**) for the FLORY-HUGGINS theory for $\alpha = 0.5$ and average composition $\bar{\phi} = 0.55$. The filled circles mark the equilibrium compositions

Bulk Phase Behavior Bulk phases in thermodynamic equilibrium are homogeneous. So, the gradient term in (8.4) vanishes, and the minimum of the free energy reduces to $\min F = \int_V \min[f\,(\phi(\mathbf{r}), T)]\mathrm{d}^3 r = \min[f\,(\phi(\mathbf{r}), T)]V$. Thus, a discussion of the free energy density allows for the prediction of the bulk behavior. For polymer blends, we insert the length ratio of the two polymers $\alpha = N_A/N_B$ and rewrite (8.16) as

$$\Delta f_{\mathrm{mix}} = \frac{k_B T}{N_A v_c}\,[\alpha\phi\ln(\phi) + (1 - \phi)\ln(1 - \phi) + \chi N_A \phi(1 - \phi)].\qquad(8.19)$$

While α is set by the choice of the sample polymers, the right hand side of (8.19) is a function of ϕ which depends on the parameter χN_A which changes with T. Figure 8.5a shows the behavior of Δf_{mix} for the case $\alpha = 0.5$ and different values of χN_A. A sample preparation with the average volume fraction $\bar{\phi} = 0.55$ is illustrated in the figure. This value of $\bar{\phi}$ does not match the minimum of Δf_{mix}, and it appears that the system can lower its free energy by a change of ϕ. However, since the average volume fraction is fixed to the value $\bar{\phi}$ by sample preparation, a lowering of ϕ in one region of space to a value ϕ_1 needs to be compensated by an enhancement of ϕ to ϕ_2 in another region of space. In order to find out if such a separation of two phases in two volumes V_1 and V_2 is thermodynamically stable or not, its averaged free energy needs to be calculated and compared with the free energy density of the mixed system. With $V_1 + V_2 = V$ and the fixed average $\bar{\phi} V = \phi_1 V_1 + \phi_2 V_2$, the averaged free energy density $\Delta \bar{f}_{\mathrm{mix}} = \{\Delta f_{\mathrm{mix}}(\phi_1)V_1 + \Delta f_{\mathrm{mix}}(\phi_2)V_2\}/V$ can be written as

$$\Delta \bar{f}_{\mathrm{mix}}\left(\bar{\phi}\right) = \Delta f_{\mathrm{mix}}(\phi_1) + \frac{\bar{\phi} - \phi_1}{\phi_2 - \phi_1}\left[\Delta f_{\mathrm{mix}}(\phi_2) - \Delta f_{\mathrm{mix}}(\phi_1)\right].\qquad(8.20)$$

This equation describes a straight line which connects the two points $(\phi_1, \Delta f_{\mathrm{mix}}(\phi_1))$ and $(\phi_2, \Delta f_{\mathrm{mix}}(\phi_2))$ which are lying on the graph of $\Delta f_{\mathrm{mix}}\,(\phi)$. For small values of χN_A, $\Delta f_{\mathrm{mix}}(\phi)$ in Fig. 8.5a is a convex function and it is not possible to find any pair (ϕ_1, ϕ_2) of compositions embracing $\bar{\phi}$ which lead to a connecting line with $\Delta \bar{f}_{\text{mix}}(\bar{\phi}) < \Delta f_{\text{mix}}(\bar{\phi})$. The system is thermodynamically stable as a mixed phase. For higher values of χN_A, on the other hand, $\Delta f_{\text{mix}}(\phi)$ becomes a concave function in the middle of the ϕ range. The total free energy of the blend is diminished by the transition from a mixed state to a two phase state.

The reduction of Δf_{mix} by phase separation is illustrated in Fig. 8.5a for the case $\chi N_A = 2.2$. Sample preparation for the selected composition $\bar{\phi} = 0.55$ is marked by the symbol $\times$. A separation of two phases of composition ϕ_1 and ϕ_2 has a lower average free energy density, since the line connecting these two compositions at $\bar{\phi}$ is below Δf_{mix}. A further reduction of Δf_{mix} is possible until the two phases reach the compositions ϕ_a and ϕ_b, where the connecting line is the common tangential line to $\Delta f_{\text{mix}}(\phi)$. The global analysis of the stability of the mixed state for different values of χ and the resulting values of ϕ_a for the A-rich phase and ϕ_b for the B-rich phase can be used to determine the binodal in a phase diagram similar to Fig. 8.1.

A local criterion if Δf_{mix} is convex or concave and thus if a mixed state is stable or unstable is the second derivative $\frac{\partial^2 \Delta f_{\text{mix}}}{\partial \phi^2}$. This quantity can be considered as a potential which provides restoring forces to composition fluctuations. Only as long as $\frac{\partial^2 \Delta f_{\text{mix}}}{\partial \phi^2} > 0$ there is a tendency to restore the previous composition ϕ. The spinodal in a phase diagram similar to Fig. 8.1 is determined by the root of $\frac{\partial^2 \Delta f_{\text{mix}}}{\partial \phi^2}$ for different values of χ, so by the point where restoring forces vanish. The binodal and the spinodal do not match. A composition might be meta-stable and withstand to small fluctuations, corresponding to the local criterion. Larger fluctuations, however, test if Δf_{mix} has convex behavior for the whole ϕ range.

From a thermodynamics point, the identical slopes at ϕ_a and ϕ_b indicate that the B polymers have the same chemical potentials μ_B in both phases. With (8.11), the definition $\mu_B = \left(\frac{\partial F}{\partial n_B} \right)_{T,P,n_B}$ can be rewritten as

$$\mu_B = v_c N_B \frac{\partial \Delta f_{\text{mix}}}{\partial \phi}, \qquad (8.21)$$

which is the slope of the tangential line in Fig. 8.5a. Similarly, the chemical potential μ_A of the A chains can be expressed as $\mu_A = v_c N_A \left(\frac{\partial \Delta f_{\text{mix}}}{\partial (1-\phi)} \right)$, which can be rewritten in the from $\mu_A = -\alpha \mu_B$. Also the A chains have the same chemical potential in both phases.

By a subtraction of the common tangential line, ϕ_a and ϕ_b become the minima of the resulting function. A graph with such functions is shown in Fig. 8.5b. For the χN_A values which describe mixed states, the tangential line at the preparation composition $\bar{\phi}$ is used for the subtraction. The subtraction of the tangential line corresponds to the LEGENDRE transformation $\Omega = F - \mu_B N$ of the free energy to the grand canonical potential Ω. The corresponding LEGENDRE transformation of Δf_{mix} to the increment in the grand canonical potential density $\Delta \omega_{\text{mix}}$ reads

$$\Delta \omega_{\text{mix}} = \Delta f_{\text{mix}} - \mu_B \phi. \qquad (8.22)$$

$\Delta\omega_{\text{mix}}$ can be considered as the starting point for the description of the interface behavior and of fluctuations. The grand potential allows for a variation in the number of particles, which is not a constant for the interface. Unfortunately, the calculation of ϕ_a and ϕ_b which determine the common tangent to Δf_{mix} require numerical calculations for most cases. So we do not have an explicit formula for the subtraction in (8.22) and thus $\Delta\omega_{\text{mix}}$. An exception is the symmetric case ($N_A = N_B$, so $\alpha = 1$), where $\mu_A = \mu_B = 0$ and thus $\Delta\omega_{\text{mix}} = \Delta f_{\text{mix}}$ with the explicit formula (8.16). Some calculations require $\frac{\partial^2 \Delta\omega_{\text{mix}}}{\partial\phi^2}$, and we can use $\frac{\partial^2 \Delta f_{\text{mix}}}{\partial\phi^2}$ instead, since the subtraction of a linear function in (8.21) does not alter the second derivative. Further, by construction $\frac{\partial \Delta\omega_{\text{mix}}}{\partial\phi} = 0$ for the equilibrium bulk composition ϕ. Thus, a second order TAYLOR expansion of $\Delta\omega_{\text{mix}}$ around the equilibrium bulk composition ϕ_{eq} reads

$$\Delta\omega_{\text{mix}}(\phi_{\text{eq}} + \Delta\phi) \approx \Delta\omega_{\text{mix}}(\phi_{\text{eq}}) + \frac{1}{2}\left.\frac{\partial^2 \Delta f_{\text{mix}}}{\partial\phi^2}\right|_{\phi=\phi_{\text{eq}}} \Delta\phi^2. \tag{8.23}$$

8.4.1 Bulk Fluctuations

Fluctuation Amplitude Fluctuations are often discussed in connection with scattering experiments. These experiments are not restricted to the investigation of the sample structure, they are also sensitive to fluctuations. Experimental parameters determine the scattering vector $\mathbf{q}$, and the experiment detects a variation of the scattering contrast with the shape of a sine wave of wavelength $2\pi/|\mathbf{q}|$. Such a contrast wave is produced by a sine concentration fluctuation in a polymer blend, and we can use $\Delta\omega_{\text{mix}}$ to determine the grand potential penalty for such a fluctuation. Based on the equipartition theorem which states that each degree of freedom is thermally excited by $\frac{1}{2}k_BT$, the root mean squared (rms) amplitude of the fluctuation results by equating this penalty with $\frac{1}{2}k_BT$. With (8.18) and (8.23), a small wave fluctuation

$$\phi((x)) = \phi_{\text{eq}} + \Delta\phi_q \cos(\mathbf{q} \cdot \mathbf{x}). \tag{8.24}$$

with amplitude $\Delta\phi_q$ and wave vector $\mathbf{q}$ around ϕ_{eq} results in the grand canonical potential increment

$$\Delta\Omega = V\Delta\omega_{\text{mix}}(\phi_{\text{eq}}) + \frac{1}{2}V\left[\left.\frac{\partial^2 \Delta f_{\text{mix}}}{\partial\phi^2}\right|_{\phi=\phi_{\text{eq}}} + \kappa q^2\right]|\Delta\phi_q|^2. \tag{8.25}$$

While the first term on the right side is the grand canonical potential increment without fluctuation, the second term describes the penalty for the fluctuation. Equating the average of this penalty to $k_BT/2$ yields

$$\left\langle \left| \Delta \phi_q \right|^2 \right\rangle = \frac{k_{\mathrm{B}} T}{V \left[\left. \frac{\partial^2 \Delta f_{\mathrm{mix}}}{\partial \phi^2} \right|_{\phi = \phi_{\mathrm{eq}}} + \kappa q^2 \right]}. \tag{8.26}$$

The neglecting of higher orders in $\left| \Delta \phi_q \right|$ in (8.26) is usually justified for small thermal fluctuations, as long as the system is not very close to the conditions of the critical point. In general, fluctuations at all wavelengths and thus all q values are present simultaneously with amplitudes $\Delta \phi_q(q)$. Thus, a modified version of (8.24) contains a sum[3] over all q values. Inserting such a sum into (8.23) results in a double sum, which looks complicated at first. However, the fluctuations for different q are orthogonal, so the cross terms vanish in the volume integration, and an equation similar to (8.26) with a single sum of squared amplitudes for all q values results.

The Bulk Correlation Length of Fluctuations In Sect. 8.1, the correlation length of bulk fluctuations ξ was discussed as the relevant length scale in the description of soft matter. For a derivation of ξ, we start from the scattering amplitude $\tilde{A}$ [17] (see Chap. 11 by A. C. Völker et al. and Chap. 12 by J. Daillant)

$$\tilde{A} = \frac{1}{V} \int_V \Delta \phi(\mathbf{r}) e^{i\mathbf{q} \cdot \mathbf{r}} \mathrm{d}^3 r. \tag{8.27}$$

The scattering intensity is proportional to the time averaged squared modulus $\left\langle \left| \tilde{A} \right|^2 \right\rangle = \left\langle \tilde{A} \tilde{A}^* \right\rangle$, where $\tilde{A}^*$ is the complex conjugate of $\tilde{A}$. An example is a light scattering experiment, where the scattering amplitude is the scattered electrical field E, and the scattering intensity results as absolute modulus $|E|^2$. With (8.27), the intensity $\left\langle \tilde{A} \tilde{A}^* \right\rangle$ results as a double integration over $\mathbf{r}$ and $\mathbf{r}'$. In an integral substitution $\mathbf{r}'$ is replaced by the difference $\Delta \mathbf{r} = \mathbf{r}' - \mathbf{r}$, and the scattering intensity becomes

$$\left\langle \left| \tilde{A} \right|^2 \right\rangle = \frac{1}{V} \int_V \mathrm{d}^3 r \frac{1}{V} \int_V \mathrm{d}^3 \Delta r \left\langle \Delta \phi(\mathbf{r}) \Delta \phi(\mathbf{r} + \Delta \mathbf{r}) \right\rangle e^{i\mathbf{q} \cdot \Delta \mathbf{r}}. \tag{8.28}$$

The integration over $\mathrm{d}^3 r$ is a spatial averaging of the starting point $\mathbf{r}$. It can be absorbed to the average $\langle . \rangle$, which becomes an average over space and time. What is left is a FOURIER transform in $\Delta \mathbf{r}$ of the correlation function $\langle \Delta \phi(\mathbf{r}) \Delta \phi(\mathbf{r} + \Delta \mathbf{r}) \rangle$. Due to the isotropy of the system, the correlation function does not depend on the direction of $\Delta \mathbf{r}$, but only on its magnitude $\Delta r = |\Delta \mathbf{r}|$. The change to polar coordinates $(\theta, \varphi, \Delta r)$ yields

$$\left\langle \left| \tilde{A} \right|^2 \right\rangle = \frac{1}{V} \int_0^\pi \sin(\theta) \mathrm{d}\theta \int_0^{2\pi} \mathrm{d}\varphi \int_0^\infty \mathrm{d}\Delta r \, \Delta r^2 \left\langle \Delta \phi(r) \Delta \phi(r + \Delta r) \right\rangle e^{iq \Delta r \cos(\theta)}. \tag{8.29}$$

---

[3]For a finite scattering volume the fluctuations form a FOURIER series, not a FOURIER integral transformation.

Here, $q = |\mathbf{q}|$ is used. The integrations over φ yields a factor 2π, and the integration over θ can be performed after the substitution $u = \cos(\theta)$. With the complex notation $\sin(x) = [\exp(ix) - \exp(-ix)]/(2i)$ of the sin function, the result reads

$$\left\langle \left| \tilde{A} \right|^2 \right\rangle = \frac{4\pi}{V} \int_0^\infty d\Delta r \, \Delta r^2 \, \langle \Delta\phi(r)\Delta\phi(r + \Delta r) \rangle \frac{\sin(q\Delta r)}{q\Delta r}. \tag{8.30}$$

For a connection between $\Delta\phi_q$ and $\tilde{A}$, we insert (8.24) in the definition of $\tilde{A}$ (8.27) and consider a scattering volume $V = L_x L_y L_z$ with $\mathbf{q}$ in x direction. With the EULER relation $\exp(iqx) = \cos(qx) + i\sin(qx)$ and $\sin^2(qx) = \frac{1}{2}[1 - \cos(2qx)]$, the integration reads

$$\tilde{A} = \frac{1}{L_y L_x L_z} \int_0^{L_x} \int_0^{L_y} \int_0^{L_z} \left[\phi_{\text{eq}} + \Delta\phi_q \cos(qx) \right] \left[\cos(qx) + i\sin(qx) \right] = \frac{1}{2}\Delta\phi_q. \tag{8.31}$$

Thus $\left\langle \left| \Delta\phi_q \right|^2 \right\rangle = 4 \left\langle \left| \tilde{A} \right|^2 \right\rangle$. In order to proceed, we use the correlation function

$$\langle \Delta\phi(r)\Delta\phi(r + \Delta r) \rangle = \left\langle \left| \Delta\phi_r \right|^2 \right\rangle \frac{\xi}{\Delta r} \exp\left[-\frac{\Delta r}{\xi} \right], \tag{8.32}$$

with the correlation length ξ. The local mean squared fluctuation amplitude $\left\langle \left| \Delta\phi_r \right|^2 \right\rangle$ at a fixed point in space in (8.32) and the mean squared amplitude $\left\langle \left| \Delta\phi_q \right|^2 \right\rangle$ of a not localized, wave-like fluctuation for selected q in (8.24)–(8.26) are distinguished by their different index. With (8.31) and (8.32), the integration in (8.30) yields

$$\left\langle \left| \Delta\phi_q \right|^2 \right\rangle = 4 \left\langle \left| \tilde{A} \right|^2 \right\rangle = \frac{16\pi\xi^3 \left\langle \left| \Delta\phi_r \right|^2 \right\rangle}{V(1 + q^2\xi^2)}. \tag{8.33}$$

A comparison of (8.33) with (8.25) reveals that these equations have the same q dependence. From this equivalence in q space one can conclude, that (8.32) has the correct form for the correlation function in real space for $\Delta\omega_{\text{mix}}$ described by (8.23), since the mapping by the FOURIER transform is unique. A comparison of the coefficients connects the thermodynamic description in (8.25) with the scattering description of (8.33). It yields

$$\xi = \sqrt{\frac{\kappa}{\left. \frac{\partial^2 \Delta f_{\text{mix}}}{\partial\phi^2} \right|_{\phi=\phi_{\text{eq}}}}} \tag{8.34}$$

$$\left\langle \left| \Delta\phi_r \right|^2 \right\rangle = \frac{k_B T}{16\pi\kappa} \frac{\left. \frac{\partial^2 \Delta f_{\text{mix}}}{\partial\phi^2} \right|_{\phi=\phi_{\text{eq}}}}{\kappa} = \frac{k_B T}{16\pi\kappa\xi}. \tag{8.35}$$

From (8.32), (8.34), and (8.35), the correlation volume V_c of a localized fluctuation results as

$$V_c = \int_V d^3r \, \langle \Delta\phi(\mathbf{r}) \Delta\phi(\mathbf{r} + \Delta\mathbf{r}) \rangle = \left\langle \left| \Delta\phi_r \right|^2 \right\rangle \xi^3 = \frac{k_B T}{16\pi \left. \frac{\partial^2 \Delta f_{mix}}{\partial \phi^2} \right|_{\phi = \phi_{eq}}}. \quad (8.36)$$

From a conceptual point, the pole at $\Delta r = 0$ in the correlation function (8.32) appears strange, and might not reflect a physical reality. On the other hand, V_c remains well defined, so fluctuations remain limited within the squared gradient theory.

The Correlation Length for the FLORY HUGGINS Theory In order to relate the previous paragraph to the example of polymer blends, we follow the discussion of Strobl for the connection of κ to the sizes of the A and B polymer chains [5]. In scattering experiments, the overall size of an object is determined in the range of small $\mathbf{q}$. This limit is called the GUINIER range, and the evaluation of the size of an object is based on either a ZIMM plot or a GUINIER plot. We use here the ZIMM presentation of the small q limit, which reads[4]

$$\left\langle \left| \tilde{A} \right|^2 \right\rangle^{-1} (\mathbf{q}^2) \approx \left\langle \left| \tilde{A} \right|^2 \right\rangle^{-1} (\mathbf{q}^2 = 0) \left[1 + \frac{1}{3} q^2 R_g^2 + O(q^4) \right]. \quad (8.37)$$

Here, R_g is the radius of gyration of an object. For an ideal polymer chain with N segments, it reads $R_g^2 = \frac{2}{3} l_{ps}^2 N$ [5]. The persistence length l_{ps} describes the decay $\exp(-l/l_{ps})$ of directional correlation when following a polymer coil. While for stiff chains l_{ps} is large, it is small for flexible polymers. For very small ϕ, a polymer blend contains only a few B chains in a background of mainly A chains. The scattering contrast to the background is thus given by the B chains, and the size of B chains R_{gB} is detected. With similar arguments the detection of the size of the A chains R_{gA} for ϕ values close to 1 is justified. So, we need to evaluate (8.37) for the Flory Huggins case, then consider the limits of vanishing B and A content and match the resulting R_g^2 values with R_{gB}^2 and R_{gA}^2, respectively. With (8.31) we can replace $\left\langle \left| \tilde{A} \right|^2 \right\rangle^{-1}$ by $\left\langle \left| \Delta\phi_q \right|^2 \right\rangle^{-1}$ and apply (8.26), which can be evaluated for the Flory Huggins case with (8.19):

$$\left\langle \left| \tilde{A} \right|^2 \right\rangle^{-1} = 4 \left\langle \left| \Delta\phi_q \right|^2 \right\rangle^{-1} = \frac{4V}{k_B T} \left\{ \frac{k_B T}{v_c} \left[\frac{1}{\phi N_B} + \frac{1}{(1-\phi) N_A} - 2\chi \right] + \kappa q^2 \right\}. \quad (8.38)$$

Beside the χ term, the q independent first part has a ϕ^{-1} contribution which diverges and thus dominates in the limit of small ϕ, and a $(1 - \phi)^{-1}$ contribution which

---

[4]For data gained from a real experiment, the measured intensity needs to be corrected by subtracting the background scattering of the solvent.

diverges and dominates in the limit $\phi \to 1$. In order to realize the required R_g^2 values in the two limits, κ is also composed of two terms with the same divergences:

$$\kappa = \frac{k_B T}{v_c} \left[\frac{1}{\phi N_B} \frac{R_{gB}^2}{3} + \frac{1}{(1-\phi)N_A} \frac{R_{gA}^2}{3} - 2\chi \frac{r_0^2}{3} \right]. \tag{8.39}$$

The interaction term $-2\chi r_0^2/3$ is added in a similar way as in (8.38), since for high values of N_A and N_B it exceeds the non-divergent term in the two limits $\phi \to 0$ and $\phi \to 1$, and thus becomes the major correction term. Strobl and Jones provide an additional justification of (8.39) based on the random phase approximation [5, 7]. Jones ascribes r_0 to the range of interactions. For a symmetric blend with $N_A = N_B$ and $R_{gA} = R_{gB}$, the only length scale to define ξ is $R_{gA} = R_{gB}$. Thus, ξ needs to become ϕ independent in this case, which is achieved for $r_0 = R_{gA} = R_{gB}$. It appears that a ϕ dependent average of R_{gA} and R_{gB} is a more suitable value for r_0.

The random phase approximation is not restricted to the small q limit, but also predicts the high q behavior, where the internal structure of a chain is resolved. We do not follow this route, since the involved higher powers in q^2 in the scattering description correspond in the real space equations either to higher powers of the gradient in ϕ or to higher order derivatives,[5] which are no longer compatible with the squared gradient approach discussed here. The restriction indicates a limitation of calculations based on the squared gradient theory: the predicted width of an interface profile scales with ξ and thus remains comparable to R_{gA} and R_{gB}. Such a situation corresponds to the weak segregation limit, discussed in the introduction. Sharper interface profiles would require higher order powers of the gradient or higher derivatives. For other systems different from polymer blends, where no additional structure is expected on a scale smaller than ξ, the limitation of the squared gradient theory might be less severe.

8.4.2 Simple Dynamics

The description of the fluctuation amplitude by (8.26) has the same form as a thermally excited harmonic oscillator, where the potential is formed by the second order approximation (8.23) of $\Delta\omega_{mix}$ and an additional q^2 dependent contribution due to the elastic constant κ. The discussion in Sect. 8.4.1 thus is a q dependent version of simple harmonic oscillator physics. The analogy can be extended to the effect of external fields h, which is considered by adding a linear term $h\phi$ to $\Delta\omega_{mix}$ in (8.23). As long as the second order approximation (8.23) holds and thus the potential of the equivalent oscillator remains harmonic, the shift of the new minimum position which describes the thermal equilibrium in the presence of h away from the original minimum position is linear in h. This description of external fields is the basis

---

[5] See the transition from (8.23) to (8.25), where the squared gradient in the real space description (8.23) transforms to the factor q^2 in the q space picture (8.25).

of the linear response theory (see e.g. [18]). It is often useful for the description of experimental results by linear response coefficients, which might be set up on a phenomenological basis.

In this brief section, the harmonic oscillator analogy is used for a simple description of the relaxation dynamics. The differential equation of an equivalent damped harmonic oscillator reads

$$m\frac{d^2x}{dt^2} + b\frac{dx}{dt} + Kx = 0. \tag{8.40}$$

Here, m is the mass of a particle, $b > 0$ its friction, and $K > 0$ the constant of the spring which forms the harmonic potential $U = \frac{1}{2}Kx^2$. Thermal fluctuations in soft matter systems are usually over-damped and inertia effects are negligible. Thus we can cancel the m term. The remaining differential equation is of first order and the solution is an exponential decay:

$$x(t) = x_0 \exp\left[-\frac{K}{b}t\right]. \tag{8.41}$$

When we apply the analogy to a soft matter system, x corresponds to the amplitude $\Delta\phi_q$ of a bulk mode with fixed wave vector q or an eigen-mode of an interface fluctuation. It is excited thermally[6] and decays exponentially. Beside the equivalent spring constant which is twice the prefactor of $\left|\Delta\phi_q\right|^2$ in (8.25), we need a friction factor b, which might also have a q dependence in general. One technique to follow the relaxation of a fluctuation is dynamic light scattering, where the exponential decay of a fluctuation turns up in the time auto correlation function as

$$\langle\Delta\phi_q(t)\Delta\phi_q(t+\Delta t)\rangle = \langle\left|\Delta\phi_q\right|^2\rangle\exp\left\{-\frac{V}{b}\left[\frac{\partial^2\Delta f_{mix}}{\partial\phi^2}\bigg|_{\phi=\phi_{eq}} + \kappa q^2\right]\Delta t\right\}. \tag{8.42}$$

A simple example are particles with no interactions under highly dilute conditions, so the thermodynamic restoring force $\frac{\partial^2\Delta f_{mix}}{\partial\phi^2}\big|_{\phi=\phi_{eq}}$ vanishes. With a q independent friction b the resulting correlation function $\exp[-q^2\Delta t\kappa/b]$ describes the characteristic q^2 dependence of diffusion with the diffusion constant $D = \kappa/b$. Generally, the effective spring constant $\frac{\partial^2\Delta f_{mix}}{\partial\phi^2}\big|_{\phi=\phi_{eq}}$ and the friction factor are complementary information of a system. While a static scattering experiment detects the mean squared excitation of a fluctuation for selected q and thus yields the effective spring constant as the static characteristic of a system, the relaxation time is extracted from a dynamic experiment. The combination of both allows the calculation of the friction coefficient. The procedures introduced in this section can be also applied to interface

---

[6]An excellent discussion of thermal excitation and time correlation function as described by a LANGEVIN equation is found in the book of Doi and Edwards [19].

fluctuations, where fluctuation eigen-modes need to be considered instead of wave like fluctuations ϕ_q.

8.4.3 Bulk Fluctuations Revisited

Role of the Translational Symmetry The discussion of fluctuations in the context of a scattering experiment over emphasizes the experimental scattering technique in the role of fluctuations. It appears fortuitous that the squared gradient in (8.18) is replaced by the simpler factor q^2, and the scattering amplitude (8.26) is calculated by simple algebraic operations, without the need to solve a differential equation. A deeper reason for this simple behavior can be traced back to the translational symmetry of the bulk system. This symmetry can be addressed by the NOETHER theorem which is introduced in mechanics, but which is usually not mentioned in connection with statistical mechanics or scattering theory. The NOETHER theorem establishes a connection between a continuous symmetry and a corresponding preserved quantity. The translational symmetry is connected to the preservation of the linear momentum. Thus, fluctuations in a homogeneous, translational invariant bulk system are necessarily eigen-modes of the momentum operator, so sine waves. The scattering vector $\mathbf{q} = \mathbf{k}_i - \mathbf{k}_s$ results as difference of the wave vectors of the incident light $\mathbf{k}_i$ and the scattered light $\mathbf{k}_s$. Since both modes $\mathbf{k}_i$ and $\mathbf{k}_s$ have well defined momenta, the difference of the two is connected to a single value of momentum transfer and thus also a sine wave. Since the eigen-modes to different momentum values or directions are orthogonal, the scattering experiment picks one eigen-mode of fluctuation with the eigen-value $\mathbf{q}$. This step is formally done by an overlap integration between the wave set by the momentum transfer of the scattering experiment and the spectrum of fluctuations. The overlap integration looks like a FOURIER transform, and selection of the mode follows directly.

For a planar interface, a similar reasoning can be set up for any direction within a plane parallel to the interface. In a direction parallel to the interface, there is translational symmetry. So any fluctuation mode is an eigen-function of the momentum operator for the direction parallel to the interface, described by the wave vector component $q_\parallel$. For the direction perpendicular to the interface, in contrast, the translational symmetry is broken. Thus there is no longer the simplifying concept of momentum conservation, and the calculation of a fluctuation mode requires the solution of a differential equation. The spectrum of interface fluctuation modes should still be orthogonal in general, since different modes have different eigen-values for the excitation energies. An interface sensitive scattering experiment, e.g. evanescent wave dynamic light scattering (EWDLS, see Chap. 13 by B. Loppinet and e.g. [1]) has a sensitivity profile different from the sine wave of a bulk scattering experiment. The overlap integration in order to calculate the sensitivity of EWDLS is now more complex than a FOURIER transform, and in general there might be overlap to different modes.

8.5   Interface Structure

Interfaces between the A-rich phase and the B-rich phase of a de-mixed state are
often not a sharp transition. Instead, a concentration profile is formed. Similarly, a
concentration profile also builds up when the mixture is in contact with a substrate.
The squared gradient theory provides predictions for these profile. Differential equations and expressions for the interface tension for a ϕ independent elastic constant
κ are derived in Sect. 8.5.1. Analytical solutions exist for the LANDAU theory, and
they are discussed in Sect. 8.5.2. For the Flory Huggins case, the ϕ dependence of
κ leads to an additional term in the differential equation for the interface profile. It
appears that this term is silently neglected in the literature [7]. The effect of this term
requires additional investigations, which will not be performed here.

8.5.1   Interface Tensions and Differential Equations

Internal Interfaces. A polymer blend in the two phase region consists of an A-
rich phase of composition ϕ_a and a B-rich phase of composition ϕ_b. The discussion
of the interface profile between these two phases starts with $\Delta\omega_{mix}$ as depicted in
Fig. 8.5b. The A-rich and the B-rich phase have the common value $\Delta\omega_{mix}(\phi_a) =
\Delta\omega_{mix}(\phi_b) = \Delta\omega_{bulk}$, so the minimum value of Fig. 8.5b. For an interface profile
$\phi(z)$ which describes a smooth transition from the A-rich to the B-rich phase, the ϕ
values in the interface interpolate between ϕ_a and ϕ_b, and thus lead to contributions to
the grand potential with $\Delta\omega_{mix}(\phi) > \Delta\omega_{bulk}$. The $\Delta\Omega$ penalty of volume elements
in the interface is thus $\Delta\omega_{mix}(\phi) - \Delta\omega_{bulk}$. The interface energy is the sum over
these contributions. Within a squared gradient theory (8.3), the interface energy for
a given profile $\phi(z)$ is calculated as

$$\gamma_{ab}[\phi] = \int_{-\infty}^{+\infty} dz \left[\Delta\omega_{mix}(\phi(z)) - \Delta\omega_{bulk} + \frac{\kappa}{2} \left(\frac{d\phi}{dz} \right)^2 \right]. \tag{8.43}$$

A system will minimize the interfacial energy (8.43) and build up $\phi(z)$ accordingly.
The interface tension results as $\gamma_{ab} = \min(\gamma_{ab}[\phi])$. Without the gradient term, so for
$\kappa = 0$, the profile minimizing (8.43) is a step profile, with a sharp transition from ϕ_a
in the A-rich phase to ϕ_b in the B-rich phase. For this case, (8.43) yields $\gamma_{ab} = 0$, since
$\Delta\omega_{mix}(\phi_a) = \Delta\omega_{mix}(\phi_b) = \Delta\omega_{bulk}$. The contribution of the gradient term for $\kappa > 0$
modifies this picture, as the step profile has an infinite gradient and thus a step profile
would have infinite interface tension. The resulting $\phi(z)$ is a compromise between the
penalties in $\Delta\omega_{mix}(\phi(z))$ and the cost for a high gradient. The discussion is restricted
to the case of a ϕ independent value of κ.

The calculus of variations for a minimization of (8.43) is briefly summarized in the appendix (Sect. 8.8). The LAGRANGE function in (8.43) reads

$$
L\left(\phi, \frac{d\phi}{dz}\right) = \Delta\omega_{mix}(\phi) - \Delta\omega_{bulk} + \frac{\kappa}{2}\left(\frac{d\phi}{dz}\right)^2,
\tag{8.44}
$$

and the EULER-LAGRANGE equation $\dfrac{\partial L}{\partial \phi} - \dfrac{d}{dz}\dfrac{\partial L}{\partial(d\phi/dz)} = 0$ becomes[7]

$$
\frac{d\Delta\omega_{mix}}{d\phi} = \kappa\frac{d^2\phi}{dz^2}.
\tag{8.45}
$$

An integration of (8.45) with respect to z is possible after multiplying it by $\frac{d\phi}{dz}$. The result reads

$$
\Delta\omega_{mix}(\phi) - \Delta\omega_{bulk} = \frac{\kappa}{2}\left(\frac{d\phi}{dz}\right)^2.
\tag{8.46}
$$

As a cross-check of this step, it might be reverted by differentiating (8.46) with respect to z to find (8.45). The integration constant is identified in (8.46) already with $\Delta\omega_{bulk}$. Formally, one can first write (8.46) with an integration constant, and then determine its value at a position z which is far away from the interface in the bulk, where $\frac{d\phi}{dz} = 0$. Re-writing (8.46) leads to a first order differential equation for the interface profile:

$$
\frac{d\phi}{dz} = \pm\sqrt{\frac{2}{\kappa}[\Delta\omega_{mix}(\phi) - \Delta\omega_{bulk}]}.
\tag{8.47}
$$

Separation of variables in (8.47) and integration leads to an implicit formula for the interface profile

$$
z = \int_{\phi_a}^{\phi(z)}\sqrt{\frac{\kappa}{2[\Delta\omega_{mix}(\phi) - \Delta\omega_{bulk}]}}\,d\phi,
\tag{8.48}
$$

with ϕ_a as starting point within the A-rich phase. With (8.47) which describes the minimum of (8.43), the interface tension is written as:

$$
\gamma_{ab} = \int_{-\infty}^{+\infty}\left[\frac{\kappa}{2}\left(\frac{d\phi}{dz}\right)^2 + \frac{\kappa}{2}\left(\frac{d\phi}{dz}\right)^2\right]dz = \int_{\phi_a}^{\phi_b}\kappa\frac{d\phi}{dz}\,d\phi
$$
$$
= \int_{\phi_a}^{\phi_b}\sqrt{2\kappa\,[\Delta\omega_{mix}(\phi) - \Delta\omega_{bulk}]}\,d\phi.
\tag{8.49}
$$

---

[7] In this step, a ϕ dependence of the elastic constant κ would lead to an additional term, which does not fit to the following manipulations in a simple way.

(8.46) indicates, that the $\Delta\omega_{mix}$ penalty and the cost for building up a gradient at the interface have the same magnitude, similar to the same magnitudes of kinetic and potential energy for a harmonic oscillator, or the same size of electric and magnetic energy in an electromagnetic wave. Thus, it is possible to express γ_{ab} by integrating (8.49) over the square root of the $\Delta\omega_{mix}$-hump in Fig. 8.5b alone, with the gradient terms eliminated.

Interface between the Mixture and a Substrate. For a discussion of the wetting properties, the variation of the interface tension between the mixture and the substrate with composition is required. Starting point are the interface energies γ_{AS} between the pure A phase ($\phi = 0$) and the substrate, and γ_{BS} between the pure B phase ($\phi = 1$) and the substrate. For a composition ϕ in between 0 and 1, the resulting interface energy to the substrate results by linear interpolation

$$\gamma_S(\phi) = (1 - \phi)\gamma_{AS} + \phi\gamma_{BS} = \gamma_{AS} + \phi[\gamma_{BS} - \gamma_{AS}]. \tag{8.50}$$

This approach is based on an addition of interactions of A molecules and B molecules which are at the interface to the substrate. So, γ_S has the meaning of a contact potential at the interface. For a calculation of interface tensions, the penalties to build up concentration gradients in interface profiles need to be considered in addition.

We use results of Sheng [20] for liquid crystals and re-write them for binary mixtures. The assumed preferential absorption of A to the interface requires[8] $\gamma_{BS} > \gamma_{AS}$, and γ_S in (8.50) increases with ϕ. Thus, the effect of γ_S alone would result in a complete coverage of the interface by A molecules with $\phi = 0$. Such a pure composition at the interface, however, deviates from the bulk composition, as defined by preparation for the mixed phase and ϕ_a or ϕ_b in the de-mixed state. The system thus has to form a concentration profile at the interface which minimizes the total interface energy, which is composed by the interface potential $\gamma_S(\phi)$, the thermodynamic contribution $\Delta\omega_{mix}(\phi)$, and the penalty for a concentration gradient, as considered by the squared gradient term. For an interface located at $z = 0$, the resulting formula reads

$$\gamma[\phi] = \int_0^{+\infty} dz \left[\Delta\omega_{mix}(\phi(z)) - \Delta\omega_{bulk} + \frac{\kappa}{2}\left(\frac{d\phi}{dz}\right)^2 \right.$$
$$\left. + \phi(z)[\gamma_{BS} - \gamma_{AS}]\delta(z) \right] + \gamma_{AS}. \tag{8.51}$$

The boundary term with the interface potential $\gamma_S(\phi)$ is taken into account by the delta-function[9] $\delta(z)$. In the same way as the transformations from (8.43) to (8.45), the differential equation for the minimum profile results as

---

[8]For a neutral substrate with $\gamma_{AS} = \gamma_{BS}$ and thus no preferential adsorption, the substrate would have no effect on the sample. This boring case does not need further discussion.

[9]The delta function is also briefly discussed in the appendix.

$$\frac{d\Delta\omega_{\text{mix}}}{d\phi} = \kappa \frac{d^2\phi}{dz^2} - \delta(z)\left(\phi[\gamma_{\text{BS}} - \gamma_{\text{AS}}] + \kappa \frac{d\phi}{dz}\right). \tag{8.52}$$

For $z > 0$, there is no contribution from the interface potential and (8.52) is identical to (8.45). Thus the integration of (8.46) and the differential equation (8.47) are derived as before. With (8.46) and (8.47) and the same integration by substitution as in (8.49), the interface tension (8.51) results as

$$\gamma = \int_{\phi_0}^{\phi_{\text{eq}}} \sqrt{2\kappa \left[\Delta\omega_{\text{mix}}(\phi) - \Delta\omega_{\text{bulk}}\right]}\, d\phi + [\gamma_{\text{BS}} - \gamma_{\text{AS}}]\phi_0 + \gamma_{\text{AS}}. \tag{8.53}$$

For the mixed state $\phi_{\text{eq}} = \bar{\phi}$. In the de-mixed phase, wetting droplets at the substrate are usually macroscopic, so their thickness is much larger than the extension of an interfacial profile, which is comparable to the bulk fluctuation length ξ. So, we can use ϕ_a or ϕ_b as bulk composition ϕ_{eq} for this case. The contact composition directly at the substrate is denoted by ϕ_0. With the assumed preferential adsorption of A to the interface, the A concentration $(1 - \phi)$ is generally enhanced compared to the bulk phase, so $\phi_0 < \phi_{\text{eq}}$. Thus the integral in (8.53) is positive. The composition at the substrate ϕ_0 is determined as the minimum[10] of (8.53):

$$0 = \frac{d\gamma}{d\phi_0} = -\sqrt{2\kappa \left[\Delta\omega_{\text{mix}}(\phi_0) - \Delta\omega_{\text{bulk}}\right]} + [\gamma_{\text{BS}} - \gamma_{\text{AS}}]. \tag{8.54}$$

The minimum condition can be transformed to

$$\Delta\omega_{\text{mix}}(\phi_0) - \Delta\omega_{\text{bulk}} = \frac{[\gamma_{\text{BS}} - \gamma_{\text{AS}}]^2}{2\kappa}. \tag{8.55}$$

There might be several values for ϕ_0 which fulfill (8.54) and (8.55), and it is required to find the value which corresponds to the absolute minimum. A distinction between minima and maxima is based on the second derivative of (8.53):

$$\frac{d^2\gamma}{d\phi_0^2} = -\frac{\kappa}{\sqrt{2\kappa \left[\Delta\omega_{\text{mix}}(\phi) - \Delta\omega_{\text{bulk}}\right]}} \frac{d\Delta\omega_{\text{mix}}}{d\phi}\Bigg|_{\phi=\phi_0}. \tag{8.56}$$

Thus, a positive second derivative (8.56) of (8.53) which indicates a minimum requires a negative slope of $\Delta\omega_{\text{mix}}(\phi)$ at $\phi = \phi_0$. With a known value of ϕ_0, the interface profile results from the implicit formula (8.48).

---

[10]The negative sign in (8.54) occurs since ϕ_0 is the lower limit in (8.53).

The Wetting Transition Triggered by the Contact Compositions The contact angle in (8.1) and the spreading coefficient S_a in (8.2) are defined by interface tensions, which are integrations with the same integrand. Inserting (8.49) and (8.53) in (8.1) and (8.2) results in

$$\cos(\Theta_a) = 1 - \frac{\int_{\phi_{0a}}^{\phi_{0b}} \left| \sqrt{2\kappa \left[\Delta\omega_{mix}(\phi) - \Delta\omega_{bulk} \right]} \right| \, d\phi - [\gamma_{BS} - \gamma_{AS}](\phi_{0b} - \phi_{0a})}{\int_{\phi_a}^{\phi_b} \left| \sqrt{2\kappa \left[\Delta\omega_{mix}(\phi) - \Delta\omega_{bulk} \right]} \right| \, d\phi}$$

(8.57)

$$S_a = [\gamma_{BS} - \gamma_{AS}](\phi_{0b} - \phi_{0a}) - \int_{\phi_{0a}}^{\phi_{0b}} \left| \sqrt{2\kappa \left[\Delta\omega_{mix}(\phi) - \Delta\omega_{bulk} \right]} \right| \, d\phi.$$

(8.58)

Here, ϕ_{0a} and ϕ_{0b} are the contact compositions for the A-rich phase and the B-rich phase, respectively. In order to emphasize the requirement that square roots with a positive sign are used, the integrants in (8.57) and (8.58) are written as absolute magnitudes. A direct consequence of (8.57) and (8.58) is that for identical contact compositions $\phi_{0a} = \phi_{0b}$ for A-rich and B-rich phases one gets $\Theta_a = 0$ and $S_a = 0$ and thus complete wetting, irrespectively of the detailed shape of $\Delta\omega_{mix}(\phi)$. As we will see below the condition $\phi_{0a} = \phi_{0b}$ is fulfilled for $T \geq T_{pw}$. On the other hand $\phi_{0a} < \phi_{0b}$, which occurs for $T < T_{pw}$, implies in general $\cos(\Theta_a) \neq 1$ in (8.57) and $S_a \neq 0$ in (8.58), as long as the integrand does not vanish. The case $\phi_{0a} > \phi_{0b}$ is excluded for the assumed preferential adsorption of A to the substrate. For $|\cos(\Theta_a)| < 1$ corresponding to a negative value of S_a, a transition of the contact compositions ϕ_{0a} and ϕ_{0b} from an equal value $\phi_{0a} = \phi_{0b}$ to different values $\phi_{0a} < \phi_{0b}$ directly induces the first order wetting transition from complete wetting $S_a = 0$ to partial wetting $S_a < 0$. For this case we have $T_w = T_{pw}$ (see Fig. 8.1).

It is also possible that (8.57) yields a value $|\cos(\Theta_a)| > 1$. This case corresponds to $S_a > 0$, so complete wetting with no contact angle defined. Since, however, we are discussing the case $\phi_{0a} \neq \phi_{0b}$, the temperature is below T_{pw}. So there is no pre-wetting for the same temperature and a composition in the one-phase region of the phase diagram. It turns out that the LANDAU grand canonical potential density $\Delta\omega_L$ discussed in the next Sect. 8.5.2 produces such a behavior. In (8.57) and (8.58), the integration term has a stronger temperature dependency than the differences' product $[\gamma_{BS} - \gamma_{AS}](\phi_{0b} - \phi_{0a})$. Roughly speaking, the integration is proportinal to the width of the integration interval $(\phi_{0b} - \phi_{0a})$, which essentially covers the temperature dependence of the differences' product, multiplied with the average height of the integrand, which contains an additional temperature dependency. At a temperature lower than T_{pw}, the integration term and the differences' product in (8.57) and (8.58) become equal, so $S_a = 0$ and $|\cos(\Theta_a)| = 1$. This temperature is thus the wetting temperature smaller than T_{pw} (see Fig. 8.1). Since $\cos(\Theta_a)$ and S_a depend steadily on ϕ_{0a}, ϕ_{0b}, and T in the absence of a bulk phase transition, the wetting transition is now continuous,[11] and so the contact angle changes steadily from 0 to a finite value.

8.5.2 Interfaces Based on the LANDAU Assumption

Analytical solutions for the integrations of 8.5.1 are available for the LANDAU theory (Sect. 8.3.1). It is convenient to switch to a description, where the composition is expressed by the deviation ϕ' from the critical composition ϕ_c:

$$\phi' = \phi - \phi_c. \tag{8.59}$$

Here, ϕ' plays the role of the order parameter in the LANDAU theory, although it does not vanish in the disordered phase (the mixed phase) like a classical order parameter. Alternately, equations can be expanded in the difference to the equilibrium composition ϕ_{eq}:

$$\phi'' = \phi - \phi_{eq} = \phi' - \phi'_{eq}. \tag{8.60}$$

In the mixed state, $\phi_{eq} = \bar{\phi}$ is set by the sample preparation, while in the de-mixed state $\phi_{eq} = \phi_a$ in the A-rich phase and $\phi_{eq} = \phi_b$ in the B-rich phase. It turns out that a discussion based on ϕ' is suitable for the two-phase region, while the one-phase region is described easier with ϕ''. The equations of Sect. 8.5.1 remain basically valid[12] when ϕ is replaced by ϕ', only the boundary terms in (8.51) and (8.53), $\phi[\gamma_{BS} - \gamma_{AS}]$ have to be replaced by $(\phi_c + \phi')[\gamma_{BS} - \gamma_{AS}]$ according to the transform (8.59). The same holds true when ϕ is replaced in these equations by ϕ'', with the boundary term $(\phi_c + \phi'_{eq} + \phi'')[\gamma_{BS} - \gamma_{AS}]$ in (8.51) and (8.53). Since these modifications are constant additional terms, they vanish after derivations or differences in (8.52) and (8.54)–(8.58).

LANDAU **Free Energy Density.** Within the LANDAU assumption, a simple free energy density is a power series in ϕ for the phase separated state:

---

[11]The mechanism of such a continuous transition is different from the discussion of Bonn and Ross [4]. They investigate conditions for the contact composition which could lead to a continuous transition based on a graphical method eqivalent to (8.55) with an additional, ϕ dependent term on the right side. Based on a discussion of this slope of the right side, they identify conditions for the contact composition where the wetting transition becomes continuous. They find continuous transitions at a higher temperature than T_{pw}. In the discussion here, in contrast, the reason for continuous wetting is a calculated value $|\cos(\Theta_a)| > 1$ at T_{pw}, which leads to continuous wetting at lower temperature than T_{pw}. The discussion of Bonn and Ross does not include a test of the magnitude of $|\cos(\Theta_a)|$ resulting from their derived boundary values.

[12]Due to the linearity of (8.59), ϕ' can simply replace ϕ in all derivatives, e.g. $\frac{d\phi}{dz} = \frac{d\phi'}{dz}$ or $\frac{d\omega}{d\phi} = \frac{d\omega}{d\phi'}$. In order to apply the equations in Sect. 8.5.1 with ϕ' instead of ϕ, one could write (8.43) with a new function $\tilde{\omega}(\phi') = \omega(\phi_c + \phi')$, and repeat all derivations of Sect. 8.5.1 with $\tilde{\omega}(\phi')$ instead of $\omega(\phi)$. In order to keep the notation traceable, we use the same symbol and write sloppily $\omega(\phi')$ for $\tilde{\omega}(\phi')$.

$$\Delta f_{\rm L}(\phi) = \frac{C}{4}(\phi - \phi_{\rm a})^2(\phi - \phi_{\rm b})^2. \tag{8.61}$$

By construction, $\Delta f_{\rm L}$ has a minimum at $\phi_{\rm a}$ and a second minimum at $\phi_{\rm b}$, corresponding to the bulk compositions of the two phases. In between, $\Delta f_{\rm L}$ has a maximum at

$$\phi_{\rm c} = \frac{\phi_{\rm a} + \phi_{\rm b}}{2}. \tag{8.62}$$

With the transformation (8.59), (8.61) becomes

$$\Delta f_{\rm L}(\phi') = \frac{C}{4}\phi_{\rm c}^4 - \frac{C}{2}\phi_{\rm c}^2\phi'^2 + \frac{C}{4}\phi'^4. \tag{8.63}$$

The important second order term in (8.63) determines the second derivative at the critical composition $\phi' = 0$. For a temperature below the critical temperature $T_{\rm c}$, in the de-mixed state, there is a maximum of $\Delta f_{\rm L}$ at $\phi' = 0$. The concave behavior of $\Delta f_{\rm L}(\phi')$ at $\phi' = 0$ indicates a negative second derivative. In order to describe the phase transition to a mixed state which requires a convex shape of $\Delta f_{\rm L}(\phi')$ with positive second derivative at $\phi' = 0$, a replacement $C\phi_{\rm c}^2 = -A(T - T_{\rm c})$ is inserted in (8.63):

$$\Delta f_{\rm L}(\phi') = \frac{A^2}{4C}(T - T_{\rm c})^2 + \frac{A}{2}(T - T_{\rm c})\phi'^2 + \frac{C}{4}\phi'^4, \tag{8.64}$$

The linear temperature dependence of the second order term is usually motivated as the lowest order of a TAYLOR series which vanishes at $T = T_{\rm c}$. For a temperature close to $T_{\rm c}$, the contribution of higher orders is small, and thus the linear description provides reasonable predictions. As discussed in Sect. 8.1, for a $T_{\rm c}$ at room temperature or higher, which is typical for soft matter systems, the relative temperature deviation $(T - T_{\rm c})/T_{\rm c}$ from the critical point remains small even for several tens of degrees temperature difference. So, the description by a linear temperature dependence usually provides a suitable description of experimental results. The coefficient $C > 0$ determines the 4th order term, which guarantees the existence of a minimum, even for $\frac{1}{2}A(T - T_{\rm c}) < 0$. The typical argument to neglect any temperature dependence of C is the weakness of its relative temperature dependence $\Delta C(T)/C$, since no change of sign in C is to be expected.

By construction, the two minima of $\Delta f_{\rm L}$ in (8.64) have the same depth and thus a horizontal common tangential line. As a consequence, the chemical potential $\mu_{\rm B}$ of (8.21) is zero in the de-mixed state $T \leq T_{\rm bin}(\phi)$. The transformation (8.22) indicates that the LANDAU free energy density $\Delta f_{\rm L}(\phi')$ is identical to the LANDAU density of the grand canonical potential $\Delta\omega_{\rm L}$ for the de-mixed phase.

$$\Delta\omega_{\rm L}(\phi') = \Delta f_{\rm L}(\phi') \quad \text{for } T \leq T_{\rm bin}(\phi'). \tag{8.65}$$

Bulk Phase Behavior. As a first application of (8.65) with (8.64), the bulk phase behavior which was discussed in Sect. 8.2 and illustrated in Fig. 8.1 is calculated. The binodal line $T_{bin}(\phi')$, which is given by the equilibrium volume fractions ϕ_a and ϕ_b of the A-rich and the B-rich phases for $T < T_c$, results from the minimum condition $\frac{d \Delta \omega_L}{d\phi} = 0$:

$$\phi'_{a,b} = \mp \sqrt{\frac{A(T_c - T_{bin})}{C}},$$

(8.66)

The sign is negative for ϕ'_a and positive for ϕ'_b. An equivalent form of (8.66) is:

$$T_{bin}(\phi') = T_c - \frac{C}{A}\phi'^2_{a,b}.$$

(8.67)

For later use, we re-write (8.65) with (8.64) inserted by using (8.67):

$$\Delta \omega_L(\phi') = \frac{C}{4}(\phi'^2 - \phi'^2_{a,b})^2 \quad \text{for } T \leq T_{bin}(\phi').$$

(8.68)

The spinodal line $T_{spin}(\phi')$ indicates the locations within the two phase region for $T < T_c$ where the restoring forces for fluctuations vanish, as described by the condition $\frac{d^2 \Delta \omega_L}{d\phi^2} = 0$. This condition leads to $\phi' = \pm\sqrt{A(T_c - T_{spin})/(3C)}$, or alternately

$$T_{spin}(\phi') = T_c - 3\frac{C}{A}\phi'^2.$$

(8.69)

Bulk Fluctuations. Based on (8.64), the amplitude of the fluctuations and the correlation length result from (8.26) and (8.34), respectively. With (8.69), the resulting expressions can be written in a form which is often used in the presentation of experimental results in the mixed phase:

$$\left\langle |\Delta\phi|^2 \right\rangle^{-1} = \frac{VA}{k_B T}\left[T - T_{spin} + \frac{\kappa}{A}q^2\right]$$

(8.70)

$$\xi^{-2} = \frac{A}{\kappa}\left[T - T_{spin}\right].$$

(8.71)

The inverse mean squared amplitude which is proportional to the inverse intensity in a scattering experiment as well as the inverse squared correlation length follow both a linear temperature dependence. The extensions of these lines become zero at the location of the spinodal temperature, which thus can be determined experimentally by such a plot. The roots of the inverse quantities correspond to poles of $\left\langle |\Delta\phi|^2 \right\rangle$ and ξ. However, in most cases these poles on the spinodal cannot be approached closely, since the system undergoes the phase transition to the de-mixed state when the binodal temperature (8.67) is reached. Thus, the increase in intensity and ξ remain limited. An exception occurs for $\phi' = 0$, so the critical composition. Here, directly at the critical point, the divergence can be observed, which is addressed as critical behavior. This increase in fluctuation amplitude and fluctuation size is the basis of the discussion of the introduction. The mean field critical exponents for $\langle |\Delta\phi|^2 \rangle$ and ξ according to (8.70) and (8.71) are 1 and $\frac{1}{2}$, respectively.

In the de-mixed phase, the compositions of the A-rich phase and the B-rich phase are set by the binodal line (8.67). Inserting $T = T_{\text{bin}}$ in (8.70) and (8.71) yields

$$\langle |\Delta\phi|^2 \rangle = \frac{k_{\text{B}}T}{V\left[2C\phi_{\text{b,a}}'^2 + \kappa q^2\right]} \tag{8.72}$$

$$\xi = \sqrt{\frac{\kappa}{2C}}\frac{1}{|\phi_{\text{b,a}}'|}. \tag{8.73}$$

Within the LANDAU assumption, the fluctuation amplitudes and the correlation lengths in the two phases are the same.

Density of the Grand Canonical Potential. For the mixed phase for $T > T_{\text{bin}}(\phi)$, $\Delta\omega_{\text{L}}$ should have a minimum for the average composition $\bar{\phi}'$ set by the sample preparation. With (8.22), this condition is achieved by subtracting the tangential line at $\bar{\phi}'$ from Δf_{L}

$$\Delta\omega_{\text{L}}(\phi'') - \Delta\omega_{\text{bulk}} = \Delta f_{\text{L}}(\phi_{\text{eq}}' + \phi'') - \left[\Delta f_{\text{L}}(\phi_{\text{eq}}') + \frac{d\Delta f_{\text{L}}}{d\phi'}\bigg|_{\phi'=\phi_{\text{eq}}'} \phi''\right]. \tag{8.74}$$

Here, $\Delta\omega_{\text{L}}$ is written in terms of the deviation ϕ'' (8.60) from the average composition ϕ_{eq}. Starting from (8.64) and with (8.69) inserted, the resulting expansion

$$\Delta\omega_{\text{L}}(\phi'') - \Delta\omega_{\text{bulk}} = \frac{A}{2}(T - T_{\text{spin}})\phi''^2 + C\phi_{\text{eq}}'\phi''^3 + \frac{C}{4}\phi''^4 \tag{8.75}$$

is valid in the mixed state $T \geq T_{\text{bin}}(\phi)$ with $\phi_{\text{eq}} = \bar{\phi}$, as well as in the A-rich and B-rich phases for $T \leq T_{\text{bin}}(\phi)$ with $\phi_{\text{eq}} = \phi_{\text{a}}$ and $\phi_{\text{eq}} = \phi_{\text{b}}$, respectively. For later use, we re-write (8.75) by adding and subtracting a term $C\phi_{\text{eq}}'^2\phi''^2$. The added term is completing the square with the 3rd and 4th order term, while the subtracted term is absorbed into the temperature dependence using (8.67) and (8.69), so T_{spin} is eliminated for T_{bin}. The result reads

$$\Delta\omega_{\text{L}}(\phi'') - \Delta\omega_{\text{bulk}} = \frac{A}{2}(T - T_{\text{bin}})\phi''^2 + C\phi''^2\left(\phi_{\text{eq}}' + \frac{1}{2}\phi''\right)^2. \tag{8.76}$$

Contact Values and Pre-wetting Temperature. In order to use (8.53) for the calculation of the interface tension between the mixture and a substrate, we first need to discuss the contact composition ϕ_0. The condition of (8.55) is illustrated in Fig. 8.6. The value of ϕ_0 results from the intersection point of the horizontal line with $\Delta\omega_{\text{L}} - \Delta\omega_{\text{bulk}}$. For a minimum of the resulting interface tension, (8.56) indicates that

Fig. 8.6 LANDAU free energy density $\Delta\omega_L$ versus composition ϕ for several temperatures. The *horizontal line* indicates the effect of the interface potential. For intersection points ϕ_0 with negative slope of $\Delta\omega_L$ the resulting interface tension has a minimum value

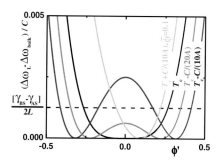

the slope of $\Delta\omega_L - \Delta\omega_{bulk}$ at the intersection has to be negative. For temperatures above and slightly below T_c, there is only one intersection point $\phi'_{0,1}$ with negative slope. The negative value of $\phi'_{0,1}$ corresponds to an A-rich composition, consistent with our assumption of preferred A adsorption to the substrate. The common value of the contact composition for an A-rich phase and a B-rich phase slightly below T_c leads to complete wetting, as discussed in connection with (8.57) and (8.58). The general occurrence of complete wetting slightly below T_c is the content of the CAHN argument [6]. Cahn argued further, that γ_{ab} vanishes with a higher power than the difference $\gamma_{bS} - \gamma_{aS}$ when T approaches T_c from below, and thus the spreading coefficient (8.2) necessarily becomes non-negative close to T_c, indicating complete wetting.

When T becomes lower, the bump in the middle of $\Delta\omega_L(\phi')$ in Fig. 8.6 grows, and at a certain temperature a second intersection point $\phi_{0,3}$ of negative slope is created. An analysis of (8.65) with (8.64) based on the roots of quadratic equations shows that 4 real valued intersection points are realized for $T \leq T_{pw}$ with

$$T_{pw} = T_c - \sqrt{\frac{2C}{\kappa}\frac{[\gamma_{BS} - \gamma_{AS}]}{A}}. \tag{8.77}$$

Since $\phi_b > \phi'_{0,3} > \phi'_{0,1}$, the integration over a positive integrand in (8.53) for the B-rich phase (upper limit $\phi_{eq} = \phi_b$) becomes smaller with the lower limit $\phi'_0 = \phi'_{0,3}$ compared to the case $\phi'_0 = \phi'_{0,1}$. So, for $T \leq T_{pw}$, the contact composition of the B-rich phase at the substrate increases to $\phi'_{0,3}$, in order to minimize the interfacial energy. The deviation of $\phi'_{0,3}$ for the B-rich phase from the contact composition $\phi'_{0,1}$ of the A-rich phase could directly lead to partial wetting for $T \leq T_{pw}$, in case (8.57) yields $|\cos(\Theta_a)| < 1$. In this case, $T_{pw} = T_w$ is the temperature of the wetting transition. The equilibrium volume fractions $\pm\phi'_{pw}$ of the A-rich and the B-rich phase at $T = T_{bin} = T_{pw}$ are calculated from (8.66) with (8.77), or from (8.67) with (8.77):

$$\phi'_{pw} = \sqrt{\frac{A(T_c - T_{pw})}{C}} = \sqrt{\sqrt{\frac{2}{\kappa C}}[\gamma_{BS} - \gamma_{AS}]}. \tag{8.78}$$

Apart from the detailed values (8.77) for T_{pw} and (8.78) for ϕ'_{pw}, the arguments for wetting do not relie on shape details of $\Delta\omega_{mix}$ but apply in general. For $\Delta\omega_L$ as special case, analytical solutions for the de-mixed state are possible. The contact compositions at the substrate result from (8.55) with (8.68) inserted. The correct choice of signs for the square roots in the solutions can be extracted from Fig. 8.6. For an A-rich region at the substrate for $T \leq T_c$, the most negative solution $\phi'_{0a} = \phi'_{0,1}$ indicates the contact composition. B-rich regions at the substrate occur only for $T \leq T_{pw}$. Here, $\phi'_{0b} = \phi'_{0,3}$ is the right choice as contact composition. The results read

$$\phi'_{0a,b} = \mp\sqrt{\phi'^2_{a,b} \pm \sqrt{\frac{2}{\kappa C}}[\gamma_{BS} - \gamma_{AS}]}. \tag{8.79}$$

In (8.79) and the following equations of this section, the upper signs apply to the A-rich phase, while the lower signs describe the B-rich phase. With (8.78), it is possible to rewrite (8.79) as

$$\phi'_{0a,b} = \mp\sqrt{\phi'^2_{a,b} \pm \phi'^2_{pw}}. \tag{8.80}$$

Interface Tensions in the Two Phase Region. The interface tension between the A-rich phase and the B-rich phase in the two phase region results from (8.49) with (8.68) inserted, where the compositions ϕ'_b and $\phi'_a = -\phi'_b$ are the binodal values (8.66):

$$\gamma_{ab} = \frac{4}{3}\sqrt{\frac{\kappa C}{2}}\phi'^3_b. \tag{8.81}$$

The interface tensions γ_{aS} and γ_{bS} between the A-rich phase or the B-rich phase and a substrate are expressed as increments $\Delta\gamma_{aS}$ and $\Delta\gamma_{bS}$ to a constant background contribution $\gamma_{AS} + [\gamma_{BS} - \gamma_{AS}]\phi_c$:

$$\Delta\gamma_{a,bS} = \gamma_{a,bS} - \gamma_{AS} - [\gamma_{BS} - \gamma_{AS}]\phi_c. \tag{8.82}$$

Due to the differences $\gamma_{bS} - \gamma_{aS}$ in (8.1) and (8.2), the constant background does not affect the contact angle and the spreading coefficient, so a discussion based on $\Delta\gamma_{a,bS}$ is sufficient. The integration of (8.53) with (8.68) inserted leads to

$$\Delta\gamma_{a,bS} - [\gamma_{BS} - \gamma_{AS}]\phi'_{0a,b} = \pm\sqrt{\frac{\kappa C}{2}}\left[\frac{\phi'^3_{a,b} - \phi'^3_{0a,b}}{3} - \phi'^2_{a,b}(\phi'_{a,b} - \phi'_{0a,b})\right]. \tag{8.83}$$

With (8.78), the prefactor can be expressed in terms of $(\gamma_{BS} - \gamma_{AS})$. The sign choice in (8.83) originates from the square root in (8.53). It is important to select the sign so the right side of (8.83) is positive: the formation of a concentration profile at a substrate has a positive contribution to the interfacial energy. In order to find the right sign, (8.83) is re-written as

$$\Delta\gamma_{a,bS} - (\gamma_{BS} - \gamma_{AS})\phi'_{0a,b} = \pm\frac{(\gamma_{BS} - \gamma_{AS})}{3\phi'^2_{pw}}\left[-(\phi'_{0a,b} + 2\phi'_{a,b})(\phi'_{0a,b} - \phi'_{a,b})^2\right].$$

(8.84)

On the A-rich side of the binodal, $\phi'_{eq} = \phi'_a \leq 0$ and $\phi'_{0a} < 0$. The bracket on the right side is positive, and the upper positive sign applies to the A-rich side. The B-rich phase neighboring the substrate is found only for $\phi'_{eq} = \phi'_b \geq \phi'_{pw} > 0$ below the pre-wetting transition. For this case, $\phi'_{0b} \geq 0$, the bracket becomes negative, and we have to use the lower negative sign for the B-rich side of the binodal. With (8.80) inserted into (8.84), $\Delta\gamma_{a,bS}$ is calculated as

$$\Delta\gamma_{a,bS} = -\frac{2(\gamma_{BS} - \gamma_{AS})}{3\phi'^2_{pw}}\left[\left(\phi'^2_{a,b} \pm \phi'^2_{pw}\right)^{\frac{3}{2}} \pm \phi'^3_{a,b}\right].$$

(8.85)

For the mixed state, an analytical integration of (8.53) with (8.76) inserted is possible. For better clarity, the mixed phase is treated with reduced variables below.

Contact Angle and Wetting Transition. An overview over the behavior of contact compositions and interface tensions is depicted in Fig. 8.7. The composition $\phi'_b = -\phi'_a$ on the x axis moves along the binodal line and thus with (8.67) implies also a change in temperature. With the chosen scaling of the y-axes, ϕ'_{0a} and ϕ'_{0b} calculated

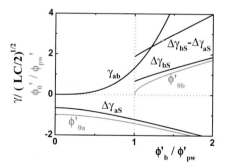

Fig. 8.7 Interface tensions (*black*) and contact compositions (*grey*) around the bulk pre-wetting composition ϕ_{pw} for the LANDAU grand canonical potential density

from (8.80) mark the contributions $(\gamma_{BS} - \gamma_{AS})\phi'_{0a,b}$ of the contact compositions to the interface tensions. Because of additional interfacial energy needed to build up the interfacial concentration profile, the actual interface tension increments $\Delta\gamma_{aS}$ and $\Delta\gamma_{bS}$ from (8.85) are slightly higher. The difference $\Delta\gamma_{bS} - \Delta\gamma_{aS} = \gamma_{bS} - \gamma_{aS}$ exceeds the interface tension γ_{ab} from (8.81) at $\phi'_b/\phi'_{pw} = 1$. Thus, there is no real solution for the contact angle Θ_a from (8.1) and the change of contact composition for a B-rich phase is not accompanied by a first order wetting transition. A continuous transition of Θ_a occurs at the intersection point of $\Delta\gamma_{bS} - \Delta\gamma_{aS}$ and γ_{ab}. When (8.81), (8.85), and (8.78) are inserted into $\Delta\gamma_{bS} - \Delta\gamma_{aS} - \gamma_{ab} = 0$, the square roots can be ellminated by two successive squaring steps. The resulting quadratic equation for $(\phi'_b/\phi'_{pw})^4$ leads to the solution for the wetting transition

$$\frac{\phi'_w}{\phi'_{pw}} = \left[1 + \sqrt{\frac{4}{3}}\right]^{\frac{1}{4}}. \tag{8.86}$$

Since ϕ'_w and ϕ'_{pw} are both on the binodal, (8.67) can be employed to transfer (8.86) to an equation for the temperatures:

$$\frac{T_c - T_w}{T_c - T_{pw}} = \sqrt{1 + \sqrt{\frac{4}{3}}}. \tag{8.87}$$

For a different ω, the morphology of a plot like Fig. 8.7 can change. With higher values of γ_{ab} an intersection point with $\Delta\gamma_{bS} - \Delta\gamma_{aS}$ can be avoided. In this case, a first order pre-wetting transition occurs right at $\phi'_b/\phi'_{pw} = 1$. The distinction of the two morphologies illustrates the discussion of first order and continuous wetting at the end of Sect. 8.5.1.

Shape of the Interface Profile. The form of (8.75) as a power series in ϕ' with only 2nd to 4th powers reminds of the shape of the LANDAU-DE GENNES theory for liquid crystals [14, 21], and so we can borrow an analytical solution from there [20]. The integration in (8.48) with the expansion (8.75) and the correlation length (8.71) inserted can be re-written as

$$\frac{z}{\xi} = \int_{\phi_0}^{\phi(z)} \sqrt{\frac{A(T - T_{spin})}{A(T - T_{spin}) + 2C\phi'_{eq}\phi''^3 + \frac{1}{2}C\phi''^4}} \, d\phi''. \tag{8.88}$$

The scaling of z with ξ shows, that it is in fact the bulk correlation length ξ which determines the length scale of the interfacial profile. A solution of (8.88) is provided by Tarczon and Miyano [22]:

$$\frac{z}{\xi} = \ln\left[\frac{R(\phi'')}{R(\phi_0'')}\right] \tag{8.89}$$

$$R(\phi'') = \frac{1}{\phi''}\sqrt{A(T-T_{\text{spin}})\left[A(T-T_{\text{spin}}) + 2C\phi_{\text{eq}}'\phi'' + \frac{1}{2}C\phi''^2\right]}$$
$$+\frac{1}{\phi''}A(T-T_{\text{spin}}) + C\phi_{\text{eq}}'. \tag{8.90}$$

Solving for ϕ'' yields the profile

$$\phi''\left(\frac{z}{\xi}\right) = \frac{2A(T-T_{\text{spin}})Z}{Z^2 - 2C\phi_{\text{eq}}'Z + C^2\phi_{\text{eq}}'^2 - \frac{1}{2}CA(T-T_{\text{spin}})}$$
$$= \frac{2A(T-T_{\text{spin}})Z}{Z^2 - 2C\phi_{\text{eq}}'Z - \frac{1}{2}CA(T-T_{\text{bin}})} \tag{8.91}$$

$$Z = R(\phi_0'')\exp\left[\frac{z}{\xi}\right]. \tag{8.92}$$

While for $z \gg \xi$ the profile approaches an exponential decay, there are deviations from an exponential for smaller z.

Profiles in the De-Mixed State. A first application of the solution (8.91) are the interface profiles in the de-mixed state. The equilibrium compositions of the A-rich and B-rich phases are on the binodal T_{bin} (8.67), and with (8.69) we get for this case $A(T_{\text{bin}} - T_{\text{spin}}) = 2C\phi_{\text{eq}}'$. This simplification reduces (8.90) and (8.92) substantially, and the profiles result as

$$\phi''(z) = \frac{2\phi_{\text{eq}}'\phi_0''}{[2\phi_{\text{eq}}' + \phi_0'']\exp\left[\frac{z}{\xi}\right] - \phi_0''}. \tag{8.93}$$

The bulk equilibrium composition ϕ_a or ϕ_b enters via ϕ_{eq}' as the deviation (8.59) from the critical composition ϕ_c. Based on (8.66) this deviation is negative for the A-rich phase and positive for the B-rich phase. The contact value ϕ_0'' is expressed as deviation from the bulk equilibrium composition. This deviation is negative for all cases due to the assumed preferential interface adsorption of A. A plot of the A enrichment $-\phi''(z)$ at the interface shows,[13] that the interface profile of an A-rich region is below the exponential decay $\exp(-z/\xi)$, while the profile of a meta-stable B-rich region is above $\exp(-z/\xi)$.

We can use (8.93) for a construction of the profile at internal flat interfaces between A-rich and B-rich phases. Based on (8.60), the deviations from the bulk equilibrium compositions $\phi_0'' = \phi_0' - \phi_{\text{eq}}'$ and $\phi''(z) = \phi'(z) - \phi_{\text{eq}}'$ in (8.93) are eliminated for the deviations from the critical composition ϕ_0' and $\phi'(z)$, respectively. From an

---

[13]For a quick and dirty check, one can plot (8.93) with example values $\phi_a' = -1$, $\phi_b' = +1$, and $\phi_0'' = -1$.

imaginary interface at $z = 0$ the B-rich phase with $\phi'_{eq} = \phi'_b$ expands to positive z, while the A-rich phase with $\phi'_{eq} = \phi'_a = -\phi'_b$ is found at negative z values. These transformations result in

$$\phi'(z) = \mp\phi'_b \frac{[\mp\phi'_b + \phi'_0]\exp\left(\mp\frac{z}{\xi}\right) - [\mp\phi'_b - \phi'_0]}{[\mp\phi'_b + \phi'_0]\exp\left(\mp\frac{z}{\xi}\right) + [\mp\phi'_b - \phi'_0]}. \tag{8.94}$$

The upper signs apply for the A-rich phase at $z < 0$, while the lower ones are for $z > 0$, where the B-rich phase is located. With the choice $\phi'_0 = 0$, so the critical composition in the middle between ϕ'_b and $\phi'_a = -\phi'_b$, (8.94) can be transformed to

$$\phi'(z) = \phi'_b \frac{\exp\left(+\frac{z}{2\xi}\right) - \exp\left(-\frac{z}{2\xi}\right)}{\exp\left(+\frac{z}{2\xi}\right) + \exp\left(-\frac{z}{2\xi}\right)} = \phi'_b \tanh\left(\frac{z}{2\xi}\right). \tag{8.95}$$

The interface extends a the distance $\xi/2$ in positive z direction and also the distance $\xi/2$ in negative direction, so the total interfacial width is again ξ. For any other choice of ϕ'_0 in (8.94) with $|\phi'_0| < \phi'_b$, a similar transformation to a hyperbolic tangents profile with shifted z-location can be performed with the substitution

$$\exp\left(\frac{\Delta z}{\xi}\right) = \frac{\pm\phi'_b + \phi'_0}{\pm\phi'_b - \phi'_0}. \tag{8.96}$$

Reduced Variables. As a preparation for the discussion of pre-wetting, we rewrite the important equations in reduced variables. Only for the critical composition ϕ_c we have $T_{bin} = T_{spin}$. For other compositions, the difference $T_{bin} - T_{spin} = 2C\phi'^2_{eq}/A$ resulting from (8.67) and (8.69) defines a suitable temperature unit which is used for the reduced temperature scale

$$\vartheta = \frac{T - T_{bin}}{T_{bin} - T_{spin}}. \tag{8.97}$$

A frequent combination of variables in the equations is reduced as $A(T - T_{spin})/(2C) = (\vartheta + 1)\phi'^2_{eq}$. A suitable length scale is the correlation length ξ_{bin} at the binodal $T = T_{bin}$. From (8.67), (8.69), and (8.71) we get

$$\xi_{bin} = \sqrt{\frac{\kappa}{2C}}\frac{1}{\phi'_{eq}} \tag{8.98}$$

$$\xi = \frac{\xi_{bin}}{\sqrt{\vartheta + 1}}. \tag{8.99}$$

The deviation ϕ'' from ϕ'_{eq} is also transformed to the reduced variable

$$\phi''_{red} = \frac{\phi''}{\phi'_{eq}},\tag{8.100}$$

and the reduced density of the grand canonical density (8.75) becomes

$$\Delta\omega_{L,red}(\phi''_{red}) - \Delta\omega_{bulk,red} = \frac{\Delta\omega_L - \Delta\omega_{bulk}}{C\phi'^4_{eq}} = (\vartheta + 1)\phi''^2_{red} + \phi''^3_{red} + \frac{1}{4}\phi''^4_{red}.\tag{8.101}$$

When used in the squared gradient expression (8.3), it needs to be combined with a reduced elastic constant $\frac{\kappa}{C\phi'_{eq}} = 2\xi^2_{bin}$. Let's get to the interface equations. The reduced form of (8.55) for the calculation of the reduced contact composition $\phi''_{red,0}$ results with (8.98)–(8.101) and (8.78)

$$\Delta\omega_{L,red}(\phi''_{red,0}) - \Delta\omega_{bulk,red} = \frac{1}{4}\phi''^4_{red,0} + \phi''^3_{red,0} + (\vartheta + 1)\phi''^2_{red,0} = \frac{1}{4}\frac{\phi'^4_{pw}}{\phi'^4_{eq}}.\tag{8.102}$$

Finally, the composition profile at the interface (8.90)–(8.92) is transformed to the reduced form

$$\phi''_{red}\left(\frac{z}{\xi}\right) = \frac{2(\vartheta + 1)Z_{red}}{Z^2_{red} - Z_{red} - \vartheta/4}\tag{8.103}$$

$$Z_{red} = \frac{Z}{2C\phi'_{eq}}$$

$$= \left[\frac{1}{\phi''_{red,0}}\sqrt{(\vartheta + 1)\left[\vartheta + \left[1 + \frac{1}{2}\phi''_{red,0}\right]\right]^2} + \frac{\vartheta + 1}{\phi''_{red,0}} + \frac{1}{2}\right]\exp\left[\frac{z}{\xi}\right].\tag{8.104}$$

With reduced variables, the only remaining free parameters are ϑ and ϕ'_{eq}. The latter will be discussed as a multiple of a scale defined by ϕ'_{pw}. These two parameters are suitable coordinates in the two dimensional phase diagram.

Pre-Wetting Profiles. The reduced equations for the interface structure show the same separation as before. The contact value $\phi''_{red,0}$ depending on ϕ'_{eq} and ϕ'_{pw} result from (8.102), without reference to the actual interface structure. The concentration profile, on the other hand, is calculated for known $\phi''_{red,0}$ based on (8.103) and (8.104), where ϕ'_{eq} and ϕ'_{pw} do not enter explicitly. The temperature enters in both cases.

Examples of interface profiles for different temperatures and two contact values $\phi''_{red,0}$ are displayed in Fig. 8.8. For $\phi''_{red,0} = -2.25$, the thickness of the adsorbed interface layer depends strongly on ϑ. Starting from the interface $z = 0$, there is first a

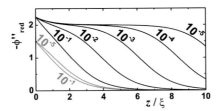

Fig. 8.8 Interface concentration profiles of the mixed phase at a substrate. Profiles with contact value $\phi_{red,0}'' = -2.25$ are drawn in *black*, while profiles for $\phi_{red,0}'' = -1.5$ are shown in *grey*. The *numbers* indicate the reduced temperatures ϑ at which the profiles were calculated

decay to $\phi_{red}'' = -2$ within the length scale ξ. The profile then remains almost flat until a certain thickness depending on ϑ, with a subsequent decay to the bulk composition with $\phi_{red}'' = 0$ within a characteristic distance 2ξ. The occurrence of an interface layer of thickness larger than ξ within the mixed phase is called *pre-wetting*. Note that ξ for the variation of ϑ in the range of small ϑ changes only marginally. While the layer thickness diverges for $\vartheta \to 0$ corresponding to $T \to T_{bin}$, the apparent point of divergence for ξ with (8.99) is $\vartheta \to -1$, or $T = T_{spin}$.

For the example profiles with $\phi_{red,0}'' = -1.5$ in Fig. 8.8, the interface is not able any more to stabilize an interface film of A enrichment $-\phi_{red}'' > 2$, which could increase in thickness. The effect of a temperature change approaching $\vartheta = 0$ is small, and the thickness of the interface profile remains comparable to ξ. The qualitative difference of an interface layer with diverging thickness for $\vartheta \to 0$ on one hand and a layer which almost indifferent thickness for $\vartheta \to 0$ on the other hand indicates, that two different interface states are possible.

The thickness of the interface layer is calculated by an integration over the interface profile

$$d = -\frac{1}{2} \int_0^\infty \phi_{red}'' \, dz. \tag{8.105}$$

The factor $-\frac{1}{2}$ is introduced in order to compensate the composition $\phi_{red}'' = -2$ in a thick layer (see Fig. 8.8). Integration[14] of (8.103) yields [23]

$$\frac{d}{\xi} = \sqrt{1+\vartheta} \ln \left[\frac{\sqrt{\vartheta + \left[1 + \frac{1}{2}\phi_{red,0}''\right]^2} + \sqrt{\vartheta+1} + \frac{1}{2}\phi_{red,0}''}{\sqrt{\vartheta + \left[1 + \frac{1}{2}\phi_{red,0}''\right]^2} + \sqrt{\vartheta+1} - \frac{1}{2}\phi_{red,0}''} \right]. \tag{8.106}$$

---

[14]For the integration of the profile, use a partial fraction decomposition of (8.103)

$$\frac{2(\vartheta+1)Z_{red}}{Z_{red}^2 - Z_{red} - \vartheta/4} = \frac{\sqrt{\vartheta+1}(1+\sqrt{\vartheta+1})}{Z_{red} - \frac{1}{2} - \frac{1}{2}\sqrt{\vartheta+1}} - \frac{\sqrt{\vartheta+1}(1-\sqrt{\vartheta+1})}{Z_{red} - \frac{1}{2} + \frac{1}{2}\sqrt{\vartheta+1}}.$$

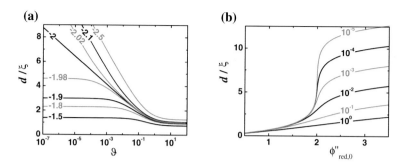

Fig. 8.9 Thickness d of the pre-wetting layer as function of the reduced temperature ϑ for several contact compositions $\phi''_{\text{red},0}$ (**a**), and as function of ϑ for several $\phi''_{\text{red},0}$ (**b**). The numbers indicate the value of the fixed parameter in the calculations

The temperature dependence of d is depicted in Fig. 8.9a. While for values $\phi''_{\text{red},0} > -2$ the layer thickness becomes constant at small ϑ, there is a logarithmic divergence for $\phi''_{\text{red},0} \leq -2$. The critical value $\phi''_{\text{red},0} = -2$ can be deduced from (8.106) with the limiting temperature $\vartheta = 0$ inserted. Only for $\phi''_{\text{red},0} > -2$ the fraction in the logarithm is positive and a solution exists. Figure 8.9b shows the change of d with $\phi''_{\text{red},0}$. At $\phi''_{\text{red},0} = -2$ there is a transition from a thin to a thick layer, which is most pronounced for small ϑ.

Pre-Wetting Contact Composition. The construction of the contact composition according to (8.102) is illustrated in Fig. 8.10. The left side of (8.102) is drawn in black for several temperatures ϑ. The right side of (8.102) indicated by the grey horizontal lines is proportional to ϕ'^{-4}_{eq}. Thus, in order to approach the critical composition ϕ_c from large ϕ'_{eq} to small ϕ'_{eq}, we have to discuss first the lower grey lines and then move upwards. As an orientation, the inset displays a part of the phase diagram Fig. 8.1, with the example compositions marked by round dots. Like in Fig. 8.6, an intersection point in the main figure with negative slope of a black curve with a grey line corresponds to a minimum of interface energy. Since the x-axis is the deviation from the bulk composition, the rightmost intersection point of negative slope is the correct one: it is the first one with increased A content relative to ϕ'_{eq} which matches the minimum condition. All intersection points of negative slope occur for $\phi''_{\text{red},0} < 0$. With (8.100), the reduced variable is negative for an A enrichment $\phi''_0 < 0$ at the interface and a positive bulk composition above the critical composition $\phi'_{\text{eq}} > 0$. Thus, pre-wetting occurs at the B-rich side of the phase diagram (see Fig. 8.1). The other negative combination $\phi''_0 > 0$ and $\phi'_{\text{eq}} < 0$ does not fit to the assumed preferential A adsorption, since it would correspond to a reduced A content at the interface. It can be excluded as an equilibrium state, however it might describe a meta-stable state.

The lowest two grey lines in Fig. 8.10 $\phi'_{\text{eq}}/\phi'_{\text{pw}} = 1.1$ and $\phi'_{\text{eq}}/\phi'_{\text{pw}} = 1$ correspond to compositions on the binodal below the pre-wetting temperature and directly at the pre-wetting temperature, respectively. The rightmost intersection point of negative

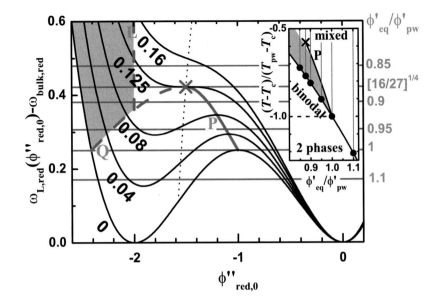

Fig. 8.10 Reduced density $\Delta\omega_{L,red}(\phi''_{red,0}) - \Delta\omega_{bulk,red}$ of the grand canonical potential for several reduced temperatures ϑ, as indicated by the *black* tilted numbers. The contact composition $\phi''_{red,0}$ at the interface results as the rightmost negative intersection of $\Delta\omega_{L,red}(\phi''_{red,0}) - \Delta\omega_{bulk,red}$ with the *grey horizontal lines*. These lines are drawn for different ratios of the bulk composition ϕ'_{eq} and the composition at the start of the pre-wetting line ϕ'_{pw}, as indicated by the *grey numbers* on the right. The pre-wetting line P marks the points where the interface phase transition to higher A contact value $-\phi''_{red,0}$ on the line Q occurs. The pre-wetting line ends at the critical pre-wetting point, marked by $\times$. For the grey area $\phi''_{red,0} \leq 2$ on the left, the logarithmic divergence of the thickness of the interface layer (see Fig. 8.9a) occurs in the limit $\vartheta \to 0$. The first part of line Q and line L mark the trace of conditions, where this logarithmic divergence starts. The *black dotted line* marks the traces of one inflection point of $\Delta\omega_{L,red}(\phi''_{red,0}) - \Delta\omega_{bulk,red}$. The inset shows a magnified section of the phase diagram Fig. 8.1. Beside the binodal, the pre-wetting temperature, the pre-wetting line P and the *grey area* of logarithmic divergence of layer thickness, the example compositions of the main figure are marked by *round points* and *grey lines*

slope moves only slightly when the temperature is increased: the contact composition $\phi''_{red,0}$ is only weakly temperature dependent. The lower value of $\phi''_{red,0}$ for $\phi'_{eq}/\phi'_{pw} = 1.1$ compared to the value for $\phi'_{eq}/\phi'_{pw} = 1$ is due to difference in ϕ'_{eq} in the normalization (8.100).

For the next two grey lines $\phi'_{eq}/\phi'_{pw} = 0.95$ and $\phi'_{eq}/\phi'_{pw} = 0.9$, the same weak temperature dependence of the rightmost negative intersection point $\phi''_{red,0}$ as before is observed for the high temperature examples $\vartheta = 0.16, 0.125$, as well as at $\vartheta = 0.08$ for $\phi'_{eq}/\phi'_{pw} = 0.95$. While for these temperatures there are two intersection points with negative slope of the black curves, there is only one such intersection point for each curve for the lower temperatures. As a consequence, there is a jump of the rightmost intersection point for $\phi'_{eq}/\phi'_{pw} = 0.95$ between $\vartheta = 0.08$ and $\vartheta = 0.04$. The contact composition jumps discontinuously from $\phi''_{red,0} \approx -1.1$ to $\phi''_{red,0} \approx -2.2$.

For $\phi'_{eq}/\phi'_{pw} = 0.9$, a similar jump occurs between $\vartheta = 0.125$ and $\vartheta = 0.08$. These jumps indicate the first order interface phase transition, the pre-wetting transition. The starting and the end points of the jump for different ϑ are marked in Fig. 8.10 by the two branches P and Q of the pre-wetting line, respectively. A qualitative difference between $\phi'_{eq}/\phi'_{pw} = 0.95$ and $\phi'_{eq}/\phi'_{pw} = 0.9$ is the location of the end-point of the jump. While for $\phi'_{eq}/\phi'_{pw} = 0.95$ it is located below $\phi''_{red,0} = -2$ and thus in a range with logarithmic thickness growth of the interface layer with ϑ, the jump for $\phi'_{eq}/\phi'_{pw} = 0.9$ ends with $\phi''_{red,0} > -2$, and thus further cooling is required until the logarithmic growth of the interface layer sets in. The range where values $\phi''_{red,0} \leq -2$ are realized is grey shaded in Fig. 8.10.

The grey line for $\phi'_{eq}/\phi'_{pw} = [16/27]^{\frac{1}{4}}$ passes through the critical pre-wetting point. In this point, the lines P and Q meet, so the jumping distance in $\phi''_{red,0}$ has become zero. The remainder of the jump is an infinite slope of $\phi''_{red,0}$ with ϑ. It indicates the critical pre-wetting transition which is now a second order interface phase transition at the end of the pre-wetting line. For lower values of ϕ'_{eq}/ϕ'_{pw}, there is a steady super-critical change of $\phi''_{red,0}$ with temperature. The curve for $\phi'_{eq}/\phi'_{pw} = 0.85$ is an example for this case.

The first section of curve Q combined with curve L in Fig. 8.10 indicates the locations where the logarithmic thickness divergence for small ϑ starts. Note, that such a divergence is possible for any choice $0 \leq \phi'_{eq} \leq \phi'_{pw}$, so any composition ϕ above ϕ_c and below the composition of the B-rich phase at the wetting line.

It remains to determine the lines P and Q and discuss the transformation of P to the phase diagram. For a selected temperature ϑ, the jump occurs at the local maximum of $\Delta\omega_{L,red}(\phi''_{red}) - \Delta\omega_{bulk,red}$ in Fig. 8.10. Thus, the abscissa value $\phi''_{red,0,P}$ is calculated as the location of the local maximum of (8.101):

$$\phi''_{red,0,P}(\vartheta) = -\frac{-3 + \sqrt{1 - 8\vartheta}}{2}. \tag{8.107}$$

The temperature range $0 \leq \vartheta < \frac{1}{8}$ where (8.107) has a real solution defines the range of the pre-wetting line. The ordinate at the maximum position results after inserting (8.107) in (8.101). With (8.102), the ordinate can be transformed to the bulk composition ϕ'_{eq}. After simplification, it becomes

$$\frac{1}{4}\frac{\phi'^4_{pw}}{\phi'^4_{eq}} = \frac{1}{8}\left[1 + 20\vartheta - 8\vartheta^2 + (1 - 8\vartheta)^{\frac{3}{2}}\right]. \tag{8.108}$$

The combination of (8.107) and (8.108) is a parametric representation of the P line in Fig. 8.10. The end-point $\vartheta = \frac{1}{8}$ inserted in (8.107) and (8.108) yields the coordinates of the critical pre-wetting point $\phi''_{red,0} = -\frac{3}{2}$ and $\phi'_{eq}/\phi'_{pw} = [16/27]^{\frac{1}{4}}$. In order to calculate also the Q line, we subtract the height of the local maximum (8.108) from (8.101). Since the compositions at the start and the end of the jump have the same ϕ'_{eq}, they have the same height in Fig. 8.10. So, the Q line results as a root of the equation after substraction. By construction, the fourth order polynomial resulting from the subtraction has a double root at the maximum position (8.107). A polynom division allows to separate these known roots, and the Q line can be determined as root of the resulting second order polynomial. These calculations require tedious book keeping of terms, but otherwise are straight forward. The abscissa $\phi''_{\text{red},0,Q}$ of the Q line reads

$$\phi''_{\text{red},0,Q}(\vartheta) = -\frac{1}{2} - \frac{1}{2}\sqrt{1 - 8\vartheta} - \sqrt{1 + \sqrt{1 - 8\vartheta}}. \tag{8.109}$$

By construction, the ordinate of the Q line is the same as for the P line (constant value of ϕ'_{eq} on both sides of the jump) and thus results also from (8.108). It can be cross-checked numerically by inserting (8.109) into (8.101). In a similar way, all relevant points in Fig. 8.10 can be assigned to values of ϑ through the temperature curve passing through the point and ϕ'_{eq} by identifying the height of a point with $\frac{1}{4}\frac{\phi'^4_{\text{pw}}}{\phi'^4_{\text{eq}}}$ according to (8.102).

For a transfer of the pre-wetting line P or Q to the phase diagram as in the inset of Fig. 8.10, one might first draw the binodal as $T_{\text{bin}} = T_c - (T_c - T_{\text{pw}})\left(\frac{\phi'_{\text{eq}}}{\phi'_{\text{pw}}}\right)^2$. According to (8.67) and (8.69), the curvature of the spinodal is 3 times the curvature of the binodal, so $T_{\text{spin}} = T_c - 3(T_c - T_{\text{pw}})\left(\frac{\phi'_{\text{eq}}}{\phi'_{\text{pw}}}\right)^2$. With $T_{\text{bin}} - T_{\text{spin}} = 2(T_c - T_{\text{pw}})\left(\frac{\phi'_{\text{eq}}}{\phi'_{\text{pw}}}\right)^2$ a connection to the reduced temperature scale ϑ (8.97) is established:

$$\vartheta = \frac{1}{2}\frac{T - T_c}{T_c - T_{\text{pw}}}\left(\frac{\phi'_{\text{pw}}}{\phi'_{\text{eq}}}\right)^2 + \frac{1}{2}. \tag{8.110}$$

The ordinate of the pre-wetting line $\frac{\phi'_{\text{eq}}}{\phi'_{\text{pw}}}(\vartheta)$ is calculated from (8.108), and it can be used to evaluate the inversion of (8.110) $T\left(\frac{\phi'_{\text{eq}}}{\phi'_{\text{pw}}}, \vartheta\right)$. Again the combination of these two equations is a parametric representation of the pre-wetting line in terms of ϑ in the range $0 \leq \vartheta \leq \frac{1}{8}$.

Interface Tension in the Mixed State. The integral (8.53) for the calculation of γ with the reduced grand canonical potential density (8.101) inserted reads

$$\gamma = \gamma_{\text{AS}} + [\gamma_{\text{BS}} - \gamma_{\text{AS}}][\phi_c + \phi'_{\text{eq}}(1 + \phi''_{\text{red},0})]$$
$$\pm \int_{\phi''_{\text{red},0}}^{0} \phi'^2_{\text{eq}}\sqrt{\frac{\kappa C}{2}}\sqrt{4\vartheta\phi''^2_{\text{red}} + (\phi''_{\text{red}} + 2)^2\phi''^2_{\text{red}}}\,(\phi'_{\text{eq}}d\phi''_{\text{red}}). \tag{8.111}$$

A substitution based on (8.100) to the reduced composition deviation ϕ''_{red} from ϕ'_{eq} is already performed, which has resulted in a differential $d\phi' = \phi'_{eq}d\phi''_{red}$ and the upper limit equal to 0. As before, it is important to select the sign of the integral in a way to keep it positive. The not reduced deviation ϕ'' from ϕ'_{eq} is generally non positive for preferential adsorption of A to a substrate. The extension of the relevant range to negative values can also be seen in Fig. 8.10. For $\phi_{eq} < \phi_c$, on the A-rich side relative to the critical composition, ϕ'_{eq} is negative and so with (8.100) $\phi''_{red} > 0$. The positive lower limit $\phi''_{red,0} > 0$ results in a factor (-1), which is compensated by the negative prefactor in the differential $(\phi'_{eq}d\phi''_{red})$. For $\phi'_{eq} > \phi_c$, on the B-rich side relative to ϕ_c, $\phi'_{eq} > 0$, so the differential prefactor is positive. With $\phi''_{red} < 0$, the lower limit is smaller than the upper limit 0, so there is also no sign contribution from the integral.

A first transformation is the extraction of a factor $\sqrt{\phi''^2_{red}} = \pm\phi''_{red}$ from the square root. Based on the discussed signs of ϕ''_{red}, we have to select the upper positive sign for the A-rich side $\phi'_{eq} < 0$, and the negative lower sign for the B-rich side $\phi'_{eq} > 0$. Another substitution $u = \phi''_{red} + 2$ leads to a sum of two integrals which are Bronstein integrable [23]. With (8.78) and (8.82), the result is expressed as increment to the constant background in terms of $(\gamma_{BS} - \gamma_{AS})$:

$$\Delta\gamma = (\gamma_{BS} - \gamma_{AS})\phi'_{eq}(1 + \phi''_{red,0})$$

$$\pm \frac{4(\gamma_{BS} - \gamma_{AS})\phi'^3_{eq}}{3\phi'^2_{pw}} \left\{ \left[1 - 2\vartheta - \frac{\phi''_{red,0}}{2} - \frac{\phi''^2_{red,0}}{2}\right]\sqrt{\left[1 + \frac{\phi''_{red,0}}{2}\right]^2 + \vartheta} \right.$$

$$\left. -[1 - 2\vartheta]\sqrt{1 + \vartheta} + 3\vartheta \ln\left[\frac{1 + \frac{1}{2}\phi''_{red,0} + \sqrt{\left[1 + \frac{1}{2}\phi''_{red,0}\right]^2 + \vartheta}}{1 + \sqrt{1 + \vartheta}}\right] \right\}. \quad (8.112)$$

The sign choice cancels with the sign of ϕ'_{eq}, so $\pm\phi'^3_{eq} = -|\phi'_{eq}|^3$.

Comparison of Binodal Values. The binodal line $T_{bin}(\phi)$ marks the boundary between the two-phase region and the one phase region. We can cross-check the equations for ϕ'_0 and γ for these two regions by a comparison on the binodal line, where the results should match. For the binodal as limiting case of the one phase region, ϕ''_0 results from (8.101) for $\vartheta = 0$. The resulting curve in Fig. 8.10 is symmetric relative to $\phi''_{red,0} = -1$. So, the transformation of (8.101) in terms of a variable $(\phi''_{red,0} + 1)$ results in a fourth order polynomial with only even powers. The roots are found by double application of the formula for quadratic equations, and $\phi''_{red,0}$ for $\vartheta = 0$ is calculated as:

$$\phi''_{red,0} = -1(\mp)_2\sqrt{1(\pm)_1\frac{\phi'^2_{pw}}{\phi'^2_{eq}}}. \quad (8.113)$$

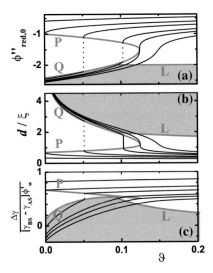

Fig. 8.11 Reduced contact composition $\phi''_{red,0}$ (**a**), layer thickness d/ξ (**b**), and reduced interface tension increment $\Delta\gamma/([\gamma_{BS} - \gamma_{AS}]\phi'_{pw})$ (**c**) plotted against reduced temperature ϑ. The *black lines* are calculated for the same compositions $\phi'_{eq}/\phi'_{pw} \in \{0.85, [16/27]^{\frac{1}{4}}, 0.9, 0.95, 1, 1.1\}$ as in Fig. 8.10. In (**a**) and (**c**), ϕ'_{eq}/ϕ'_{pw} decreases when going from *upper lines* to the *lower lines*, while in (**b**) it increases. The *shaded regions* limited by the lines Q and L indicate the areas of logarithmic divergence of layer thickness. The *grey thick lines* P and Q mark start and end of the jump in a first order pre-wetting transition

Here, $(\pm)_1$ and $(\mp)_2$ indicate the sign choices in the first and second application of the formula for quadratic equations. For $|\phi'_{eq}| < \phi'_{pw}$, there are only two real solutions of (8.113), and we have to select the negative one for preferential A adsorption to the substrate: $(\pm)_1 = +, (\mp)_2 = -$. For $|\phi'_{eq}| \geq \phi'_{pw}$, there are four real solutions. For an A-rich region close to the substrate the previous solution remains valid. For a B-rich region at the substrate, the right choice of signs in (8.113) is $(\pm)_1 = -, (\mp)_2 = +$, in order to obtain the negative solution of smallest magnitude. This solution corresponds to the selection of intersection point $\phi_{0,3}$ in Fig. 8.10 for $\phi'_{eq}/\phi'_{pw} > 1$. The signs in (8.113) are again written in a form where the upper signs describe the A-rich side, while the lower signs stand for the B-rich side of the phase diagram. As a result, the limit of (8.113) for $\vartheta = 0$ from the side of the mixed phase is the reduced form of (8.80), which indicates the contact values in the two phase region.

When (8.113) is inserted into (8.112) for the one-phase limiting case $\vartheta = 0$ and the prefactor expressed in terms of $(\gamma_{BS} - \gamma_{AS})$ based on (8.78), one recovers the two-phase formula (8.85). The consistent and steady behavior of ϕ'_0 and γ at the binodal has been used as a cross check of signs for the formulas.

How to see the Pre-Wetting Transition in an Experiment? Experimental parameters which might serve for the detection of the pre-wetting transition are the layer thickness d and the interface tension γ. A direct experimental access to the contact composition $\phi''_{red,0}$ appears more difficult. The temperature dependence of these three parameters is shown in Fig. 8.11 for the same selection of bulk compositions ϕ'_{red}/ϕ'_{pw} as in Fig. 8.10. The contact composition in Fig. 8.11a was determined from (8.102) by a simple numeric NEWTON-RAPHSON Method [24]. The behavior is already discussed in connection with Fig. 8.10: for $\phi'_{eq}/\phi'_{pw} = 1.1$ and $\phi'_{eq}/\phi'_{pw} = 1$ in the upper two lines, there is only a marginal increase of $\phi''_{red,0}$ when ϑ is decreased. For $\phi'_{eq}/\phi'_{pw} = 1$, the critical composition $\phi''_{red,0} = -1$ is reached for $\vartheta = 0$. The two examples $\phi'_{eq}/\phi'_{pw} = 0.95$ and $\phi'_{eq}/\phi'_{pw} = 0.9$ show the jump in $\phi''_{red,0}$ at the first order pre-wetting transition. For $\phi'_{eq}/\phi'_{pw} = [16/27]^{\frac{1}{4}}$, the pre-wetting transition becomes second order: there is no longer a jump, but an infinite slope of $\phi''_{red,0}(\vartheta)$. In the supercritical example $\phi'_{eq}/\phi'_{pw} = 0.85$, there is a continuous transition from strongly to a moderately negative values of $\phi''_{red,0}$.

Based on the numerical solution for $\phi''_{red,0}$, the temperature dependencies of d (Fig. 8.11b) and γ (Fig. 8.11c) are calculated with (8.106) and (8.112), respectively. Also the P, Q, and L lines were transferred to lines in Figs. 8.11b, c. Similar to the behavior of $\phi''_{red,0}$, γ and d have jumps in the first order pre-wetting transition. For γ, the jumps become very small at higher ϑ. The infinite slopes in the second order pre-wetting critical point is found only for d. The interface tension is thus less suitable for a determination of the pre-wetting transition. With the known contact compositions on the P line (8.107) and the Q line (8.109) before and after the jump, on can use (8.106) and (8.112) to determine the mean field cirtical exponents when ϑ approaches the the critical pre-wetting value $\vartheta = \frac{1}{8}$. As a result, $[\phi''_{red,0,P} - \phi''_{red,0,Q}]$ and $[d(\phi''_{red,0,P}) - d(\phi''_{red,0,Q})]$ vanish with an exponent $\frac{1}{2}$, while $[\gamma(\phi''_{red,0,P}) - \gamma(\phi''_{red,0,Q})]$ approaches zero with an exponent 2.

The occurrence of a thick pre-wetting layer is accompanied by a reduction of the interface tension. So, the A component acts like a surfactant on the B-rich side of the one phase region. The possibility to switch this surfactant in the first order pre-wetting transition by a small temperature change might be of technological interest. Due to the scaling with the interface tension difference $[\gamma_{BS} - \gamma_{AS}]$, this surfactant has different strength on different substrates, which might be used for selectivity. An example could be a separation of a mixture of colloids or nano-particles of different material as substrates within a binary mixture as dispersion medium.

Interface Fluctuations A discussion of interface fluctuations for a binary system similar to the treatment of interface fluctuations in liquid crystals [25], which is an interesting field of study in its own right, needs to be postponed to a separate publication. Experimental results for the fluctuation amplitude and the fluctuation dynamics of a liquid crystalline pre-wetting layer are found in [26, 27], respectively.

LANDAU **Theory: Summary.** The LANDAU theory, where all quantities except the contact composition in the mixed phase can be calculated analytically, is a very well suited didactic tool for an illustration of the interface calculations based on the squared gradient approach. For a grand canonical density different from $\Delta\omega_L$, the same calculations can be performed numerically. The general morphology of the phase diagram should remain similar in this case.

8.6 Summary and Outlook

The squared gradient approach allows a description of interfaces and the bulk phases with a common set of parameters. A combination of interactions and entropy contribution determines the phase behavior of the bulk, as well as the contact composition at a substrate. The correlation length of concentration fluctuations in the bulk phase sets the length scale of an interfacial concentration profile. A number of parameters for the description of interfaces can be extracted already from bulk results. Thus, interfaces are no longer exceptionally complicated exotic objects in an otherwise perfect bulk world. The complication in real interface experiments stems to a large part from the highly demanding purity requirements in preparation.

The squared gradient theory is limited, since interfaces with a too steep profile cannot be described. This limitation is less severe for truly soft systems, where structural changes expand over a correlation length. So, the theory is well adapted to the focus of SOMATAI. Beside the derivation of the equilibrium state, the theory allows the calculation of fluctuation amplitudes and restoring forces, which give access to a basic understanding of relaxation dynamics, if a suitable friction constant is known. We touched the distinction between meta-stable states and thermodynamically stable states in the two phase region of the bulk between the binodal and the spinodal only briefly here. An access to meta-stable states at the interface could be constructed on similar grounds. Such meta-stable states appear to be highly relevant for interface rheology, where often the interface structure is frozen and not at all in a thermodynamic equilibrium. In a recent workshop "Dynamics of complex fluid-fluid interfaces" organized by Peter Fischer and others from ETH Zürich at Monte Verita in Ascona, a majority of interface rheology contributions appeared to have frozen non equilibrium structures. It might be even speculated, that modern applications of interface science, e.g. the creation of new food products with low fat and sugar content in food science require frozen non-equilibrium structures at the internal interfaces of the product. Here, the squared gradient approach could provide an access to a theoretical description of meta-stable states, which might be able to predict food live-times based on breakdown of frozen structures at internal interfaces by relaxation processes.

8.7 Exercises

The SOMATAI initial training network and also its summer school to which this contribution was delivered intend to train young researchers. An essential point of training is to practice the concepts under study. Exercises of the tutorial during the SOMATAI summer school are listed below.

1. Consider the Figure showing partial and complete wetting of the vapor phase by the A-rich phase, which has a higher density than the B-rich phase and thus mainly is at the bottom of the container.

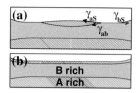

a. Which phase has the higher interface tension against the container wall? Hint: look at the contact angle at the container wall.

b. Write down the force balance for the three phase contact line for partial wetting and express the interface energies of A and B against the vapor phase by the contact angles and the interface tension γ_{AB} of the A-rich phase versus the B-rich phase.

2. Assume that the binodal $T_{bin}(\phi)$ of an A and B mixture is characterized by the equation

$$\frac{T_{bin}(\phi)}{K} = 1600\phi^2 - 1600$$

a. Determine the critical point.

b. A mixture 20 mL of A and 40 mL of B is prepared at the temperature 300 K. Is the mixture in a one phase state, or in a two phase state?

c. Determine the compositions of the two phases.

d. Determine the volumes of the two phases.

3. Consider a blend of polystyrene PS (C_8H_8, $\rho = 1.04\,\text{g/cm}^3$) of molecular mass $M_w = 25\,000\,\text{g/mol}$ and polyisoprene PI (C_5H_8, $\rho = 0.93\,\text{g/cm}^3$) of molecular mass $M_w = 20\,000\,\text{g/mol}$. Use the polystyrene monomer volume as cell volume v_c.

a. Determine the degrees of polymerization N_{PS} and N_{PI}', and the effective degree of polymerization for N_{PI} to be used in a FLORY-HUGGINS description.

b. For an interaction parameter $\chi = 0.1$, do you expect a mixed (one phase) or a de-mixed (two phase) state at the temperature $T = 150\,^\circ C$?

4. Derive (8.36).

5. Linear response theory: start from the quadratic approximation (8.23) for $\Delta\omega_{mix}$ and consider the effect of an external field h by an additional linear term $h\phi$. Determine the new equilibrium position $\phi_{eq}(h) = \phi_{eq} + \Delta\phi$ with h and show that the response $\Delta\phi = \phi_{eq}(h) - \phi_{eq}(h = 0)$ is linear in h. Determine the response coefficient $\frac{d\phi_{eq}}{dh}$.

6. Derive (8.70) and (8.71) starting from (8.75).

8.8 Appendix

8.8.1 Calculus of Variations

A brief summary of functionals and their derivatives clearly remains insufficient from
a mathematical point of view. However, it provides the techniques to perform the
calculus of variations employed in this contribution. We follow the presentation of
Großmann, and recommend his book for a detailed discussion on solid mathematical
grounds [28].

Starting point is a space $\mathscr{F}$ of functions $f : \mathscr{M} \mapsto \mathbb{R}$, which can be chosen with
sufficiently tame properties. Here $\mathscr{M} \subset \mathbb{R}^n$ is the space of arguments for the functions, for our purposes the real axis $\mathscr{M} = \mathbb{R}$, the 3D space $\mathscr{M} = \mathbb{R}^3$ for bulk problems, or $\mathscr{M} = \mathbb{R}^2 \times \mathbb{R}_+$ for interface problems. The latter describes the half space
with only non-negative z values. The tame functions $f, g \in \mathscr{F}$ are assumed to be
steady and sufficiently often differentiable. For two functions $f, g \in \mathscr{F}$ the inte-
gration of the product fg over $\mathscr{M}$ is considered as scalar product $\langle f|g \rangle$ over $\mathscr{F}$. A
functional

$$F : \mathscr{F} \to \mathbb{R} \qquad\qquad (8.114)$$
$$f \mapsto r \in \mathbb{R}$$

maps $\mathscr{F}$ to the real numbers $\mathbb{R}$ and assigns to each $f \in \mathscr{F}$ a real number r.

As a first example, a fixed function $g \in \mathscr{F}$ can be chosen, which forms the kernel
in an integral

$$F[f] = \int_{\mathscr{M}} g(\mathbf{x}) f(\mathbf{x}) \, \mathrm{d}^n x = \langle g|f \rangle . \qquad\qquad (8.115)$$

So, $f(\mathbf{x})$ is mapped by the functional F to the real number calculated as the integral over the product $g(\mathbf{x}) f(\mathbf{x})$, so the scalar product mentioned above. The squared
brackets of $F[.]$ which were used in (8.3), (8.7), (8.43), and (8.51) indicate the dependence of the functional F on the function f. A more general functional can take
also derivatives of $f(x)$ into account, and the squared gradient theories in the form
(8.3) are examples for such functionals. A second example to assign a real value to
a function $f(x)$ defined over an interval $\mathscr{I}$ is to select a specific point $x_0 \in \mathscr{I}$ and to
evaluate $f(x_0)$. Formally, the corresponding functional $\delta_{x_0}[f] = f(x_0)$ can be written
in a similar form as in (8.115):

$$\delta_{x_0}[f] = \int_{\mathscr{I}} \delta(x - x_0) f(x) \, \mathrm{d}x. \qquad\qquad (8.116)$$

The kernel $\delta(x - x_0)$ of $\delta_{x_0}[.]$ is usually called a delta function, although it cannot
be defined in a mathematically consistent way as a function. The interpretation of
(8.116) as a formal notation for the functional which assigns to each $f \in \mathscr{F}$ its value

$f(x_0)$ gives a more consistent picture. The concept of a distribution which forms the basis for $\delta(x - x_0)$ is also discussed by Großmann [28].

The calculation of an equilibrium interface profile is traced back to a minimization of a functional (8.43) which yields the interfacial energy for arbitrary interfacial profiles. Similar to a discussion of the minimum of a function, the minimum is found by setting the derivative equal to zero. However, now we have to use a functional derivative, not the normal derivative. The derivative of the functional F at the 'position' $f \in \mathscr{F}$ in the 'direction' $h \in \mathscr{F}$ is calculated as:

$$\frac{\partial F}{\partial f}[h] = \lim_{\varepsilon \to 0} \frac{F[f + \varepsilon h] - F[f]}{\varepsilon}. \tag{8.117}$$

Under suitable conditions for a steady behavior of $\frac{\partial F}{\partial f}[h]$ in f and h, the existence of functional derivations in specific directions (8.117) implies the existence of a unique functional derivative

$$\frac{\delta F}{\delta f}[h] = \int_{\mathscr{M}} \frac{\delta F}{\delta f(\mathbf{x})} h(\mathbf{x}) \, d^n x = \left(\left. \frac{\delta F}{\delta f} \right| h \right). \tag{8.118}$$

The resulting functional derivative $\frac{\delta F}{\delta f}$ is again a functional in h. It can be written as an integral over the kernel $\frac{\delta F}{\delta f(x)}$, which might be either a regular integration, as in (8.115), or a formal integral similar to (8.116). While $F[h]$ might contain also derivatives of h, they can be removed by partial integration in order to reach the representation of $\frac{\delta F}{\delta f}[.]$ as an integral which contains h only and no derivatives of h. In this way, it can be expressed as a scalar product in the second form of (8.118). Removing any derivative of h is important when the first functional derivative is set equal to zero in order to find a function with an extremal value. For the mathematically sloppy discussion we are bound to in this brief appendix, we can assume the existence of the functional derivative $\frac{\delta F}{\delta f}$ of (8.118), if the limit in (8.117) leads to an expression which is linear in h, so a presentation as scalar product in (8.118) is possible with a kernel $\frac{\delta F}{\delta f(x)}$ independent of h.

A well known example in physics is the functional derivative of the action

$$F[x] = \int_{t_1}^{t_2} L(x, \dot{x}, t) \, dt. \tag{8.119}$$

of the differentiable trace $x(t)$ with speed $\dot{x} = \frac{dx}{dt}$ of a particle which starts at time t_1 and ends at time t_2. Here, $L(x, v, t)$ is a two times differentiable LAGRANGE-function. For the potential energy $V(x)$ it reads $L(x, v, t) = \frac{1}{2}mv^2 - V(x)$. The functional derivative in the 'direction' $h(t)$ is now calculated according to (8.117) as

$$\frac{\partial F}{\partial x}[h] = \lim_{\varepsilon \to 0} \frac{1}{\varepsilon} \left\{ \int_{t_1}^{t_2} L(x(t) + \varepsilon h(t), \dot{x}(t) + \varepsilon \dot{h}(t), t)\, dt - \int_{t_1}^{t_2} L(x(t), \dot{x}(t), t)\, dt \right\}$$

$$= \int_{t_1}^{t_2} \left[\frac{\partial L}{\partial x} h(t) + \frac{\partial L}{\partial v} \dot{h}(t) \right] dt. \tag{8.120}$$

The arguments of the LAGRANGE-function $L(x(t), \dot{x}(t), t)$ are not written explicitly in (8.120) and will also be omitted in the following equations, for facility of inspection. The time derivative $\dot{h}(t)$ can be removed by integration by parts, which leads to the form

$$\frac{\partial F}{\partial x}[h] = \int_{t_1}^{t_2} \left[\frac{\partial L}{\partial x} - \frac{d}{dt}\frac{\partial L}{\partial v} \right] h(t)\, dt + \left[h(t)\frac{\partial L}{\partial v} \right]_{t_1}^{t_2}$$

$$= \int_{t_1}^{t_2} \left[\frac{\partial L}{\partial x} - \frac{d}{dt}\frac{\partial L}{\partial v} + \frac{\partial L}{\partial v}[\delta(t - t_2) - \delta(t - t_1)] \right] h(t)\, dt. \tag{8.121}$$

Since (8.121) is linear in h, the functional derivative in the direction h, $\frac{\partial F}{\partial x}$, represents the functional derivative $\frac{\delta F}{\delta x}$, which does not depend on the choice of h. Its kernel can be extracted from the integral of (8.121) and reads

$$\frac{\delta F}{\delta x(t)} = \frac{\partial L}{\partial x} - \frac{d}{dt}\frac{\partial L}{\partial v} + [\delta(t - t_2) - \delta(t - t_1)]\frac{\partial L}{\partial v}. \tag{8.122}$$

The test function $h(t)$ describes the variation of the path $x(t)$, see (8.117). For fixed starting and end points of a path only variations with $h(t_1) = h(t_2) = 0$ are considered, and the boundary terms in (8.122) indicated by the delta functions do not contribute. For a path of extremal action, the functional derivative (8.121) needs to vanish for any such variation h. Thus, the kernel (8.122) needs to vanish, and we arrive at the EULER- LAGRANGE equation

$$0 = \frac{\delta F}{\delta x(t)} = \frac{\partial L}{\partial x} - \frac{d}{dt}\frac{\partial L}{\partial v}. \tag{8.123}$$

Inserting the arguments of $L(x(t), \dot{x}(t), t)$ into (8.123) leads to a differential equation, the solution of which is the path of the particle.

A second example adapted to interface problems starts from the functional

$$F[\phi] = \int_{-\infty}^{+\infty} dx \int_{-\infty}^{+\infty} dy \int_{0}^{+\infty} dz \left[\omega(\phi) + \frac{\kappa}{2}\left[\left[\frac{\partial \phi}{\partial x}\right]^2 + \left[\frac{\partial \phi}{\partial y}\right]^2 + \left[\frac{\partial \phi}{\partial z}\right]^2 \right] \right]. \tag{8.124}$$

While x and y range from $-\infty$ to $+\infty$, z extends only from 0 which is the location of the interface to $+\infty$. The LAGRANGE function is extracted from (8.124):

$$L\left(\phi, \frac{\partial \phi}{\partial x}, \frac{\partial \phi}{\partial y}, \frac{\partial \phi}{\partial z}\right) = \omega(\phi) + \frac{\kappa}{2}\left[\left[\frac{\partial \phi}{\partial x}\right]^2 + \left[\frac{\partial \phi}{\partial y}\right]^2 + \left[\frac{\partial \phi}{\partial z}\right]^2\right]. \quad (8.125)$$

We consider only variations h which vanish for $x, y \to \pm\infty$ and $z \to +\infty$. A calculation similar to (8.120) leads to the kernel of $\frac{\delta F}{\delta \phi}$:

$$\frac{\delta F}{\delta \phi(x, y, z)} = \frac{\partial \omega}{\partial \phi} - \kappa\left[\frac{\partial^2 \phi}{\partial x^2} + \frac{\partial^2 \phi}{\partial y^2} + \frac{\partial^2 \phi}{\partial z^2} - \delta(z)\frac{\partial \phi}{\partial z}\right]. \quad (8.126)$$

Since the squared gradients in x and y in (8.124) give positive contributions, they need to vanish for the equilibrium interface profile. So (8.124) and (8.126) effectively depend on the gradient in z direction only and thus the problem becomes a one dimensional. Apart from the contact potential, (8.124) corresponds to (8.51), and (8.126) is similar to (8.52). Also bulk fluctuations are described by (8.126), if the kernel is integrated over the full $\mathbb{R}^3$. The boundary term at the interface with $\delta(z)$ vanishes, if we consider variations with $h(z) = 0$, so for a profile with fixed value at the interface. When $\phi(z = 0)$ is not fixed but determined by an interface potential, the gradient in the boundary term affects the boundary condition for fluctuation modes in the second functional derivative.

For the calculation of the second functional derivative $\frac{\delta^2 F}{\delta f^2}$, one considers the first derivative $\frac{\delta F}{\delta f}[h]$ as a functional in f for fixed 'direction' h. For this functional in f, the functional derivative is calculated with the same rules. The second derivative can be written either as an integral, or as a bi-linear mapping, similar to the scalar product (8.115):

$$\frac{\delta^2 F}{\delta f^2}[h, k] = \int_{\mathcal{M}} h(\mathbf{x})\frac{\delta^2 F}{\delta^2 f(\mathbf{x})}k(\mathbf{x})\, dx = \left\langle h \left| \frac{\delta^2 F}{\delta f^2} \right| k \right\rangle. \quad (8.127)$$

We do not look for an extremum with the second derivative, and so we do not set the kernel (8.127) equal to zero, as we did in (8.123). So, it is not required and often not possible to remove the derivatives of k by partial integration, and thus $\frac{\delta^2 F}{\delta^2 f(x)}$ might contain derivative operators acting on k. In case the derivative of a product of a term included in $\frac{\delta^2 F}{\delta^2 f(x)}$ and $k(x)$ is encountered, it can be resolved with the product rule of derivation for improved clarity. A function f which results in an extremum as determined by the first functional derivative and its derivative can be inserted to the LAGRANGE Function of (8.127), in order to build up a bi-linear form for fluctuations around the equilibrium profile. This bi-linear form allows the definition of orthogonality of fluctuation modes. Eigen-fluctuations are the eigen-modes of the bi-linear form, and if these eigen-fluctuations have different eigen-values they are necessarily orthogonal. Higher derivatives can be calculated along the same lines, and a TAYLOR expansion for a functional reads

$$F[f + h] \approx F[f] + \frac{\delta F}{\delta f}[h] + \frac{1}{2}\frac{\delta^2 F}{\delta f^2}[h, h] + \ldots + \frac{1}{n!}\frac{\delta^n F}{\delta f^n}[h, h, \ldots, h]. \quad (8.128)$$

As an illustration of the second derivative, we use the second example (8.124), start from the first derivative (8.118) with (8.126) inserted, and perform a limit calculation with a second variation with respect to $k(t)$ similar to (8.120). The kernel operator of the resulting bilinear form (8.127) results as

$$\frac{\delta^2 F}{\delta f^2}[h, k] = \frac{\partial^2 L}{\partial x^2} - \left(\frac{d}{dt}\frac{\partial^2 L}{\partial x \partial v}\right) - \left(\frac{d}{dt}\frac{\partial^2 L}{\partial v^2}\right)\frac{d}{dt} - \frac{\partial^2 L}{\partial v^2}\frac{d^2}{dt^2}$$
$$+ [\delta(t - t_2) - \delta(t - t_1)]\left[\frac{\partial^2 L}{\partial x \partial v} + \frac{\partial^2 L}{\partial v^2}\frac{d}{dt}\right]. \quad (8.129)$$

8.8.2 List of Important Symbols

Symbol	First Occurrence	Meaning
$\tilde{A}$	(8.27)	scattering amplitude
α	(8.19)	length ratio of the polymers of a blend
χ	(8.13)	FLORY-HUGGINS interaction parameter
$F, \Delta F, \Delta f$	(8.4), (8.3), (8.8)	free energy, free energy increment and its density
$\Delta F_{\text{mix}}, \Delta f_{\text{mix}}$	(8.12), (8.16)	Free energy of mixing and its density, FLORY-HUGGINS theory
Δf_{L}	(8.61)	free energy density LANDAU theory
$\Delta\Omega, \Delta\omega_{\text{mix}}$	(8.3), (8.22)	grand canonical potential increment and its density
$\Delta\omega_{\text{L}}, \Delta\omega_{\text{L,red}}$	(8.65), (8.101)	density grand canonical potential LANDAU theory, reduced form
$\Delta\omega_{\text{bulk}}, \Delta\omega_{\text{bulk,red}}$	(8.43), (8.101)	minimum density grand canonical potential, reduced form
ϕ, ϕ', ϕ''	Fig. 8.1, (8.59), (8.60)	volume fraction of B, deviation from ϕ_c, deviation from ϕ_{eq}
ϕ_a, ϕ_b	(8.20)	compositions on the A-rich and B-rich side on the binodal line
$\phi_0, \phi_{0a}, \phi_{0b}$	(8.53), (8.57)	contact compositions at a substrate, for A-rich and B-rich phases
ϕ_c	(8.59)	critical composition
$\bar{\phi}$	Fig. 8.5, (8.20)	average composition set by sample preparation
ϕ_{eq}	(8.23)	equilibrium composition
ϕ_{pw}	(8.78)	bulk composition at the start of the pre-wetting line
$\phi_{\text{red}}, \phi_{\text{red,0}}$	(8.100), (8.102)	reduced composition, contact value of it
$\phi_{\text{red,0,P}}, \phi_{\text{red,0,Q}}$	(8.107), (8.109)	reduced contact composition on the P line and the Q line

(continued)

(continued)

Symbol	First Occurrence	Meaning
ϕ_w	Fig. 8.7	bulk composition at the wetting transition
κ	(8.3)	elastic constant of a squared gradient theory
μ_A, μ_B	(8.21)	chemical potential of A polymers and B polymers
N_A, N_B, N'_B	(8.11)	degree of polymerization of the A and of the B molecules
n_A, n_B	(8.11)	number of A polymers and number of B polymers in the volume
R_g, R_{gA}, R_{gB}	(8.37)	radius of gyration, for A and B chains
S_a	(8.2)	spreading coefficient of the A-rich phase
γ_{ab}	(8.1)	interface tension between the A-rich phase and the B-rich phase
γ_{AS}, γ_{BS}	(8.50)	interface tensions between the pure A phase or the pure B phase and a substrate
γ_{aS}, γ_{bS}	(8.1)	interface tensions between the A-rich phase or the B-rich phase and a substrate
$\gamma_S(\phi)$	(8.50)	contact potential at a substrate
T_{bin}, T_{spin}	(8.66), (8.69)	binodal and spinodal temperature
Θ_a	(8.1)	contact angle of the A-rich phase
T_{pw}	Fig. 8.1, (8.77)	temperatures at the start of the pre-wetting line
T_w	Fig. 8.1, (8.87)	bulk composition at the wetting transition
T_c	(8.64)	temperature of the bulk critical point
V_A, V_B	(8.11)	total volumes of A and B
v_A, v_B	(8.11)	volumes of an A monomer and a B monomer
v_c	(8.11)	cell volume in the FLORY-HUGGINS theory
ξ, ξ_{bin}	(8.32), (8.98)	correlation length of fluctuations, ξ on the binodal line
z_{eff}	(8.11)	number of neighboring cells in the FLORY-HUGGINS theory

Acknowledgments The author thanks Helgard Sigel for her hospitality during summer 2014 in Markdorf, when the first part of this contribution was written.

References

1. R. Sigel, Curr. Opin. Colloid Interface Sci. **14**, 426–437 (2009)
2. E.H.A. de Hoog, H.N.W. Lekkerkerker, Measurement of the interfacial tension of a phase-separated colloid-polymer suspension. J. Phys. Chem. B **103**, 5274–5279 (1999)
3. J.N. Israelachvili, *Intermolecular and Surface Forces* (Academic Press, Orlando, 2010)
4. D. Bonn, D. Ross, Wetting transitions. Rep. Prog. Phys. **64**, 1085–1163 (2001)
5. G. Strobl, *The Physics of Polymers* (Springer, New York, 1996)
6. J.W. Cahn, Critial point wetting. J. Chem. Phys. **31**, 3667–3672 (1977)
7. R.A.L. Jones, R.W. Richards, *Polymers at Surfaces and Interfaces* (Cambridge University Press, Cambridge, 1999)
8. A. Erbe, K. Tauer, R. Sigel, Ion distribution around electrostatic stabilized poly(styrene) latex particles studied by ellipsometric light scattering. Langmuir **23**, 452–459 (2007)
9. R. Okamoto, A. Onuki, Charged colloids in an aqueous mixture with a salt. Phys. Rev. E **84**, 051401 (2011)
10. J.W. Cahn, J.E. Hilliard, Free energy of a nonuniform system. I. Interfacial free energy. J. Chem. Phys. **28**, 258–267 (1958)
11. J. Cahn, Free energy of a nonuniform system. II. Thermodynamic basis. J. Chem. Phys. **30**, 1121–1124 (1959)

12. J.W. Cahn, J.E. Hilliard, Free energy of a nonuniform system. III. Nucleation in a two component incompressible fluid. J. Chem. Phys. **31**, 688–699 (1959)
13. A.G. Lamorgese, D. Molin, R. Mauri, Phase field approach to multiphase flow modeling. Milan J. Math. **79**, 597–642 (2011)
14. P. Sheng, E.B. Priestley, *The Landau-de Gennes Theory of Liquid Crystal Phase Transitions*, in Introduction to liquid crystals ed. by E.B. Priestley, P. Wojtowicz, P.J. Sheng (Plenum Press, New York, 1974), pp. 143–201
15. M.L. Huggins, Solutions of long chain compounds. J. Chem. Phys. **9**, 440 (1941)
16. P.J. Flory, Thermodynamics of high polymer solutions. J. Chem. Phys. **9**, 660–661 (1941)
17. J.S. Higgins, H.C. Benoitl, *Polymers and Neutron Scattering* (Clarendon Press, Oxford, 2002)
18. S. Dattagupta, *Relaxation Phenomena in Condensed Matter Physics* (Academic Press, Orlando, 1987)
19. M. Doi, S.F. Edwards, *The Theory of Polymer Dynamics* (Clarendon, Oxford, 1988)
20. P. Sheng, Boundary layer phase transitions in liquid crystals. Phys. Rev. A **26**, 1610–1617 (1982)
21. E.F. Gramsbergen, L. Longa, W.H. de Jeu, Landau theory of the nematic-isotropic phase transition. Phys. Rep. **135**, 196–257 (1986)
22. J.C. Tarczon, K. Miyano, Surface induced ordering of a liquid crystal in the isotropic phase. II. 4methoxybenzylidene-4-n-butylaniline (MBBA). J. Chem. Phys. **73**, 1994–1998 (1980)
23. G. Grosche, V. Ziegler, D. Ziegler, *Bronstein-Semendjajew: Taschenbuch der Mathematik* (Verlag Harri Deutsch, Thun, 1985)
24. W.H. Press, S.A. Teukolsky, W.T. Vetterling, B.,P. Flannery, *Numerical Recipes Third Edition* (Cambridge University Press, Cambridge, 2007)
25. R. Sigel, Untersuchung der nematischen Randschicht eines isotropen Flüssigkristalls mit evaneszenter Lichtstreuung. PhD thesis, Freiburg (1997)
26. R. Sigel, G. Strobl, Light scattering by fluctuations within the nematic wetting layer in the isotropic phase of a liquid crystal. J. Chem. Phys. **112**, 1029–1039 (2000)
27. R. Sigel, G. Strobl, Static and dynamic light scattering from the nematic wetting layer in an isotropic liquid crystal. Prog. Coll. Polym. Sci. **104**, 187–190 (1997)
28. S. Großmann, *Funktionalanalysis*, 4th edn. (AULA-Verlag Wiesbaden, 1988)

Chapter 9
Polymer Physics at Surfaces and Interfaces

Jens-Uwe Sommer

Abstract The aim of this chapter is to give a brief introduction to the physical concepts for polymers interacting with surfaces and interfaces. In particular we introduce scaling and mean-field concepts for polymers in confinement, at adsorbing interfaces and surfaces and show how these simple models can be used to understand the diversity of surface and interface phenomena of polymers. After a short introduction to concepts of polymer physics, self-similarity of polymer conformations is introduced as the basics for scaling arguments. Then, mean-field concepts as a thermodynamic approach are presented as a versatile tool in polymer theory. Using these concepts the physics of polymer adsorption is explored. After consideration of adsorption of single chains, concentration and saturation at the substrate is discussed. Polymer brushes represent a particular class of substrate-fixed polymers displaying new features due to the highly stretched conformations. Mean-field concepts are used to explore the physics of charged brushes under different solvent conditions.

9.1 Flexible Polymer Chains, Scaling, and Mean-Field Concepts

9.1.1 Ideal Chains

The universal property of long polymer chains is their huge number of conformational, microscopic states. This can be illustrated by random paths having a given length L. In Fig. 9.1 sketches of such conformations are drawn for a polymer chain with fixed ends defining the end-to-end distance vector R. In the simplest

J.-U. Sommer ($\boxtimes$)
Theory of Polymers, Leibniz-Institut für Polymerforschung Dresden, Hohe Strasse 6, 01067 Dresden, Germany
e-mail: sommer@ipfdd.de

J.-U. Sommer
Institut für Theoretische Physik, Technische Universität Dresden, Zellescher Weg 17, 01067 Dresden, Germany

© Springer International Publishing Switzerland 2016
P.R. Lang and Y. Liu (eds.), *Soft Matter at Aqueous Interfaces*,
Lecture Notes in Physics 917, DOI 10.1007/978-3-319-24502-7_9

Fig. 9.1 Sketch of polymer
conformations with fixed end
points located at x_0 and x. The
conformational entropy is
given by all possible paths of
a given length which connect
the end-points

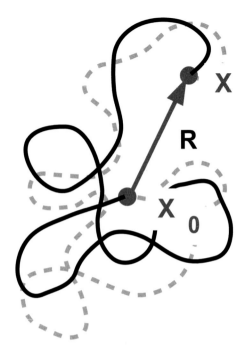

model one assumes that monomers are grouped into flexibly joint units of length l,
called the statistical or Kuhn-segment with random orientations. The number of
such segments is given by $N = L/l$. This model is called the *ideal chain model* and
can be envisioned as the trace of a *random walk* where all steps of length l are taken
randomly and uncorrelated. The total number of conformations without any
restriction on the chain is thus given by

$$Z_0 = z_0^N, \tag{9.1}$$

where z_0 denotes the number of independent states (for instance orientations) of
each segment. Note that the number of conformations for a single chain is huge:
Taken $z_0 = 6$ and $N = 100$ we obtain: $Z_0 \simeq 5 \cdot 10^{77}$! Consequently the *conformational entropy* of the free chain is given by $S_0 = k_B N \ln(z_0) \sim N$, which is much
larger than the entropy associated with a thermal fluctuation ($S_{th} \simeq k_B$), where k_B is
the Boltzmann-constant. Thus, a long polymer chain can be considered as a small
thermodynamic system. When applying any constraint to the chain the conformational entropy is reduced and a thermodynamic force has to be applied. Let us
consider a chain fixed by its endpoints at distance R. Since the segments are
uncorrelated one can easily prove the relation

$$\langle R^2 \rangle = l^2 N. \tag{9.2}$$

Here, the brackets $\langle\ldots\rangle$ denote averaging over all conformations. The probability to realize this state due to the statistical independence of the individual segments is given by a Gaussian distribution according to

$$P(R,N) = \left(\frac{3}{2\pi l^2 N}\right) \exp\left(-\frac{3}{2}\frac{R^2}{l^2 N}\right) = \frac{Z(R,N)}{Z_0}. \tag{9.3}$$

From this we can calculate the free energy to fix the end-points as compared to the state of the unconstrained chain:

$$F = -k_B T \ln(Z/Z_0) = \frac{3}{2}k_B T \frac{R^2}{l^2 N} + k_B T \ln N, \tag{9.4}$$

where we have suppressed constant terms. Thus, an ideal polymer chain behaves as a harmonic spring with an entropic spring constant, $3k_B T/N$, being proportional to the temperature.

We can rewrite Eq. (9.2) in the following form

$$R \sim N^{1/2} \text{or } N \sim R^2. \tag{9.5}$$

Here and in the following we use the short hand notation $R := \langle R^2\rangle^{1/2}$. Because the states of the segments are statistically independent all these results are also valid for any sub-chain consisting of g monomers and thus having an average extension

$$\xi \sim g^{1/2}. \tag{9.6}$$

We note that the last relation in Eq. (9.5) can be read as the dependence of the mass, N, of an object as a function of its size, R. From this we would conclude that the random coil, which represents the conformation of the chain has the (fractal) dimension $d_f = 2$ regardless of the fact that the chain has the chemical dimension, $d_c = 1$.

9.1.2 Self-Similarity, Scale-Invariance

Equation (9.6) indicates a very important feature of polymer conformations; their statistical *self-similarity*. This means that any sub-chain behaves statistically the same as the whole chain. An observer who investigates the chain conformations cannot determine the absolute scale of the system: A shorter chain in higher magnifications is indistinguishable from a longer chain observed in lower magnification. This has important consequences for the physics of polymer chains. In Fig. 9.2 we show two polymers with different chain length between two plates with different distance in between them. Since we cannot tell the absolute scale of the system the physical properties can only depend of the dimensionless ratio

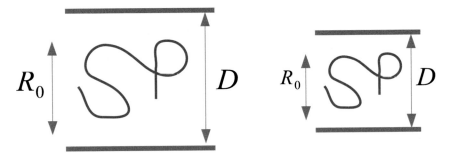

Fig. 9.2 Illustration of scale invariance: the two polymers of different length between two plates are identical up to rescaling of the absolute length

$$y = R_0/D,$$

where D denotes the distance between the plates and R_0 denotes the size of a free chain. The fact that the absolute scale for self-similar objects can not be important for their physical behavior is called *scale-invariance*. Then, only combinations of relevant length scales control the physical properties. In our example all geometric and thermodynamic properties of chains in a slit are controlled by the *scaling variable* y which represents the only possible combination of relevant length scales here. For an ideal chain Eq. (9.2) yields $y = lN^{1/2}/D$ and hence only the combination l^2N/D^2 is relevant instead of the two formally independent variables D and N. This is already a great reduction of complexity for the present problem! In Sect. 9.1.4 we will continue the discussion of polymers in the slit geometry based on this consideration.

9.1.3 Good Solvent, Real Chains and Excluded Volume

The model of ideal chains is a good starting point for understanding the physics of flexible polymers. There are real systems where this model is also fully appropriate such as dense polymer melts. Moreover, this model can be also relevant on larger scales quite generally for overlapping polymers such as in semi-dilute solutions. On the other hand, let us consider a single chain in a dilute polymer solution in a good solvent. If the ideal chain statistics as expressed in Eq. (9.3) would be valid the number of pair contacts of two monomers would grow with chain length according to $V \sim v_T \cdot N^{2-d/2}$ (d—dimension of space), where

$$v_T = v_0 \frac{T - \Theta}{T} . \qquad (9.7)$$

denotes the effective repulsion free energy between two monomers. Here, v_0 corresponds to the excluded volume of a monomer in the athermal case ($T \gg \Theta$).

In the above equation this defines the so-called Θ-temperature at which the pair-wise interaction vanishes and the solvent becomes marginal, i.e. $v_T = 0$. At the Θ-point, chain statistics is nearly ideal (perturbed only by higher order interactions). Above the Θ-point, the interaction energy per chain, V, increases with chain length without limit for $d < 4$. On the other hand, swelling of the chain conformation requires only an effort of the order of $k_B T$, see Eq. (9.4). Therefore, ideal chain statistics cannot be preserved and we expect an increase of the chain size with $R > N^{1/2}$. Here and in the following we will often denote the scaling relations only and ignore prefactors and constant length scales in order to focus on the essential result. In the following, we set the statistical segment size, l, to unity (i.e. defines the length unit) to achieve a simplified notation.

To understand polymers in good solvent, $v_T > 0$, one can use the concept of self-avoiding walks. In generalization of the picture of a random walk for an ideal chain this refers to a stochastic path which a priori does not has any self-intersections. In contrast to the random walk the statistical features of self-avoiding walks are complex and mathematically challenging. It could be shown using concepts of statistical field theory and intriguing analogies between polymer statistics and magnetic systems [1] that one essential result for chains with excluded volume interactions is given by a generalization of Eqs. (9.2) and (9.6) according to

$$R \sim N^\nu \quad \text{and} \quad \xi \sim g^\nu \quad \text{with} \quad \nu = \nu_3 \simeq 0.588. \tag{9.8}$$

This result for the exponent ν_3 in three dimension is obtained numerically and using the methods denoted above. We add that in 2 dimension the exact result is given by

$$\nu_2 = 3/4 . \tag{9.9}$$

The most important consequence is that also the self-avoiding walk (real chain in a good solvent) displays self-similarity and scale-free behavior. Referring again to the example of the chain between two plates the scaling variable can be now written as $y = N^\nu/D$. This includes the case of an ideal chain with $\nu = 1/2$.

For $v_T < 0$ the segments of the chain attract each other (poor solvent conditions). Using the same argument as above the free energy of contacts is exceeding $k_B T$ for $g^{1/2} v_T \sim 1$ and chains longer than $g \sim v_T^{-2}$ tend to collapse. Much longer chains $(N \gg g)$ will form a compact globule with $R \sim N^{1/3}$. It is important to note that chains in poor solvent are not self-similar for $N \gg g$ but display the behavior of ideal chains confined in a droplet.

9.1.4 Scaling and Blobs

Let us now consider the example of a chain in a slit in more detail. Since only the scaling variable can control the physics we can write

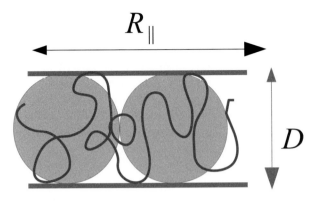

Fig. 9.3 Chain squeezed in a narrow slit, $R_{\parallel} \gg D$

$$R_{\parallel}/R_0 = f(y), \qquad (9.10)$$

where $R_{\parallel}$ denotes the (average) extension of the chain in the lateral direction and R_0 denotes again the extension of the free chain. The function of the scaling variable, $f(y)$, is called *cross-over* or *scaling function*. For $y \to 0$, (very wide slits) we have $R_{\parallel} = R_0$, and $f(0) = 1$. More interesting is the opposite case where the chain is squeezed in a narrow silt, see Fig. 9.3. In this case the chain is two-dimensional for $R_{\parallel} \gg D$ and we expect: $R_{\parallel} \sim N^{\nu_2} \sim N^{3/4}$ using Eqs. (9.8, 9.9). But how can we smoothly change from the behavior $R_{\parallel} \sim N^{0.588}$ to $\sim N^{3/4}$ This is only possible if the cross-over function displays a power-law behavior in the limit $y \gg 1$ of the general form y^m with an unknown exponent m. Substituting this into Eq. (9.10) we arrive at: $R_{\parallel} \sim N^{\nu} N^{\nu \cdot m}/D^m \sim N^{\nu_2}$ with the solution $m = (\nu_2 - \nu)/\nu$. We can conclude that the real chain in good solvent expands laterally due to compression according to

$$R_{\parallel} \sim D^{1-\nu_2/\nu} N^{3/4} \sim D^{-1/4} N^{3/4}. \qquad (9.11)$$

In the last relation we have used the approximation $\nu \simeq 3/5$. The cross-over function must change smoothly from the asymptotic values $f(0) = 1$ to $f(y \gg 1) \sim y^m$. This is expected to happen in the cross-over region $y \simeq 1$. The general functional form including the cross-over region cannot be deduced from simple arguments. However, if we plot results for different chains lengths and slit widths in a scaling plot using Eq. (9.10) we expect that all data obey a single master-curve for all values of y until the self-similarity is broken. This takes place if the slit width reaches the statistical segment size $D \simeq 1$. Further squeezing will then depend of the local chain properties and chemical details of the polymer and shall be not of interest here. Besides, this would require a substantial squeezing force. Quite generally, scaling and self-similarity will be broken for real chains at a certain minimal length scale, usually at the statistical segment size.

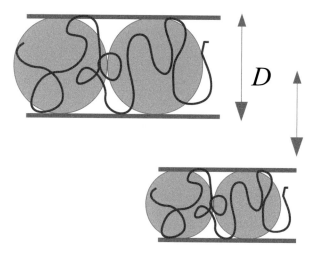

Fig. 9.4 Scale transformation of a chain squeezed in a narrow slit. Without knowing the absolute scale, scenarios are indistinguishable, if they display the same number of blobs

These result can be also obtained using a different approach. Let us redefine the scaling variable[1] according to

$$y' = y^{1/\nu} = \frac{N}{g} \quad \text{with} \quad g \sim D^{1/\nu}. \tag{9.12}$$

We observe that g is the number of segments which just cross the slit without being strongly perturbed, see Fig. 9.3. We call this sequence a *blob*. For $N \gg g$ we can imagine a chain forming a 2D self-avoiding walk made of blobs:

$$R_\| \sim \left(\frac{N}{g}\right)^{\nu_2} \cdot D, \tag{9.13}$$

where D is the step-length of the 2D-chain. Using Eq. (9.12) we reobtain the result of Eq. (9.11).

The blob concept yields another clue, namely the access to thermodynamic observables. Consider again a scale transformation of the squeezed chain as displayed in Fig. 9.4. By reducing the slit width and the chain length simultaneously so that the number of blobs is invariant the two physical situations cannot be distinguished. Thus, the thermodynamic properties shall be invariant (up to possible rescaling of length scale). In particular the free energy should only depend on the number of blobs: $F = f_F(N/g)$. The idea of this confinement-blob is apparently that a chain sequence of length g is not yet squeezed but just fixed in the slit, i.e. it looses

---

[1]Every monotonic function, in particular a power-law function of the scaling variable is again a scaling variable.

its translational degree of freedom in the direction perpendicular to the slit. Thus, we can associate the free energy loss of a blob by

$$F_{blob} \simeq k_B T.$$

Then, the total free energy effort to squeeze a chain is given by

$$F \sim k_B T \left(\frac{N}{g}\right) \sim k_B T \cdot N \cdot D^{-1/\nu}.$$

Based on this result we can now calculate the force necessary to keep the chain in the squeezed state:

$$f_D = -\frac{\partial F}{\partial D} \sim k_B T \cdot N \cdot D^{-1/\nu - 1} \sim \frac{F}{D}.$$

The force exerted by a single chain will be rather week. To estimate its value let us consider ten blobs squeezed in a slit of width 4 nm. Using $k_B T = 4.1 \times 10^{-21}$ J at ambient temperature we obtain $f_D \simeq 1 pN$. We can extend our calculation if we consider M chains squeezed in the slit without overlapping each other and having the segment concentration c. The osmotic pressure is then given by

$$\Pi = \frac{M \cdot f_D}{A} = \frac{f_D}{N} \frac{M \cdot N}{A \cdot D} \cdot D = \frac{f_D}{N} \cdot D \cdot c \sim k_B T \cdot c \cdot D^{-1/\nu} \simeq k_B T \cdot c \cdot D^{-5/3},$$

where in the last relation we have used again the approximation for ν noted above. This result can be also used for ideal chains (for instance under Θ-conditions) which leads to $\Pi \sim k_B T \cdot c \cdot D^{-2}$. We note that the expression for ideal chains can be derived exactly including the numerical prefactors unknown in the scaling theory. We note further that the similarity relation ($\sim$) in scaling expressions can be replaced by an equality with a fixed (but unknown) numerical constant. However, this constant can be easily obtained from computer simulations by applying a scale free representation such as given in Eq. (9.10). Here, also the scaling function itself including the crossover region can be obtained.

In general analytical calculations for real chains in three spatial dimensions are cumbersome and not available for all situations. Moreover, such calculations are approximate (mostly based on methods of statistical field theory) and need additional assumptions to find an explicit solution.

Let us consider a further example for scaling arguments which is important to understand polymer solutions with overlapping chains of concentration c. The overlap threshold where chains are in contact to each other in d dimensions is given by

$$c^* \sim \frac{N}{R^d} \sim N^{1 - d\nu_d}. \tag{9.14}$$

In 3D we obtain $c^* \sim N^{1-3v} \sim N^{-4/5}$. For $c \gg c^*$ chains are strongly overlapping and interpenetrating and we call this state a *semi-dilute solution*. The conformational and thermodynamic properties in the semi-dilute state will be strongly effected by the interactions between the chains. We define a blob for $c > c^*$ as a sequence of g monomers which is just on the verge of overlapping as sketched in Fig. 9.5. One can argue that for length scales smaller then the correlation length, $\xi \sim g^v$, chain sequences are only weakly perturbed by other segments and thus follow the statistics of single chains in good solvent, denoted by "EV" in Fig. 9.5. In this case we can assume that the extension of the blob is given by Eq. (9.8). Thus, we obtain:

$$g \sim c^{1/(1-v_d d)} \quad \text{and} \quad \xi \sim c^{v_d/(1-v_d d)}. \tag{9.15}$$

Chain sequences much larger than the correlation length are strongly interacting. Here, one can assume that the information whether a given segment is interacting with a segment of the same chain or with one from any other chain is lost. As a consequence the statistics can be assumed to be Gaussian again. This conclusion can be obtained for dense melts—the case of the so called Flory theorem—and *a semi-dilute solution can be considered as a melt of correlation blobs*. This conclusion can

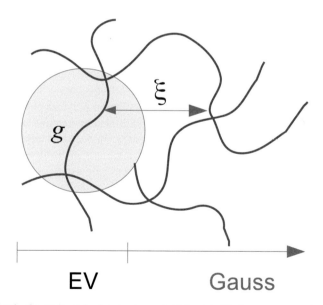

Fig. 9.5 Sketch of a mesh-work of chains in semi-dilute state. Chain sequences up to the length of g monomers are below their overlap. Thus, up to the correlation length, ξ, the chain statistics in good solvents is only weakly perturbed. For length scales larger then ξ chains interact strongly and excluded volume correlations along a single chain are destroyed. The *arrow* denotes the region of validity of excluded volume statistics (EV) and ideal statistics (Gauss) as a function of the length scale

be further obtained using arguments from statistical field theory and the interested reader is referred to the textbook by de Gennes [1].

Above the size of the blobs the excluded volume interaction is screened and the interaction between monomers of different chains dominates. The free energy per blob can be considered as $k_B T$. Thus, we obtain for the total free energy of the semi-dilute solution in three dimensions $F \sim k_B T \cdot V/\xi^3$, and the osmotic pressure is

$$\Pi = -\frac{\partial F}{\partial V} \sim \xi^{-3} \sim c^{3\nu/(\nu d - 1)} \sim c^{9/4}, \tag{9.16}$$

where in the last relation again the approximation $\nu \simeq 3/5$ has been used. The result $\Pi \cdot \xi^3 \sim k_B T$ is called the law of Des Cloizeaux [1].

Because of the screening of excluded volume interactions on a scale larger than ξ, the chain in a semi-dilute solution can be described as a Gaussian chain made of blobs:

$$R \sim \left(\frac{N}{g}\right)^{1/2} \xi \sim N^{1/2} c^{(\nu - 1/2)/(1-3\nu)} \sim N^{1/2} c^{-1/8}. \tag{9.17}$$

Hence the chain extension swells with $c^{-1/8}$ if the concentration is decreased. For $c \to c^*$ the value of $R_0 \sim N^\nu$ is reached.

9.1.5 Mean-Field Concepts

Besides the concept of self-similarity which leads to scaling variables and the blob-picture, mean-field concepts are widely applied in polymer physics. Here one usually refers to the ideal chain statistics and additional interactions and constraints are included as additive terms to the free energy. In a certain sense, mean-field concepts use thermodynamic separability of the various interactions. For a single chain in a solvent we can write the free energy per chain using Eqs. (9.4) and (9.7) in the following way:

$$F = a \cdot \frac{3 R^2}{2 N} + v_T \cdot \frac{N^2}{R^d} + w \cdot \frac{N^3}{R^{2d}}. \tag{9.18}$$

Here and the in the following we use the convention $k_B T = 1$ as the unit of energy (note that we also use $l = 1$ as the unit of length). The first term stems from Eq. (9.4) and represents the elasticity of the chain. Since the ends are freely fluctuating the numerical prefactor is not necessarily equal to 3/2 why we introduced a factor a. Within the mean-field approach, R is considered as an averaged, thermodynamic observable which is given by the minimum of the total free energy. The second term in the above equation is the mean-field expression of the pairwise interactions between the monomers: The average density within the chain volume is given by $c = a' \cdot N/R^3$ and the thus the average interactions of all monomers is $N\,c$. Note that the assumption of constant density within the chain volume brings about another constant which can be, however, just included in the definition of v_T, see Eq. (9.7). The third term represents the triple-monomer interactions in mean field constructed similarly to the second term (probability of finding two monomers in the same place is given by c^2). The prefactor w of the three-monomer interactions is considered as strictly positive and to be not temperature-dependent. The equilibrium state is given by $\partial F/\partial R = 0$. For $v_T > 0$ (good solvent) we can ignore the third term and obtain

$$R = A_d \cdot v_T^{1/(d+2)} \cdot N^{3/(d+2)}$$

Here, A_d denotes a numerical constant. In $d = 3$ we obtain $R = A_3 v_T^{1/5} N^{3/5}$. This result is well know as Flory's result and following our notation above we obtain

$$v_d^F = \frac{3}{d+2}, \tag{9.19}$$

and in particular $v_3^F = v^F = 3/5$, the approximation we have used already in some of the formulas above. We note that although this result is very close to the value of v- exponents discussed above[2] its meaning is a bit different: The polymer considered here is a stretched Gaussian chain and not self-similar. Furthermore, the free energy is given by $F_{eq} \simeq N^{1/5}$ (in $d = 3$) and is overestimated by the mean-field approach.

In the case of poor solvent, $v_T < 0$, the third term is necessary to obtain a stable solution. In this case, the second and the third term in Eq. (9.18) balance each other with the solution: $R \sim (w/v_T)^{1/d} \cdot N^{1/d}$, which corresponds to a compact globule with a temperature-dependent density.

We can reconsider the chain squeezed in a slit under good solvent conditions using mean-field concepts. The corresponding free energy reads:

$$F = b \cdot \frac{R_\parallel^2}{N} + v_T \cdot \frac{N^2}{R_\parallel^2 \cdot D} + F(N, D). \tag{9.20}$$

The first term corresponds to the stretching of the chain in the lateral direction, the second term represents the mean-field contribution of the pairwise interactions and the third term reflects contributions which do not depend on $R_\parallel$. The prefactor of the first term corresponds to two-dimensional stretching and, therefore, may differ numerically from that in Eq. (9.18). The solution for the equilibrium value of the chain extension reads

---

[2]In fact it is exact in $d = 1$, 2 and 4.

$$R_\parallel \sim N^{3/4} \cdot D^{-1/4}. \tag{9.21}$$

Note that this result coincides with that obtained from scaling in Eq. (9.11) if the Flory-approximation for v is used. We note that the thermodynamics can be obtained from the mean-field result from Eq. (9.20). However, here the dominating contribution stems from the third term in Eq. (9.20) but we will not outline calculations in detail here. Let us only mention that the free energy for the squeezed chain is again overestimated as compared to the scaling result.

9.2 Polymer Adsorption

The effective interaction between polymers and surfaces or interfaces causes depletion or adsorption effects. The most interesting and practically relevant case is adsorption. As we will see, weak attraction between monomers and adsorbing surface causes strong adsorption effects for the whole chain: Even in a very dilute solution a weakly adsorbing surfaces can be saturated with polymers.

9.2.1 Adsorption of Single Chains at Interfaces and on Surfaces

9.2.1.1 Single Chains at Penetrable Interfaces

In Fig. 9.6 we have sketched a polymer chain which is adsorbed at a penetrable interface. We assume that the segments experience the same energetic environment at both sides of the interface and experience a certain attraction only if they are crossing the interface. The adsorbing strength should be ε per segment. Apparently a sequence of monomers is strongly adsorbed if the number of contacts formed with the interfaces times the strength of interaction is larger then $k_B T$. Without adsorption the number of interface contacts for the chain is given by

$$M_0 \sim N^\phi,$$

where ϕ defines the *cross-over exponent*. For an ideal chain one can use Eq. (9.3) to derive: $\phi_{Gauss} = 1/2$. In general (for self-similar chain conformations) one can show: $\phi = 1 - v$ [2]. If $M_0 \cdot \varepsilon \simeq k_B T$ we expect a crossover from a desorbed state into an adsorbed state. In the adsorbed state the chain conformations will be strongly perturbed by the interface and one expects a flat, "pancake-like", state, if $M_0 \cdot \varepsilon \gg k_B T$. In this case let us repeat the argument for a small sequence of

Fig. 9.6 Sketch of a polymer chain adsorbed at an interface with the energy per monomer $\varepsilon \ll k_B T$. Adsorption blobs comprising g monomers are formed which have an effective interaction energy of $k_B T$. The adsorbed chain can be considered as 2D chain of adsorption blobs. The *thickness* of the adsorbed chain is given by the size of a single blob, $D \sim g^\nu$

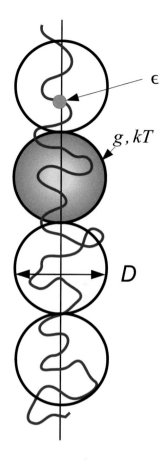

g monomers. This sequence will be on the verge of adsorption if the following condition is fulfilled:

$$g^\phi \varepsilon \simeq k_B T \text{ or } g \simeq \chi^{-1/\phi} \,,$$

where the dimensionless parameter $\chi = \varepsilon/k_B T$ has been introduced. We call this sequence an *adsorption blob*. Each adsorption blobs has an effective interaction energy of the order of $k_B T$, just enough to compensate for the lost translational degree of freedom. Since we can assume that inside the blob the chain is self-similar and Eq. (9.8) is valid, all chains having the same number of blobs cannot be distinguished if the absolute scale is not known. Thus, we can introduce the scaling variable

$$y' = \frac{N}{g} \sim N \cdot \chi^{1/\phi} \text{ or, alternatively } y = N^\phi \chi. \tag{9.22}$$

The extension of the chain perpendicular to the interface, D, can then be written as

$$\frac{D}{R_0} = f_D(y) .$$

(9.23)

In the limit of $y \ll 1$ obviously we have $f_D(0) = 1$. In the limit of strong adsorption the chain is localized and D should not depend on N. Thus, we obtain

$$D \sim \chi^{-\nu/\phi} \sim \chi^{-3/2} .$$

(9.24)

In the second relation we have used the Flory-exponent. We note that $D \sim \xi \sim g^\nu$, i.e. the perpendicular extension is given by the unperturbed size of the blob. The free energy in the adsorbed state is given by

$$F_{ad} \sim \frac{N}{g} \cdot k_B T \sim N \cdot \chi^{-1/\phi} \cdot k_B T,$$

(9.25)

a result which could also be obtained from a scaling function using $F_{ad} = k_B T \cdot f_F(y)$ and requesting the asymptotic limit to be extensive in the number of monomers, i.e. $F_{ad}(y \gg 1) \sim N$.

Next, we consider the number of monomers in contact with the interface. Using the blob concept in the adsorbed state we can directly write

$$N_M \sim g^\phi \left(\frac{N}{g}\right) \sim N \cdot g^{\phi-1} \sim N \cdot \chi^{1/\phi-1} .$$

In the case $y \ll 1$ we have $M = M_0$. Again a scaling function of y defines the cross-over from the non-adsorbed to the adsorbed state. We can define the order parameter for single chain adsorption as

$$m = \frac{N_M}{N} ,$$

being the ratio of the adsorbed segments. For $N \gg 1$, we obtain $m \to 0$ in the non-adsorbed state, and $m \sim g^{\phi-1} \sim \chi^{1/\phi-1} > 0$ in the adsorbed state, which associates single chain adsorption with a continuous phase transition, see Ref. [3] for more details.

Finally, in some analogy to the chain squeezed in the slit, the extension parallel to the interface in the adsorbed regime is given by

$$R_\| \sim \left(\frac{N}{g}\right)^{\nu_2} D \sim N^{\nu_2} g^{\nu-\nu_2} \sim N^{\nu_2} \chi^{(\nu_2-\nu)/\phi} \sim N^{3/4} \chi^{3/8},$$

where in the last relation we have used $v = v^F = 3/5$ as well as the exact result $v_2 = 3/4$. Due to adsorption, a chain in a good solvent is squeezed to a localization length, D, independent of N, and swells in the direction parallel to the interface. We note that for ideal chains there is not swelling in the direction parallel to the interface. Here we have, $v_{ideal} = 1/2$ in all dimensions and thus, $v - v_2 = 0$.

9.2.1.2 Some Notes About Polymer-Polymer Interfaces

When talking about penetrable interfaces, common examples are interfaces between two immiscible polymers, A and B. Let us denote the effective interaction between the segments ε_{AB} and let us define $\chi = \varepsilon_{AB}/k_BT$. First, one can show that the interface has a finite width even if the immiscibility is very strong, i.e. $\chi N \gg 1$. Let us consider a sequence of g monomers forming a loop into the other phase, see lhs of Fig. 9.7. If the total repulsion of this loop given by $g \cdot \varepsilon_{AB}$ is of the order of k_BT, it can be easily formed by thermal fluctuations. Much larger loops, however, will be suppressed. Since polymer chains in a dense melt obey Gaussian statistics the width of the interface can be estimated by $D \sim g^{1/2}$, and with $g \cdot \chi \sim 1$ we obtain

$$D \sim \chi^{-1/2} . \tag{9.26}$$

Thus the polymer-polymer interface width increases with the temperature as $D \sim T^{1/2}$.

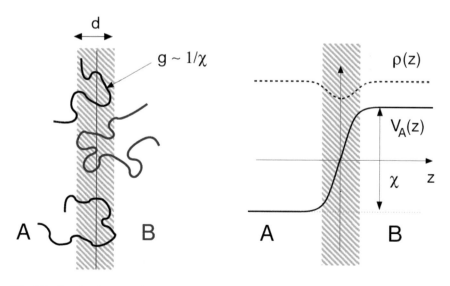

Fig. 9.7 Sketch of an interface between two immiscible polymers A and B. On the rhs the interface potential, V_A, with respect to the A-phase and the density profile are shown along the coordinate perpendicular to the interface, z

The free energy stored in the interface can be estimated by the number of loops (each stores a free energy of $k_B T$). Thus we get for the interface tension

$$\gamma \sim D/g \sim \chi^{1/2} . \tag{9.27}$$

These results can be confirmed by a local mean-field approach. Here one can furthermore predict a smooth interface profile as sketched in the rhs of Fig. 9.7. Moreover, a real polymer melt has a finite compressibility which results in a minimum of the total polymer density in the interface region. This becomes important for interactions with a third component such as a third polymer species, copolymers or nano-particles. We stop, to note that interfaces can dominate the physical properties of materials such as for block-copolymers or lipid-bilayers.

9.2.1.3 Single Chains Adsorbed at Hard Surfaces

Perhaps the most common example for polymer adsorption is the adsorption onto solid substrates. Let us denote the energy gain of a segment in direct contact with the surface again by ε and $\chi = \varepsilon/k_B T$. In contrast to penetrable interfaces, a polymer chain looses conformational degrees of freedom, if it gets close to an impenetrable wall. Thus, there is an entropic penalty for chains approaching the surfaces to a distance of the order of the chain extension, R_0, which must be compensated before adsorption can take place. One can show that there exists a critical adsorption strength, χ_c, independent of chain length which divides surface depletion, $\chi < \chi_c$, from adsorption, $\chi > \chi_c$. In fact for $N \to \infty$ this becomes a sharp transition and χ_c is a critical point in the thermodynamic sense [4]. For finite chains, adsorption is a cross-over phenomena as for the penetrable interface. Because of the definition of χ, the critical point can be either a critical temperature, T_c (for a fixed interaction energy), or a critical interaction energy, ε_c, (for fixed temperature). The latter can be practically realized by mixing two solvents with different affinity to the surface. The distance to the critical point of adsorption can be defined by (Fig. 9.8)

$$\kappa = \frac{\varepsilon - \varepsilon_c}{\varepsilon_c} . \tag{9.28}$$

It can be shown that κ plays a similar role for the adsorbing surface as χ for the penetrable interface. The relevant scaling variable can now be noted as

$$y = N^\phi \kappa. \tag{9.29}$$

The important difference to the penetrable interface is that the crossover-exponent is not a priori related to ν, but an independent, so-called surface critical exponent. As for ν in 3D we do not know its exact numerical value. From scaling analysis of simulation data at least two different predictions have been made. Direct Monte Carlo simulations are in good agreement with $\phi \simeq 0.59$.

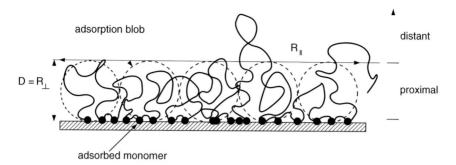

Fig. 9.8 The structure of chains adsorbed onto a solid substrate. As for the adsorption of chains at penetrable interfaces, an adsorbed chain can be subdivided into adsorption blobs which define the thickness, D, of the adsorbed chain. Its extension in the lateral directions, $R_\parallel$, is given by a 2D self-avoiding walk of adsorption blobs. The region inside the layer of adsorption blobs defines the "proximal" region

Conformational sampling methods led to the conjecture $\phi = 0.5$. Note that that both values are very different from the value of $\phi = 1 - v \simeq 0.4$ for the penetrable interface. The true value of ϕ is still an open question. Part of the problem here is that the critical point of adsorption, χ_c depends of the polymer system (and on the simulation model) and has to be estimated at the same time. In fact, very small changes of χ_c have a major effect on the estimate of ϕ. In simulations, for instance, only finite chain lengths can be considered and different values of ϕ in the range of $\phi = 0.5\ldots0.6$ are correlated with a best estimate of χ_c of the order of one percent. Such an accurate determination of the critical point of adsorption is quite impossible for chains up to a length of $N = 500$. The correlation between the apparent value of ϕ and the uncertainty in the estimation of χ_c explains the yet unresolved problem of the best approximation of ϕ. However, it can be shown that for chain lengths up to several hundred monomers the correlated estimate of both parameters yields to excellent data collapse of observables using the scaling variable of Eq. (9.29). In full analogy with the penetrable interface, scaling predicts

$$\frac{D}{R_0} = f_\perp (N^\phi \kappa) \rightarrow D \sim \kappa^{-v/\phi} \tag{9.30}$$

$$\frac{R_\parallel}{R_0} = f_\parallel (N^\phi \kappa) \rightarrow R_\parallel \sim N^{v_2} \kappa^{(v_2-v)/\phi} \tag{9.31}$$

$$\frac{M}{N^\phi} = f_M (N^\phi \kappa) \rightarrow M \sim N \kappa^{(1-\phi)/\phi} \tag{9.32}$$

Here, the relation on the rhs give the asymptotic results for large values of the scaling variable, i.e. in the adsorbed state. In the desorbed state, i.e. for $\kappa < 0$ the number of adsorbed monomers should not depend on N asymptotically, thus one obtains $M/N^\phi \sim N^{-\phi}$. In Fig. 9.9 we show an example which has been obtained by

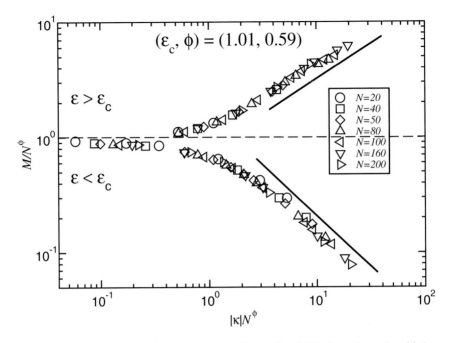

Fig. 9.9 Scaling plot of the order parameter according to Eq. (9.32) above ($\varepsilon > \varepsilon_c$) and below ($\varepsilon < \varepsilon_c$) the critical point of adsorption. Here the temperature is taken constant. The data have been obtained by Monte Carlo simulations using the bond fluctuation model (BFM) [16]. The set of critical parameters is chosen as $\varepsilon_c = 1.01$ and $\phi = 0.59$. Very good scaling is obtained. The *solid lines* indicate the asymptotic behavior both in the adsorbed and in the desorbed state, see Eq. (9.32) and text. Reproduced with permission from Ref. [5]

computer simulations using the bond fluctuation model (BFM) [5]. Using the best estimate of $\chi_c = 1.01$ and $\phi = 0.59$ a collapse of data into a single master curve, see Eq. (9.32), can be obtained. Also the asymptotic slopes are well reproduced. Similar results are obtained for the other observables according to Eqs. (9.30) and (9.31) [5]. Alternative critical parameters are tested in Ref. [5] where either scaling or the asymptotics is less satisfying. We can conclude that for moderate chain length as used in the simulations (corresponding to a molar mass for polyethylene up to approximately 14 kg/mol) the scaling approach to polymer adsorption provides very good results.

In full analogy to the case of a penetrable interface we can calculate the free energy of adsorption per chain in the well adsorbed state, where several adsorption blobs are formed, $N^\phi \kappa \gg 1$. As in Eq. (9.25) it can be written as the number of adsorption blobs per chain and we obtain

$$F_{ad} \sim \frac{N}{g} \cdot k_B T \sim N \cdot \kappa^{1/\phi}. \tag{9.33}$$

If κ is not much smaller than unity this leads to an extremely large free energy gain per chain ($\sim N$) which can be of the order of $100 k_B T$, which is of the order of a chemical bond energy!

9.2.1.4   Adsorption Isotherm and Saturation

The understanding of single chain adsorption is mandatory for going further towards the thermodynamics of adsorption. In reality a finite concentration of polymers in the bulk, c, is in contact with an adsorbing surface at constant temperature as sketched in Fig. 9.10. The amount of adsorbed polymer at the surface per unit area is denoted by Γ (surface excess). On the molecular level we consider a chain in the adsorbed state if at least one monomer is touching the surface. In the following we will consider only the adsorbed case, i.e. $N^\phi \kappa \gg 1$, where the asymptotic scaling relations of Eqs. (9.30)–(9.32) and (9.33) hold.

In order to derive the universal behavior of the adsorption isotherm we start with the dilute surface state where the adsorbed chains do not overlap. This is given for $\Gamma < \Gamma^* \sim N/R_\parallel^2$ and $R_\parallel$ given by Eq. (9.31). Here, we can assume Boltzmann statistics and write

$$\Gamma = A \cdot e^{N\kappa^{1/\phi}} \cdot c . \tag{9.34}$$

where A is a numerical prefactor. Since $N\kappa^{1/\phi} \gg 1$ the initial slope of the adsorption isotherm is very large and an exponentially small value of c is sufficient to reach overlap at the surface ($\Gamma \to \Gamma^*$). Thus, the bulk is in the very dilute state of the polymer solution when the surface begins to be crowded by chains.

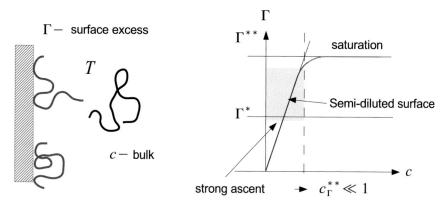

Fig. 9.10 Polymer solution in equilibrium with an adsorbing surface. The surface excess is given by the amount of polymer attached to the surface by at least one monomer. On the rhs a sketch of the adsorption isotherm is shown where the characteristic concentration regimes are indicated

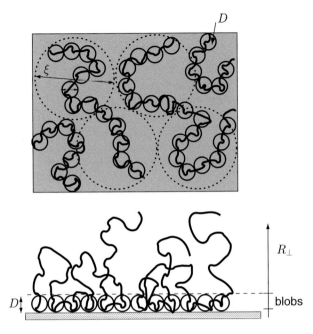

Fig. 9.11 The semi-dilute surface state. The formation of 2D concentration blobs formed by adsorption blobs is illustrated in the upper part. Due to crowding of chains for $\Gamma > \Gamma^*$ loops and dangling ends can extend into a distance $R_\perp$ much larger than the adsorption blob size as sketched in the lower part. By approaching the saturated state the layer thickness reaches the size of the chains in the bulk

Next, we consider the case $\Gamma > \Gamma^*$. Here, a 2D semi-dilute state on the surface is formed as discussed in Sect. 9.1.4, see Eqs. (9.14)–(9.15). This is sketched in Fig. 9.11(top). Note that the natural unit for the 2D chain conformation is the adsorption blob, given by $D \sim \kappa^{\nu/\phi}$, see Eq. (9.30), and $g \sim D^{1/\nu} \sim \kappa^{1/\phi}$. Following the arguments given in Sect. 9.1.4, the effort to form a concentration blob is of the order of $k_B T$. Thus, if the concentration blob is much larger then the adsorption blob, i.e. $\xi \gg D$, this corresponds to a weak perturbation of the adsorption and only to a slight deviation from the Boltzmann result of Eq. (9.34). In a first approximation we can consider Eq. (9.34) to be valid up to $\xi \simeq D$. Corrections to this behavior will be important to explain the crossover to saturation, see rhs of Fig. 9.10 [6].

If the surface excess is increased further, the 2D concentration blobs become of the size of the adsorption blobs, $\xi \simeq D$, and further increase of the surface concentration by adsorption is not possible. In the saturated surface state, the free energy gain by adsorption is balanced by the effort to pack the adsorption blobs close to each other. The saturated surface state can thus be considered as closely packed adsorption blobs. This defines the limiting surface excess by

$$\Gamma^{**} \sim \frac{g}{D^2} \sim B \cdot \kappa^{(2\nu-1)/\phi} \sim B \cdot \kappa^{0.3} . \tag{9.35}$$

In the last relation we used $\phi \simeq 0.59$ and B is a numerical prefactor. At the same time the corresponding bulk concentration can be extremely small. This can be understood by extrapolating the linear regime of the adsorption isotherm to Γ^{**}, see Fig. 9.10, and we obtain: $c_\Gamma^{**} = 1/A \cdot \exp(-N\kappa^{1/\phi}) \cdot B \cdot \kappa^{0.3}$. Due to the large exponent we get $c_\Gamma^{**} \ll c^*$, where c^* again denotes the overlap density in the solution. Thus, the surface can be saturated by polymers already in a highly dilute solution. This is one reason why polymer adsorption is technologically important: Even weakly adsorbing surfaces are fully covered by polymers in dilute solution, and it is difficult to clean surfaces by exposing to solvent once they were in contact with adsorbing polymers. A practical example is protein adsorption on metal surfaces, related with fouling.

If the saturation density is reached, chains form longer loops into the solution. The thickness of the adsorption layer approaches again the extension of free chains in the solvent. Using the argument that in this state the loops (and tails) are in a semi-dilute state and the correlation length is given by the distance to the wall (the density is self-organizing this way) one obtains using Eq. (9.15):

$$c(z) \sim z^{-4/3} ,$$

where the Flory-exponent has been used and z denotes the distance to the wall [1].

9.3 Polymer Brushes

As we have seen, adsorption of polymers leads to saturation which limits the amount of adsorbed polymers to Γ^{**} and the size of the adsorbed layer to the size of the polymer coil in solution, $R_0 \sim N^\nu$. However, this limit can be overcome by adsorbing or chemically grafting polymers by one end only. A sketch is shown in Fig. 9.12. Such a *polymer brush* is characterized by the number of chains, M, grafted on the surface area, A:

$$\sigma = \frac{M}{A} = \frac{1}{\xi^2} , \tag{9.36}$$

where ξ denotes the distance between the grafting points. Because of volume conservation the thickness of the brush, L, must increase proportional to the degree of polymerization provided that N is large enough and for large enough N we always obtain

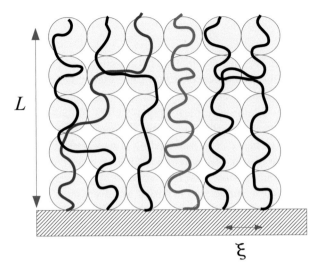

Fig. 9.12 Sketch of a polymer brush. The distance between the grafting points is denoted by ξ and the thickness of the brush is given by L. In a simple model the brush can be understood as a column of blobs of size ξ

$$L \sim N \gg R_0 \sim N^\nu. \tag{9.37}$$

The polymer surface excess (given the bulk is diluted) is $\Gamma = \sigma N$. End-grafting chains can lead to film thicknesses much larger than obtained by adsorption and usually much larger than long-range van der Waals interactions between bare substrates with useful application for emulsion stabilization and for coatings.

Let us consider first a polymer brush in poor solvent (or under vacuum in the absence of liquid solvent). The volume of the dense polymer phase controls the brush thickness and we obtain $L/\sigma = N \cdot v_0$, where we denote the monomer volume by v_0, and thus

$$L \sim \sigma \cdot N. \tag{9.38}$$

This is sometimes called the dry brush limit. We note that a dense film is only defined for $\sigma \gg \sigma^* \sim N^{-1}$, where we have introduce the overlap density σ^*. At σ^* the polymer layer has the thickness of the monomer size and is thus much thinner then R_0, the chains are collapsed on the substrate. The chains are stretched with respect to their conformation in the melt state only if $\sigma \gg \sigma^{**} \sim N^{-1/2}$. We call σ^{**} the stretching threshold of the brush in poor solvent. We note that in the region $\sigma < \sigma^{**}$ inhomogeneous surface patterns are predicted which result from the formation of micelles which reduce the contact area with the poor solvent [7].

In many cases brushes under good solvent conditions are of interest. Good solvents swell the chains beyond the prediction of Eq. (9.38). To understand this we can first apply a mean-field model, see Sect. 9.1.5. The total free energy for the brush reads: $F_M = M \cdot a\frac{3}{2}\frac{L^2}{N} + v_T \frac{M^2 N^2}{L \cdot A}$. The free energy per chain is then given by[3]

$$F = \frac{F_M}{M} = a\frac{3}{2}\frac{L^2}{N} + v_T \cdot \sigma \frac{N^2}{L}. \tag{9.39}$$

From $\partial F/\partial L = 0$ we obtain the equilibrium height of the brush as

$$L \sim v_T^{1/3}\sigma^{1/3}N, \tag{9.40}$$

which differs from Eq. (9.38) by the exponent of σ. Because of $\sigma \ll 1$ we have $L_{good-solvent} \gg L_{poor-solvent}$.

This result can be also explained using a scaling approach. Here, we assume that the grafting density determines the blob size of chains in a good solvent, see Fig. 9.12. The brush is then considered as a dry brush made of blobs each containing $g \sim \xi^{1/\nu}$ monomers and we obtain

$$L \sim \left(\frac{N}{g}\right)\xi \sim N \cdot \sigma^{(1/\nu-1)/2}, \tag{9.41}$$

which corresponds to Eq. (9.40) by using Flory's value of ν^F. In this picture the polymer brush corresponds to a semi-dilute polymer solution with $c \sim g/\xi^3 \sim \sigma^{(3-1/\nu)/2} \sim \sigma^{2/3}$ and an the osmotic pressure given by the law of Des Cloizeaux, see Eq. (9.16) according to $\Pi \sim \xi^{-3} \sim \sigma^{3/2}$. This can be applied if we want to calculating the free energy effort to insert a particle of volume δV into the brush which is thus given by $\delta F = \Pi \delta V \sim \delta V \sigma^{3/2}$. We note that the crossover from non-interacting grafted chains at low grafting density (so-called mushroom regime) to the brush state is given by $\sigma^* \sim N^{-2\nu}$ and the above obtained results can be re-derived using the scaling variable σ/σ^*.

9.3.1 Density Profile, Surface Instabilities, and the Role of End-Group Modification

The mean-field solution and the scaling model, commonly denoted as the Alexander-de-Gennes model [8, 9], implicitly assumes a square-shape density profile with a density jump at the brush surface. While this concept is sufficient to explain most relevant properties of polymer brushes it leaves open the question about the driving force for stretching of the chains. A gradient of the density profile

---

[3]We note that the numerical prefactor a is different for different applications of the mean-field concept depending of the way the ideal elastic free energy is parametrized (radius of gyration, brush height etc.

is necessary to explain the stretching effect based on pair interactions, i.e. the immersion of monomers in the solution of other monomers. This can be taken into account by considering a continuous stretching of a chain in an external force field, ignoring fluctuations of the chain conformations. Assuming Gaussian chain statistics the density profile is parabolic and can be written in the following form

$$c(z) = (L^2 - z^2)\pi^2/8N^2 v_T , \qquad (9.42)$$

where $L \sim \sigma^{1/3} N$ and the average monomer positions are related by $\langle z \rangle = 3L/8$. The stretching force for each monomer is here provided by the gradient of the density profile, if the brush forms a semi-dilute solution where the pair interaction between the monomers dominates. As a consequence of this solution the distribution of end-monomers is also smooth and given by $c_e \sim z(L^2 - z^2)^{1/2}$. While in computer simulations Eq. (9.42) can be rather well approximated, the calculation of the end-point distribution usually fails dramatically. A reason for this is the finite extensibility of chains which often plays an important role because of the high stretching. Taking this into account leads to a more box-like behavior of the density profile which can have important consequences for applications of polymer brushes.

In Fig. 9.13 the role of sharp interface for the conformations of a chain in the brush is sketched. For the parabolic form all monomers experience a force and fluctuations of the chains are stable around the mean extension. For a box-like profile the gradient zone is located in the surface region of the brush which are occupied by the chain ends. Here, only the chain ends experience a force which is strong enough to substantially stretch the whole chain. If, for some reason the chain end is diving into the brush there is only a weak force acting which pushes the chain back into the stretched state. Thus, a box-like profile creates a *surface instability*. At least two cases can be considered where this instability can have important consequences.

Let us consider a minority of chains of length $N + \delta N$ in a majority of chains of length N. For $\delta N < 0$ the chains cannot easily reach the gradient zone and are thus collapsed in the brush profile. Longer chains, $\delta N > 0$, by contrast have a stronger anchor in the gradient zone and are thus overstretched. It can be shown in computer simulations [10] that already a change of only one monomer leads to a jump in the

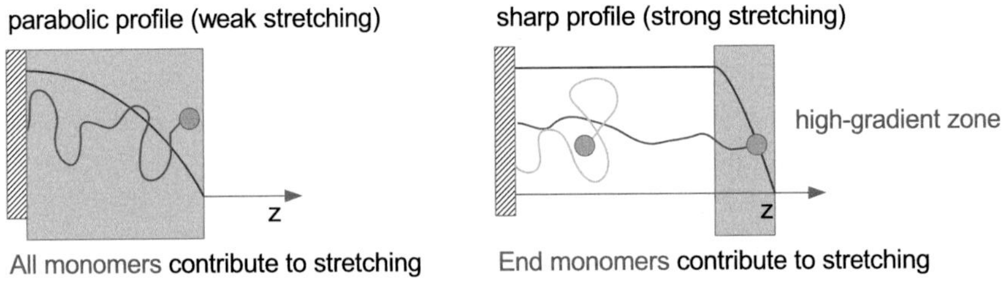

parabolic profile (weak stretching)   sharp profile (strong stretching)

high-gradient zone

z   z

All monomers contribute to stretching   End monomers contribute to stretching

Fig. 9.13 Sketch of the behavior of a single chain inside the density profile of a polymer brush. *Left* parabolic profile with a smooth gradient distributed throughout the brush. *Right* Sharp profile with a strong gradient zone localized in the surface region of the brush

extension of the chain, see Fig. 9.14, which results from the collapse of the slightly shorter chain. This illustrates nicely the instability of the strongly stretched chain conformations.

Next, we consider the modification of the end-monomer by changing its size as illustrated in Fig. 9.15. The excluded volume interaction of the end-monomer is

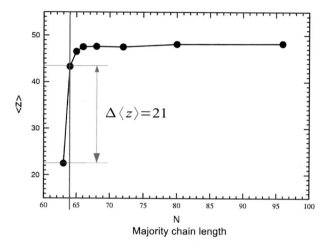

Fig. 9.14 Extension of a single chain inside a brush of length $N = 64$ as obtained in Monte Carlo simulations. For details see Ref. [10]. Here, we have considered a single chain with length N in a brush with chain length $N = 64$ (see the *vertical line*). One missing monomer leads already to a collapse of the chain in the brush

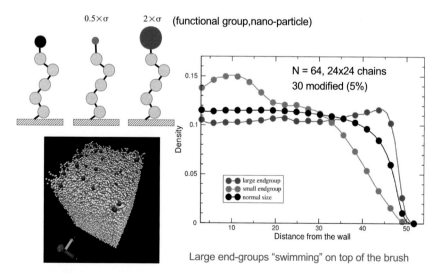

Fig. 9.15 The role of modified end-groups in a highly stretched polymer brush. For more details see Ref. [10]

proportional to its volume $v_T \sim v_0$. Therefore, bigger monomers interact stronger in the density gradient and vice versa. One would therefore expect that bigger end-groups provide a stronger stretching of the chains and smaller endgroups lead to collapse of the chains in the brush. This has been tested in computer simulations [10]. In Fig. 9.15 the monomer density profile of a minority of 5 % of chains with modified endgroups is plotted. The curve for not modified chains displays the rather box-like profile deviating from the prediction of the parabolic profile due to high grafting density and strong stretching of chains. This profile changes dramatically if only the size of the end-groups is modified: For endgroups having two-times larger radius the profile becomes non-monotonous with a peak in the surface region of the brush. Chains with smaller endgroups collapse with a peak of the density profile close to the substrate.

As the snapshot in Fig. 9.15 shows, the bigger endgroups are floating on top of the brush and are strongly localized there. Such modified endgroups usually will have different properties such as different interactions with the solvent and the other monomers in the brush. Again, computer simulations have shown that repulsion of endgroups with respect to solvent leads to a switching of the behavior from stretched to collapsed [11]. Such switching effects can be tuned by external conditions such as a change in temperature or pH (in the case of charged groups) and have potential applications for stimuli-responsive surfaces: If the endgroups are formed by biologically active units such as enzymes, switching would lead to a transition of an exposed (active) to an hidden (passive) state of the surface with respect to the enzymatic activity.

9.3.2 Charged Polymer Brushes

In particular in aqueous solutions some polymers can become charged by releasing counter-ions from monomer units. Charged polymers are called polyelectrolytes or poly-ions. If the charging is complete and nearly not pH-dependent we call this a strong polyelectrolyte, if the charging can be changed strongly by changing pH, this is called a weak polyelectrolyte. In general the effect of charges creates a formidable problem for the theoretical understanding, because of the long-range character of charge effects. However, for rather dense polymer systems such as networks, brushes or dendrimers there is a simple argument which in most cases leads to a general understanding of the effects of charges: If counterions leave the polymer phase (a brush in our case) they are attracted by the oppositely charged polymer phase. In the case of a brush the electric field due to charging is constant and thus the potential of the counterions increases without limit with increasing height. This is similar to the barometric effect of a gas in the gravitational field. The typical length scale within which the counterions can be distributed above an oppositely charged surface is called the Gouy Chapman length, $\lambda_{GC} \sim k_B T \varepsilon_r / q^2 \omega$, where ω denotes the density of charges of the polymer per surface unit, q, is the charge of

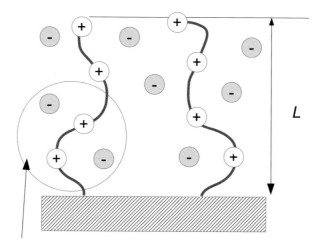

Fig. 9.16 Sketch of a charged polymer brush with a thickness L. Negative counterions are trapped inside the positively charged brush and exert an osmotic pressure

counterions and ε_r denotes the relative dielectric constant of the solution. One can understand λ_{GC} as the length scale at which the energy of the charge, $\lambda q^2 \omega / \varepsilon_r$ is of the order of $k_B T$, although it is usually introduced in the context of the Poisson-Boltzman equation leading to a power-law decay of the ion-density profile over the charged surface.

If the height of the brush, L, is much larger than λ_{GC} the counterions are trapped inside the brush but are not localized to the charged monomers.[4] Thus we have *local charge neutrality* inside the brush as indicated by the circle in Fig. 9.16. In a first approximation counterions are freely moving inside the volume of the brush $V = A \cdot L$ and thus exert an osmotic pressure given by the ideal gas law

$$\Pi^0_{CI} = N \cdot M \cdot f \cdot k_B T / A \cdot L = \sigma N \cdot f \cdot k_B T / L,$$

where f denotes the fraction of monomers charged. The osmotic pressure tends to stretch the chains such as the gas pressure in a balloon stretches its elastic surface. The total free energy per chain of the charged brush in the mean-field model extends the expression of Eq. (9.39) into

$$F = a \frac{3 L^2}{2 N} + v_T \cdot \sigma \frac{N^2}{L} - fN \ln\left(\frac{L}{fN\sigma}\right), \tag{9.43}$$

---

[4]Such localization is called counterion condensation but shall be not considered in the following.

where we have again used the convention $k_B T = 1$. The equilibrium condition is given by

$$\frac{\partial F}{\partial L} = 0 = \Pi_{el} + \Pi_{EV} + \Pi_{CI} = 3a\frac{L}{N} - v_T\sigma\frac{N^2}{L^2} - \frac{fN}{L}, \qquad (9.44)$$

where Π_{el}, Π_{EV} and Π_{CI} are defined according to the three terms of the free energy (elasticity, excluded volume and counterion pressure respectively in Eq. (9.43). If we ignore the excluded volume interactions and consider the osmotic pressure of the counterions as the dominating force we obtain

$$L \sim f^{1/2}N.$$

Interestingly this result is independent of the grafting density. This can be understood by the fact that by doubling the number of chains per unit area we double the number of counterions but also the number of elastic units. The lowest grafting density where this behavior can be expected is given by $\lambda_{GC} \simeq L$, using $\omega = \sigma N f$. The region where the counterion pressure is dominating the chain extension is called the osmotic brush regime [12].

Frequently, the solution contains further ion-pairs due to a finite salt-concentration, c_s. The exchange by like-charge salt ions makes it possible that counterions leave the brush volume as indicated in Fig. 9.17. This reduces the osmotic effect of the counterions. At high concentration of salt the counterion effect will vanish and only the interplay of the solvent quality (now also controlled by the salted solution) and elasticity remains—thus we recover the situation of effectively uncharged chains.

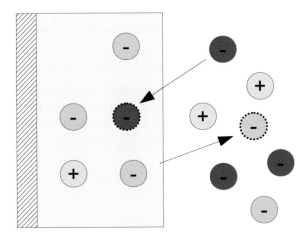

Fig. 9.17 Effect of salt on the polymer brush. Counter ions are *blue* and salt ions are *red* and *yellow*. The charged brush is indicated as a *box*. Exchange of counterions by like-charged salt ions as indicated by the *arrows* does not break the charge neutrality but effectively decreases the osmotic pressure

For monovalent salt ions one can show that the salt effect reduce the osmotic pressure as follows:

$$\Pi_{CI} = \Pi_{CI}^0 \left[\sqrt{1 + y_S^2} - y_S \right],$$

(9.45)

with the scaling variable

$$y_S = 2\frac{c_s}{c_{CI}} = 2\frac{c_s}{f\sigma}r \text{ with } r = \frac{L}{N}.$$

(9.46)

Here, we have introduced the stretching ratio, r, and y stands for the ratio of concentrations of salt ions and counterions. Using this expression in Eq. (9.44) we obtain an equation for the equilibrium stretching ratio with the abbreviations $s = 2c_s/\sigma$ and $r_0^3 = \sigma v_T$ [13]:

$$a \cdot r \cdot f \left\{ \left[1 + \left(\frac{sr}{f}\right)^2 \right]^{1/2} - \frac{sr}{f} \right\} + r_0^3 - r^3 = 0,$$

where the prefactor a can be extracted for the unknown prefactor in elasticity. Let us note that we repeatedly use the symbol "a" to note a constant prefactor in different problems where it does not need to be same value. This equation for $r(s, t, r_0)$ can be solved numerically only. However, for the case of high salt concentration, $y \gg 1$ we obtain from Eq. (9.45) $\Pi \simeq \frac{f^2}{4c_s}c^2$ which corresponds to a second virial coefficient of $v_s = f^2/4c_s$. Thus we can approximate the highly salted brush with an effective excluded volume coefficient $v_{eff} = v_T + v_s$ and thus we obtain from Eq. (9.40)

$$\frac{L}{N} \simeq \left[\frac{\sigma}{3a} \left(v_T + \frac{f^2}{4c_s} \right) \right]^{1/3}.$$

(9.47)

The first observation from this result is that the effective Θ-point is now reached in the poor-solvent state, i.e. for $v_T = -f^2/4c_s$. Thus, a charged brush can appear to be in good solvent, even if the backbone is already under poor solvent conditions. On the other hand, for $v_T \to 0$, we obtain $L \sim c_s^{-1/3}$. This behavior has been taken as an indication of the validity of the osmotic brush model. As we have shown, this depends on the solvent quality of the backbone. For hydrophilic backbones, v_T will strongly limit the $c_s^{-1/3}$- behavior for large values of c_s and leads to a plateau-value of L. For weakly hydrophobic backbones, high salt concentrations lead to collapse which, however, might degenerate into a smooth crossover for finite chain lengths and thus is rather difficult to delineate from the salt-effect. Note that high salt concentrations are typical under physiological conditions in living systems.

9.3.3 Charged Brushes in Poor Solvents

Poor solvent conditions of the backbone are more effectively compensated if the salt concentration is low. In this case we can reconsider Eq. (9.43) for $v_T < 0$. To avoid a singular behavior in the collapsed state at least the third virial coefficient (better would be the full equation of state) has to be taken into account. Then the free energy per monomer reads

$$F_m = a\frac{3}{2}r^2 + v_T \cdot \sigma\frac{1}{r} + w\sigma^2\frac{1}{r^2} - fN\ln\left(\frac{r}{f\sigma}\right). \tag{9.48}$$

The free energy can have two minima at low temperatures which are dominated by the first and the fourth term (swollen osmotic state), and the second and the third term (collapsed state) respectively. The general form of the free energy as a function of the stretching factor is sketched in Fig. 9.18. At the transition temperature, T_c, both minima have the same height and a discontinuous collapse transition is predicted which is related to a jump of the stretching ratio.

The equation $\partial F_m/\partial r = 0$ can be implicitly solved in the form: $T(r, \sigma, f; v_0, w)$ as $1 - \Theta/T = (ar^4 - fr^2 - 2w\sigma^2)/v_0\sigma r$, where we have used the definition of v_T according to Eq. (9.7).

In Fig. 9.19 we display the solution of the mean-field equation. The function T (r) is not monotonous at high charge fractions. This corresponds to a discontinuous phase transition from the swollen to the collapsed state which is expected far below the Θ-point in salt-free solutions. The arrow indicates the Maxwell-construction, $T_{col}(f, \sigma)$, where the jump from the swollen to the collapsed state is expected.

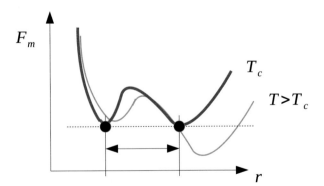

First-order collapse transition

Fig. 9.18 Sketch of the free energy function of Eq. (9.48) as a function of the streching ratio above and at the point of the collapse transition, T_c. At T_c the stretching ratio jumps as indicated by the *double-arrow*

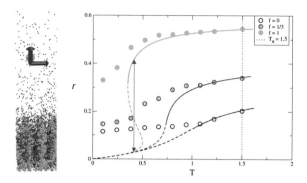

Fig. 9.19 Charged brush in poor solvent. The stretching ratio, r, calculated as described in text (*lines*), and obtained from computer simulations (*points*). The *arrow* indicates the first-order transition (jump) of the stretching ratio at high charge fractions as predicted from the mean-field model. On the left a snapshot of the simulated system is displayed. The fraction of charged monomers is indicated by f and $f = 0$ corresponds to the non-charged system. For more details see Ref. [14]

Computer simulations do not display this behavior as can be also seen from Fig. 9.19 (data points). While a certain fit can be achieved in the swollen state (however, quite below the Θ-point), the collapse is much less pronounced, and in fact only a rather smooth crossover is displayed at low temperatures. Nevertheless, the temperatures where a shrinkage in height is pronounced coincide roughly with the prediction of the mean-field model. We note that for lower charge densities also the mean-field model does not display a discontinuous transition. We note that the large deviations at low temperatures (very poor solvent) are related with missing higher order terms of the equation of state of dense polymers. This can be improved be considering a model equation of state such as the Flory-Huggins-equation [14].

The snapshot on the left of Fig. 9.19 indicates the reason for the failure of the mean-field approach: At low temperatures, the brush forms bundles (upright cylinders) which are surrounded by the counterions [15]. Spontaneous formation of this nano-structure increases the entropy of the counterions since the available volume (height) is much larger as compared to the homogeneously collapsed state. The failure of the mean-field model is to assume an homogenous phase from the beginning.

9.4 Conclusions

Polymers can be understood as flexible chains and typical polymer behavior is related to conformational entropy. To consider the influence of geometrical restrictions and interactions two types of simplified concepts have been established: scaling and mean-field models. In many situations both models make similar predictions and are to some extend exchangeable. While scaling concepts are based on the statistical self-similarity of polymer conformations, mean-field approaches use the superposition of independent contributions to the free energy such as elasticity, mean interactions and counterion entropy. In general scaling models should be considered as more realistic since correlation effects are taken into account by the assumption of self-similarity in the case of good solvents. While many problems can be considered with both approaches, more sophisticated situations where many different effects play a role are usually better accessible in mean-field models. The simplicity provided by these approaches allows to consider more complex polymer systems such as adsorbed layers, brushes and even charge effects. Since both approaches rely on rather strong assumptions, testing of the predictions by simulations is a most suitable pathway to check the consistency and validity of the predictions. This is all the more important since polymer adsorption experiments cannot access easily details such as chain conformations and density profiles.

Scaling and mean-field models give a rich and detailed picture of polymers at surfaces and interfaces. This ranges from critical adsorption and the blob-concept for single chain adsorption to the properties of surface saturation and adsorption isotherms. Brushes formed by end-grafted polymer chains are another example of the diversity of polymer systems which can be achieved by variation of their components and parameters. The stretched conformations in the brush are unstable if the stretching is high such that slight differences in chain length or end-group volume of a minority species can lead to collapse and switching of conformation under change of external parameters.

Charge effects are important for aqueous systems in particular for biological systems. Often, charging of monomers by dissociation of counterions is the only way to achieve hydrophilic behavior of the polymers. Charged brushes can be theoretically understood by considering the entropy of counterions which are rather freely moving inside the brush. The osmotic pressure of counterions even overcomes the effect of poor solvent conditions which are otherwise often present for the uncharged backbone of the chain in water. Mean-field theory predicts a collapse transition of the brush when the poor solvent effect is strong enough at low temperatures. Interestingly polymer brushes avoid the collapse to a large extend by lateral structure formation which allows for higher entropy of the counterions.

Acknowledgments The author thanks Jaroslav Paturej for critical reading of the manuscript.

References

1. P. de Gennes, *Scaling Concepts in Polymer Physics* (Cornell University Press, Ithaca, 1979)
2. P. de Gennes, Scaling theory of polymer adsorption. Journ. Physique **37**(12), 1445 (1976)
3. E. Eisenriegler, K. Kremer, K. Binder, Adsorption of polymer chains at surfaces: Scaling and monte carlo analyses. J. Chem. Phys. **77**(12), 6296–6320 (1982)
4. E. Eisenriegler, K. Kremer, K. Binder, Adsorption of polymer chains at surfaces: Scaling and monte carlo analyses. J. Chem. Phys. **77**, 6296–6320 (1982)

5. R. Descas, J.-U. Sommer, A. Blumen, Grafted polymer chains interacting with substrates: Computer simulations and scaling. Macrom. Theor. Sim. **17**, 429–453 (2008)
6. R. Descas, J.-U. Sommer, A. Blumen, Concentration and saturation effects of tethered polymer chains on adsorbing surfaces. J. Chem. Phys. **125**, 214702 (2006)
7. D. Williams, Grafted polymers in bad solvents: Octupus surface micelles. J. Phys. II France **3** (9), 1313–1318 (1993)
8. S. Alexander, J. Phys. (France) **38**, 977 (1977)
9. P. de Gennes, Macromolecules **13**, 1069 (1980)
10. H. Merlitz, G.-L. He, C.-X. Wu, J.-U. Sommer, Surface instabilities of monodisperse and densely grafted polymer brushes. Macromolecules **41**, 5070 (2008)
11. H. Merlitz, G.-L. He, C.-X. Wu, J.-U. Sommer, Nanoscale brushes: How to build a smart surface coating. PRL **102**(11), 115702 (2009)
12. P. Pincus, Colloid stabilization with grafted polyelectrolytes. Macromolecules **24**(10), 2912 (1991)
13. L. Chen, H. Merlitz, S.-Z. He, C.-X. Wu, J.-U. Sommer, Polyelectrolyte brushes: Debye approximation and mean-field theory. Macromolecules **44**, 3109–3116 (2011)
14. G.-L. He, H. Merlitz, J.-U. Sommer, Molecular dynamics simulations of polyelectrolyte brushes under poor solvent conditions: Origins of bundle formation. J. Chem. Phys. **140**, 104911 (2014)
15. J.M.Y.C.D.J. Sangberg, A.V. Dobrynin, Langmuir **23**, 12716 (2007)
16. I. Carmesin, K. Kremer, The bond fluctuation method: A new effective algorithm for the dynamics of polymers in all spatial dimensions. Macromolecules **21**(9), 2819–2823 (1988)

Chapter 10
Colloidal Hydrodynamics and Interfacial Effects

Maciej Lisicki and Gerhard Nägele

Abstract Interfaces and boundaries play an important role in numerous soft matter and biological systems. Apart from direct interactions, the boundaries interact with suspended microparticles by altering the solvent flow field in their vicinity. Hydrodynamic interactions with walls and liquid interfaces may lead to a significant change in the particle dynamics in (partially) confined geometry. In these lecture notes we review the basic concepts related to colloidal hydrodynamics and discuss in more detail the effects of geometric confinement and the hydrodynamic boundary conditions which an interface imposes on a suspension of microparticles. We start with considering the general characteristic features of low-Reynolds-number flows, which are an inherent part of any colloidal system, and discuss the appropriate boundary conditions for various types of interfaces. We then proceed to develop a proper theoretical description of the friction-dominated, inertia-free dynamics of colloidal particles. To this end, we introduce the concept of hydrodynamic mobility, and analyse the solutions of the Stokes equations for a single spherical particle in the bulk and in the presence of a planar solid-fluid, and fluid-fluid interfaces. Both forced and phoretic motions are considered, with a particular emphasis on the principles of electrophoresis and the associated fluid flows. Moreover, we discuss the hydrodynamic interactions of self-propelling microswimmers, and the peculiar motion of bacteria attracted to slip and no-slip walls.

M. Lisicki (✉)
Faculty of Physics, Institute of Theoretical Physics, University of Warsaw, Warsaw, Poland
e-mail: mklis@fuw.edu.pl

G. Nägele
Institute for Complex Systems ICS-3, Forschungszentrum Jülich, and Institut für Theoretische Physik, Heinrich-Heine-Universität Düsseldorf, Jülich, Germany
e-mail: g.naegele@fz-juelich.de

© Springer International Publishing Switzerland 2016
P.R. Lang and Y. Liu (eds.), *Soft Matter at Aqueous Interfaces*,
Lecture Notes in Physics 917, DOI 10.1007/978-3-319-24502-7_10

10.1 Introduction

Mesoscale particles suspended in a viscous fluid are found in numerous technological processes and products, including paints, cosmetics, pharmaceuticals, and
food stuff. They are encountered also in biological processes involving complex
fluids, and animalcules such as eukaryotic cells and bacteria. Understanding the
dynamics of such systems, often referred to as passive or active soft matter systems,
is of importance not only for industrial development, materials science, microbi-
ology and health care, but also from the point of view of fundamental scientific
problems such as in dynamic phase transitions. The importance of soft matter
systems derives also from their diversity, and the variety of tunable particle interactions giving rise to a plethora of phenomena that are partially still unexplored. An
inherent feature of such systems is the presence of a viscous solvent which trans-
mits mechanical stresses through the fluid, affecting in this way the motion of
suspended particles. These solvent-mediated particle interactions are known as
hydrodynamic interactions (HIs). The presence of HIs affects the dynamic properties of soft matter systems: In colloidal suspensions, e.g., they change the diffusion and rheological suspension properties [1], and play an important role in the
dynamics of DNA helices and proteins in solution [2]. Moreover, HIs modify the
characteristics of the coiling-stretching transition in polymers [3], influence the
pathways of phase separation in binary mixtures [4], alter the kinetics of macromolecules adsorption on surfaces [5] and cell adhesion [6], and are at the origin of
the flow-induced polymer migration in microchannels [7].

There has been a growing interest in the physics of soft matter systems, particularly triggered by the development of experimental techniques allowing for
probing soft matter on smaller length and time scales. The widespread use of
advanced optical microscopy and light scattering techniques in scientific and
industrial laboratories has fostered the insight in the structure and dynamics of soft
matter systems, and has boosted the development of theoretical and numerical tools
used in tackling emerging problems. The complexity of the studied systems has
considerably grown over the past years. Yet, the underlying physical principles
remain rather simple, so that if not fully quantitative then at least qualitative predictions of dynamic properties can be made.

Quite interestingly, many relevant hydrodynamic processes take place under
(partial) confinement, such as in a vessel or channel, close to a cell wall, inside
droplets, in the presence of bubbles, or near macroscopic fluid interfaces. Since the
confining boundaries or interfaces can have a dominant effect on the system
dynamics, it is important to analyse in detail their effect on the fluid flow in their
relative vicinity, and on the motion of suspended particles.

The aim of these lecture notes is to give an elementary introduction into
hydrodynamic effects occurring in colloidal systems, with a particular emphasis on
interfacial effects. There are various mathematical subtleties showing up in the
theoretical and computer simulation modelling of colloidal hydrodynamics. In this

more elementary introduction, however, we leave these subtleties aside, focusing instead on the physical principles without attempting to be mathematically rigorous.

There exists a large number of overview articles and textbooks on the hydrodynamics of soft matter systems, on different levels of complexity. As introductory texts on colloid hydrodynamics, we recommend the textbooks by Dhont [1] and Guazzelli and Morris [8], the lecture notes by Nägele in [9, 10], and the overview articles by Hinch [11], Pusey [12], and Pusey and Jones [13]. More advanced topics related to slow viscous flows are addressed in the excellent textbooks of Kim and Karrila [14], Happel and Brenner [15], and Zapryanov and Tabakova [16]. Standard textbooks on general hydrodynamics are the ones by Batchelor [17], and Landau and Lifshitz [18]. We further recommend the textbook by Guyon et al. [19]. A set of classical videos by G. I. Taylor [20] is recommended as an enjoyable illustration of the general features of low-Reynolds-number hydrodynamics discussed in the present notes.

Outline We start by introducing in Sect. 10.2 the Stokes (creeping flow) equations governing the low-Reynolds-number quasi-incompressible motion of a viscous fluid on colloidal time and length scales. The linear Stokes equations are a special case of the non-linear Navier-Stokes equations of incompressible flow, under the conditions where inertial effects are negligible and the particle motion is viscosity dominated. We show that these equations apply to flows related to the motion of suspended colloids and unicellular animalcules. The Stokes equations are amended by boundary conditions (BCs) on particle surfaces, confining interfaces and container walls. In this context, we discuss as important examples the no-slip and Navier partial-slip BCs for the fluid at a rigid surface, and the fluid-fluid BCs at a clean fluid-fluid interface. In Sect. 10.3, we explain salient generic features of Stokes flows, namely linearity, instantaneity, and kinematic reversibility. These features are used subsequently to infer some general knowledge on the motion of rigid microparticles in a viscous liquid. In Sect. 10.4, we analyse the bulk hydrodynamics of an unbounded colloidal suspension and the associated microparticles motion. For this purpose, we introduce the important concept of hydrodynamic friction and mobility tensors. Moreover, we discuss a versatile set of elemental solutions of the Stokes equations from which the flow profiles in simple situations are readily constructed. As examples, we discuss the motion of a slender particle (a rod) where the shape anisotropy results in anisotropic friction, and a spherical particle driven by body forces (i.e. gravitational settling), or by external fields such as temperature or electric potential gradient (phoretic motion). We introduce the notion of many-body hydrodynamic interactions (HIs) between microparticles, and outline how these interactions can be accounted for theoretically. The section is concluded by the lubrication analysis of the motion of two nearly touching spheres, and of a sphere near a flat no-slip wall.

Section 10.5 is dedicated to single-particle dynamics in the presence of a flat interface. We show how the solutions of the Stokes equations in (partially) confined geometry can be constructed using a superposition of the previously introduced elemental flow solutions, and discuss the implications of various interfacial boundary conditions on the dynamics of a suspended colloidal particle. In particular, we discuss the translational and rotational motion of a spherical particle near a no-slip wall, and comment on generalizations of this system to elastic particles and deformable interfaces. In Sect. 10.6, we explore the self-propulsion of microswimmers such as bacteria and spermatozoa, both in a bulk fluid and near to a confining surface. Our concluding remarks are contained in Sect. 10.7.

10.2 Fluid-Particle Dynamics on Microscale

In this Section, we elucidate some of the basic features of microscale flows. Due to the typical small sizes and velocities of microparticles, the flow on these scales can be treated as inertia-free and dominated by viscous effects. The neglect of inertia in the Navier-Stokes equations of hydrodynamics leads to the linear Stokes equations. These equations need to be supplemented with appropriate boundary conditions at interfaces confining the fluid. We introduce and discuss the BCs for a no-slip rigid wall, a clean fluid-fluid interface, and a partial slip surface.

10.2.1 Low-Reynolds-Number Flow

On length and time scales where continuum mechanics applies, the flow of an incompressible Newtonian fluid of shear viscosity η and constant mass density ρ_f is governed by the Navier-Stokes equations,

$$\rho_f \left(\frac{\partial \mathbf{u}(\mathbf{r},t)}{\partial t} + \mathbf{u}(\mathbf{r},t) \cdot \nabla \mathbf{u}(\mathbf{r},t) \right) = -\nabla p(\mathbf{r},t) + \eta \nabla^2 \mathbf{u}(\mathbf{r},t) + \mathbf{f}(\mathbf{r},t), \quad (10.1)$$

$$\nabla \cdot \mathbf{u}(\mathbf{r},t) = 0, \quad (10.2)$$

where $\mathbf{u}(\mathbf{r},t)$ is the velocity field at a point $\mathbf{r}$ at time t, and $p(\mathbf{r},t)$ is the pressure field. The shear viscosity η and the fluid mass density ρ_f are constant for a Newtonian fluid. The second equation follows from the continuity equation for a fluid of constant mass density, and is referred to as the incompressibility condition. The pressure in an incompressible fluid is determined only up to an additive constant, for p appears in the Navier-Stokes equations in the form of its gradient only. The external body force field per unit volume acting on the fluid is denoted by $\mathbf{f}(\mathbf{r},t)$. It can be due, e.g., to an applied electric or magnetic field, and to particle surfaces or system boundaries confining the fluid. For the latter two cases, the body forces are singularly concentrated on two-dimensional surfaces. For surface hydrodynamic boundary conditions (BCs) involving velocities only, the effect of a constant gravitational field on the fluid can be included conveniently by redefining the pressure according to $p \to p + \rho_f \mathbf{g} \cdot \mathbf{r}$ where $\mathbf{g}$ is the gravitational acceleration. A particle of uniform mass density ρ_p and volume $\Delta\Omega_p$ experiences in the fluid the buoyancy-corrected gravitational force $(\rho_p - \rho_f)\Delta\Omega_p\mathbf{g}$ acting at its center-of-mass (Archimedes principle).

Consider now an ensemble of rigid, impermeable microparticles immersed in the fluid (Fig. 10.1). In many but not all cases, the particles have no-slip surfaces. This means that the velocity of the fluid at every point of a particle surface must match the velocity of the particle at this point. The motion of the material on the surface and inside a rigid particle i can be described by

$$\mathbf{u}(\mathbf{r}) = \mathbf{V}_i + \boldsymbol{\Omega}_i \times (\mathbf{r} - \mathbf{R}_i), \tag{10.3}$$

where $\mathbf{V}_i$ and $\boldsymbol{\Omega}_i$ are the particle's translational and angular velocity vectors, respectively, and $\mathbf{R}_i$ is a body-fixed reference point which can be taken, e.g., to be the centre-of-mass position. The particles influence the flow outside through the boundary conditions applied to their surfaces. Another influence on the fluid flow is caused by the boundary conditions on external boundaries such as container walls, or at infinity.

The Navier–Stokes Eqs. (10.1) include both inertial effects, represented by the two terms on the left-hand side proportional to ρ_f, and fluid viscosity effects which are included in the viscous force density term $\eta\nabla^2\mathbf{u}$ on the right-hand side. The relative importance of these effects can be read off from the dimensionless Reynolds number Re. Suppose a sphere of radius a translates through the fluid with a velocity of magnitude V. The Reynolds number associated with the fluid flow caused by the sphere's motion is

$$Re = \frac{\rho_f V a}{\eta} \sim \frac{\rho_f |\mathbf{u} \cdot \nabla\mathbf{u}|}{|\eta\nabla^2\mathbf{u}|}. \tag{10.4}$$

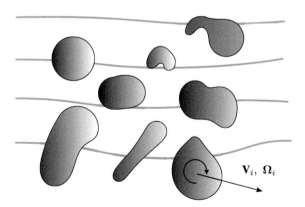

Fig. 10.1 Sketch of fluid flow around a microparticle labelled i with translational and rotational velocities $\mathbf{V}_i$ and $\boldsymbol{\Omega}_i$, respectively, in the presence of other immersed particles. Even if the rigid particles are free to translate and rotate in response to an externally imposed flow (freely advecting particles), they still modify the flow pattern through the BCs on their surfaces

For nano- to micrometer-sized particles, including in particular colloidal systems, the Reynolds number is typically of the order of 10^{-3} or even smaller [12]. This implies an important feature of the so-called low-Reynolds-number flows: Inertial effects can be neglected as compared to the viscous ones, so that the non-linear convective term $\rho_f \mathbf{u} \cdot \nabla \mathbf{u}$ in Eq. (10.1) can be taken as zero. In the absence of intrinsic time scales originating from high-frequency oscillatory or ultra-strong forcing of particles, the linear time derivative term on the left-hand-side of the Navier-Stokes equation can be likewise neglected.

The description of microparticle-induced hydrodynamics in a Newtonian fluid reduces then to the Stokes equations,

$$-\nabla p(\mathbf{r}) + \eta \nabla^2 \mathbf{u}(\mathbf{r}) + \mathbf{f}(\mathbf{r}) = 0, \tag{10.5}$$

$$\nabla \cdot \mathbf{u}(\mathbf{r}) = 0, \tag{10.6}$$

also referred to as creeping flow equations. These equations have no explicit time dependence and are linear in the velocity and pressure fields.

Equation (10.5) expresses the balance, at any instant of time and for every fluid element, of pressure gradient, viscous and external force densities. In the absence of external force density, the instantaneous values of velocity and pressure, and consequently the fluid stress field, depend solely on the momentary configuration and shape of particles and system boundaries, and on the surface boundary conditions taken at the particle surfaces and system boundaries. There is thus no dependence on the earlier flow history. Note that motion under Stokes flow conditions can be unsteady, with the velocities of particles and surrounding fluid changing as a function of time. An important example illustrating this fact is the settling of a spherical particle towards a stationary wall in its vicinity. This settling is discussed in Sect. 10.4.7 in relation to the effect of lubrication. At any instant, however, the net force and torque on each particle and each fluid element are zero, with accordingly instantaneous linear force-velocity relations characteristic of non-inertial fluid *and* immersed microparticles motions. The flow and pressure fields pattern readjust quasi-instantaneously to the moving system boundaries and particle surfaces.

In consequence, the hydrodynamic drag force $\mathbf{F}^h$ and torque $\mathbf{T}^h$ acting on a particle due to its surface friction with the surrounding fluid are exactly balanced, according to

$$\begin{aligned} \mathbf{F}^h + \mathbf{F} &= 0, \\ \mathbf{T}^h + \mathbf{T} &= 0, \end{aligned} \tag{10.7}$$

by a non-hydrodynamic 'external' force $\mathbf{F}$ and torque $\mathbf{T}$, respectively, caused by direct interactions with other particles and system boundaries, and by external force fields. Only a force-free and torque-free particle will move quasi-inertia-free. There is in addition a so-called thermodynamic force contribution to $\mathbf{F}$ and $\mathbf{T}$ proportional to the system temperature T which accounts for the on average isotropic thermal bombardment of a microparticle by the surrounding fluid molecules. If viewed on the time and length scales where creeping flow applies this bombardment leads to an erratic Brownian motion of the particles which persists even in the absence of additional force contributions to $\mathbf{F}$ and $\mathbf{T}$.

The strength of the Brownian motion of a particle can be characterized by the diffusion time τ_D which is the time required by a particle to diffuse by Brownian motion over a distance comparable to its size. For a spherical particle of radius a, this characteristic diffusion time is

$$\tau_D = \frac{a^2}{D^0} \propto \eta \frac{a^3}{T}, \tag{10.8}$$

where

$$D^0 = \frac{k_B T}{C \eta a}, \tag{10.9}$$

is the single-sphere Stokes-Einstein translational diffusion coefficient. This coefficient decreases with increasing particle size and fluid viscosity, and it increases with increasing temperature T. The numerical coefficient C depends on the hydrodynamic boundary condition for the flow at the sphere surface. According to [1]

$$\left\langle [\mathbf{R}(t) - \mathbf{R}(0)]^2 \right\rangle = 6D^0 t, \tag{10.10}$$

where D^0 quantifies the magnitude of the mean-squared displacement, after the time span t, of the position vector $\mathbf{R}$ of an isolated Brownian particle immersed in an unbounded fluid. The brackets denote here an average over an equilibrium ensemble of non-interacting Brownian particles.

The diffusion time grows strongly with increasing particle size. For water at room temperature as the suspending fluid, it increases from $\tau_D \sim 5$ ms for $a = 0.1$ μm to $\tau_D \sim 0.3$ h for $a = 5$ μm. Brownian motion is thus negligibly small for particles of several micrometers in size or larger. These particles are therefore referred to as non-Brownian. A dispersion of non-Brownian particles requires external driving agents to keep them in motion. This agent can be gravity, provided some of the particles are lighter or heavier than the fluid, or an applied electric, magnetic or temperature gradient field. Additionally, the particles are hydrodynamically moved by incident flows created by moving system boundary parts (e.g., in cylindrical Couette cell flow) or applied pressure gradients (e.g., in pipe flow).

The distinguishing and to some extent surprising properties of fluid flows described by the Stokes equations, and of the associated microparticles motions, are an important theme of the present lecture notes, in addition to interfacial effects related to the fluid dynamics. In our discussion, we will make ample use of streamlines pattern in order to visualize Stokes flow fields formed around particles in the bulk fluid and at interfaces. A streamline is tangential to the local velocity field at any fluid point, and for stationary flow it agrees with the pathway of a fluid element. For each streamline segment $d\mathbf{r}$, we have thus

$$d\mathbf{r} \times \mathbf{u}(\mathbf{r}) = 0. \qquad (10.11)$$

The three Cartesian components of this vectorial equation form a coupled set of differential equations, for given $\mathbf{u}(\mathbf{r})$, from which the streamlines can be determined.

10.2.1.1 Hydrodynamic Stresses

To every solution, $\{\mathbf{u},p\}$, of the Stokes equations, referred to as a Stokes flow solution, one can associate a fluid stress field described in terms of a stress tensor $\boldsymbol{\sigma}$. This symmetric second-rank tensor consists of nine elements σ_{ij} with $i,j \in \{1,2,3\}$ which at a given fluid position $\mathbf{r}$ have values depending on the considered (rectangular) coordinate system spanned by its three basis unit vectors $\{\mathbf{e}_1,\mathbf{e}_2,\mathbf{e}_3\}$. The stress tensor has the following physical meaning: Imagine a small planar surface element dS in the fluid with the unit normal vector $\mathbf{n}$. The hydrodynamic drag force, $d\mathbf{F}$, exerted by the fluid on this surface element, located on the side where $\mathbf{n}$ points to, is then given by $d\mathbf{F} = \boldsymbol{\sigma} \cdot \mathbf{n}dS$. The tensor (matrix) element σ_{ij} is therefore the hydrodynamic force component per unit area (referred to as stress) acting in the direction $\mathbf{e}_i$ on a surface element with the normal vector equal to $\mathbf{e}_j$ [17]. The stress field of an incompressible Newtonian fluid is given in terms of the flow fields $\mathbf{u}$ and p by

$$\boldsymbol{\sigma}(\mathbf{r}) = -p(\mathbf{r})\mathbf{I} + \eta\mathbf{E}(\mathbf{r}), \qquad (10.12)$$

where $\mathbf{I}$ is the unit tensor, and

$$\mathbf{E}(\mathbf{r}) = [\nabla\mathbf{u}(\mathbf{r})] + [\nabla\mathbf{u}(\mathbf{r})]^T \qquad (10.13)$$

is the symmetric fluid rate-of-strain tensor, with the superscript T denoting the transposition operation.

While the polyadic tensor expression for $\boldsymbol{\sigma}(\mathbf{r})$ in Eqs. (10.12) and (10.13) applies to all coordinate systems, the explicit form of its elements depends on the selected coordinates [15]. In Cartesian coordinates where the orthonormal basis vectors $\{\mathbf{e}_1, \mathbf{e}_2, \mathbf{e}_3\} = \{\mathbf{e}_x, \mathbf{e}_y, \mathbf{e}_z\}$ are constant, the stress tensor elements are simply given by

$$\sigma_{ij} = -p\delta_{ij} + \eta\left[\frac{\partial u_i}{\partial r_j} + \frac{\partial u_j}{\partial r_i}\right], \qquad (10.14)$$

where $\{r_1,r_2,r_3\} = \{x,y,z\}$ are the Cartesian components of the fluid element position vector $\mathbf{r}$.

The hydrodynamic stress field depends on the properties of the fluid flow which in turn is influenced by the characteristics of the particles and confining walls, namely their porosity and fluid permeability, and other non-hydrodynamic surface properties such as surface charge density, van der Waals attraction etc. The knowledge of stresses in the fluid is of importance, since it allows for the calculation of hydrodynamic drag forces and torques acting on bodies immersed in the fluid. It is also of key importance for the calculation of rheological properties such as the effective suspension viscosity of a fluid with immersed microparticles [21]. Once the hydrodynamic stresses are known, the hydrodynamic drag force and torque, $\mathbf{F}^h$ and $\mathbf{T}^h$, acting on a particle can be calculated as the sum (integral) of the local surface force and torque contributions, respectively, according to

$$
\begin{aligned}
\mathbf{F}^h &= \int_S dS \boldsymbol{\sigma}(\mathbf{r}) \cdot \mathbf{n}(\mathbf{r}) \\
\mathbf{T}^h &= \int_S dS (\mathbf{r} - \mathbf{R}) \times \boldsymbol{\sigma}(\mathbf{r}) \cdot \mathbf{n}(\mathbf{r})
\end{aligned}
\tag{10.15}
$$

The surface S can be replaced by any fluid surface S^* enclosing the considered particle without intersecting another one, provided there is no body force density acting on the enclosed fluid part, since the hydrodynamic force and torque on a particle are transmitted loss-free through the fluid [9]. The vector $\mathbf{n}$ is normal to the surface of the particle and points into the fluid. See here Fig. 10.2.

It is important to realize that the arguments used to neglect the effects of inertia and the explicit time dependence of the fluid velocity and pressure fields are not applicable (i) if one studies processes on very short time and length scales where the time-dependence of $\mathbf{u}$ and p becomes essential, such as sound propagation in the fluid, and (ii) for processes which occur at different length scales so that the effective Reynolds number becomes large as compared to one. For a detailed discussion of the involved time and length scales of fluid and immersed microparticles, we refer to [1, 8, 12].

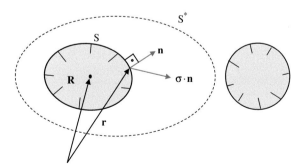

Fig. 10.2 Surface stress or traction (force per area), $\boldsymbol{\sigma}(\mathbf{r}) \cdot \mathbf{n}(\mathbf{r})$, exerted by the fluid on a particle surface element dS at position $\mathbf{r}$. The surface normal vector $\mathbf{n}$ points into the fluid. The vector $\mathbf{R}$ points to a particle-fixed reference point. For the calculation of drag force and torque, the surface S^* enclosing the particle can be chosen rather arbitrarily (see text)

10.2.2 Boundary Conditions

Fluid flows close to interfaces are strongly influenced by the interfacial properties. There exists an abundance of soft matter systems in which interfacial effects are highly influential on the dynamics. Notable examples of interfaces include: A smooth solid wall or particle surface; an engineered nano-structured surface; an interface between two immiscible fluids such as water and oil; liquid-gas free interfaces such as for gas bubbles in a liquid; surfactant-covered interfaces, and polymer-coated and grafted particle surfaces. To describe and understand the effect of interfaces on the flow behaviour, one needs to consider the appropriate boundary conditions imposed on the fluid at the surface or interface. The no-slip boundary condition for rigid impermeable surfaces noted in Eq. (10.3), first described by Navier in 1823, has been given considerable attention over the past two centuries, concerning in particular its applicability and validity [22]. It is generally accepted as the proper boundary condition for smooth solid hard walls, and for rigid particles with smooth non-permeable surfaces and sizes exceeding ~ 30 nm.

We discuss in the following two additional types of boundary conditions which are also frequently applied to soft matter systems. The first one concerns a clean liquid-liquid interface between two immiscible Newtonian fluids of viscosity ratio

$$\lambda = \eta_2/\eta_1, \tag{10.16}$$

with the associated near-interface flow sketched in the left part of Fig. 10.3. The appropriate boundary conditions are here the continuity of tangential velocities and tangential (shear) stresses of the two fluids at the interface, and the impermeability of the interface. The latter condition implies the equality of the normal velocity components of both fluids. Assuming a planar interface stretching out in the $x - y$ plane at $z = 0$, these continuity conditions read

$$u_z^{(1)} = u_z^{(2)}, \quad u_{x,y}^{(1)} = u_{x,y}^{(2)}, \quad E_{xz}^{(1)} = \lambda E_{xz}^{(2)}, \quad E_{yz}^{(1)} = \lambda E_{yz}^{(2)}, \tag{10.17}$$

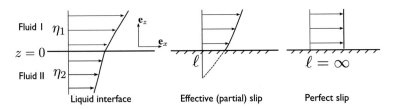

Fig. 10.3 Schematic flow profiles close to different interfaces as discussed in the notes. *Left* A clean interface between two immiscible fluids of different viscosities $\eta_1 < \eta_2$, where du_x/dz changes discontinuously. *Middle* Flow above a stationary rigid partial-slip surface, characterized by the Navier slip length ℓ equal to the distance to an apparent no-slip plane inside the stationary wall. *Right* Plug-like flow above a stationary (perfect) slip wall where $\ell = \infty$. The same flow is observed near an ideal liquid-gas interface where $\eta_2/\eta_1 = 0$, and where the gas phase is situated in the half-space $z < 0$

where the $E_{\alpha\beta}^{(i)}$ are the Cartesian elements of the rate-of-strain tensor $\mathbf{E}$ of the fluid introduced in Eq. (10.13). The liquid-liquid BCs include as limiting cases firstly a free interface (e.g., a water-air interface) where the viscosity of the second fluid is negligible, so that $\lambda = 0$, and secondly a fluid above a rigid no-slip wall with the latter described as a fluid of infinite viscosity, so that $\lambda \to \infty$. In fact, $\lambda = 0$ implies a (perfect) slip surface of zero tangential stress while $\lambda \to \infty$ implies the no-slip condition $\mathbf{u}(z = 0,x,y) = 0$ for a stationary wall.

The equality of tangential stresses of both fluids at their interface is valid for uniform interfacial tension γ only, i.e. for constant free energy per area gone into the formation of the interface. Any mechanism creating a gradient, $\nabla\gamma$, in the interfacial tension breaks the shear stress continuity and drives motions in the two fluids. These motions are referred to as Marangoni flows [23]. One possible way to cause a Marangoni flow is to establish a sufficiently strong temperature gradient along the interface [19].

For a planar liquid-liquid interface with zero motion of both fluids, the hydro-static pressure on both sides is the same in order to maintain a stationary interface. However, for a stationary spherical droplet of radius a in a stationary fluid, the hydrostatic pressure in its interior exceeds the outside fluid pressure by the capillary (Laplace) pressure contribution γ/a. Any deformation of the droplet away from its equilibrium spherical shape of constant curvature and minimal surface free energy will cause flows and associated droplet motion which tend to re-establish its spherical shape. See Ref. [23] for a lucid discussion of droplet motions and Marangoni flow effects.

The second type of a boundary condition describes the partial-slip of fluid along the surface of a fluid-impermeable solid material, as illustrated in the middle and right parts of Fig. 10.3 where the solid extends to $z < 0$ with the fluid residing on top. The so-called Navier BCs for a stationary partial-slip surface demand, in addition to a vanishing normal velocity component at the surface, the proportion-ality of surface-tangential fluid velocity and shear stress according to

$$\mathbf{t} \cdot \mathbf{u} = \frac{\ell}{\eta} \mathbf{t} \cdot \boldsymbol{\sigma} \cdot \mathbf{n}, \quad \mathbf{n} \cdot \mathbf{u} = 0. \tag{10.18}$$

Here, ℓ is the Navier slip length, and $\mathbf{t}$ and $\mathbf{n}$ are tangential and normal unit vectors at a surface point. For the planar stationary surface at $z = 0$ depicted in Fig. 10.3, the partial-slip BC for the surface-tangential velocity part simplifies to

$$u_{x,y} = \ell \frac{\partial u_{x,y}}{\partial z}, \quad u_z = 0. \tag{10.19}$$

The slip length ℓ is here the distance into the interior of the wall for which the near-surface flow linearly extrapolates to zero, defining in this way an effective no-slip plane at $z = -\ell$. In the limit $\ell = 0$, the no-slip BC with zero surface slip velocity is recovered. In the opposite limit $\ell \to \infty$, the free-surface boundary condition of zero tangential stress is obtained, with fluid slipping perfectly along the surface in a plug-flow-like manner.

The Navier partial-slip BCs can serve as an effective description for a hydrophobic wall, a rigid particle with surface roughness or corrugations [24], and to some extent also for a wall grafted with polymer brushes acting as depletants [25]. Moreover, it can be used for a fluid-solid interface with free polymers in the fluid, and a polymer depletion layer at the interface [26]. An effective (apparent) fluid slip is also found in electrokinetic [21] and other phoretic flows where the no-slip boundary condition holds right at the wall and the particle surfaces. Outside a thin fluid boundary layer with viscous flow, however, flow slip is observed [22]. In Sect. 10.4.5, effective slip is discussed in relation to phoretic motion of a microsphere.

10.3   Generic Features of Stokes Flows

Creeping flows have interesting generic properties which appear counter-intuitive from the perspective of our macroscopic world experience where inertia and high-Reynolds-number effects prevail, with the flow governed by the non-linear Navier-Stokes equations. The three generic features of the Stokes equations are linearity, kinematic reversibility, and instantaneity. In this section, their implications for the colloidal dynamics are described.

10.3.1   Linearity

The Stokes equations are linear in contrast to the underlying Navier-Stokes equations. This means that the pressure, velocity and stress field are linearly related. The consequences of linearity are far-reaching. For instance, in a slow viscous channel flow, on doubling the applied pressure gradient, a doubling of the flow rate is obtained. Moreover, a twofold increase in the rate of flow of viscous fluid through a porous medium will result in an unchanged pattern of streamlines of the flow, but with the magnitude of the fluid elements' velocities doubled. For a sphere settling in a viscous liquid, doubling the settling velocity gives rise to a correspondingly doubled hydrodynamic drag force. The fact that the hydrodynamic force on a particle and the associated velocity (increment) are linearly related is exploited further in Sect. 10.4.1, where we discuss the hydrodynamic friction and mobility coefficients in many-particle dispersions.

For linear evolution equations such as the Stokes equations, the superposition principle is valid: If $\mathbf{u}_1$ and $\mathbf{u}_2$ are two velocity solutions of the Stokes equations, then

$$\mathbf{u} = \lambda_1 \mathbf{u}_1 + \lambda_2 \mathbf{u}_2 \tag{10.20}$$

$$\nabla p = \lambda_1 \nabla p_1 + \lambda_2 \nabla p_2 \tag{10.21}$$

are likewise solutions with coefficients λ_1 and λ_2. Here, ∇p_i is the pressure gradient field solution to the Stokes equations associated with $\mathbf{u}_i$. For a given flow boundary value problem, the unique velocity field $\mathbf{u}$ can be obtained from the linear super-position of two (simpler) flows with unchanged geometry, provided the velocity BCs of the two partial flows superimpose correspondingly, with the same coefficients, to the BCs of the full flow solution.

The linearity of the Stokes flow solutions can lead to rather unexpected con-clusions; Consider a particle moving through the fluid with the velocity $\mathbf{V}$ with Cartesian components V_i, $i = 1,2,3$. The particle experiences then the drag force $-\mathbf{F}$ which we can decompose into forces acting along the axes of the coordinate system according to $\mathbf{F} = \{F_i\}$. From linearity, we conclude that the force F_1 acting on a particle moving with velocity $(V_1,0,0)$ must be of the form $F_1 = \alpha V_1$, with α being a positive constant. Imagine now that the particle is a cube with its edges aligned along the coordinate axes. Then, from symmetry, $F_2 = \alpha V_2$ and $F_3 = \alpha V_3$, and in general $\mathbf{F} = \alpha \mathbf{V}$. Hence the drag force experienced by a cube does not depend on its orientation, and it is collinear with the velocity (see Fig. 10.4). As everyday experience teaches us, this is obviously not valid any more for large Re. In fact, a more general statement is true in Stokes dynamics: Any homogeneous body with three orthogonal planes of symmetry (such as spheroids, rods, cylinders, disks, or rings), will translate under the action of force without rotating, although in general with a sidewise velocity component perpendicular to the driving force. The sidewise motion is absent only if the force is acting along the rotational symmetry axis of the particle. In addition, force and velocity are collinear independently of the particle orientation for highly symmetric particles, namely for a homogeneous sphere and the five regular polyhedra (tetrahedron, cube, etc.), and also for homogeneous bodies made from the polyhedra by equally rounding off their cor-ners, provided the hydrodynamic BCs are homogeneous [9, 11]. For this statement to be true, the particle centre must be selected as the reference point.

As noted earlier, linearity can be used to decompose a complex flow problem into a number of simpler ones: one can for instance consider the problem of a spherical particle translating and rotating in a viscous fluid as the two separate problems of sole rotation and sole translation of a sphere, provided a corresponding linear decomposition of the surface boundary conditions in Eq. (10.3) is used. Such a decomposition proves useful in various numerical schemes for the calculation, e.g., of the hydrodynamic drag forces on an ensemble of spherical particles at a given fixed configuration. One has to bear in mind, however, that the imposed BCs must be simultaneously satisfied at the surfaces of all the particles. For more than two particles, this requires in general a complicated numerical analysis.

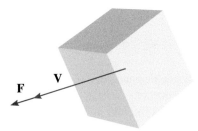

Fig. 10.4 A cube translating through a viscous fluid with velocity **V** under the influence of force **F** acting on its centre. For highly symmetric particles, linearity of Stokes equation implies that the force and velocity are collinear, with the drag force being independent of the particle orientation

10.3.2 Instantaneity

On the time and length scales where significant motion of colloidal microparticles is observed, the accompanying viscous flows are described by the quasi-stationary linear Stokes equations which have no explicit time dependence. As noted before, this means that the pressure and velocity fields adjust themselves instantaneously, on the coarse-grained colloidal time and length scales, to changes in the driving forces. The flow disturbances propagate in the fluid with an (apparently) infinite speed. A slight change in a particle's position or velocity is instantaneously communicated to the whole system. The fluid flow $\{\mathbf{u},p\}$ at a given time is therefore fully determined by the instantaneous positions and velocities of the particle surfaces and wall boundaries, independently of how the momentary boundary values have been reached (history independence). In particular, the instantaneous fluid flow pattern does not depend on whether the boundary velocities will stay constant in the future or change, such as in oscillatory motions.

This feature of Stokes flows appears counter-intuitive on the first sight. Yet, there exist nice demonstrations highlighting its validity, provided the frequency of oscillatory boundary motions and the probed distances are not too large. Otherwise, hydrodynamic retardation effects come into play reflecting the actually non-instantaneous spreading of flow perturbations by pressure (sound) waves, and by the diffusional spreading of flow vorticity in the viscous fluid with an associated vorticity diffusion coefficient, η/ρ_f, equal to the kinematic viscosity [19, 27].

10.3.3 Kinematic Reversibility

Kinematic reversibility is a remarkable feature of viscosity-dominated flows. The linearity of the Stokes equations in the flow fields $\{\mathbf{u},p\}$ and the applied forces, including the ones due to the fluid boundaries, implies that under the reversal of the driving forces, the flow fields are also reversed to $\{-\mathbf{u},-p\}$. Moreover, if the forces and also the history of their application is reversed, all fluid elements retrace their motion in the opposite direction along the unchanged streamlines.

Kinematic reversibility was beautifully demonstrated in G. I. Taylor's video [20] from 1966, where a drop of coloured ink is immersed in highly viscous glycerine, to maintain low-Reynolds-number flow, filling the gap between two concentric cylinders (Couette cell geometry). See here Fig. 10.5. On rotating the inner cylinder, the drop is smeared out along concentric streamlines into a thin filament. When the rotation is reversed subsequently by the same number of turns, the original droplet is reconstituted up to a small amount of blurring originating from the irreversible residual Brownian motion of the dye particles. The length of the filament depends on the number of turns only, independently of the rate at which the inner cylinder is rotated. This nicely illustrates the earlier discussed instantaneity of Stokes flows.

The kinematic reversibility in combination with specific symmetries puts general constraints on the motion of a microparticle in a viscous fluid. A classical example is a spherical rigid microparticle settling under gravity near a stationary vertical hard wall (see Fig. 10.6). While the particle is rotating clockwise during settling, owing to the larger wall-induced hydrodynamic friction on its semi-hemisphere facing the wall (see Sect. 10.5.2 for details), a question arises whether it will approach the wall or recede from it. Given that gravity acts vertically downwards parallel to the wall, assume for the time being that the sphere approaches the wall while settling (see Fig. 10.6a). Kinematic reversibility requires that once the direction of the motion-driving gravitational force is reversed, the Stokes flow pattern remains unchanged except for the directional reversal of the fluid elements motion, provided the translational and angular particle velocities are likewise reversed. According to Fig. 10.6b, this implies that the sphere sediments upwards while receding from the wall. On rotating Fig. 10.6b by 180° around the horizontal symmetry axis line going through the sphere centre, Fig. 10.6c is obtained in conflict with Fig. 10.6a wherein the sphere had been assumed to approach the wall. A contradiction is avoided only if the sphere remains at a constant distance from the wall while settling, as in Fig. 10.6d. An analogous reasoning can be employed to

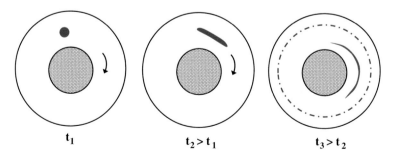

t_1 $t_2 > t_1$ $t_3 > t_2$

Fig. 10.5 Ink-spreading experiment by G. I. Taylor. An ink droplet inserted in a high-viscosity Newtonian fluid at time t_1 is smeared out in a thin concentric filament when the inner cylinder is rotated subsequently. The initial droplet shape is recovered after reversal of the rotation, independent of the rotation rate

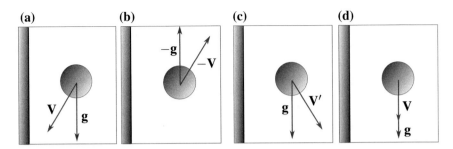

Fig. 10.6 A non-Brownian sphere settling with translational velocity **V** under gravity of field strength **g** near a vertical hard wall. The principle of kinematic reversibility in conjunction with the flow geometry leads to the conclusion that the sphere maintains a fixed distance from the wall while settling and rotating clockwise. See the text for details

show that in a Poiseuille channel flow, a non-Brownian microsphere translates along the flow streamline, without any cross-flow velocity component.

As discussed in Sect. 10.5.3, a non-spherical rigid particle, such as a rod, can move sidewise while settling and so approach the vertical wall. The wall-induced rotation of the particle can lead to a subsequent motion away from the wall. A deformable liquid droplet settling close to a vertical wall will deform into a shape which makes it glide away from the wall.

While non-spherical rigid particles and deformable particles can migrate across streamlines under Stokes flow conditions, this is not the case for an isolated non-Brownian spherical particle. However, the non-linear hydrodynamic coupling of the motions of three or more nearby spheres in a driven system such as in the pipe flow of a suspension, can lead to irregularly looking trajectories which depend sensitively on the initial particle configuration. Any reversibility-breaking slight perturbation of the initial particle configuration caused, e.g., by direct particle interactions in the form of surface roughness, flexibility or electric charge, or residual Brownian motion and inertia effects, becomes exponentially amplified, giving rise to chaotic trajectories causing cross-stream migration and the mixing of the particles. A macroscopic manifestation of this so-called (anisotropic and temperature-independent) hydrodynamic diffusion is the flow-induced migration of spherical particles in concentrated suspensions from the regions of high to low shear rates. The shear-induced diffusion has an application, e.g., in the inside-out microfiltration (enrichment) of a particle suspension pumped through a microfluidic filter pipe where it reduces the formation of irreversible particle deposits at the filter membrane (fouling reduction).

Already on the particle pair-interaction level, the kinematic reversibility of rigid particles is broken by physical processes modifying the Stokes equations or the evolution of the particle trajectories such as non-Newtonian terms in the stress-shear relation (cf. Eq. (10.14)) occurring in viscoelastic media (e.g., in polymer solutions and melts), short-range repelling forces, significant Brownian motion and non-inertial hydrodynamic effects. For example, for Reynolds numbers significantly larger than zero, a particle immersed in a pipe flow experiences an inertia-induced lift force driving it away from the pipe wall. This so-called tubular pinch or Segre-Silberberg effect, named after its discoverers, appears already on the single-particle level and should be distinguished from the many-particle shear-induced diffusion effect which takes place under Stokes flow conditions of zero inertia.

10.4   Colloidal Hydrodynamics in a Bulk Fluid

On the time and length scales of colloidal dynamics, the fluid flow is described by the Stokes equations supplemented by appropriate boundary conditions at interfaces and surfaces of suspended particles. Since the dynamics of these particulates is often of interest, one needs to construct a description of their interaction including the solvent-mediated hydrodynamic effects. In this Section, we introduce the notion of friction and mobility, and show how the linearity of the Stokes equations can be used to construct relations between forces and velocities of particles in a many-body system in the case where the interfaces are far away and the fluid may be regarded to be unbounded. We then proceed to explore the basic solutions of the creeping flow equations for point forces which are the simplest approximation to the flow field generated by the immersed particles. The set of solutions is then extended by multipole expansion to include more subtle flow effects. We apply this formalism to investigate the motion of shape-anisotropic slender bodies, such as rod-like colloids, and later on to construct a solution for a spherical particle moving through the fluid as a result of a force, or by a phoretic motion. We conclude this Section by a discussion of more advanced approaches to hydrodynamic interactions and of the lubrication effects which are essential when the particles are very close together.

10.4.1   Friction and Mobility of Microparticles

We outline here the theoretical framework for the description of the dynamics of a dispersion consisting of N rigid microparticles of basically arbitrary shape evolving under Stokes flows conditions [14]. Consider the particles to be at the instantaneous configuration $\mathbf{X} = (\mathbf{R}, \boldsymbol{\Theta}) = (\mathbf{R}_1, \ldots, \mathbf{R}_N, \boldsymbol{\Theta}_1, \ldots, \boldsymbol{\Theta}_N)$, with body-fixed particle position vectors $\{\mathbf{R}_i\}$ and orientations $\{\boldsymbol{\Theta}_i\}$. Here, $\boldsymbol{\Theta}_i$ abbreviates the three Euler angles characterizing the orientation of the particle i.

Suppose now that the particles are subjected to external forces $\mathbf{F} = (\mathbf{F}_1, \ldots \mathbf{F}_N)$ and torques $\mathbf{T} = (\mathbf{T}_1, \ldots, \mathbf{T}_N)$ where we have introduced $3N$-dimensional super-vectors $\mathbf{F}$ and $\mathbf{T}$ for notational convenience. As a consequence of this forcing, motion of the particles and the fluid is induced, and the particles acquire quasi-instantaneously the translational velocities $\mathbf{V} = (\mathbf{V}_1, \ldots, \mathbf{V}_N)$ and the rotational velocities $\boldsymbol{\Omega} = (\boldsymbol{\Omega}_1, \ldots, \boldsymbol{\Omega}_N)$. We have assumed a quiescent fluid for simplicity, meaning that the fluid would be at rest in the absence of particles. This implies, in particular, that there is no ambient flow caused, e.g., by confining boundary parts in relative motion. In the inertia-free Stokes flow system under consideration, each external force and torque are balanced by a hydrodynamic drag force and torque. Owing to the linearity of the Stokes equations and the hydrodynamic boundary conditions, the forces (torques) and translational (rotational) velocities are linearly related according to

$$\begin{pmatrix} \mathbf{V} \\ \boldsymbol{\Omega} \end{pmatrix} = \mu(\mathbf{X}) \cdot \begin{pmatrix} \mathbf{F} \\ \mathbf{T} \end{pmatrix}, \tag{10.22}$$

where the $6N \times 6N$ hydrodynamic mobility matrix μ has the four $3N \times 3N$ submatrices

$$\mu(\mathbf{X}) = \begin{pmatrix} \mu^{tt}(\mathbf{X}) & \mu^{tr}(\mathbf{X}) \\ \mu^{rt}(\mathbf{X}) & \mu^{rr}(\mathbf{X}) \end{pmatrix}. \tag{10.23}$$

The superscripts tt and rr label the purely translational and rotational mobility matrix parts, respectively. The off-diagonal matrices with superscripts tr and rt describe the hydrodynamic coupling between translational and rotational particle motions. The tensor elements of these matrices have a straightforward physical meaning. To give an example, the tensor $[\mu^{tt}(\mathbf{X})]_{ij}$ relates the instant force $\mathbf{F}_j$ on particle j with the translational velocity $\mathbf{V}_i$ of particle i, in a situation where particles different from j are all force- and torque-free. The coupling tensor $[\mu^{rt}(\mathbf{X})]_{ij}$, on the other hand, relates the force $\mathbf{F}_j$ on particle j to the resulting angular velocity $\boldsymbol{\Omega}_i$ of particle i. It is important to note here that the mobility matrix μ and its $4N^2$ mobility tensor elements depend on the configuration of the whole system, i.e. the instant positions and orientations of all particles, as well as on the particle shapes and sizes, and the surface boundary conditions. Finding the mobility tensor is therefore a very difficult problem which for arbitrary particle shapes can be addressed only numerically for a small number of particles.

It should be further noted that the form of the mobility matrix depends also on the selection of reference points $\mathbf{R}$ inside the particles. For these points, the so-called center of mobility of each particle should be selected which in Stokes flow dynamics plays a similar role as the center-of-mass position in Newtonian dynamics. For an axisymmetric homogeneous rigid body, the center-of-mobility and the center-of-mass are both located on the symmetry axis but they do not necessarily coincide. They coincide, however, for a homogeneous sphere. Different from the center-of-mass, the center-of-volume is depending on the shape of the particle surface only, for uniform surface BC, independent of the mass distribution inside the particle. For a more detailed discussion of this important issue, see [14, 28].

In the simplest case of hydrodynamically non-interacting spherical particles of equal radius a, the tt and rr tensors reduce to the 3×3 unit matrices,

$$[\boldsymbol{\mu}^{tt}(\mathbf{X})]_{ij} = \mu_0^t \delta_{ij}, \qquad [\boldsymbol{\mu}^{rr}(\mathbf{X})]_{ij} = \mu_0^r \delta_{ij} \tag{10.24}$$

describing the free translation and rotation of isolated spheres. This limiting case is approached for an ultra-dilute dispersion where the mean distance between two particles is very large compared to their sizes. The single-particle mobility coefficients of a no-slip sphere are explicitly (see Sect. 10.5.2)

$$\mu_0^t = \frac{1}{6\pi\eta a}, \qquad \mu_0^r = \frac{1}{8\pi\eta a^3} \tag{10.25}$$

with $\mathbf{V}_i = \mu_0^t \mathbf{F}_i$ and $\boldsymbol{\Omega}_i = \mu_0^r \mathbf{T}_i$. The tr and rt mobility tensors are here zero implying that there is no coupling between the translational and rotational motion of the particles.

Equation (10.22) describes the so-called mobility problem where the forces and torques acting on the particles are given, and the translational and rotational velocities are searched for. The inverse problem where the velocities are given and the forces are searched for, referred to as the friction problem, is straightforwardly formulated by introducing the $6N \times 6N$ friction matrix

$$\boldsymbol{\zeta} = \boldsymbol{\mu}^{-1} \tag{10.26}$$

defined as the inverse of the mobility matrix. That this inverse exists is due to the fact that $\boldsymbol{\mu}$ is symmetric and positive definite for all physically allowed particle configurations. This follows from general principles of the Stokes flows, and it implies physically that the power supplied to the particles by external forces is completely and quasi-instantaneously dissipated by heating the fluid. We quantify this statement for the motion of N torque-free microparticles in an infinite quiescent fluid where the rate of change of the particles kinetic energy, $W(t)$, instantaneously dissipated into heat by friction is given by

$$0 < \frac{dW(t)}{dt} = \begin{pmatrix} \mathbf{F} \\ \mathbf{T} \end{pmatrix} \cdot \begin{pmatrix} \mathbf{V} \\ \boldsymbol{\Omega} \end{pmatrix} = \begin{pmatrix} \mathbf{F} \\ \mathbf{T} \end{pmatrix} \cdot \boldsymbol{\mu}(\mathbf{X}) \cdot \begin{pmatrix} \mathbf{F} \\ \mathbf{T} \end{pmatrix}. \tag{10.27}$$

Since the $6N$-dimensional supervector with the particles forces and torques as elements is arbitrary, the second equality expresses the positive definiteness of the $6N \times 6N$ symmetric mobility matrix $\boldsymbol{\mu}^{tt}$. Any violation of the positive definiteness of this matrix would imply thus the violation of the second law of thermodynamics. In specializing Eq. (10.27) to torque-free and force-free particles, respectively, it follows readily the positive definiteness likewise of the $3N \times 3N$ symmetric sub-matrices $\boldsymbol{\mu}^{tt}$ and $\boldsymbol{\mu}^{rr}$ for all physically allowed particle configurations $\mathbf{X}$.

The knowledge of the configuration-dependence of $\boldsymbol{\mu}$, or likewise that of $\boldsymbol{\zeta}$, allows for the exploration of the microparticles' dynamics using numerical simulations, without having to address explicitly the accompanying fluid flow. For torque-free particles large enough for their Brownian motions to be negligible, the $3N$ coupled first-order equations of motion for the particles centre-of-mobility positions, in the presence of external and also non-hydrodynamic particle interaction forces all subsumed in $\mathbf{F}$, are given by

$$\frac{d\mathbf{R}(t)}{dt} = \boldsymbol{\mu}^{tt}(\mathbf{R}(t)) \cdot \mathbf{F}(t). \tag{10.28}$$

Integration of these evolution equations gives the positional trajectories of the particles. This is referred to as Stokesian dynamics [29]. Due to the non-linearity of the Stokesian dynamics evolution equations in Eq. (10.28), originating from the non-linear positional dependence of the mobility matrix, the trajectories are highly sensitive to the initial particle configuration: A slight change in the initial configuration can lead to large differences in the trajectorial evolution. Deterministic chaos in the trajectories of as little as three hydrodynamically interacting non-Brownian particles settling under gravity has been discussed theoretically first in the point-particle limit [30] and later also for extended spheres [31].

For smaller Brownian particles, on the other hand, the mobility matrix is needed as input not only for the generation of Stokesian particle displacements, but also for the generation of additional stochastic displacements caused by the thermal fluctuations of the solvent. These displacements are the essential ingredients of the so-called Brownian dynamics numerical scheme for the generation of Brownian stochastic trajectories [32]. For a pedagogical introduction to Brownian dynamics simulations, see [33]. From the generated trajectories, quantities such as the particle mean-squared displacement in Eq. (10.10) can be calculated, for the general case of interacting microparticles. The positive definiteness of the mobility matrix plays a key role for Brownian particles. It guarantees that a perturbed suspension evolves towards thermodynamic equilibrium, in the absence of external forcing and ambient flow.

Complementary to the Stokesian dynamics and Brownian dynamics simulation schemes, the evolution of microparticle dispersions is studied theoretically also in terms of the probability density distribution function $P(\mathbf{X},t)$, where $P(\mathbf{X},t)d\mathbf{X}$ is the probability of finding N particles at time t in a small $6N$-dimensional neighbourhood $d\mathbf{X}$ of the configuration $\mathbf{X}$. The evolution equations for $P(\mathbf{X},t)$ for Brownian and non-Brownian particles under Stokes-flow conditions are, respectively, the many-particle Smoluchowski diffusion equation and the Stokes-Liouville equation. An introductory discussion of these equations is given in Ref. [9].

10.4.2 *Method of Singularity Flow Solutions*

Linearity of the Stokes equations allows for the representation of the fluid velocity and pressure in dispersions of microparticles in terms of a discrete or continuous superposition of elementary flow solutions. We discuss in the following a very useful set of singularity incompressible flow solutions for an unbounded quiescent fluid which decay all to zero far away from a specified fluid point where they exhibit a pole singularity [14, 34, 35]. For simple geometries, this set can be profitably used to obtain, with little effort, exact Stokes flow solutions by linear superposition. We will exemplify this for the forced and phoretic motions of a microsphere, and for the velocity field of a point force in front of a fluid-fluid interface (see Sect. 10.5.1). To solve the latter problem, an image method is used similar to that in electrostatics [36]. For more complicated geometries such as for a complex-shaped particle, the singularity method remains useful to gain information about the flow at far distances from the particle, in the form of a multipolar series. We shall demonstrate this in our discussion of the swimming trajectories of a self-propelling microswimmer near a surface.

The important observation is that for a given solution, $\{\mathbf{u}, p\}$, of the homogeneous Stokes equations, its derivatives are likewise flow solutions. We can thus construct a complete set of singularity solutions by taking derivatives of increasing order, of two fundamental flow solutions, namely those due to a point force and a point source.

We should add that for dispersions of spherical particles, specialized elementary sets of Stokes flow solutions can be constructed, which are different from the singularity set discussed below, and which account for the high symmetry of spheres. These specific sets are used in numerically precise methods [37, 38] of calculating the many-sphere hydrodynamic mobility and friction coefficients required in Brownian and Stokesian dynamics simulations.

Point-force solution and Oseen tensor: The fundamental flow solutions, $\{\mathbf{u}_{St}, p_{St}\}$, due to the body force density $\mathbf{f}(\mathbf{r}) = \delta(\mathbf{r} - \mathbf{r}_0)\mathbf{F}$ of a point force $\mathbf{F} = F\mathbf{e}$, directed along the unit vector $\mathbf{e}$ and acting on a quiescent, infinite fluid at a position $\mathbf{r}_0$, can be obtained in several ways (see [39]). We only quote here the result

$$\mathbf{u}_{St}(\mathbf{r}) = \mathbf{T}(\mathbf{r} - \mathbf{r}_0) \cdot \mathbf{F} \tag{10.29}$$

$$p_{St}(\mathbf{r}) = \frac{1}{4\pi} \mathbf{U}_S(\mathbf{r} - \mathbf{r}_0) \cdot \mathbf{F}. \tag{10.30}$$

The second-rank Oseen tensor, $\mathbf{T}(\mathbf{r})$, has the form

$$\mathbf{T}(\mathbf{r}) = \frac{1}{8\pi\eta} \frac{1}{r} (\mathbf{1} + \hat{\mathbf{r}}\hat{\mathbf{r}}), \tag{10.31}$$

where $\mathbf{1}$ is the unit tensor, $\mathbf{r} = r\hat{\mathbf{r}}$, and $\hat{\mathbf{r}}\hat{\mathbf{r}}$ is a dyadic tensor formed with the positional unit vector $\hat{\mathbf{r}}$. In Cartesian coordinates, the Oseen tensor elements read

$$T_{ij} = \frac{1}{8\pi\eta} \left(\frac{\delta_{ij}}{r} + \frac{r_i r_j}{r^3} \right). \tag{10.32}$$

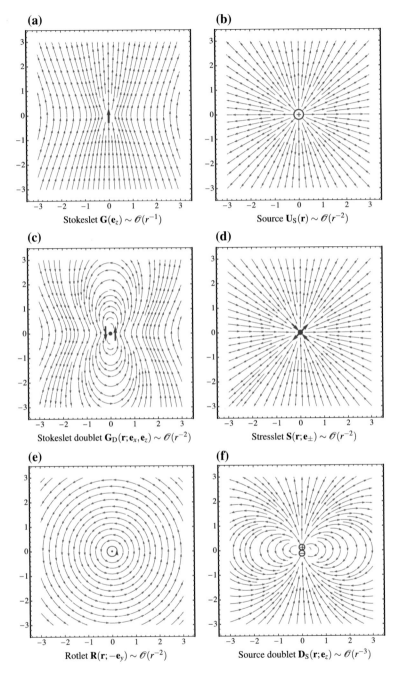

Fig. 10.7 A few elemental singularity solutions used in constructing specific Stokes flow solutions. The x-axis points horizontally to the right, and the z-axis vertically upwards. The singularities are marked in *red* with pictograms reflecting their structure. Note here that $\mathbf{G}_D(\mathbf{r}; \mathbf{e}_x, \mathbf{e}_z) = -\mathbf{G}_D(\mathbf{r}; \mathbf{e}_z, \mathbf{e}_x)$, and $\mathbf{e}_\pm = (\mathbf{e}_x \pm \mathbf{e}_z)/\sqrt{2}$ span the stretching and compression axes. The elemental source, $-\mathbf{U}_S(\mathbf{r})$, is marked as a *circle* with a minus sign inside

The pressure field $p_{St}(\mathbf{r})$ due to the point force at $\mathbf{r}_0$ is expressed here in terms of the elementary source vector field

$$\mathbf{U}_S(\mathbf{r}) = \frac{\hat{\mathbf{r}}}{r^2} = -\nabla \frac{1}{r}. \tag{10.33}$$

If multiplied by a constant $c > 0$ with the dimension of volume per time, $c\mathbf{U}_S(\mathbf{r})$ describes the radially directed outflow of fluid from the source point $\mathbf{r}_0 = 0$. The flow rate through a surface S enclosing the source point is thus equal to

$$c \int_S dS\, \mathbf{U}_S \cdot \mathbf{n} = 4\pi c. \tag{10.34}$$

The elementary velocity field $\mathbf{u}_{St}(\mathbf{r})$ of a point-force is called a Stokeslet of strength F in the direction of $\mathbf{e}$, and with the centre at $\mathbf{r}_0$ where it has a simple pole singularity. Note here that T_{ij} is the i-th component of the Stokeslet velocity field generated by a unit force acting in the j direction. The streamlines of the Stokeslet are drawn as dashed lines in the left part of Fig. 10.9, together with those generated by a spherical no-slip particle subjected to the same force. The hydrodynamics of a translating sphere is discussed in detail further down. A significant difference between the two streamlines pattern exterior to the impermeable sphere is visible only near its surface. The streamlines generated by the translating sphere are further out indistinguishable from those of the point-force Stokeslet.

The Stokeslet velocity field decays like $1/r$ at far distances from the point force. This slow decay can be ascribed to the conservation of momentum injected into the fluid by the point force, which is spread out quasi-instantaneously. It creates major difficulties in dealing theoretically with the hydrodynamics of suspensions, since forced velocity disturbances influence even well-separated particles. An additional difficulty is that the hydrodynamic interactions between three and more non-point-like particles are not pairwise additive, i.e. the hydrodynamic interactions of two nearby particles are changed in a rather complicated way if a third one is in their vicinity.

According to Eq. (10.33), the pressure field of a point force decays faster than the velocity field by a factor of $1/r$. Note that the pressure itself, and not just its gradient, has been uniquely specified by demanding $p \to 0$ for $r \to \infty$. Employing Eq. (10.14), the stress on a fluid surface element at position $\mathbf{r}$ and normal $\mathbf{n}$, due to a point force at the coordinate system origin, is

$$\boldsymbol{\sigma}_{St}(\mathbf{r}) \cdot \mathbf{n}(\mathbf{r}) = -\frac{3}{4\pi} \mathbf{F} \cdot \left(\frac{\hat{\mathbf{r}}\hat{\mathbf{r}}\hat{\mathbf{r}}}{r^2} \right) \cdot \mathbf{n} = -\frac{3}{4\pi} \frac{(\hat{\mathbf{r}} \cdot \mathbf{F})(\hat{\mathbf{r}} \cdot \mathbf{n})}{r^2} \hat{\mathbf{r}}. \tag{10.35}$$

On integrating the stress over a surface enclosing the point force, the expected result $\mathbf{F}^h = -\mathbf{F}$ is obtained.

That the pressure field decays by the factor $1/r$ faster than the associated velocity field is a general rule. It follows from the homogeneous Stokes equation written in the form $\nabla p = \eta \nabla^2 \mathbf{u}$, where the first-order derivatives of p are expressed by the second-order derivatives of $\mathbf{u}$. It can be also noticed here that the pressure in Stokes flows is a subsidiary quantity, fully determined by the velocity field for BCs invoking velocities only. The velocity field can be calculated without reference to the pressure as a solution of the bi-harmonic differential equation

$$\nabla^2 \nabla^2 \mathbf{u}(\mathbf{r}) = \mathbf{0}, \tag{10.36}$$

which readily follows from the application of the divergence operation to the homogeneous Stokes equation, using in addition the flow incompressibility constraint.

For completeness, consider also the vorticity field, $\nabla \times \mathbf{u}(\mathbf{r})$, associated with a velocity field $\mathbf{u}(\mathbf{r})$. The vorticity is twice the angular velocity of a fluid element at $\mathbf{r}$. The vorticity due to a point force at position $\mathbf{r}_0$ is

$$\nabla \times \mathbf{u}_{St}(\mathbf{r}) = -\frac{1}{4\pi\eta} \mathbf{U}_S(\mathbf{r} - \mathbf{r}_0) \times \mathbf{F}, \tag{10.37}$$

identifying the Stokeslet as an incompressible rotational flow solution.

The Oseen tensor for an unbounded infinite fluid is of key importance not only in generating higher-order elemental force singularity solutions (see below), but also for the so-called boundary integral method of calculating the flow around complex-shaped bodies. The disturbance flow, i.e. the flow taken relative to a given ambient flow field $\mathbf{u}_{amb}(\mathbf{r})$, observed in the exterior of a rigid no-slip particle in infinite fluid is given by the integral

$$\mathbf{u}(\mathbf{r}) - \mathbf{u}_{amb}(\mathbf{r}) = \int_{S_p} dS' \mathbf{T}(\mathbf{r} - \mathbf{r}') \cdot \boldsymbol{\sigma}(\mathbf{r}') \cdot \mathbf{n}(\mathbf{r}'), \tag{10.38}$$

over the particle surface S_p, i.e. by a continuous superposition of surface-located Stokeslets of vectorial strength $\boldsymbol{\sigma} \cdot \mathbf{n}$. We emphasize here that if the fluid at S_p is tangentially mobile such as for a rigid particle with Navier partial-slip BC, and a liquid droplet or gas bubble, there is an additional surface integral contribution to the exterior flow. The form of this additional contribution is discussed in detail in textbooks on low-Reynolds-number fluid dynamics [14, 35, 40].

The ambient velocity field $\mathbf{u}_{amb}(\mathbf{r})$ is a Stokes flow caused by sources exterior to the considered particle. In a non-quiescent situation it can be, e.g., a linear shear or a quadratic Poiseuille flow. The ambient flow can be also the flow due to the motion of other rigid or non-rigid particles. If the considered particle was not present, the ambient flow would be measured in the system.

Integrating Eq. (10.38) with respect to $\mathbf{r}$ over the particle surface, and using the no-slip BC in Eq. (10.3) for its left-hand side, results in a linear surface integral equation for the surface stress field $\boldsymbol{\sigma} \cdot \mathbf{n}$ in terms of the given translational and rotational particle velocities $\mathbf{V}$ and $\boldsymbol{\Omega}$, and the ambient flow field (friction problem). The integral equation can be solved numerically by an appropriate surface discretization (triangulation). For given particle force and torque (mobility problem),

and given ambient flow, the velocities are determined from substituting the cal-
culated stress field into the likewise discretized Eq. (10.15) for $\mathbf{F}^h$ and $\mathbf{T}^h$. See here
[8, 9, 35] for details on the boundary integral method which has the main advantage
of requiring only a two-dimensional surface mesh for a three-dimensional flow
calculation.

Force multipoles solutions: Singularity solutions of increasing multipolar order are
obtained from derivatives of the fundamental flow solution $\mathbf{u}_{St}(\mathbf{r})$. They also show
up in the expansion of $\mathbf{u}_{St}$ in a Taylor series about the force placement (singularity)
point $\mathbf{r}_0$. Recall now that a point force $\mathbf{F} = F\mathbf{e}$ oriented along the direction $\mathbf{e}$ generates the velocity field

$$\mathbf{u}_{St}(\mathbf{r} - \mathbf{r}_0) = \frac{F}{8\pi\eta}\mathbf{G}(\mathbf{r} - \mathbf{r}_0; \mathbf{e}), \qquad (10.39)$$

with

$$\mathbf{G}(\mathbf{r}; \mathbf{e}) = 8\pi\eta\mathbf{T}(\mathbf{r}) \cdot \mathbf{e} = \frac{\mathbf{e}}{r} + \frac{\mathbf{e} \cdot \mathbf{r}}{r^3}\mathbf{r}. \qquad (10.40)$$

We select $\mathbf{G}(\mathbf{r}; \mathbf{e})$ as the starting element of the singularity set, quoting it as the
fundamental $\mathbf{e}$-directed Stokeslet. It is actually equal to a Stokeslet of unit force in
the direction $\mathbf{e}$, made non-dimensional by multiplication with $8\pi\eta$ and division by
the force unit. The first two singularity solutions obtained from directional
derivatives of the fundamental Stokeslet are the Stokeslet doublet $\mathbf{G}_D$, and the
Stokeslet quadrupole $\mathbf{G}_Q$ [26, 41]

$$\mathbf{G}_D(\mathbf{r} - \mathbf{r}_0; \mathbf{d}, \mathbf{e}) = (\mathbf{d} \cdot \nabla_0)\mathbf{G}(\mathbf{r} - \mathbf{r}_0; \mathbf{e}) \sim \mathcal{O}(r^{-2}), \qquad (10.41)$$

$$\mathbf{G}_Q(\mathbf{r} - \mathbf{r}_0; \mathbf{c}, \mathbf{d}, \mathbf{e}) = (\mathbf{c} \cdot \nabla_0)\mathbf{G}_D(\mathbf{r} - \mathbf{r}_0; \mathbf{d}, \mathbf{e}) \sim \mathcal{O}(r^{-3}), \qquad (10.42)$$

where the gradient operator ∇_0 acts on the singularity placement $\mathbf{r}_0$, and $\mathbf{d}$ and $\mathbf{c}$ are
arbitrary vectors. We have indicated here the decay of these velocity fields far away
from the singularity point. Higher-order singularity flow solutions with an $\mathcal{O}(r^{-4})$
asymptotic decay are obtained accordingly by repeated differentiation. For later use,
we explicitly quote the Stokeslet doublet,

$$\mathbf{G}_D(\mathbf{r}; \mathbf{d}, \mathbf{e}) = \mathbf{d} \cdot \frac{1}{r^2}\left[\hat{\mathbf{r}}\mathbf{1} - \mathbf{1}\hat{\mathbf{r}} - {}^T(\hat{\mathbf{r}}\mathbf{1}) + 3\,\hat{\mathbf{r}}\hat{\mathbf{r}}\hat{\mathbf{r}}\right] \cdot \mathbf{e} \qquad (10.43)$$

$$= \frac{1}{r^2}\left[\mathbf{e}(\hat{\mathbf{r}} \cdot \mathbf{d}) - \mathbf{d}(\hat{\mathbf{r}} \cdot \mathbf{e}) - (\mathbf{d} \cdot \mathbf{e})\hat{\mathbf{r}} + 3(\hat{\mathbf{r}} \cdot \mathbf{e})(\hat{\mathbf{r}} \cdot \mathbf{d})\hat{\mathbf{r}}\right]. \qquad (10.44)$$

where the pre-transposition symbol T implies the interchange of the first two
Cartesian indices to its right. The Stokes doublet $\mathbf{G}_D(\mathbf{r}; \hat{\mathbf{d}}, \mathbf{e})$, with $\hat{\mathbf{d}}$ denoting a unit
vector, has the following physical interpretation: It is the velocity field times $8\pi\eta$, of

two opposing Stokelets of vector strengths $\pm F\mathbf{e}$ and singularity locations at $\mathbf{r}_0 \pm \mathbf{d}$ (with $\mathbf{d} = d\hat{\mathbf{d}}$), in the limits $d \to 0$ and $F \to \infty$ with the force dipole moment $p = 2Fd$ kept constant equal to one. The unit vector $\hat{\mathbf{d}}$ points from the Stokeslet of strength $-F\mathbf{e}$ to the one of strength $F\mathbf{e}$. This interpretation is obviated from the explicit calculation of the flow field,

$$
\begin{aligned}
\mathbf{u}_D(\mathbf{r}) &= [\mathbf{T}(\mathbf{r} - \mathbf{r}_0 - \mathbf{d}) - \mathbf{T}(\mathbf{r} - \mathbf{r}_0 + \mathbf{d})] \cdot \mathbf{e}F = 2dF(\hat{\mathbf{d}} \cdot \nabla_0)\mathbf{T}(\mathbf{r} - \mathbf{r}_0) \cdot \mathbf{e} + \mathcal{O}(d^2) \\
&= \frac{p}{8\pi\eta} \mathbf{G}_D(\mathbf{r} - \mathbf{r}_0; \hat{\mathbf{d}}, \mathbf{e}) + \mathcal{O}(d^2).
\end{aligned}
$$

$$(10.45)$$

The force doublet provides the far-field behaviour of flows caused by force-free microparticles. It is named asymmetric when $\hat{\mathbf{d}}$ is not co-linear with the force direction $\pm\mathbf{e}$, and referred to as symmetric otherwise. The symmetric force doublet $\mathbf{G}_D(\mathbf{r} - \mathbf{r}_0; \mathbf{e}, \mathbf{e})$ is also called a linear force dipole. It plays a major role in the discussion of the flow created by many autonomous microswimmers, including various types of prokaryotic bacteria and eukaryotic unicellular microorganisms. Microswimmers in the bulk fluid and near interfaces are discussed in Sect. 10.6.

The force doublet can be split into an anti-symmetric part, named Rotlet $\mathbf{R}$, and a symmetric part named Stresslet $\mathbf{S}$, each of which has a direct physical meaning. We exemplify this for the Rotlet and in the special situation where the force strengths of the two opposing Stokeslets are orthogonally displaced, and aligned with the z-axis and x-axis, respectively. Then, $\mathbf{F} \cdot \mathbf{d} = 0$ and the dipole moment $T = 2dF$ has the meaning of an applied torque. The Rotlet at the singularity point $\mathbf{r}_0 = \mathbf{0}$ is in this case

$$
\mathbf{R}(\mathbf{r}; -\mathbf{e}_y) = \frac{1}{2}[\mathbf{G}_D(\mathbf{r}; \mathbf{e}_x, \mathbf{e}_z) - \mathbf{G}_D(\mathbf{r}; \mathbf{e}_z, \mathbf{e}_x)] = -\mathbf{e}_y \times \frac{\hat{\mathbf{r}}}{r^2}, \qquad (10.46)
$$

and after division by the factor $8\pi\eta$ it describes the rotational flow field due to a unit point torque aligned with the negative y-axis.

The symmetric Stresslet part reads

$$
\begin{aligned}
\mathbf{S}(\mathbf{r}; \mathbf{e}_\pm) &= \frac{1}{2}[\mathbf{G}_D(\mathbf{r}; \mathbf{e}_x, \mathbf{e}_z) + \mathbf{G}_D(\mathbf{r}; \mathbf{e}_z, \mathbf{e}_x)] = \frac{3\hat{\mathbf{r}}}{r^2}(\hat{\mathbf{r}} \cdot \mathbf{e}_x)(\hat{\mathbf{r}} \cdot \mathbf{e}_z) \\
&= \mathbf{G}_D(\mathbf{r}; \mathbf{e}_+, \mathbf{e}_+) - \mathbf{G}_D(\mathbf{r}; \mathbf{e}_-, \mathbf{e}_-).
\end{aligned}
$$

$$(10.47)$$

It describes a straining fluid motion [14] originating from the superposition of two linear force dipoles oriented along the diagonal stretching axis $\mathbf{e}_+$ and the anti-diagonal compression axis $\mathbf{e}_-$, respectively, where $\mathbf{e}_\pm = (\mathbf{e}_x \pm \mathbf{e}_z)/\sqrt{2}$. The streamlines of a linear force dipole are discussed in Sect. 10.6, and are drawn in Fig. 10.27.

The key point to notice here is that the stresses $\sigma_S \cdot \mathbf{n}$ and $\sigma_R \cdot \mathbf{n}$, associated with the Stresslet and Rotlet force doublet parts, respectively, are decaying as $\mathcal{O}(1/r^3)$. When these stresses are integrated over a surface enclosing the singularity point, according to Eq. (10.15) they do not contribute a hydrodynamic drag force.

Differently from the Stresslet which is torque-free, the Rotlet contributes a hydrodynamic torque of magnitude equal to $8\pi\eta$ times the torque unit. The stress fields of all the higher-order force singularity solutions including the one by the force quadrupole $\mathbf{G}_Q$ are all of $\mathcal{O}(1/r^4)$, so that they contribute neither a drag force nor a torque [8, 9].

Source multipoles solutions: Elementary singularity solutions in addition to the force singularities are obtained from derivatives of the source vector field $\mathbf{U}_S(\mathbf{r} - \mathbf{r}_0)$ in Eq. (10.33) with respect to the singularity (source) point $\mathbf{r}_0$. The two leading-order flows obtained in this way are the source doublet (dipole) and quadrupole,

$$\mathbf{D}_S(\mathbf{r} - \mathbf{r}_0; \mathbf{e}) = (\mathbf{e} \cdot \nabla_0)\mathbf{U}_S(\mathbf{r} - \mathbf{r}_0) \sim \mathcal{O}(r^{-3}) \tag{10.48}$$

$$\mathbf{Q}_S(\mathbf{r} - \mathbf{r}_0; \mathbf{d}, \mathbf{e}) = (\mathbf{d} \cdot \nabla_0)\mathbf{D}_S(\mathbf{r} - \mathbf{r}_0; \mathbf{e}) \sim \mathcal{O}(r^{-4}). \tag{10.49}$$

The source doublet multiplied by a constant c of dimension volume per time describes the flow due to a source flow with outflow rate $4\pi c$, and a sink flow of the same inflow rate. The source at $\mathbf{r}_0 + d\mathbf{e}$ and the sink of the doublet at $\mathbf{r}_0 - d\mathbf{e}$ are an infinitesimal vector distance $2d\mathbf{e}$ separated from each other and have the moment $2dc$ equal to one. Explicitly,

$$\mathbf{D}_S(\mathbf{r}; \mathbf{e}) = \frac{1}{r^3}[3\hat{\mathbf{r}}\hat{\mathbf{r}} - \mathbf{1}] \cdot \mathbf{e} = \frac{1}{r^3}[3(\hat{\mathbf{r}} \cdot \mathbf{e})\hat{\mathbf{r}} - \mathbf{e}]. \tag{10.50}$$

The source singularity solutions are related to the force singularity solutions by

$$\mathbf{D}_S(\mathbf{r} - \mathbf{r}_0) = -\frac{1}{2}\nabla_0^2\mathbf{G}(\mathbf{r} - \mathbf{r}_0), \tag{10.51}$$

and its derivatives. Equation (10.51) identifies the source doublet as a degenerate force quadrupole, which explains its faster decay than that of the force doublet. The stress fields of the source multipoles decay as $\mathcal{O}(1/r^4)$ or faster, except for the source flow $\mathbf{U}_S$ itself, implying that they make no force and torque contributions. As the derivatives of the Coulomb-type potential $1/r$ (see Eq. (10.33)), the source multipoles belong to the class of irrotational potential flows (where $\nabla \times \mathbf{u} = \mathbf{0}$) with associated constant pressure fields. To understand the pressure constancy, note with $\mathbf{u} = \nabla\psi$ for some scalar (potential) function ψ that incompressibility implies $\Delta\psi = 0$. It follows from the Stokes equation that $\nabla p = \eta\Delta(\nabla\psi) = \eta\nabla(\Delta\psi) = \mathbf{0}$.

Superposition of singularity solutions: Linear superposition of fundamental singularity solutions, appropriately selected and positioned to conform with the system symmetry and BCs under consideration, can be profitably used to construct (approximate) flow solutions. The coefficients in the superposition series can be determined from the prescribed BCs.

As an example of such a superposition, in Sect. 10.4.3 we discuss the gravitational settling of a slender body whose flow field can be described in decent approximation by a continuous distribution of Stokeslets placed along the body's center line.

For a particle axisymmetric along the direction $\mathbf{e}$, and in a flow situation sharing this axial symmetry, the appropriate superposition describing the far-distance velocity field is

$$\mathbf{u}(\mathbf{r}) = c_1\big(a\,\mathbf{G}(\mathbf{r};\mathbf{e})\big) + c_2\big(a^2\mathbf{G}_D(\mathbf{r};\mathbf{e};\mathbf{e})\big) + c_3\big(a^3\mathbf{S}_D(\mathbf{r};\mathbf{e})\big) + \mathcal{O}(r^{-4}), \quad (10.52)$$

with scalar coefficients $\{c_i\}$ having the physical dimension of a velocity. The centre $\mathbf{r}_0$ of the particle is placed here in the origin, and a can be taken as the lateral length of the particle. The Rotlet part of the symmetric force doublet $\mathbf{G}_D$ is zero here, since a torque-free, non-rotating particle is required by the symmetry of the flow problem. If the particle moves force-free along its axial direction, as it is the case for a self-propelling microswimmer, there is no Stokeslet contribution so that $c_1 = 0$. On the other hand, if the particle is sedimenting along its axis, the co-linear driving force is given by

$$\mathbf{F} = c_1(8\pi\eta a)\mathbf{e}, \quad (10.53)$$

with the coefficient c_1 determining the strength of the Stokeslet depending on the particle BCs. In Sect. 10.4.4, we show that the flow created by a sphere translating through a quiescent fluid, is exactly represented by the superposition of a Stokeslet and a source dipole, owing to the high symmetry of this flow problem. If the sphere is placed in an ambient linear shear flow where the stress distribution on its surface becomes non-uniform, then a more general superposition of singularity solutions must be used including a source quadrupole, which in addition accounts for all relevant Cartesian directions, to obtain the exact flow solution [42]. Note also that for a translating spheroid, a line distribution of Stokeslets and source doublets extending between the two focal points must be used [34]. As it is explained in Sect. 10.5.1, an appropriate placement of elemental singularity solutions at a reflection point provides an analytic solution for the velocity field of a point force in the vicinity of a fluid-fluid interface.

10.4.3 Slender Body Motion

As a first application, we use the force singularity method to determine the hydrodynamic friction experienced by a settling rigid slender body, that is a particle without sharp corners whose contour length, L, is large compared to its thickness d. Examples of such bodies include rod-like particles, and elongated (prolate) spheroids. Owing to the shape anisotropy, the friction force depends on the orientation of the body relative to the direction of motion. The slenderness of the body renders it possible, in place of having to solve a complicated boundary integral problem for a no-slip particle on the basis of Eq. (10.38), to describe approximately the disturbance flow field caused by its motion as that of a line of Stokeslets, uniformly distributed along the axis of the particle. The surface integration reduces then to a one-dimensional integral over a line of Stokeslets. This concept is originally due to Batchelor [43].

To demonstrate this, consider the sedimentation of a thin rigid rod of length L and diameter $d \ll L$ translating with velocity $\mathbf{V}$ through an unbounded quiescent fluid, in response to an external force $\mathbf{F}$ due, e.g., to gravity. We approximate the flow created by the particle by a sum of $2n + 1$ Stokeslets, placed in the centres of spherical beads of diameter d building up the rod in the form of a necklace (see Fig. 10.8). Each Stokeslet is assumed to have the same strength $(d/L)\mathbf{F}$, disregarding the end effects which are small for long rods. The rod is oriented along the unit vector $\mathbf{e}$ as depicted in the figure. On identifying, according to $\mathbf{V} = \mathbf{u}(\mathbf{0})$, the rod velocity with the velocity of the central bead under the hydrodynamic influence of the $2n$ other ones, the superposition of the $2n$ Stokeslets fields at the central bead position gives

$$\mathbf{V} = \frac{d}{\zeta^0_{bead}L}\mathbf{F} + \sum_{\substack{i = -n, \\ n \neq 0}}^{n} \mathbf{T}(id\mathbf{e}) \cdot \frac{d}{L}\mathbf{F} \approx \frac{1}{4\pi\eta L}\left(\sum_{i=1}^{n}\frac{1}{i}\right)(\mathbf{1}+\mathbf{ee}) \cdot \mathbf{F}, \qquad (10.54)$$

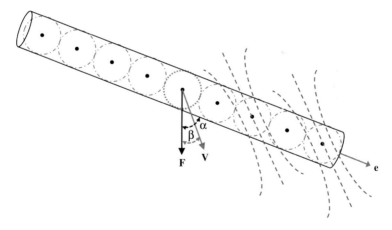

Fig. 10.8 A thin rod-shaped particle translating with a velocity $\mathbf{V}$ under the action of an applied force $\mathbf{F}$ pointing downwards. The stress distribution on the rod surface is approximated by a line of equally-spaced and equally strong Stokeslets placed in the centres of the large number of spherical beads building up the rod. The flow lines of two of the superposing Stokeslets are sketched. The tilt angle α of the rod with respect to the applied force is in general different from the sedimentation angle β, resulting in a sidewise velocity component. Reproduced from the COMPLOIDS book [9] with kind permission of the Societa Italiana di Fisica

where the sum behaves like $\log n \approx \log L/d$ for large n, and where $\zeta^0_{bead} = 3\pi\eta d$ is the no-slip single-bead friction coefficient. On noting that $\mathbf{F}^h = -\mathbf{F}$, the result of this summation is the force-velocity friction problem relation

$$\mathbf{F}^h = -\left[\zeta^{tt}_\parallel \mathbf{ee} + \zeta^{tt}_\perp (\mathbf{1} - \mathbf{ee})\right] \cdot \mathbf{V}, \tag{10.55}$$

and likewise the inverse relation,

$$\mathbf{V} = -\left[\mu^{tt}_\parallel \mathbf{ee} + \mu^{tt}_\perp (\mathbf{1} - \mathbf{ee})\right] \cdot \mathbf{F}^h, \tag{10.56}$$

for the mobility problem of given force. The friction and mobility coefficients for the translation of a thin rod parallel and perpendicular to its axis $\mathbf{e}$ have been obtained here as

$$\zeta^{tt}_\parallel = \frac{1}{\mu^{tt}_\parallel} = \frac{2\pi\eta L}{\log(L/d)}, \qquad \zeta^{tt}_\perp = \frac{1}{\mu^{tt}_\perp} = 2\zeta^{tt}_\parallel. \tag{10.57}$$

Corrections to this asymptotic result from a refined hydrodynamic calculation for a cylinder with end effects included are provided in [44]. Note that the application, i.e. dot-multiplication, of the dyadic $(\mathbf{1} - \mathbf{ee})$ to a vector gives the component of this vector perpendicular to $\mathbf{e}$.

Remarkably, the friction coefficient for the broadside motion of a thin rod is only twice as large as that for the axial motion. Both coefficients scale essentially with the length L of the rod, so that the drag force acting on a thin rod is not far less than that experienced by a sphere of diameter L enclosing it. It should be noted here that from general properties of Stokes flows it follows that the magnitude of the drag force on an arbitrarily-shaped body is always in between those for the inscribing and enclosing spheres [8, 14].

It is interesting to analyse the effect of the friction anisotropy on the direction of sedimentation. Denoting as α the angle between the rod axis $\mathbf{e}$ and the applied force $\mathbf{F}$, and as β the angle between rod velocity and applied force, we can relate the two angles by decomposing the external force into its components along and perpendicular to the rod axis, with the accompanying components of $\mathbf{V}$ determined by the mobility coefficients. In this way, one obtains

$$\beta = \alpha - \arctan\left(\frac{1}{2}\tan \alpha\right). \tag{10.58}$$

For a vertically or horizontally oriented rod and the applied force pointing downwards (see again Fig. 10.8), there will be no sidewise rod motion due to symmetry. Except for these special configurations, however, the tilt angle α of the rod is different from its sedimentation angle β, although both will remain constant during the motion. The maximum settling angle $\beta_{\max} = \arctan(\sqrt{2}/4) \approx 19.5°$ corresponds to $\alpha \approx 54.7°$.

Kinematic reversibility in conjunction with the system symmetry (no nearby walls are present here) commands that the rod is settling without rotation. The friction asymmetry of rod-shaped particles discussed here is a key ingredient in the swimming strategy of microswimmers with helical flagellar propulsion.

An elementary introduction to the concept of slender body motion is contained in the works [11, 14]. For a general discussion of slender bodies, which may also be curved, see [45, 46].

10.4.4 Forced Translation of a Microsphere

We consider here a microsphere of radius a with Navier partial slip BCs, which translates with constant velocity $\mathbf{V}_0 = V_0 \mathbf{e}$ and without rotation through an unbounded quiescent fluid. The origin of the coordinate system is placed at the momentary sphere center. We attempt to describe the exterior fluid velocity ($r \geq a$) by the linear superposition of a Stokeslet and source doublet in accord with Eq. (10.52), and with the coefficients c_1 and c_3 determined by the BCs. It is convenient to express $\mathbf{u} = u_r \hat{\mathbf{r}} + u_\theta \mathbf{e}_\theta$ and $\mathbf{V}_0 = V_{0,r} \hat{\mathbf{r}} + V_{0,\theta} \mathbf{e}_\theta$ in polar coordinates with components

$$
\begin{aligned}
u_r(r,\theta) &= 2\cos\theta \left[c_1 \left(\frac{a}{r}\right) + c_3 \left(\frac{a}{r}\right)^3 \right], \quad V_{0,r} = V_0 \cos\theta, \\
u_\theta(r,\theta) &= -\sin\theta \left[c_1 \left(\frac{a}{r}\right) - c_3 \left(\frac{a}{r}\right)^3 \right], \quad V_{0,\theta} = -V_0 \sin\theta,
\end{aligned}
\tag{10.59}
$$

where $\mathbf{e} \cdot \hat{\mathbf{r}} = \cos\theta$ and $\mathbf{e} \cdot \mathbf{e}_\theta = -\sin\theta$ have been used, with θ denoting the angle between polar axis $\mathbf{e}$ and fluid position vector $\mathbf{r}$. The two coefficients can be determined from enforcing the BC of zero normal velocity difference between the sphere and the fluid at the surface, $u_r(a,\theta) = V_{0,r}$, in conjunction with the Navier partial slip condition in Eq. (10.18) for the tangential velocity part. The latter is formulated in terms of the fluid velocity in the particle rest frame,

$$
\mathbf{u}'(\mathbf{r}) = \mathbf{u}(\mathbf{r}) - \mathbf{V}_0,
\tag{10.60}
$$

labelled by the prime, where the fluid far away from the stationary sphere is moving with uniform velocity $-\mathbf{V}_0$. The Navier BC reads then

$$
u_\theta'(r,\theta) = \frac{\ell}{\eta} \left(\frac{\partial u_\theta'}{\partial r} - \frac{u_\theta'(r,\theta)}{a} \right), \quad (r=a)
\tag{10.61}
$$

The two coefficients are determined from the BCs as

$$c_1 = \frac{3V_0}{4}\left(\frac{1+2\ell^*}{1+3\ell^*}\right), \quad c_3 = -\frac{V_0}{4}\left(\frac{1}{1+3\ell^*}\right), \tag{10.62}$$

where $\ell^* = \ell/a$.

Using Eq. (10.53), the hydrodynamic drag force opposing the motion of the sphere follows from the Stokeslet contribution as

$$\mathbf{F}^h = -(8\pi\eta a c_1)\mathbf{e} = -6\pi\eta a\left(\frac{1+2\ell^*}{1+3\ell^*}\right)\mathbf{V}_0. \tag{10.63}$$

Note that the single-sphere friction coefficient, $\zeta_0^t = 1/\mu_0^t$, relating the velocity of an isolated sphere to its drag force according to

$$\mathbf{F}^h = -\zeta_0^t \mathbf{V}_0, \tag{10.64}$$

reduces from $6\pi\eta a$ for a no-slip sphere ($\ell^* = 0$) to $4\pi\eta a$ for a perfect-slip sphere with stress-free surface ($\ell^* = \infty$). For a perfect-slip sphere such as a gas bubble with clean surface without adsorbed contaminants, the drag force is entirely due to the pressure changes in the fluid without viscous stress contributions. Since $c_3 = 0$ for $\ell^* = \infty$, the flow exterior to a translating gas bubble is that of a Stokeslet of strength $\mathbf{F} = 4\pi\eta a\mathbf{V}_0$ placed in its center. The relation $F^h = Ca\eta V_0$ with undetermined constant C follows readily from linearity of the Stokes equations and BCs, and a dimensional analysis using the sphere radius as the only physical length scale. The difficult part is the determination of C which as we have shown requires an elaborate calculation.

The tangentially oriented slip velocity, $\mathbf{u}'_{\text{slip}}(\theta) = u'_\theta(a,\theta)\mathbf{e}_\theta$, relative to the sphere surface is

$$\mathbf{u}'_{\text{slip}}(\theta) = \frac{1}{2}\left(\frac{3\ell^*}{1+3\ell^*}\right)V_0\sin\theta\,\mathbf{e}_\theta = -\frac{1}{2}\left(\frac{3\ell^*}{1+3\ell^*}\right)[\mathbf{1}-\hat{\mathbf{r}}\hat{\mathbf{r}}]\cdot\mathbf{V}_0. \tag{10.65}$$

It is zero at the poles, where $\theta = \{0,\pi\}$ and $\hat{\mathbf{r}} = \pm\mathbf{e}$, and maximal in magnitude on the equator, where $\theta = \pi/2$ and $\hat{\mathbf{r}}\perp\mathbf{e}$. On the equator, the slip velocity points oppositely to $\mathbf{V}_0$ since $\mathbf{e}_\theta(\theta = \pi/2) = -\mathbf{e}$. For perfect slip, the result $\mathbf{u}_{\text{slip}} = -\mathbf{V}_0/2$ is obtained at the sphere equator.

We proceed by discussing the lab frame velocity field of a translating no-slip microsphere, given in dyadic notation by

$$\mathbf{u}(\mathbf{r}) = \left[\left(\frac{a}{r}\right)(\mathbf{1}+\hat{\mathbf{r}}\hat{\mathbf{r}}) - \frac{1}{3}\left(\frac{a}{r}\right)^3(3\hat{\mathbf{r}}\hat{\mathbf{r}}-\mathbf{1})\right]\cdot\frac{\mathbf{F}}{8\pi\eta a}, \tag{10.66}$$

with $\mathbf{F} = 6\pi\eta a\mathbf{V}_0$, and with r measuring the distance from the center of the sphere. This result has been first calculated by Stokes more than 150 years ago [47], in a way different from the one described above. The velocity field in Eq. (10.66) has two interesting features: Firstly, far away from the particle it reduces to a Stokeslet flow field. The far-distance form is also recovered from taking the limit $a \to 0$ with the applied force kept constant. Secondly, the shorter-ranged source doublet potential flow contribution accounting for non-zero sphere volume is of importance in the near-distance region of the sphere only. The streamlines for a no-slip sphere and for an equal-force Stokeslet are drawn in Fig. 10.9. They are shown both in the lab frame where the sphere is moving and the fluid is quiescent, meaning that $\mathbf{u} \to \mathbf{0}$ for $r \to \infty$ (left part), and in the rest frame of the sphere where $\mathbf{u} \to -\mathbf{V}_0$ for $r \to \infty$ (right part). The vorticity field $\nabla \times \mathbf{u}$ around a translating sphere describing the local rotation of fluid elements is due to the Stokeslet part only, since the potential part has no vorticity. It is given by Eq. (10.37) restricted to the exterior of the sphere. Differently from $\mathbf{u}$, the associated vorticity field is invariant under a Galilean change of the reference frame.

10.4.5 Phoretic Particle Motion

So far we have dealt with the translational motion of an isolated microsphere and a rod settling subjected to an external body force such as the buoyancy-corrected gravitational force. The $\mathcal{O}(1/r)$ far-distance decay of the velocity field outside the moving particle is due to the momentum imparted to the fluid by the applied force.

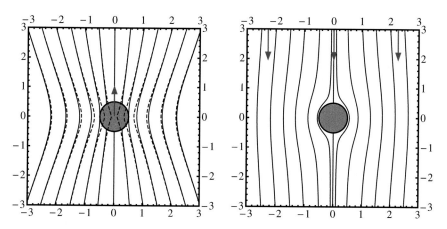

Fig. 10.9 *Left* Comparison of flow streamlines generated by an upward point force centred at the origin (a Stokeslet), marked by *dashed lines*, and a spherical no-slip particle (*solid lines*) moving upwards with velocity V_0 (*red arrow*) as viewed in the laboratory frame. *Right* The streamlines *circle* around the *sphere* if viewed in its rest frame, owing to its experiencing of a uniform ambient flow equal to $-V_0$. Note the fore-aft symmetry of the flow field related to the kinematic reversibility of Stokes flows. Reproduced from the COMPLOIDS book [9] with kind permission of the Societa Italiana di Fisica

Another mechanism for creating motion of a suspended microparticle is due to an imposed field gradient, such as an electric field $\mathbf{E}_\infty = -\nabla\phi_\infty$, a temperature gradient ∇T, or a concentration gradient, ∇c, in the concentration field c of small solutes surrounding the colloidal particle. Depending on the physical field, one refers to electrophoresis, thermophoresis or diffusiophoresis, respectively. The field gradients drive a so-called phoretic fluid flow relative to the particle surface in an interfacial region surrounding the particle. The effective hydrodynamic slip associated with the relative motion of the particle and the surface-phoretic flow results in a non-zero phoretic velocity, $\mathbf{V}_{\mathrm{phor}}$, of the particle in the lab frame where the fluid velocity is zero at infinity (quiescent fluid). The phoretic particle motion occurs even though the total direct force and torque on the particle plus its interfacial layer are zero, i.e.

$$\mathbf{F}_T = \mathbf{0}, \quad \mathbf{T}_T = \mathbf{0}, \tag{10.67}$$

with the balancing hydrodynamic drag force and torque being likewise equal to zero. There is no Stokeslet involved in the velocity field exterior to the particle and its boundary layer, referred to as the outer flow region. The velocity field in the outer region decays thus asymptotically like $\mathscr{O}(r^{-2})$ or faster, with r measuring the radial distance to the sphere center. As we are going to show in the context of the electrophoresis of a charged colloidal sphere, the outer velocity field is actually a source doublet potential flow with the characteristic $\mathscr{O}(r^{-3})$ far-distant decay. A classical review of phoretic motions is given in [48]. For recent lecture notes on electrophoresis, and the dynamics of charge-stabilized suspensions in general, see [10].

Consider now an insulating, charged colloidal sphere of radius a immersed in an infinite electrolyte solution. For the matter of definiteness, the sphere is assumed to carry a uniform negative surface charge (see Fig. 10.10).

In thermal equilibrium without an externally applied electric field, the charged sphere is surrounded by a diffuse spherical layer of mainly oppositely charged electrolyte ions which screen its electric effect to the outside fluid region. On the length scale of the colloidal sphere, this interfacial region can be considered as an oppositely (here: positively) charged fluid of charge density $\rho_{\mathrm{el}}(r)$. This density decays exponentially with outgoing radial distance from the sphere surface at $r = a$. The outer electrolyte fluid in the region $r > a + \lambda_{\mathrm{D}}$ is practically electroneutral, since the influence of the surface charge is screened out across the interface. Here, λ_{D} is the Debye screening length characterizing the thickness of the charged interfacial layer. It describes, at least for larger distances, the exponential radial decay of the interfacial charge density

$$\rho_{\mathrm{el}}(r > a) \propto \exp\{-r/\lambda_{\mathrm{D}}\}. \tag{10.68}$$

The charged sphere and its neutralizing interfacial fluid region form an electric double layer (EDL) sphere of radius $a + \lambda_{\mathrm{D}}$ whose net charge content is zero. The total electric force and torque on the EDL sphere are consequently zero. The thickness, λ_{D},

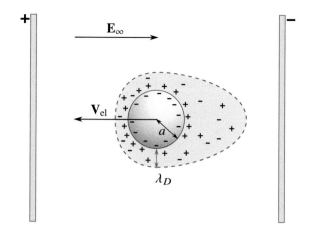

Fig. 10.10 Electrophoretic migration, with velocity $\mathbf{V}_{el}$, of a negatively charged colloidal sphere in a uniform electric field of strength $\mathbf{E}_\infty$. The oppositely charged interfacial layer of electrolyte ions with extension $\sim \lambda_D$ is slightly distorted (polarized) from the spherical shape at zero external field. The counterions in the near-surface interfacial region drive an electro-osmotic flow. Figure redrawn after [49]

of the interfacial region decreases with increasing concentration of electrolyte ions. This can be triggered experimentally by the addition of salt, or through osmotic contact with an electrolyte reservoir (buffer). For an aqueous strong 1-1 electrolyte at 25 °C, the Debye length in nanometres is

$$\lambda_D = \frac{0.304}{\sqrt{n_s[M]}}, \qquad (10.69)$$

where $n_s[M]$ is the concentration of salt ion pairs in mol per litre. For a 1 M solution $\lambda_D = 0.3$ nm. An upper bound is set by the self-dissociation of water requiring that $\lambda_D < 961$ nm.

If exposed to an uniform electric field, $\mathbf{E}_\infty = E_\infty \mathbf{e}$, created by distant sources such as a pair of electrode plates with an applied voltage, the negatively charged sphere migrates with constant electrophoretic velocity

$$\mathbf{V}_{el} = \mu_{el} \mathbf{E}_\infty, \qquad (10.70)$$

opposite to the direction of the applied electric field. The field-independent electrophoretic mobility, μ_{el}, characterizing the phoretic particle motion, has a negative sign for a negatively charged particle. The electrophoretic drift velocity, $\mathbf{V}_{el}$, is determined by two retarding electrokinetic effects, termed the electro-osmotic flow and the ion cloud polarization effect, respectively. Both effects lower the magnitude of the electrophoretic velocity below the limiting value, $V_{el}^0 = |QE_\infty|/(6\pi\eta a)$, for a sphere of surface charge Q without an electrolyte boundary layer, which is determined by the balance of the electric and hydrodynamic friction forces on the sphere.

This limiting velocity is reached by a charge-stabilized colloidal sphere with an ultra-extended, and therefore ultra-dilute, diffuse layer where $\lambda_D \gg a$.

The osmotic flow effect represents the hydrodynamic drag exerted on the sphere surface by the counter-flowing (since oppositely charged) fluid forming the diffuse layer. The electro-osmotic counterflow is driven by the electric body force density,

$$\mathbf{f}(\mathbf{r}) = \rho_{el}(\mathbf{r})\mathbf{E}(\mathbf{r}). \tag{10.71}$$

in the inhomogeneous Stokes Eqs. (10.5). Here, $\mathbf{E}(\mathbf{r})$ is the local electric field inside the diffuse layer which in general differs from the externally applied field.

The charge polarization effect, on the other hand, describes the field-induced slight distortion of the interfacial EDL zone from its spherically symmetric equilibrium shape at zero field. This distortion or polarization sets up diffusion currents of electrolyte ions which tend to equilibrate the EDL system back to spherical symmetry, with the net effect of slowing the sphere motion. The polarization effect becomes stronger with increasing particle charge and electric surface potential, and with decreasing mobilities of the electrolyte ions.

We restrict ourselves in the following to a charged colloidal sphere with an ultrathin interfacial layer for which $\lambda_D \ll a$. We further assume a weak surface charge density. The electrophoresis problem in this limiting situation was first treated in some detail by Smoluchowski (1903). It is therefore referred to as the Smoluchowski limit. Since the diffuse interfacial region reduces now to a thin boundary layer coating the sphere, the fluid can be mentally divided into an extended outer region $r > a + \lambda_D$ where $\rho_{el} = 0$, and in a thin boundary layer region $a \leq r < a + \lambda_D$ which is locally flat. The flows in the two regions are first determined independently and afterwards matched (asymptotically) to fix the remaining integration constants. We just outline here the major steps of this calculation. For details see, e.g., [10, 50]. For the assumed low surface charge density, ρ_{el} remains radially symmetric (i.e., unpolarized) in the presence of the external field. The sphere plus its boundary layer have the appearance in the outside fluid region of a neutral, non-conducting sphere of radius $a + \lambda_D$ (the EDL sphere). This gives rise to a dipolar electric field, $\mathbf{E}_{out}(\mathbf{r})$, in the outer region whose field lines must bend tangentially near the sphere surface, for otherwise there would be an electric current perpendicular to the non-conducting sphere surface. The surface-tangential electric field acts on the charged fluid inside the boundary layer through the body force term in Eq. (10.5), creating a surface-tangential flow profile as depicted in Fig. (10.11) which sketches the locally flat interfacial region. While the fluid sticks to the actual sphere surface S_a (i.e., the no-slip surface BC is used here), outside the enclosing surface $S_{a+\lambda_D}$ of the EDL sphere a plug-like local flow is observed, reminiscent of perfect slip. Incidentally, this plug-like electro-osmotic flow outside the Debye layer region is used in microfluidic devices to drive ion-containing aqueous media through narrow micro-channels where the Stokes flow conditions apply. In micro-channels, electro-osmotic flow transport is far more efficient than using an imposed pressure gradient along the channel [49]. See Fig. 10.12 for the sketch of such a device.

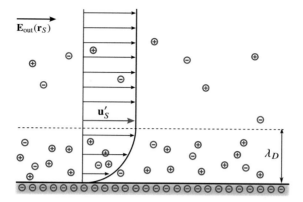

Fig. 10.11 Electro-osmotic flow profile of an electrolyte solution flowing tangentially past the locally flat surface of a negatively charged microsphere, viewed in its rest frame. The flow outside the thin boundary layer of thickness λ_D is plug-like with slip velocity $\mathbf{u}'_s(\mathbf{r}_s) = -\mu_{\text{el}}\mathbf{E}_{\text{out}}(\mathbf{r}_s)$. Redrawn after [9]

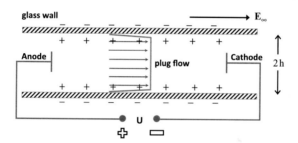

Fig. 10.12 Sketch of electro-osmotic plug flow in an open microcapillary tube with negatively charged glass walls. The non-dissipative plug flow is driven by the field-induced migration of counterions accumulated in the thin Debye interfacial layer of thickness $\lambda_D \ll 2h$ at the glass walls. In a tube closed at both ends, a pressure difference is created along the channel which drives a Poiseuille-type backflow of fluid in the central region of the tube [49]

The velocity field at the slip surface, in the rest frame of the sphere, is calculated as [10, 48]

$$\mathbf{u}'_s(\mathbf{r}_S) = -\frac{\varepsilon\zeta}{4\pi\eta}\mathbf{E}_{\text{out}}(\mathbf{r}_S) = -\frac{3\varepsilon\zeta}{8\pi\eta}(\mathbf{1} - \hat{\mathbf{r}}\hat{\mathbf{r}}) \cdot \mathbf{E}_\infty, \qquad (10.72)$$

where $\mathbf{r}_s \in S_{a+\lambda_D}$, and ε is the static dielectric constant of the fluid. Since $\lambda_D \ll a$, we were allowed in the second equality to identify the slip surface $S_{a+\lambda_D}$ with the actual particle surface S_a. The so-called zeta potential, ζ, is likewise identified with the electric potential at the actual sphere surface. The surface potential decays exponentially to zero in going outwards radially from a to $a + \lambda_D$.

Since the outer fluid is uncharged, the outer velocity field in the lab frame is found from a purely hydrodynamic consideration, namely from the solution of the homogeneous Stokes equations with inner and outer BCs,

$$\mathbf{u}_{out}(\mathbf{r}_S) = \mathbf{V}_{el} + \mathbf{u}'_s(\mathbf{r}_S), \quad \mathbf{u}_{out}(\mathbf{r} \to \infty) = \mathbf{0}. \tag{10.73}$$

respectively, where $\mathbf{r}_s \in S_{a+\lambda_D} \approx S_a$, and the sphere located momentarily at the coordinate frame origin. The electrophoretic velocity $\mathbf{V}_{el}$ is still unknown to this point. The boundary value problem has a unique flow solution if one demands in addition the EDL sphere to be force- and torque free, in accord with its overall electroneutrality.

To determine $\mathbf{u}_{out}$ using the singularity method, we notice first that the Stokeslet and Rotlet are ruled out as flow contributions since they would result in a non-zero drag force and torque. Therefore, we use the ansatz,

$$\mathbf{u}_{out}(\mathbf{r}) = c_3 a^3 \mathbf{D}_S(\mathbf{r}; \mathbf{e}). \tag{10.74}$$

for the out flow in the lab frame, with a source doublet in direction $\mathbf{e}$ of the applied electric field. With this ansatz, the inner and outer BCs are fulfilled with c_3 determined as $c_3 = \varepsilon \zeta E_\infty / (8\pi\eta)$, and the electrophoretic velocity as

$$\mathbf{V}_{el} = \mu_{el} \mathbf{E}_\infty, \quad \mu_{el} = \frac{\varepsilon\zeta}{4\pi\eta}, \tag{10.75}$$

respectively. The so-called Smoluchowski electrophoretic mobility μ_{el} scales linearly with the electric zeta potential, and it is independent of the particle radius. The outer velocity field in the lab frame is thus

$$\mathbf{u}_{out}(\mathbf{r}) = \frac{1}{2} \left(\frac{a}{r}\right)^3 [3\hat{\mathbf{r}}\hat{\mathbf{r}} - \mathbf{1}] \cdot \mathbf{V}_{el} = \mu_{el} \mathbf{E}_{out}(\mathbf{r}) + \mathbf{V}_{el}. \tag{10.76}$$

In the second equality, we have expressed the velocity field in terms of the outer electric dipole field.

Being an incompressible potential flow field, $\mathbf{u}_{out}$ satisfies $\Delta\mathbf{u}_{out} = \mathbf{0}$ with $\nabla p_{out} = 0$, so that the outer pressure field is uniform. The source doublet streamlines of the outer velocity field are shown in the right part of Fig. 10.13. They diverge in front of the moving sphere and curve back at its rear side. There are no closed streamlines, which are not allowed for a potential flow where $\nabla \times \mathbf{u} = \mathbf{0}$. Since $\mathbf{u}_{out}$ decays asymptotically like $\mathcal{O}(r^{-3})$ there is indeed no hydrodynamic torque exerted on the EDL sphere. Moreover, the absence of a $\mathcal{O}(r^{-2})$ Stresslet contribution to $\mathbf{u}_{out}$ is consistent with the fact that a non-uniform ambient flow was not considered. The streamlines, shown in the left part of Fig. 10.13, are different if viewed in the particle rest frame where distant from the sphere the fluid moves with uniform velocity $-\mathbf{V}_{el}$.

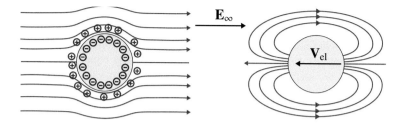

Fig. 10.13 Streamlines of the velocity field, $\mathbf{u}_{\text{out}}(\mathbf{r})$, outside the thin boundary layer of a negatively charged microsphere in electrophoretic motion. *Left* Streamlines in the rest frame of the sphere for which $\mathbf{u}_{\text{out}}(\mathbf{r} \to \infty) = -\mathbf{V}_{\text{el}}$. *Right* Streamlines in the lab frame where the fluid is at rest at infinity. The *sphere* migrates in the direction opposite to the external field $\mathbf{E}_\infty$

Without requiring the outer flow solution, the electrophoretic velocity for a spherical colloidal particle can be obtained directly as the surface average of the slip velocity field in the particle rest frame. In fact, it follows from Eq. (10.72) that

$$\frac{1}{4\pi a^2} \int_{S_a} dS \mathbf{u}'_s(\mathbf{r}_S) = -\mathbf{V}_{\text{el}}. \tag{10.77}$$

This result implies that the surface average of the outer velocity field in the lab frame is zero. This should be contrasted with the lab-frame Stokeslet velocity field for $r \geq a$,

$$\mathbf{u}(\mathbf{r}) = \frac{1}{2} \left(\frac{a}{r}\right) [1 + \hat{\mathbf{r}}\hat{\mathbf{r}}] \cdot \mathbf{V}_0, \tag{10.78}$$

due to a Navier perfect-slip sphere translating with velocity $\mathbf{V}_0$, which was discussed earlier in relation to Eq. (10.65). On taking the surface average of this Stokeslet field, the non-zero result $\mathbf{V}_0/3$ is obtained. This highlights the difference between a phoretically moving sphere with an *effective* slip, and a forced perfect-slip sphere without the phoretic boundary layer. The electrophoretic mobility becomes different from the Smoluchowski result in Eq. (10.75), if in place of the no-slip BC at S_a the Navier partial-slip BC is used. The Smoluchowski mobility is then enhanced by the factor $(1 + \ell/\lambda_D)$, for the Navier slip length small compared to the sphere radius [51]. Measured slip lengths are typically of the order of nanometres.

For the general case of a phoretically moving microsphere, it can be shown that its drift velocity is given by the surface average [52, 53]

$$\mathbf{V}_{\text{phor}} = -\frac{1}{4\pi a^2} \int_{S_a} dS \mu_S(\mathbf{r}_S)[1 - \hat{\mathbf{r}}\hat{\mathbf{r}}] \cdot \nabla \phi(\mathbf{r}_S), \tag{10.79}$$

where ϕ stands likewise for the potential of the applied electric field, temperature or solute concentration. Here, $\mu_s(\mathbf{r}_s)$ is a local surface mobility coefficient allowed to vary over the sphere surface. In our discussion of electrophoresis the potential gradient and mobility are taken as constant, with the latter equal to $(3/2)\mu_{el}$.

On recalling the steps which led to Eq. (10.75), one notices that owing to the local surface flatness on the scale of λ_D, the inner boundary value problem for the interfacial region is independent of the global shape of the microparticle. Therefore, the second equality in Eq. (10.76) remains valid for arbitrarily shaped, non-conducting rigid particles, provided the local radius of curvature at all surface points is large compared to λ_D. What varies with the particle shape is only the near-distance form of $\mathbf{E}_{out}(\mathbf{r})$. The latter is determined by the electrostatic boundary value problem of a non-conducting particle in an external field, with its field lines bending tangentially close to the particle surface to satisfy the non-conductance condition. The crucial point here is that the explicit form of $\mathbf{E}_{out}(\mathbf{r})$ is not required for the determination of the electrophoretic mobility. In fact, on noting $\mathbf{E}_{out} \rightarrow \mathbf{E}_\infty$ and $\mathbf{u}_{out} \rightarrow \mathbf{0}$ for $r \rightarrow \infty$, the Smoluchowski mobility result in Eq. (10.75) is readily recovered from the second equality in Eq. (10.76), but now for arbitrary particle shape and size, under the proviso that the particle has *uniform* surface zeta potential ζ. This remarkable result is supplemented by the likewise remarkable feature that the particle migrates phoretically without rotating. This follows from the fact that the far-distant form of $\mathbf{E}_{out}(\mathbf{r})$ is a dipolar electric field of $\mathcal{O}(r^{-3})$, independent of the particle shape. This in turn implies that $\mathbf{u}_{out}(\mathbf{r})$ is a potential flow field with a likewise $\mathcal{O}(r^{-3})$ far-distance decay. Such a flow exerts no hydrodynamic torque on the particle which consequently is non-rotating while translating.

However, the remarkable finding of shape-independent mobility has been obtained based on various restrictions. We quote here the most important ones: Firstly, the surface zeta potential is assumed to be weak enough for the polarization of the diffuse layer to be negligible, i.e. $|\zeta e|$ is small compared to k_BT, with e denoting the elementary charge. Secondly, the diffuse boundary layer must be very thin. Moreover, the applied electric field should be uniform on the scale of the particle size, and the small excluded volume of the electrolyte ions should not matter. The ions must be monovalent, and the dielectric constant and electrolyte viscosity should not change across the boundary layer. See [54] for a quantitative theory for the concentration dependence of the electrolyte viscosity.

For an isolated charge-stabilized sphere, surface uniformity of zeta potential and surface charge density go hand in hand. However, this is not the case for a non-spherical particle which has surface regions of varying curvature. A particle with uniform surface charge density has a larger zeta potential in the regions of higher curvature. This causes the particle to reorient while translating.

We proceed with the interesting generalization that a force $\mathbf{F} = F\mathbf{e}$ co-linear with the applied field is acting on the electrophoretically moving sphere. The sphere velocity, $\mathbf{V} = V\mathbf{e}$, and the flow for this situation can be constructed by the addition of a Stokeslet field of strength $\mathbf{F}$ to the source doublet, i.e.

$$\mathbf{u}(\mathbf{r}) = \left\{ \left(\frac{a}{r}\right)[\mathbf{1} - \hat{\mathbf{r}}\hat{\mathbf{r}}]\left(\frac{F}{8\pi\eta a}\right) + \frac{c_3}{2}\left(\frac{a}{r}\right)^3[3\hat{\mathbf{r}}\hat{\mathbf{r}} - \mathbf{1}] \right\} \cdot \mathbf{e}. \tag{10.80}$$

The two conditions determining c_3 and the sphere velocity are

$$\mathbf{u}(\pm a\mathbf{e}) = \mathbf{V}$$
$$\mathbf{u}(a\hat{\mathbf{r}}) - \mathbf{V} = -\frac{3}{2}(\mathbf{1} - \hat{\mathbf{r}}\hat{\mathbf{r}}) \cdot \mathbf{V}_{\mathrm{el}}. \tag{10.81}$$

The first condition assures that the fluid velocity at the poles of the sphere agrees with its velocity. The second condition demands the phoretic surface flow in the rest frame of the sphere to be the same as in unforced electrophoresis. This is a reasonable requirement, since for negligible EDL polarization the local osmotic flow in the boundary layer does not depend on whether the sphere is forced or not. To evaluate the second condition, one selects a unit vector $\hat{\mathbf{r}} = \mathbf{e}_\perp$ perpendicular to $\mathbf{e}$. The result is

$$V = V_{\mathrm{el}} + \frac{F}{6\pi\eta a} = \mu_0^t F + \mu_{\mathrm{el}} E_\infty$$
$$c_3 = V_{\mathrm{el}} - \frac{F}{12\pi\eta a}. \tag{10.82}$$

The first equation expresses the expected linear superposition of the particle velocities of the two problems of a phoretically moving sphere without body force, and a forced no-slip sphere without phoresis. It can be likewise formulated as a mobility problem for given forces F and eE_∞. The second force contribution, however, is not a body force.

The sphere becomes stationary, with $V = 0$, when the body force is equal to

$$F = (6\pi\eta a)V_{\mathrm{el}}, \tag{10.83}$$

i.e. equal to the hydrodynamic drag force on a no-slip and non-phoretic sphere moving with velocity V_{el}. Stationarity can be achieved experimentally, up to the inevitable undirected Brownian motion for smaller particles, using optical tweezers. The flow field of a stationary phoretic sphere is thus

$$\mathbf{u}(\mathbf{r}) = -\frac{3}{4}\left\{ \left(\frac{a}{r}\right)[\mathbf{1} - \hat{\mathbf{r}}\hat{\mathbf{r}}] - \left(\frac{a}{r}\right)^3[3\hat{\mathbf{r}}\hat{\mathbf{r}} - \mathbf{1}] \right\} \cdot \mu_{\mathrm{el}}\mathbf{E}_\infty. \tag{10.84}$$

The surface average of this velocity field yields $\mu_{\mathrm{el}}\mathbf{E}_\infty$, which expresses again the body-force independence of the surface osmotic flow.

The flow in Eq. (10.84) describes an interesting situation: Even though the sphere is held stationary in a quiescent fluid, it maintains a non-zero flow driven by the osmotic current along its surface. For a negatively charged sphere where μ_{el} is negative, the flow is oriented in direction of the external electric field. The flow

lines of the stationary sphere are shown in Fig. 10.14. At large distances from the
sphere, the Stokeslet term dominates and the streamlines converge in the rear and
diverge in front. If the sphere is placed close to an interface, using optical tweezers,
a sufficiently strong temperature gradient, gravity, or electric wall attraction, the
streamlines facing the wall will bend along the interface. This sets up an attractive
hydrodynamic force which favors the formation of particle clusters at the interface
[55].

10.4.6 Many-Particle Hydrodynamic Interactions

As explained in Sect. 10.4.1, when a colloidal particle moves in a viscous liquid in
the presence of other particles, it creates a flow pattern which affects not only the
motion of neighbouring particles, but through hydrodynamic back reflection also
the motion of the particle itself, and this even in the (hypothetical) absence of direct
interactions between the particles. The velocity field generated by a moving particle
is quasi-instantaneously transmitted through the fluid, inducing forces and torques
on all the other particles. These solvent-mediated interactions are referred to as
hydrodynamic interactions (HIs) [1]. For a quiescent dispersion of N microparticles,
these interactions are characterized by the $6N \times 6N$ mobility matrix, $\boldsymbol{\mu}(\mathbf{X})$, in
Eq. (10.23), or likewise by the associated $6N \times 6N$ friction matrix, $\boldsymbol{\zeta}(\mathbf{X})$, which both
linearly relate translational and angular particles velocities with drag forces and
torques. There are three major distinguishing features of HIs which render them
difficult to deal with. Firstly, they are long-ranged, i.e. the velocity field due to a

Fig. 10.14 Streamlines
around an electrophoretically
driven negatively charged
sphere in quiescent fluid, held
stationary by an applied body
force according to
Eq. (10.83). The flow is
maintained by the
electro-osmotic counterions
current in the boundary layer
region near the sphere surface

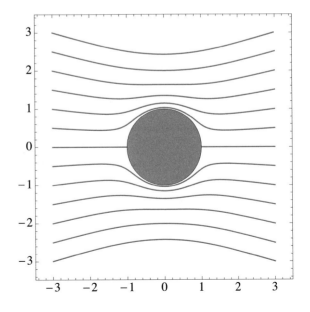

moving particle decays for forced motion with the distance r as $1/r$. Secondly, they are of genuinely many-body character, meaning that the HIs between a pair of particles are changed through the presence of a third one in their vicinity. To account for the non-additivity requires the consideration of multiple flow reflections by neighbouring microparticles, with the reflections being responsible for the deviations of the exact HIs from the approximate form in terms of a superposition of pair contributions. The approximate pairwise additivity treatment of the HIs where flow reflections are disregarded altogether can be justified for dilute dispersions only, for conditions where all particles are mutually well separated on the scale of their sizes. Non-pairwise additive higher-order HIs effects are important in particular in the dynamics of concentrated dispersions. The third distinguishing feature of the HIs is that, differently from direct interactions, they do not affect the equilibrium microstructure (i.e., the configurational distribution function $P_{eq}(\mathbf{X})$), in dispersions without external fields or imposed macroscopic flows. This is reflected by the fact that owing to their interrelation with hydrodynamic friction forces, HIs are not derivable from a many-particle conservative potential energy function determining the equilibrium distribution function.

Since diffusion transport properties of suspensions are expressible as configurational averages of certain tensor elements of $\boldsymbol{\mu}(\mathbf{X})$, and since $\boldsymbol{\mu}(\mathbf{X})$ is a key ingredient in the many-particle Smoluchowski and Stokes-Liouville equations governing the evolution of the configurational distribution function $P(\mathbf{X},t)$ of Brownian and non-Brownian microparticles, respectively, it is essential to be able to calculate this matrix for a many-particle system. Over the years, various numerical methods have been developed for this purpose, and a number of review articles [29, 37] and books [1, 14, 21] dealing with HIs are available.

If the microparticles are mutually well separated, with relative distances large compared to their size, the flow induced by their motion can be regarded approximately as originating from point forces concentrated at their centres. This liberates one from the severe complications of having to invoke the prescribed hydrodynamic BCs simultaneously on the particle surfaces. The tensorial mobility coefficients describing the linear relations between applied forces and resulting velocities are then straightforwardly approximated by the superposition of Stokeslets at the particle centres according to (point-particles model)

$$\boldsymbol{\mu}_{ii}^{tt} \approx \mu_0^t \mathbf{1} \tag{10.85}$$

$$\boldsymbol{\mu}_{ij}^{tt} \approx \mathbf{T}(\mathbf{R}_i - \mathbf{R}_j) \tag{10.86}$$

for $i \neq j$. Here, μ_{0i}^t is the single-particle translational mobility coefficient of particle i, and $\mathbf{1}$ is the unit tensor. Only the leading-order long-distance behaviour of the flow field is included in the point-particles model, with the off-diagonal mobility tensors decaying as r^{-1} in the inter-particle distance r. While the pairwise additive point-particles model is applicable to well-separated particles, it usually fails when particles are close to each other or to a confining wall, as indicated by the possible violation of the positive definiteness of the mobility matrix approximation in

Eqs. (10.85) and (10.86). At large concentrations and for smaller inter-particle distances, more refined theoretical methods and numerical schemes are required to account for the near-distance and non pairwise additive contributions to the HIs.

The next important step in going beyond the point-particles model is to account for the non-zero volumes of the particles while still maintaining the pairwise additive single-particle flow superposition. For spherical no-slip particles this is referred to as the Rotne-Prager (RP) mobility matrix approximation [56], and with an appropriate extension to overlapping no-slip spheres also as the Rotne-Prager-Yamakawa (RPY) approximation [57, 58]. The RP approximation is still pairwise additive, with the RP mobility tensors including terms with the long-distance decay of $\mathcal{O}(r^{-3})$. All the hydrodynamic back reflections, which give rise to more steeply decaying flow field contributions of $\mathcal{O}(r^{-4})$ are hereby disregarded. The main merit of the RP approximation, in addition to its convenient simplicity, is that is leads to a translational mobility matrix approximation, $\boldsymbol{\mu}^{tt,RP}$, which is positive definite for all physically allowed configurations of non-overlapping spheres. Moreover, $\boldsymbol{\mu}^{tt,RP}$ is an upper bound to the exact translational mobility matrix, $\boldsymbol{\mu}^{tt}$, in the sense that $\mathbf{b} \cdot (\boldsymbol{\mu}^{tt,RP} - \boldsymbol{\mu}^{tt}) \cdot \mathbf{b} > 0$ for all non-zero $3N$-dimensional vectors $\mathbf{b}$ and all allowed configurations. This property of the RP approximation can be used for constructing upper bounds to certain transport properties of concentrated dispersion such as the (short-time) mean sedimentation velocity. The RP approximation can be profitably used for dilute to moderately concentrated dispersions of no-slip microspheres which strongly repel each other over larger distances. Examples in case are like-charged colloidal particles, and lower-salinity aqueous solutions of globular proteins.

The derivation of the translational mobility matrix in RP approximation proceeds as follows: Consider first an isolated no-slip microsphere j of radius a in an infinite quiescent fluid, translating without rotation under the action of the force $\mathbf{F}_j$ acting at its centre at $\mathbf{R}_j$. From Eq. (10.38), one notices that $\boldsymbol{\sigma} \cdot \mathbf{n} = \frac{\mathbf{F}_j}{4\pi a^2}$ is consistent with the no-slip BC on the sphere surface, i.e. the surface shear stress in this special situation is constant. Consequently,

$$\mathbf{u}_j^{(0)}(\mathbf{r}) = \int_{S_j} dS' \mathbf{T}(\mathbf{r} - \mathbf{r}') \cdot \frac{\mathbf{F}_j}{4\pi a^2} = \left(1 + \frac{a^2}{6}\nabla^2\right)\mathbf{T}(\mathbf{r} - \mathbf{R}_j) \cdot \mathbf{F}_j \qquad (10.87)$$

where for the second equality, the mean-value theorem for bi-harmonic functions has been used [59], on recalling that $\nabla^2\nabla^2\mathbf{T}(\mathbf{r}) = \mathbf{0}$ for all non-zero vectors $\mathbf{r}$. From performing the second-order differentiation, one verifies that $\mathbf{u}_j^{(0)}(\mathbf{r})$ is equal to the velocity field in Eq. (10.66) which we had obtained earlier using the singularity flow solutions.

Consider next the incident flow, $\mathbf{u}_{\text{inc}}^{(N-1)}(\mathbf{r})$, created by the motion of $(N–1)$ no-slip spheres at the position $\mathbf{R}_i$ of another sphere i. It should be noticed here that $\mathbf{u}_{\text{inc}}^{(N-1)}(\mathbf{r})$ is determined by the BCs on all N spheres, including the singled-out sphere i. In the RP treatment, the incident flow is crudely approximated by the superposition of the single-sphere flow fields,

$$\mathbf{u}_{\text{inc}}^{(N-1)}(\mathbf{r}) \approx \sum_{j \neq i}^{N} \mathbf{u}_j^{(0)}(\mathbf{r}). \tag{10.88}$$

In this incident flow field approximation, the $(N-1)$ no-slip BCs are only approximately fulfilled for configurations of mutually well separated spheres.

The velocity, $\mathbf{V}_i$, of a no-slip microsphere i of radius a in an infinite fluid, subjected to a force $\mathbf{F}_i$ and an incident flow field $\mathbf{u}_{\text{inc}}(\mathbf{r})$, is given by the exact translational Faxén law

$$\mathbf{V}_i = \mu_0^t \mathbf{F}_i + \left(1 + \frac{a^2}{6}\nabla_i^2\right)\mathbf{u}_{\text{inc}}(\mathbf{R}_i), \tag{10.89}$$

where μ_0^t is the translational single-particle mobility of a no-slip sphere. The only restriction on the form of the incident flow is that it has to be a Stokes flow solution, created by sources located outside the volume of sphere i. These sources could be, e.g., other microparticles or confining walls, whereby the wall influence can be described by an image system as discussed in the following section. A derivation of the presented Faxén theorem is given, e.g., in [14, 35]. According to the Faxén theorem, freely advected, i.e. force- and torque-free, small particles can be used to trace out the streamlines caused by the motion of big ones. For an example, consider a small tracer sphere i placed in the (incident) flow of, say, a phoretically moving big sphere. Since on the small length scale a of the tracer the curvature contribution in Eq. (10.89) described by the Laplacian can be neglected, one has

$$\mathbf{V}_i \approx \mathbf{u}_{\text{inc}}(\mathbf{r} = \mathbf{R}_i) \tag{10.90}$$

for the velocity of the tracer dragged along in the flow field of the big particle. Small tracer particles have been used experimentally, e.g., to visualize the thermophoretic quasi-slip flow around a big polystyrene or silica sphere near a surface, driven by a temperature gradient in the fluid oriented perpendicular to the surface [60].

We use now the Faxén theorem to obtain the velocity of a no-slip sphere i of radius a in the incident flow of $(N-1)$ other ones of equal radii, using the single-spheres flow superposition. The result is

$$\mathbf{V}_i \approx \mu_0^t \mathbf{F}_i + \sum_{j \neq i}^{N} \left(1 + \frac{a^2}{3}\nabla_i^2\right)\mathbf{T}(\mathbf{R}_i - \mathbf{R}_j) \cdot \mathbf{F}_j, \tag{10.91}$$

where $\nabla_i^2 \nabla_j^2 \mathbf{T}(\mathbf{R}_i - \mathbf{R}_j) = 0$ for $i \neq j$ has been used in the derivation. The translational mobilities in the RP approximation follow readily from this result as

$$\boldsymbol{\mu}_{ii}^{tt,RP} = \mu_0^t \mathbf{1} \tag{10.92}$$

$$\boldsymbol{\mu}_{ij}^{tt,RP} = \mu_0^t \left[\frac{3}{4} \left(\frac{a}{r} \right) (\mathbf{1} + \hat{\mathbf{r}}\hat{\mathbf{r}}) + \frac{1}{2} \left(\frac{a}{r} \right)^3 (\mathbf{1} - 3\hat{\mathbf{r}}\hat{\mathbf{r}}) \right] \tag{10.93}$$

with $\mathbf{r} = \mathbf{R}_i - \mathbf{R}_j$. The positive definiteness of the RP translational mobility matrix can be shown, e.g., using the double surface integral representation of its tensor elements,

$$\boldsymbol{\mu}_{ij}^{tt,RP} = \frac{1}{(4\pi a^2)^2} \int_{S_i} dS \int_{S_j} dS' \mathbf{T}(\mathbf{r} - \mathbf{r}'), \tag{10.94}$$

which is valid also for $i = j$. It has been assumed for the validity of this representation for $i \neq j$ that the two spheres do not overlap. For the details on how the positive definiteness is shown using this representation we refer to [58].

The simplicity of the RP approximation renders it very attractive for practical applications. It has been generalized to no-slip spheres of different radii (see, e.g., [58]), and to overlapping particles configurations [61]. Moreover, the RP approximation has been employed for the calculation of the mobilities of complex-shaped particles using bead-modelling of the particles [62].

For problems where particles are close to each other, like in concentrated dispersions, more precise methods than the simple RP approximation are required. Various advanced numerical methods have been developed for this purpose. A very powerful and versatile one is the force multipoles method advanced by Cichocki and coworkers. It is based on a multipolar expansion of the flow field for a system of spherical particles which is used to construct a set of equations determining the a priori unknown stresses at the sphere surfaces. The addition of lubrication corrections for nearly-touching particles has led to an efficient numerical algorithm allowing for a controlled high-precision calculation of the many-spheres mobility matrix [37]. A review of this method is given in [38]. The force multipoles method has been extended to account for HIs with a planar wall [63], two parallel walls [64], and a cylindrical channel [65]. A particularly intriguing feature of the method is its facile adaptability to different hydrodynamic BCs such as the Navier partial-slip and liquid interface BCs, with has opened the possibility to study the dynamics of permeable particles, droplets, and surfactant-covered particles.

10.4.7 Lubrication Effects

Hydrodynamic lubrication plays an important role when two or more particles are brought close to each other. This is sketched in Fig. 10.15 for two no-slip smooth spheres, and for three types of relative translational motions, namely squeezing and receding motions along the line of centres, and shearing motion.

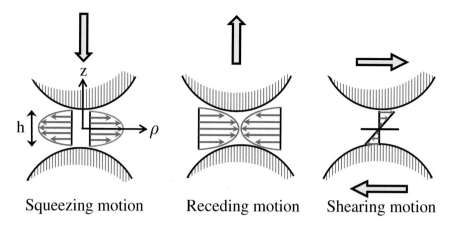

Fig. 10.15 A sketch of the flows in the thin gap region between two smooth spheres near contact, for relative translational motions as indicated

Since the gap between the particles is small, large pressure gradients build up in the gap region in order to squeeze out (or push in) the fluid in the form of an in-to-out Poiseuille flow, as illustrated in Fig. 10.15. This leads to strong friction forces slowing down the motion of close particles and thus has the effect of strongly slowing the relative motion of nearly touching particles. As an example consider the squeezing motion of two no-slip spheres of radii a_1 and a_2, respectively (with sphere 1 on top) approaching each other with constant relative velocity V_{rel}. The lubrication flow analysis leads to the following expression [66]

$$p(\rho) - p_\infty \approx -\frac{3\eta a_1 a_2}{(a_1 + a_2)h^2}\left[1 + \left(\frac{a_1 + a_2}{2a_1 a_2 h}\right)\rho^2\right]^{-2} V_{rel}, \qquad (10.95)$$

for the pressure distribution inside the narrow gap with the minimal surface-to-surface distance $h \ll \{a_1, a_2\}$. Here, $\rho \ll \sqrt{ah}$ is the distance from the symmetry axis, and p_∞ is the inconsequential pressure far away from the gap region. The strong $\mathcal{O}(1/h^2)$ divergence of the pressure in the gap along the vertical symmetry axis dominates the less severe $\mathcal{O}(1/h)$ divergence of the viscous contribution, $\sim \eta|\nabla\mathbf{u}|$, to the stress tensor. Thus only the pressure contribution is required in calculating the hydrodynamic drag force, $\mathbf{F}^h$, on the upper sphere 1 on the basis of Eq. (10.96). The result is [8, 21, 67]

$$\mathbf{F}^h \approx -\frac{6\pi\eta V_{rel}}{h}\left(\frac{a_1 a_2}{a_1 + a_2}\right)^2 \mathbf{e}_z, \qquad (10.96)$$

for the dominant near-contact lubrication part of the drag force which diverges like $\mathcal{O}(1/h)$. The main effect of lubrication is therefore a dramatic increase of hydrodynamic friction between closely spaced no-slip rigid surfaces. This effect can lead in concentrated suspensions to the formation of transient hydrodynamic clusters of particles with consequential shear thickening which has applications, e.g., in manufacturing of protective clothing.

The singular pressure concentration in the gap region of two no-slip spheres is less severe for shearing motion where it gives rise to weaker logarithmic singularity of the drag force at contact. Due to their mobile surfaces, two spherical drops slip past each other in shearing motion even when in contact, experiencing therefore non-divergent drag forces. For squeezing motion, however, there is still a $\mathcal{O}(1/\sqrt{h})$ singularity for two droplets with fully mobile interfaces. It is assumed here that the droplet interfacial tension is so large that the flow-induced deviations from the spherical shape are negligible. For more information about lubrication effects under different surface BCs see [14, 66].

As an illustration of the dynamic effect of lubrication consider a no-slip sphere of radius a approaching a stationary, horizontal no-slip wall in squeezing motion (see Fig. 10.16). The sphere is driven by the *constant* external force $\mathbf{F} = -|F|\mathbf{e}_z = -\mathbf{F}^h$. This is a limiting case of Eq. (10.96), for $a = a_1$ and $a_2 \rightarrow \infty$ with the lower sphere 2 expanded into a planar wall. Here, h denotes now the distance from the wall to the closest surface point of the sphere. It follows that

$$\frac{dh(t)}{dt} = V_{rel}(t) \approx -\frac{\mu_0^t|F|}{a} h(t), \tag{10.97}$$

with the solution

$$h(t) \approx h_0 \exp\left\{-\left(\frac{\mu_0^t|F|}{a}\right)t\right\}. \tag{10.98}$$

Here, h_0 is the near-contact starting distance with $h_0/a \sim 0.01$, and μ_0^t is the single-sphere translational mobility defined in Eq. (10.25). The exponentially slow vertical approach of the sphere to the wall is a good description in reality only until the gap distance h becomes comparable to the surface roughness of the sphere and plane which in fact leads to plane-sphere contact after a finite time. A finite contact

Fig. 10.16 Squeezing motion of a no-slip sphere towards a near-contact planar wall, driven by the constant force $\mathbf{F} = -|F|\mathbf{e}_z$

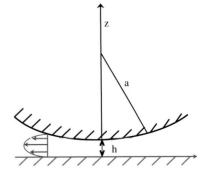

time would be reached also due to the van der Waals attraction force between the planar wall and the sphere which in the distance range where lubrication applies scales as $\mathcal{O}(1/h^2)$ [68].

10.5 Single-Particle Dynamics Near a Flat Interface

The preceding section was devoted to the effects of the solvent flow on the dynamics of suspended particles in an unbounded fluid, except for the discussion in Fig. 10.16 of the near-wall settling of a sphere. Most realistic situations involve the presence of a boundary that may considerably change the flow character, and by reflecting the flow incident upon it, may modify the hydrodynamic interactions between the particles. A pronounced example is the effect of sedimentation, where the backflow of fluid due to the presence of a container bottom, however far distant it may be, cannot be neglected in order to correctly determine the sedimentation velocity of dispersed microparticles [9]. Soft matter systems are very often bounded, especially in biophysical flows and in a number of technical applications. In this section, we investigate the effects of the presence of an interface on the fluid flow, and in consequence on the suspended particles. We start by discussing a general solution for a point force in the presence of a boundary, and its dependence on the BCs at the interface. This allows to elucidate how the physical character of the interface may influence the flow close to it. We then consider in more detail the motion of an isolated spherical particle close to a wall, with the BCs at the particle surface and lubrication effects taken into account. We discuss the commonly used approximations for the mobility of a microsphere near a no-slip wall, and present examples of experimental techniques capable of grasping the dynamic behaviour. Finally, we explore the motion of flexible particles and the effect of deformable boundaries, in situations where the elasticity of the surface or particles may not be neglected.

10.5.1 Point Force Near an Interface

To demonstrate the effects of partial confinement on the hydrodynamics of a single particle, we will present explicit solutions for the velocity fields for a point force acting on a fluid bounded by a planar free surface, and a planar no-slip wall. These are limiting cases for a point-force near a planar liquid-liquid interface, in the limit that the viscosity ratio, $\lambda = \eta_2/\eta_1$, is reaching infinity and zero, respectively. The point force $\mathbf{F}$ is located at the point $\mathbf{r}_0 = (x_0, y_0, h)$ in the upper half space at vertical distance $h > 0$ above the $(x-y)$ interface at $z = 0$. Searched for is the flow field $\mathbf{u}(\mathbf{r})$ in the upper half space $z \geq 0$. We take advantage of the linearity of the Stokes equations to construct the flow field using the method of images. Akin to electrostatics [36], a number of hydrodynamic problems of higher symmetry can be solved by this elegant method. In this method, the fluid in the upper half-space of viscosity $\eta = \eta_1$ is mentally extended into the lower half-space, with the BCs imposed by the taken-out interface now accounted for in conjunction with the Stokeslet at $\mathbf{r}_0$ by an image system of elemental singularity solutions placed at appropriate positions in the lower half-space. The flow field constructed in this way is in the upper half-space identical to the original flow problem where the interface is present.

For a point force in front of a planar liquid-liquid interface, the image system multipoles are all located at the same image point

$$\mathbf{r}_0^* = \mathbf{P}_z \cdot \mathbf{r}_0 = (x_0, y_0, -h). \tag{10.99}$$

Here

$$\mathbf{P}_z = \mathbf{1} - \mathbf{e}_z \mathbf{e}_z \tag{10.100}$$

is the $(x - y)$ plane reflection matrix. This matrix acting on an arbitrary vector turns the z-component of this vector into its negative.

Free surface: We explain the image method first for the simplest case of a Stokeslet above a free surface, e.g., a non-contaminated air-water interface or a Navier perfect-slip wall. The BCs for such an interface is that the surface flow has only an in-plane tangential component, and that there is no tangential stress across the interface. This is expressed by Eq. (10.17) in the limit that the viscosity, η_2, of the fluid in the lower half space vanishes, that is for $\lambda \to 0$. We shall assume the free surface not to deform in response to the motion of the fluid. This is justified when the surface tension of the interface is large enough.

The image system is here simply the mirror Stokeslet of strength $\mathbf{F}^* = \mathbf{P}_z \cdot \mathbf{F}$ at the mirror location $\mathbf{r}_0^*$. The flow in the fluid occupying the upper half-space is thus

$$\mathbf{u}(\mathbf{r}) = [\mathbf{T}(\mathbf{R}) + \mathbf{T}(\mathbf{R}^*) \cdot \mathbf{P}_z] \cdot \mathbf{F}, \tag{10.101}$$

where $\mathbf{R} = \mathbf{r} - \mathbf{r}_0$ and $\mathbf{R}^* = \mathbf{r} - \mathbf{r}_0^*$ are the vectors to the observation point from the Stokeslet and image Stokeslet locations respectively. The Cartesian frame can be oriented here such that $\mathbf{F} = F_\parallel \mathbf{e}_x + F_\perp \mathbf{e}_z$. The flow components tangential and vertical to the interface are then

$$\mathbf{u}_x(\mathbf{r}) = \frac{F_\parallel}{8\pi\eta} [\mathbf{G}(\mathbf{e}_x) + \mathbf{G}^*(\mathbf{e}_x)]$$

$$\mathbf{u}_z(\mathbf{r}) = \frac{F_\perp}{8\pi\eta} [\mathbf{G}(\mathbf{e}_z) + \mathbf{G}^*(\mathbf{e}_z)], \tag{10.102}$$

where we use $\mathbf{G}(\mathbf{e}) = \mathbf{G}(\mathbf{R}; \mathbf{e})$ and $\mathbf{G}^*(\mathbf{e}) = \mathbf{G}(\mathbf{R}^*; \mathbf{e})$ as abbreviations to shorten the notation.

The image system for a planar free surface, consisting simply of the mirror reflection of the actual flow singularity, is sketched in Fig. 10.17 for the two basic situations of the Stokeslet directed along and perpendicular to the plane, respectively. We use here the elemental singularity solution pictograms introduced in Fig. 10.7.

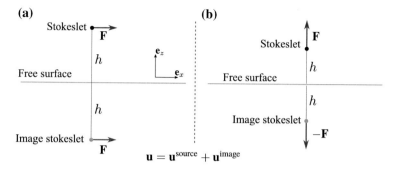

Fig. 10.17 The image system satisfying the boundary conditions of a free surface is simply the mirror Stokeslet of strength $\mathbf{F}^* = \mathbf{P}_z \cdot \mathbf{F}$. **a** For a point-force oriented parallel to the surface, the image has the same force direction, while **b** for the force pointing towards or away from the surface, the direction of the image force is reversed

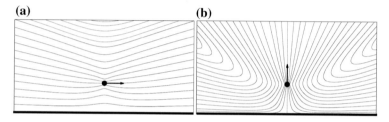

Fig. 10.18 Flow streamlines **a** due to a point-force oriented parallel to the free surface, and **b** due to a perpendicularly oriented point-fore. Reproduced from [69] with kind permission

The streamlines for the two cases, based on the flow fields in Eq. (10.102), are plotted in Fig. 10.18. By the reflection symmetry of the present image system, it is obvious that the velocity field at the interface is purely tangential. Moreover, since $\mathbf{u}(x, y, z) = \mathbf{u}(x, y, -z)$ by construction, the tangential stress at the interface proportional to $\partial u_{x,y}/\partial z$ is zero.

The flow solution for a Stokeslet in front of a planar interface can be used to construct, by superposition, the flow field due to a rigid no-slip body near the interface. For a free interface, the flow field is determined by

$$\mathbf{u}(\mathbf{r}) = \int_{S_p} dS' \mathbf{T}(\mathbf{r} - \mathbf{r}') \cdot (\boldsymbol{\sigma}(\mathbf{r}) \cdot \mathbf{n}') + \int_{S_p^*} dS' \mathbf{T}(\mathbf{r} - \mathbf{r}') \cdot \mathbf{P}_{z'} \cdot (\boldsymbol{\sigma}(\mathbf{r}') \cdot \mathbf{n}'),$$

(10.103)

where S_p^* is the surface of the mirror body, and $\mathbf{P}_{z'} \cdot (\boldsymbol{\sigma}(\mathbf{r}') \cdot \mathbf{n}')$ is the mirror stress field on this surface. This is sketched in Fig. 10.19.

As an interesting problem related to Eq. (10.103), consider the tangential motion of an isolated no-slip sphere along a planar free interface, in the extreme situation that the sphere is permanently touching the interface in a single point. This

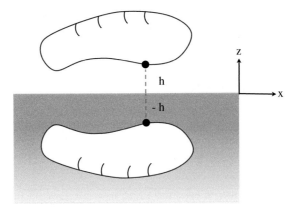

Fig. 10.19 To determine the flow due to a no-slip particle moving above a planar free surface at $z = 0$, one mentally replaces the surface by an image particle, with the fluid continued into the lower half-space. The surface stress and thus the translational and angular velocities of the image particle are the surface reflections of those of the real particle

quasi-two-dimensional motion is realized to good accuracy in an experimentally well studied system of super-paramagnetic microspheres at the air-water interface of a hanging drop. To determine the single-sphere mobilities starting from Eq. (10.103), it is crucial to account for lubrication effects in the zero gap case. Lubrication forbids the rotation of the surface-touching sphere along an axis parallel to the surface [70], i.e. $\mu_\parallel^r = 0$. Rotation is allowed only around the perpendicular sphere axis. While particle surface roughness and remnant undulations in the planar surface should give rise to a non-zero mobility $\mu_\parallel^r$, the actual effect of lubrication is to lower its value significantly below the rotational mobility, μ_0^r, of a sphere in the bulk fluid. A precise calculation of the mobility tensors of hydrodynamically interacting spheres touching a free planar interface was made in [70]. The special result for the mobilities of an isolated sphere is

$$\mu_\parallel^t/\mu_t^0 \approx 1.380, \quad \mu_\perp^r/\mu_r^0 \approx 1.109. \tag{10.104}$$

Owing to the smaller friction experienced by the hemisphere facing the free surface, the translational mobility for the motion parallel to the surface is raised above its bulk value by 38 percent, and the rotational mobility $\mu_\perp^r$ by mere 11 percent. Note that the sphere is not rotating while translating, since it can be considered as moving side by side in contact with its twin image sphere located just below the free interface. Each of the two twin spheres experiences the same hydrodynamic drag force parallel to the free surface. A microsphere above a no-slip wall which has the freedom to move away from the wall is discussed further down in relation to Fig. 10.22.

Liquid-liquid interface: The image system for a point force in the presence of a rigid no-slip wall, and more generally of a (clean) fluid-fluid interface was given by Blake in the 1970's [71, 72].

For a Stokeslet parallel, $\mathbf{G}(\mathbf{e}_x)$, and perpendicular, $\mathbf{G}(\mathbf{e}_z)$, to the $(x - y)$ fluid-fluid interface, the image system is given as a function of the viscosity ratio λ by [26]

$$\mathbf{G}_{\text{Im},x}(\mathbf{r}) = \frac{1 - \lambda}{1 + \lambda} \mathbf{G}^*(\mathbf{e}_x) + \frac{2\lambda h}{\lambda + 1} \mathbf{G}_D^*(\mathbf{e}_x, \mathbf{e}_z) - \frac{2\lambda h^2}{\lambda + 1} \mathbf{D}_S^*(\mathbf{e}_x), \qquad (10.105)$$

$$\mathbf{G}_{\text{Im},z}(\mathbf{r}) = -\mathbf{G}^*(\mathbf{e}_z) + \frac{2\lambda h}{\lambda + 1} \mathbf{G}_D^*(\mathbf{e}_z, \mathbf{e}_z) - \frac{2\lambda h^2}{\lambda + 1} \mathbf{D}_S^*(\mathbf{e}_z), \qquad (10.106)$$

where $\mathbf{G}(\mathbf{e}) = \mathbf{G}(\mathbf{R}; \mathbf{e})$ and $\mathbf{G}^*(\mathbf{e}) = \mathbf{G}(\mathbf{R}^*; \mathbf{e})$ are used as abbreviations, with analogous abbreviations used for the other elemental singularities. The superscript (*) indicates that the considered singularity is located at the position, $\mathbf{r}_0^*$, of the image. The Cartesian longitudinal and transversal components of the flow field in the upper half-space $z > 0$ are expressed in terms of this image system as

$$\mathbf{u}_x(\mathbf{r}) = \frac{F}{8\pi\eta} [\mathbf{G}(\mathbf{r}; \mathbf{e}_x) + \mathbf{G}_{\text{Im},x}(\mathbf{r})], \qquad (10.107)$$

$$\mathbf{u}_z(\mathbf{r}) = \frac{F}{8\pi\eta} [\mathbf{G}(\mathbf{r}; \mathbf{e}_z) + \mathbf{G}_{\text{Im},z}(\mathbf{r})], \qquad (10.108)$$

with the force $\mathbf{F}$ of magnitude $F > 0$ pointing along the x-axis and z axis, respectively. In the limit $\lambda = 0$ of zero viscosity of the lower half space fluid, the result for a Stokeslet above a free surface is recovered.

No-slip wall: The system of image multipoles at $\mathbf{r}_0^*$ for a no-slip rigid surface, obtained by taking the limit $\lambda \to \infty$ in Eq. (10.105), has more components than that for the free surface. The reason for this is that in addition to a zero normal velocity component at the interface, also the tangential fluid velocity component must be zero.

The pictogram representation of the image system for a no-slip wall is shown in Fig. 10.20, for the strength $\mathbf{F}$ of the Stokeslet oriented parallel (a) and (b) perpendicular to the wall. The image system for the general case of a tilted Stokeslet is obtained by linear superposition. The image system has now two members in addition to the image Stokeslet in the free surface case, namely a Stokes doublet of strength modulus $2hF$, and a source doublet of strength modulus $2h^2F$. The streamlines for a Stokeslet parallel and perpendicular to the no-slip wall are shown Fig. 10.21. Notice the pronounced flow vortices in the latter case, with fluid dragged in behind the up-pointing Stokeslet. For a detailed discussion of the image system solution we refer to Blake's original work. Quite interestingly, the presence of a no-slip wall changes the asymptotic behaviour of the flow field in the upper half-space at distances far from the Stokeslet and the wall. Although the image Stokes doublet and source doublet contributors to the velocity field decay faster than the two Stokeslet contributors, there is a long-distance flow cancellation. As a

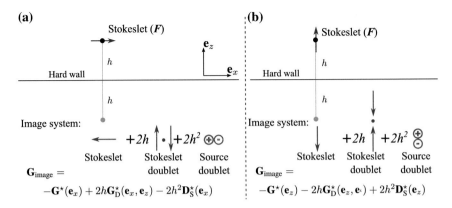

Fig. 10.20 The image system satisfying the no-slip BC at a rigid wall is more complex than that for a free surface. The three different flow singularities of the image system and their respective strengths are expressed as pictograms defined in Fig. 10.7a–f. The image singularity solutions are all located at the mirror point of the Stokeslet in the actual upper fluid. Note that in **a** the mirror Stokeslet is oppositely oriented. Redrawn after [71]

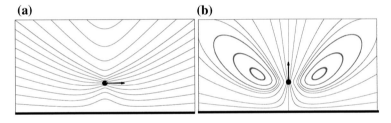

Fig. 10.21 Flow field streamlines for a point force located above a stationary no-slip wall, with the wall represented by the *thick bottom line*. Parallel (**a**) and perpendicular (**b**) orientations of the Stokeslet are considered. Reproduced from [69] with kind permission

result, the velocity field decays asymptotically as $\mathcal{O}(r^{-2})$, for the Stokeslet oriented parallel to the wall, and an even faster $\mathcal{O}(r^{-3})$ decay is found for the perpendicularly oriented Stokeslet [71]. This is an example of hydrodynamic screening induced by a stationary no-slip boundary which takes momentum out from the fluid.

The method of images can be successfully applied also to the case of higher-order elemental singularities near a planar wall such as a Stokes dipole, Rotlet, Rotlet dipole and source doublet [26, 73]. These solutions are quite useful, e.g., to describe the hydrodynamic attraction of a bacterial microswimmer by an interface, and its resultant circular motion.

Partial-slip wall: The image system for a planar rigid wall with Navier partial-slip boundary conditions is more complicated than that for a liquid-liquid interface, except for the zero and infinite slip length limits where likewise the no-slip wall and free surface are recovered. As shown by Lauga and Squires in [74], the image system for the perpendicularly oriented Stokeslet contains the same set of elemental singularity solutions as that for a liquid-liquid interface. However, these are now continuously distributed along a line extending from the reflection point $\mathbf{r}_0^*$ into the negative $\mathbf{e}_z$ direction, with magnitudes that decay exponentially downside this line over the slip length ℓ. For $\ell \to 0$, Blake's solution for a no-slip wall is recovered. The image system for a Stokeslet parallel to the partial-slip wall involves a larger set of elemental singularity solutions, likewise distributed along the aforementioned singularity line. The system includes now also a Rotlet and a Rotlet dipole. Here, not all singularity solution magnitudes decay exponentially in going downside along the line.

10.5.2 Motion of a Spherical Particle Near a No-Slip Wall

The hydrodynamic problem of the motion of a spherical particle in a viscous liquid bounded by a planar no-slip wall has been studied since more than a century. The difficulty of the problem relates to the fact that BCs must be satisfied both at the wall and the sphere surface.

Anisotropic mobilities: Owing to the BCs both for the sphere and the wall, the 6×6 mobility matrix characterizing the translational and rotational motion of the sphere near the wall is of an anisotropic character, with scalar elements (mobility coefficients) depending on the distance z of the sphere centre to the wall extending into the $x - y$ plane. Five independent mobility coefficients are required to characterize the sphere motion, as depicted in Fig. 10.22. The four coefficients $\mu_\parallel^t, \mu_\perp^t, \mu_\parallel^r, \mu_\perp^r$ characterize the translational and rotational sphere motion parallel

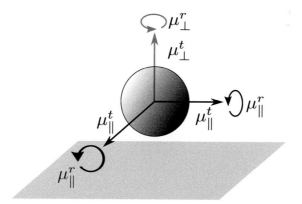

Fig. 10.22 Schematic representation of mobility coefficients describing the near-wall translational (superscript t) and rotational (superscript r) motions of a rigid microsphere parallel and perpendicular to a planar wall. Not depicted is an additional mobility coefficient, $\mu^{tr} = \mu^{rt}$, characterizing the wall-induced translation-rotation coupling of the sphere motion. See the text and [75] for details

and perpendicular to the wall, while the additional coefficient μ^{tr} describes the wall-induced coupling of translational and rotational motion. A torque-free sphere translating parallel to a near-distant wall is rotating. This should be contrasted to the motion of a sphere far distant from the wall which is fully characterized by the two bulk coefficients μ_0^t and μ_0^r given in Eq. (10.25), without any translational-rotational coupling.

The 6×6 mobility matrix for a sphere near a planar wall is of the form

$$\boldsymbol{\mu}(z) = \begin{pmatrix} \boldsymbol{\mu}^{tt}(z) & \boldsymbol{\mu}^{tr}(z) \\ \boldsymbol{\mu}^{rt}(z) & \boldsymbol{\mu}^{rr}(z) \end{pmatrix}, \tag{10.109}$$

with all coefficients depending on the sphere-wall distance z. The components of the tt submatrix have in the selected coordinate frame the simple structure

$$\boldsymbol{\mu}^{tt}(z) = \begin{pmatrix} \mu_\perp^{tt}(z) & 0 & 0 \\ 0 & \mu_\parallel^{tt}(z) & 0 \\ 0 & 0 & \mu_\parallel^{tt}(z) \end{pmatrix}, \tag{10.110}$$

with a similar structure for the rotational rr submatrix. The structure of the rt and tr coupling tensors is different (see, e.g., [75, 76]).

Numerous works have been devoted to the evaluation of the z-dependence of the mobility coefficients, dating back to Lorentz [77] and Faxén [78] more than a century ago who calculated the first terms in the expansion of the two translational coefficients in terms of the reciprocal sphere-wall distance, $t = a/z$, in units of the sphere radius, given by

$$\mu_\parallel^t(t) \approx 1 - \frac{9}{8}t, \tag{10.111}$$

$$\mu_\perp^t(t) \approx 1 - \frac{9}{8}t + \frac{1}{8}t^3 - \frac{45}{256}t^4 - \frac{1}{16}t^5. \tag{10.112}$$

These expressions provide a crude approximation of the exact translational coefficients for large sphere-wall distances $z/a \geq 10$. Subsequent refined calculations by Brenner et al. [15, 79–81] and Dean and O'Neill [82, 83] have led to formally exact expressions for part of the mobility coefficients in terms of infinite series. While frequently quoted, these series expressions are of limited practical importance owing to their slow convergence at near-contact distances. More recently, numerically precise and convenient inverse distance series results for all five mobility coefficients have been obtained, using a high-precision numerical scheme based on the force multipoles method by Cichocki and Jones [76] combined with a Padé approximation and with near-contact lubrication effects taken into account.

For the presentation of these numerical results, we introduce dimensionless mobilities by division through respective bulk mobility coefficients according to

$$\tilde{\mu}^t_{\parallel,\perp} = \frac{\mu^t_{\parallel,\perp}}{\mu^t_0}, \quad \tilde{\mu}^r_{\parallel,\perp} = \frac{\mu^r_{\parallel,\perp}}{\mu^r_0}, \quad \tilde{\mu}^{tr} = \frac{\mu^{tr}}{a\mu^r_0}. \tag{10.113}$$

The dimensionless mobility coefficients for a no-slip sphere near a planar no-slip wall are plotted in Fig. 10.23, as functions of the inverse distance parameter t. Significant deviations from the bulk mobility values are observed for distances $z/a < 5$. The slowing hydrodynamic effect of the wall is in general more pronounced for translational than rotational motion. Physical processes where this can be of importance are cellular adhesion [6], and channel flows where translation is hindered but rotation is still strong enough to allow for the reorientation of particles in external fields [84]. All mobility coefficients except for $\mu^r_\perp$ tend to zero in a non-analytical way when the contact distance $t = 1$ is approached. It follows from lubrication theory that the asymptotic behaviour of the mobility coefficients close to the wall can be expressed in terms of the dimensionless gap width $\varepsilon = (z - a)/a$ [76]. In the case of translational coefficients, one finds

$$\tilde{\mu}^t_\perp \sim \varepsilon + \frac{1}{5}\varepsilon^2 \log \varepsilon, \qquad \tilde{\mu}^t_\parallel \sim -2(\log \varepsilon)^{-1}. \tag{10.114}$$

Wall lubrication effects imply also that $\mu^r_\parallel(t \to 1) = 0$, whereas the sphere rotation with the angular velocity oriented perpendicular to the wall is possible even at contact where the related mobility coefficient $\mu^r_\perp$ is reduced by about 18 percent below the isotropic bulk value. The coefficient μ^{tr} relating translational motion to

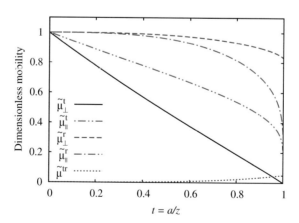

Fig. 10.23 Dimensionless mobility coefficients, defined in Eq. (10.113), of a no-slip microsphere near a planar no-slip wall, as functions of the dimensionless reciprocal sphere-wall distance $t = a/z$. At sphere-wall contact where $t = 1$, all mobilities except for $\mu^r_\perp$ are vanishing. The effect of translation-rotation coupling, absent both at contact and far away from the wall, is strongest near sphere-wall contact where $\tilde{\mu}^{tr}$ attains its largest value of about 0.05. Mobility coefficients have been obtained using the method in [76]

applied torque is zero at sphere-wall contact since lubrication implies zero translational velocity of the sphere at wall contact.

Mobility measurement by light scattering: Theoretical predictions for the distance dependence of the mobility coefficients of an isolated microsphere near a wall have been scrutinized in experimental studies, for sphere sizes ranging from about 100 nm up to several microns, using various optical techniques. These techniques include optical trap microscopy [85], nano-PIV [86, 87], dynamic light scattering (DLS) in presence of two parallel walls [88], low-coherence DLS [89], resonance-enhanced DLS [90, 91], and evanescent wave dynamic light scattering (EWDLS) in a system bounded by one or two walls [27, 92–95]. However, only recently has it been possible to determine both the translational and rotational diffusion of a colloidal sphere in the vicinity of a planar wall [75, 96], using EWDLS from optically anisotropic spherical particles. In many experiments such as in EWDLS, the relation

$$\mathbf{D} = k_B T \boldsymbol{\mu}, \tag{10.115}$$

between the hydrodynamic mobility matrix and the diffusion matrix, $\mathbf{D}$, of a dispersed Brownian particle at system temperature T is used, on measuring the diffusion matrix coefficients instead of the associated hydrodynamic mobility coefficients.

Light scattering is a powerful tool to investigate the properties of sub-micron soft matter systems [97]. In evanescent wave scattering experiments, a colloidal suspension is typically illuminated by a monochromatic laser beam that is totally reflected from a planar glass surface bounding the sample, so that no refracted light enters into the suspension (which is placed above the glass surface in Fig. 10.24) except for an evanescent wave whose intensity decays exponentially in going away from the glass surface into the suspension. Thus, only a colloidal particle close to the glass surface scatters enough of the incident evanescent light to be detected. The penetration depth, $2/\kappa$, of the evanescent wave can be changed to probe the particle diffusion at different glass surface-particle distances. For a review of the EWDLS method including experimental details, see [98].

The key quantity determined in (EW)DLS experiments is the scattered light intensity time-autocorrelation function. Of particular significance is the short-time (initial) decay rate, Γ, of the intensity autocorrelation function referred to as the first cumulant. This quantity can be theoretically predicted on basis of the generalized Smoluchowski equation determining the evolution of the configurational probability density function of Brownian particles under Stokes flow conditions [1, 99]. Inside the bulk region of a very dilute suspension of colloidal hard spheres far away from confining walls, the first cumulant is proportional to $\Gamma = q^2 D^0$, where q is the modulus of the scattering vector $\mathbf{q}$, and $D^0 = k_B T \mu_0^t$ is the translational (Stokes-Einstein) diffusion coefficient of an isolated Brownian sphere.

In the data analysis gained from a typical EWDLS set-up such as the one sketched in Fig. 10.24, one conveniently decomposes the scattering vector into its components parallel and perpendicular to the wall, $q_\parallel$ and $q_\perp$, respectively. The first

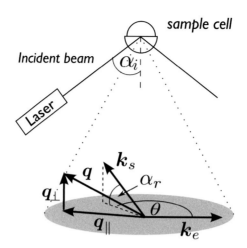

Fig. 10.24 Schematics of an EWDLS set-up. The wave vectors of the incident evanescent and scattered light beams are $\mathbf{k}_e$ and $\mathbf{k}_s$, respectively, with their difference defining the scattering vector $\mathbf{q} = \mathbf{k}_s - \mathbf{k}_e$. Independent experimental variation of the components $q_\parallel$ and $q_\perp$ of the scattering vector parallel and perpendicular to the confining glass wall allows for the determination of wall-distance averaged diffusion coefficients. See the text for details. Redrawn after [95]

cumulant for the translational motion of a Brownian sphere near the glass wall is then given by

$$\Gamma = q_\parallel^2 \left\langle D_\parallel^0 \right\rangle_\kappa + \left(q_\perp^2 + \frac{\kappa^2}{4} \right) \left\langle D_\perp^0 \right\rangle_\kappa, \tag{10.116}$$

where $D_\parallel^0(z) = k_B T \mu_\parallel^t(z)$ and $D_\perp^0(z) = k_B T \mu_\perp^t(z)$, and $\langle \cdots \rangle_\kappa$ denotes a κ-dependent weighted average of the z-dependent diffusion coefficients over all sphere - glass wall separations z, for a given evanescent wave penetration parameter κ. The average diffusion coefficients in Eq. (10.116) for translational diffusion parallel and perpendicular to the glass wall are not purely statistical mechanical properties but are dependent on the value of κ selected in the optical setup: The smaller κ is the larger are the resulting average diffusion coefficients [95]. For a smaller κ, diffusion is detected in EWDLS over a larger distance from the glass wall, and $\mu_\parallel^t(z)$ and $\mu_\perp^t(z)$ are increasing with increasing wall-sphere distance z.

By determining Γ as a function of $q_\parallel$ for fixed $q_\perp$ and vice versa, the two average translational diffusion coefficients are obtained using Eq. (10.116). Moreover, on the basis of an analytic expression for the first cumulant generalized to optically anisotropic spherical particles, the distance-averaged rotational diffusion coefficients are obtained in a light polarization-sensitive EWDLS experiment in addition to the translational ones [75, 96].

Hydrodynamic radius models: Transport properties of colloidal suspensions such as the viscosity and translational and rotational diffusion coefficients depend in principle on the details of the hydrodynamic particle structure, e.g. on the particle surface BC, and the fluid permeability profile in the case of fluid-permeable particles. Important examples of particles with internal structure are micro- and nanogels, non-permeable rigid particles with surface corrugation, and core-shell particles consisting of a dry core coated by a polymer brush. The hydrodynamic effect of the in general quite complicated intra-particle structure can be characterized under surprisingly general conditions by a single parameter, namely the effective hydrodynamic radius a_{eff}. For globular particles, this radius can be determined experimentally in a DLS experiment on using the single-sphere Stokes-Einstein relation for the diffusion coefficient D^0. The hydrodynamic radius is the radius of an effective no-slip sphere with the same diffusion coefficient as the one of the actual internally structured particle. It has been shown in recent theoretical work [100] that the error introduced by simply using a_{eff} for the particle structure characterization can be well controlled.

10.5.3 Near-Wall Dynamics of Anisotropic and Flexible Particles

The presence of a nearby wall or interface drastically affects the hydrodynamics experienced by a microparticle. This can be easily understood qualitatively using symmetry arguments. Since the hydrodynamic friction experienced by a particle is in general larger on the side of the particle facing the wall, there is translational-rotational coupling even for a highly symmetric particle such as a sphere with uniform surface BC. For a non-spherical particle, there is an additional dependence of the friction coefficients on the particle orientation giving rise to interesting dynamic effects.

For an example, consider the sedimentation of a rod-like rigid particle near a vertical rigid wall. As we have discussed earlier, an inclined rod in an unbounded fluid has a horizontal side drift while settling but it does not reorient its body axis. This absence of rod reorientation/rotation does not hold any more in sedimentation close to a vertical wall. It has been experimentally observed and numerically calculated by Russel et al. [101] for a no-slip rod sedimenting near to a no-slip vertical wall that there are two possible sedimentation scenarios. The first one is a glancing motion, where one tip of the rod always points downwards while the rod is reorienting close to the wall, and the second one a reversing motion, where the rod tumbles while approaches its closest distance to the wall. In both scenarios, the wall to centre-of-rod distance decreases initially during sedimentation, increasing again subsequently. Which of the two scenarios is taking place depends on the initial wall distance and inclination angle of the rod. To gain a quantitative understanding of the rod Stokesian dynamics, a slender-body analysis for the motion of an elongated microparticle close to a flat fluid-fluid interface has been made for determining the drag force and torque acting on it for a fixed spatial orientation [102].

The motion of a flexible microparticle is more complicated than that of a rigid one, owing to inevitable hydrodynamically induced particle surface deformations which invalidate the simple kinematic reversibility arguments for the associated Stokes flows. As an illustration, consider in Fig. 10.25a, b, a flexible particle (say a droplet or vesicle) moving away or towards a rigid planar wall, respectively. During its motion, the particle will deform in a way controlled by the interplay of fluid stresses and particle surface tension. The latter tends to restore the spherical particle shape of minimal surface free energy which it would have if the fluid and particle were stationary. The particle deformations are different for (a) and the oppositely directed motion in (b), since the distribution of friction forces (stresses) along the particle surface is different in the two cases. Since symmetry is here obviously broken, kinematic reversibility arguments can not be used to gain information on the particle motion. The changing shape of an elastic body approaching a rigid flat wall has been numerically calculated, e.g., for a liquid droplet sedimenting in another fluid [103]. Elasticity effects can lead to cross-stream migration of flexible particles [40] which is of importance in blood flows, where the suspended cor-puscles are often highly elastic, in cell adhesion problems in shear flows [104], and for industrial processes involving macromolecules or polymer flows [105].

So far, we have assumed that the interface in the proximity of a particle is rigid. This is a valid assumption for non-deformable container walls or liquid-liquid interfaces of large interfacial tension. Deformable interfaces give rise to additional

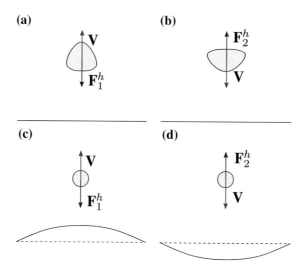

Fig. 10.25 Sketch of two physical situations where kinematic reversibility is not applicable owing to hydrodynamically induced particle (in **a–b**) or interface (in **c–d**) deformations. In **a–b**, a flexible particle is deformed differently if moving towards or away from a rigid flat interface. In **c–d**, the flexible interface experiences a local deformation different for a rigid particle moving towards or away from the interface. The frictional force, F^h, experienced by the particle is likewise dependent on the motion direction, since the distance to the deformed interface in (**c**) is smaller than in (**d**)

effects. The motion of a rigid particle towards or away from a flexible interface causes a motion-dependent local deformation of the interface. This is related to a special type of particle-interface interaction through the geometric coupling of local interface shape and flow [106]. A rigid particle moving away from the interface sketched in Fig. 10.25c pulls the fluid along with it which in turn causes a local deformation in the flexible interface extending in the direction of the particle motion. This results in an enlarged hydrodynamic friction force on the particle as compared to the flat-surface situation, i.e. $F_1^h > F_{\text{flat}}^h$. The geometric coupling for the oppositely oriented particle motion in figure part (d) leads to a frictional force on the particle smaller than in the flat interface case so that $F_2^h < F_{\text{flat}}^h$. This is reminiscent of the lift force acting on deformable particles in flow, and is of interest for biological system flows under elastic confinement. Quite interestingly, the vicinity of a soft interface or object can be used by microorganisms in their propulsion even if performing reciprocal motions [107], since the micro-scallop theorem discussed in Sect. 10.6 does not apply near a flexible surface [108].

10.6 Self-propelling Microswimmers

The locomotion and transport of autonomous (self-propelling) biological and artificial microswimmers under low-Reynolds-number flow conditions has generated a lot of interest over the past years. See here [108, 109] for two very informative overview articles, and the classical book by Lighthill [110]. The motility of microorganisms such as bacteria, sperm cells, and algae affects many biological processes including reproduction and infection. Appropriate swimming strategies are essential for microorganisms in their search for food or the avoidance of toxic environments (chemotaxis), the reaction to light (phototaxis), and the orientation under gravity (gravitaxis). Theoretical design and fabrication of synthetic (robotic) microswimmers who can transport cargo or remove toxins bears the perspective of highly useful applications in medicine, biology and environmental science.

Autonomous swimmers are characterized by the absence of an external forcing agent driving their translational and rotational motion. For microswimmers under Stokes-flow conditions, this means

$$\mathbf{F}^h(t) = \mathbf{0}, \quad \mathbf{T}^h(t) = \mathbf{0}, \tag{10.117}$$

expressing that the total hydrodynamic drag force and torque exerted on the swimmers are zero at any instant of time. The long-distance decay of the disturbance fluid velocity field caused by the swimmer's motion is thus of $\mathcal{O}(r^{-2})$ or faster.

There are a variety of autonomous propulsion mechanisms. For instance, a microobject could swim by self-diffusiophoresis, by creating through a surface-active site a small gradient, $\nabla \phi$, in the concentration ϕ of a dissolved species (solute). This self-created gradient, in turn, propels the microobject through the phoretic osmotic solute flow in its interfacial region [48, 52], with the self-phoretic velocity, $\mathbf{V}_{\text{phor}}$, of the object determined by Eq. (10.79). The potential flow outside the self-phoretic object has thus the characteristic $\mathcal{O}(r^{-3})$ far-distance decay.

10.6.1   Purcell's Micro-scallop Theorem

We focus here on purely mechanical microswimmers which self-propel by continuously changing their body shape in a periodic way. After one cycle, the microswimmer returns to its initial body shape.

Two conditions have to be fulfilled to achieve a net translational displacement after one cycle of body shape deformations. The first one known as Purcell's micro-scallop theorem [22, 111] is related to the kinematic reversibility of Stokes flows. It reads

> In the absence of inertia, the periodic sequence of body shape configurations must be non-reciprocal, that is it must be different when viewed in a time-reversed way.

This excludes in particular cyclic shape changes depending on a single parameter only. For the scallop theorem to apply, it is understood that the single-parametric microobject is far away from a flexible interface which invalidates kinematic reversibility arguments (c.f. Sect. 10.5.3). As an illustrative example of the no-go scallop theorem consider with Purcell in Fig. 10.26 a (one-hinge) symmetric micro-scallop in the bulk fluid periodically opening and closing its legs. While shaking back and forth, the net displacement after one cycle is zero. The opening angle φ is here the only parameter characterizing different shapes, and the sequence of different shapes is thus necessarily reciprocal.

A non-reciprocal sequence of cyclic shape deformations is not sufficient to achieve a net propulsion under Low-Reynolds-number conditions. An additional requirement, applying also to non-small Reynolds number locomotion, is:

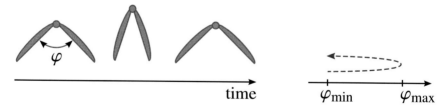

Fig. 10.26 *Left* A symmetric micro-scallop shakes back and forth during one motion cycle without a net vertical translation. *Right* The sequence of shape configurations in the one-dimensional parameter space is necessarily reciprocal

Successful self-propulsion requires an anisotropy or asymmetry in the fluid friction experienced by the moving swimmer.

The net displacement of an isolated microswimmer does not depend on the rate at which a given non-reciprocal sequence of shape configurations occurs (rate independence) but only on their geometry.

Many microorganisms such as bacteria and spermatozoa have an elongated body with a distinct body axis or polarity which dictates the direction of their motion. Roughly speaking, such a microorganism consists of a passive head part and a slender active filament (flagellum). Animalcules such as sperms are propelled by wavelike beating of their flagellum which causes them to move in the direction opposite to that of the flagellar wave travelling away from the head. Unlike eukaryotic flagella, bacterial flagella are passive fibers incapable of active bending. They use instead a helical wave propulsion, in the form of a rotating rigid helical bundle of flagella driven by a molecular motor at the cell body. If viewed from behind in swimming direction, the flagellar bundle rotates counter-clockwise. The torque introduced by this rotation is balanced by the clockwise rotation of the cell body.

In both propulsion mechanisms, swimming is possible because of the friction anisotropy and non-reciprocal travelling wave deformations of the flagellum. The slender flagellum can be mentally subdivided, at any instant, into rod-like segments characterized by two segmental friction coefficients $\zeta_\perp \approx 2\zeta_\parallel$. The force $\Delta\mathbf{F}_s$ exerted on the fluid by a segment moving with an instantaneous velocity $\mathbf{v}$ is

$$\Delta\mathbf{F}_s = \zeta_\parallel \mathbf{v}_\parallel + \zeta_\perp \mathbf{v}_\perp, \tag{10.118}$$

where $\mathbf{v}_\parallel$ and $\mathbf{v}_\perp$ are the velocity projections parallel and perpendicular to the segment, determined by the instant shape and rate of change of shape of the swimmer. Integration over the filament contour leads to the momentary propulsion (thrust) force,

$$\mathbf{F}_{\text{prop}} \propto (\zeta_\perp - \zeta_\parallel)\mathbf{e}, \tag{10.119}$$

along the body axis $\mathbf{e}$ of the swimmer who is kept stationary for this first calculation step. The propulsion force is proportional to the difference of the segment friction coefficients and points from the filament to the head. See [108] for the details of such a calculation, e.g., for a simplified sperm model with a two-dimensional wave-like beating pattern. The propulsion force must be balanced at any instant by a hydrodynamic drag force, $F^h(t) = -F_{\text{prop}}(t)$, exerted by the fluid on the instantaneously frozen-in shape of the swimmer, which is moving in this second calculation step with the searched for instantaneous axial swimming velocity $V(t)$. From a decent estimate of the axial friction coefficient of the frozen-in swimmer appearing in the relation $F^h(t) = -\zeta_{\text{froz}}(t)V(t)$, the instantaneous swimming velocity is approximately obtained. The here outlined procedure is once again a direct consequence of the additivity of Stokes flow solutions, and of the associated particle velocities. Since both F_{prop} and F^h are proportional to η, the swimming velocity is independent of the fluid viscosity. However, the rate of dissipated energy caused by swimming depends on the viscosity.

10.6.2   Dipole Swimmers

The flow around an isolated axial microswimmer with flagellar propulsion is in general well approximated by a linear force dipole (linear Stresslet), $\mathbf{G}_D(\mathbf{r};\mathbf{e},\mathbf{e})$, in the direction of the swimmer's body axis $\mathbf{e}$. The two infinitesimally distant Stokeslets pointing away from each other represent the balance of propulsive and drag forces discussed above. This far-distance flow model is valid for distances larger than the axial extension L of the swimmer. Higher-order elemental multipoles containing additional information on the shape and near-distance motion come into play when microswimmers get close to each other or to a boundary [26, 41, 108, 109].

With the dipole singularity positioned at the origin, and the dipole orientation along $\mathbf{e} = \mathbf{e}_z$, the dipole flow field is

$$\mathbf{u}_D(\mathbf{r};\mathbf{e}) = \frac{p}{8\pi\eta}\mathbf{G}_D(\mathbf{r};\mathbf{e}_z,\mathbf{e}_z) = \frac{p}{8\pi\eta r^2}[3(\hat{\mathbf{r}}\cdot\mathbf{e})^2 - 1]\hat{\mathbf{r}}. \tag{10.120}$$

Note that $(\hat{\mathbf{r}}\cdot\mathbf{e}) = \cos\psi$, with ψ denoting the angle between the dipole (swimmer) axis $\mathbf{e}$ and the fluid observation point at $\mathbf{r}$. The streamlines of the dipole velocity field are depicted in Fig. 10.27 for $p > 0$, where the two Stokeslets are pointing away from each other. The dipole strength scales as $|p| \sim \eta|V|L^2$ where

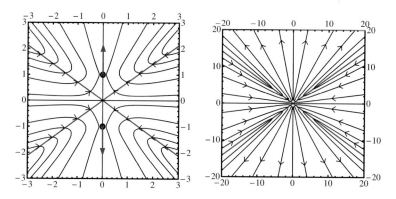

Fig. 10.27 Streamlines of a linear Stokeslet dipole (force dipole) oriented along the z axis with positive dipole moment $p > 0$ (pusher). Lengths are scaled in units of d, and $p/(8\pi\eta d^2) = 1$. *Left* Near-field streamlines of the oppositely oriented Stokeslets. *Right* Streamlines of the ideal force dipole. The two separatrix lines $z = \pm x/\sqrt{2}$ separate the sectors of in- and outflowing fluid. Reproduced from the COMPLOIDS book [9] with kind permission of the Societa Italiana di Fisica

V is the swimming speed and L the axial length of the elongated swimmer. Notice here the distinct differences in the linear Stresslet field of a dipolar microswimmer, and the source doublet potential flow field of an auto-phoretic microswimmer. The Stresslet velocity field is in particular longer-ranged than the potential flow velocity field of a phoretic swimmer.

Microswimmers with a positive dipole moment are called *pushers*. They have their active propelling part on their rear side, and as seen in Fig. 10.28 they push the fluid out along the long (swimming) axis (repulsive flow) and draw fluid in on their side (attractive flow field). The aforementioned microrganisms are all pushers in addition to many types of bacteria. Microswimmers with a negative dipole moment ($p < 0$) are termed *pullers*. Their streamline pattern is the same as that for pushers, however with the flow direction reversed since the two Stokeslets are now pointing towards each other. Pushers pull in fluid along their long swimming sides (attractive flow) and push it out at their side (repulsive flow). An example of a puller is the green algae *Chlamydonamas rheinhardtii* which swims with two head-sided flagella in a breast-stroke-like motion.

10.6.3 Hydrodynamic Interactions Between Swimmers

Two pushers swimming side-by-side attract each other, while being repulsive if swimming one behind the other into the same direction. While this behaviour can be obviated qualitatively from the dipole flow pattern for $p > 0$ depicted in Fig. 10.27, it can be more quantitatively discussed by considering the motion of one force dipole in the incident dipolar flow field of the other one. On employing the

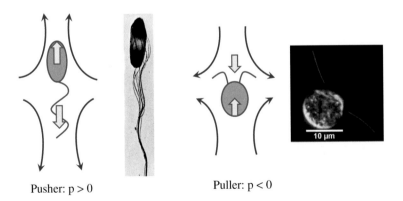

Pusher: p > 0   Puller: p < 0

Fig. 10.28 *Left* Schematics of the far-distant flow and associated linear force dipole (*arrows*) of a pusher ($p > 0$) with the *Salmonella* bacterium as an example. *Right* Schematics of a puller ($p < 0$) with the green algae *Chlamydomonas* as an example. The swimming direction (unit vector) **e** is here in the upward direction. Reproduced from the COMPLOIDS book [9] with kind permission of the Societa Italiana di Fisica

translational Faxén theorem for point-like torque- and force-free objects (see
Eq. (10.90)), the velocity increments, $\mathbf{V}_i$, of the two equal-moment dipoles at
positions $\mathbf{R}_i$ are

$$\mathbf{V}_1 = \mathbf{u}_{\mathrm{D}}(\mathbf{R}_{12}; \mathbf{e}_2) = \frac{p}{8\pi\eta(R_{12})^2}[3(\hat{\mathbf{R}}_{12}\cdot\mathbf{e}_2)^2 - 1]\hat{\mathbf{R}}_{12} \tag{10.121}$$

$$\mathbf{V}_2 = \mathbf{u}_{\mathrm{D}}(\mathbf{R}_{21}; \mathbf{e}_1), \tag{10.122}$$

respectively, with $\mathbf{R}_{ij} = \mathbf{R}_i - \mathbf{R}_j$ and $i, j \in \{1,2\}$.

If the two swimmers move momentarily side-by-side in the same direction so
that $\hat{\mathbf{R}}_{12}\cdot\mathbf{e}_i = 0$, they acquire the relative velocity increment

$$\mathbf{V}_1 - \mathbf{V}_2 = -\frac{p}{4\pi\eta(R_{12})^2}\hat{\mathbf{R}}_{12}, \tag{10.123}$$

expressing that pusher 1 is hydrodynamically attracted to pusher 2, and vice versa.
To see this just view the relative motion in the rest frame of pusher 2 where $\mathbf{V}_2 = \mathbf{0}$.
For pusher 2 following pusher 1, we have $(\hat{\mathbf{R}}_{12}\cdot\mathbf{e}_i)^2 = 1$, and the velocity difference describes now the hydrodynamic attraction of the two swimmers. The opposite
trends apply to two pullers owing to their negative force dipole moments.

This is not the whole story for arbitrary orientations and positions of the two
swimmers. The flow field $\mathbf{u}_{\mathrm{D}}(\mathbf{r};\mathbf{e}_2)$ created by swimmer 2 at the centre position $\mathbf{R}_1$ of
swimmer 1 has rotating and straining parts, $(\nabla \times \mathbf{u}_{\mathrm{D}})$ and $\mathbf{E}_{\mathrm{D}} = \frac{1}{2}(\nabla\mathbf{u}_{\mathrm{D}} + (\nabla\mathbf{u}_{\mathrm{D}})^T)$,
respectively. Swimmer 1 exposed to the flow field of swimmer 2 has thus the
tendency to align itself with the principal axis (dilation axis) of the strain field part
$\mathbf{E}_{\mathrm{D}}(\mathbf{R}_{12}; \mathbf{e}_2)$ of swimmer 2. For a quantitative analysis, let us model swimmer 1
geometrically overall as a force-free and torque-free prolate spheroid of aspect ratio
$\Gamma > 1$, with the long-axis orientation unit vector $\mathbf{e}_1$. A general rotational Faxén
theorem for a no-slip spheroid states [14] that if the spheroid with its centre at
position $\mathbf{R}_1$ is subjected to an arbitrary incident Stokes flow field $\mathbf{u}_{\mathrm{inc}}(\mathbf{r})$, it will rotate
with an angular velocity $\boldsymbol{\Omega}_1$ given by

$$\boldsymbol{\Omega}_1 = \frac{1}{2}(\nabla \times \mathbf{u}_{\mathrm{inc}})(\mathbf{R}_1) + \left(\frac{\Gamma^2 - 1}{\Gamma^2 + 1}\right)\mathbf{e}_1 \times (\mathbf{E}_{\mathrm{inc}}(\mathbf{R}_1)\cdot\mathbf{e}_1) + \cdots. \tag{10.124}$$

The dots annotate that higher order derivative contributions of the incident flow
field are neglected. A freely advecting sphere for which $\Gamma = 1$ rotates thus with half
the vorticity of the incident flow $\mathbf{u}_{\mathrm{inc}}$, taken at the sphere center. An elongated body
has an additional angular velocity part due to the shear strain part of the incident
flow. This additional angular velocity part is oriented perpendicular to the long
body axis unit vector $\mathbf{e}_1$ of swimmer 1.

Substitution of $\mathbf{u}_{\mathrm{D}}(\mathbf{R}_{12};\mathbf{e}_2)$ for the incident flow into Eq. (10.124) reveals that
two nearby pushers on a converging course reorient each other hydrodynamically
into a parallel side-by-side configuration. As depicted in the left part of Fig. 10.29,

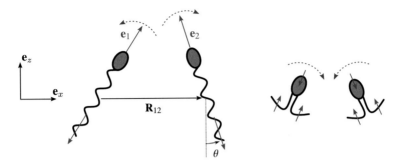

Fig. 10.29 *Left* Two pushers ($p > 0$) on a not too diverging course attract each other hydrodynamically, and reorient each other into a parallel side-by-side motion. *Right* Two pullers ($p < 0$) on a diverging course reorient each other towards an antiparallel configuration, swimming subsequently away from each other in anti-parallel direction (see also [108])

if two pushers located in the $x - z$ plane are separated by the distance $h = R_{12}$, and symmetrically oriented with inclination angles $\pm\theta$ relative to the z axis, their reorientation into a parallel configuration takes place with the angular velocities $\Omega_{y,12} \sim \mp p\theta/(\eta h^3)$ [108]. In contrast, and owing to the opposite flow fields, two pullers on a diverging course align each other in an antiparallel configuration, swimming subsequently away from each other. This is illustrated in the right part of Fig. 10.29.

It should be recalled that the analysis presented here is based on the leading-order singularity flow solutions. It applies in principle to inter-swimmer distances h large compared to the elongational swimmer size L only, although it is often found to be quite accurate even for distances comparable to L [109]. For two closely moving swimmers, the details of their shapes and propulsion mechanisms play a role. This requires then a refined hydrodynamic modeling and more elaborate methods to determine the swimmer dynamics. These methods include multiparticle collision dynamics (MPCD) simulations of bacteria and sperm cells [109], bead-modeling of complex-shaped swimmers combined with Stokesian dynamics simulations [112], and numerical boundary integral equation methods invoking particle surface triangularization [89].

10.6.4 Swimming Near a Surface

For a dipolar swimmer above a planar wall or surface ($x - y$ plane at $z = 0$), the flow field is a superposition of its dipolar flow field $\mathbf{u}_D$ and an image flow field, $\mathbf{u}_{Im}$, generated by a system of singularities located below the surface. The image flow contribution is required to satisfy the surface BC (recall Sect. 10.5). In Fig. 10.30, this situation is illustrated for the simplest case of a pusher swimming above a free surface. The only BC here is the fluid-impermeability of the surface which can be

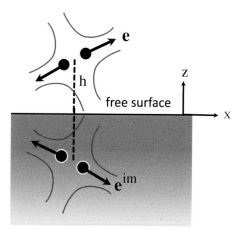

Fig. 10.30 Swimming with an image: A free surface attracts a puller and reorients its axis parallel to the surface ($\theta = 0$ with $\mathbf{e} \cdot \mathbf{e}_x = \cos \theta$). A puller on the other hand is reoriented perpendicular to the surface, swimming subsequently either away or head-on into the surface. This qualitative behaviour remains valid for a Navier partial-slip wall, a no-slip wall, and a liquid interface [26]

fulfilled by considering the swimmer in the half-space $z > 0$ to move along with its mirror image in the fluid extended to the lower half-space $z < 0$. This is akin to the symmetric side-by-side motion of two swimmers in bulk fluid discussed earlier in relation to Fig. 10.29.

Using again the translational Faxén theorem for a point-like freely advected particle, the vertical velocity component induced on the swimmer at $\mathbf{R}_0 = (0,0,z_0)$ with $z_0 = h > 0$ is

$$V_z(\theta, h) = u_{\mathrm{D},z}(\mathbf{R}_0 - \mathbf{R}_0^*; \mathbf{e}^{im}) = -\frac{p}{32\pi\eta h^2}\left[1 - 3\sin^2\theta\right], \qquad (10.125)$$

where θ is the tilt angle of the swimmer with respect to the surface, so that $\sin\theta = \mathbf{e} \cdot \mathbf{e}_z = -\mathbf{e}^{im} \cdot \mathbf{e}_z$. Here, $\mathbf{R}_0^*$ and $\mathbf{e}^{im}$ are the position and orientation vectors of the mirror dipole, respectively. Provided the tilt angle is not too large so that $\theta < \arcsin(1/\sqrt{3})$, the dipole is attracted by the surface. To reveal the influence of the free surface (i.e. of the image dipole flow part) on the swimmer's orientation, we employ the rotational Faxén law where the swimmer is described as an elongated spheroid. The result is [26]

$$\Omega_y(\theta, h) = \frac{3p\sin(2\theta)}{128\pi\eta h^3}\left[1 + \left(\frac{\Gamma^2 - 1}{\Gamma^2 + 1}\right)\sin^2(\theta)\right], \qquad (10.126)$$

according to which for all values of θ, the pusher always aligns parallel to the free surface, with the induced velocity in this aligned configuration being equal to

$V_z(0,h) = -p/(32\pi\eta h^2)$. Note that $\Gamma \gg 1$ for a typical bacterium such as *E. coli*, owing to its extended flagella bundle.

Using the respective image flows, the calculations outlined above for a free surface are rather straightforwardly extended to a clean liquid interface and a Navier partial-slip wall, respectively, with the no-slip wall included as a limiting case [26, 113]. For a clean liquid-liquid interface, e.g., the corresponding result is

$$\Omega_y(\theta,h) = \frac{3p\sin(2\theta)}{128\pi\eta_1 h^3}\left[1 + \frac{1}{2}\left(\frac{\Gamma^2 - 1}{\Gamma^2 + 1}\right)\left(\frac{\lambda + (2+\lambda)\sin^2(\theta)}{1+\lambda}\right)\right] \qquad (10.127)$$

$$V_z(0,h) = -\frac{p(2+3\lambda)}{32\pi\eta_1(1+\lambda)h^2}, \qquad (10.128)$$

with the free and no-slip surface results recovered as the limiting cases for $\lambda = \eta_2/\eta_1 = 0$ and $\lambda \to \infty$, respectively. A pusher is always attracted by and aligned to a nearby liquid interface, for any value of the viscosity ratio. The pusher is likewise attracted and aligned parallel by a partial-slip wall, for any value of the Navier slip length [26].

The attraction by and the accumulation of biological microswimmers near surfaces is indeed observed in many biological experiments. According to the above analysis, pullers are oriented hydrodynamically in the direction perpendicular to the surface ($\theta = \pm\pi/2$), swimming either head-on away or right into the surface.

The simple modeling of pushers as linear force dipoles gives the prediction that they should move in straight trajectories along a surface, owing to the rotational symmetry of the aligned dipole with respect to the surface normal. However, it is known experimentally that bacteria such as *E. coli* do swim in clockwise (CW) traversed circles along a glass surface [114], as viewed from inside the fluid, whereas near a clean free surface the spherical trajectories are traversed counter-clockwise (CCW). The CW circular motion near the glass plate can be changed to a CCW motion if a sufficient amount of free polymers is added to the fluid. Likewise, the CW circular motion at a clean free surface can change to a CCW motion upon the addition of detergents accumulating at the surface. The circular motion of a bacterium is interrupted when a tumbling event occurs.

This surface-specificity of the trajectories can be attributed to the chiral bacterial propulsion mechanism not resolved in the simple force dipole model. It becomes essential when the swimmer gets close to a surface. To understand qualitatively the CCW circular motion of a pusher near a free surface, recall that this is equivalent to a pusher swimming in the flow field of its image. The image moves with the same speed as the actual pusher, but the helical flagellar bundle (cell body) of the image rotates opposite to that of the pusher. Recall that the bundle of a bacterium rotates counterclockwise if viewed from behind, and suppose that the bacterium is oriented momentarily along the *y*-axis. The counter-rotating flagellar helices (cell bodies) of the swimmer and its image create then a disturbance flow directed in the positive (negative) *x*-direction. Since the swimmer is freely advecting in this flow, it will perform a CCW circular motion [115]. Qualitative arguments similar to the present ones can be given to explain the CW circular motion of a bacterium at a no-slip surface where an advection flow oppositely directed to that near a free surface is created [108].

As Lopez and Lauga have recently shown [26], the surface-specific circular motion of a bacterium is quantitatively explainable on the singularity method level by adding a so-called rotlet dipole singularity solution, $\mathbf{u}_{RD}(\mathbf{r}) \sim \mathcal{O}(r^{-3})$, to the linear force dipole solution of $\mathcal{O}(r^{-2})$. The rotlet dipole part accounts on the long-distance level for the counter-rotating flagellum and cell body parts of the torque-free pusher. Using this flow singularity model in conjunction with surface-specific image systems, interesting quantitative predictions regarding critical parameter values for the CW to CCW motion transition for partial-slip walls and surfactant-covered interfaces have been made [26].

10.7 Concluding Remarks

The present notes give an introduction into the world of low-Reynolds-number flows and associated passive and active (i.e. self-propelling) microparticle motions, with the focus on surface and interfacial effects. As the topics treated in the notes should have amply illustrated, studying low-Reynolds-number phenomena is important, and often leads to surprising findings. Our aim has been to provide the reader with basic background knowledge which facilitates further reading of more advanced research texts on processes involving microparticles and animalcules suspended in viscous fluids. We have presented the governing equations of Stokes flows and their essential properties in a rather descriptive way, avoiding detailed calculations which can be found in more specialized textbooks and overview papers. While the material is presented in a rather systematic way, many important phenomena such as Marangoni surface flows [23], hydrodynamic screening near boundaries [116], wall-induced apparent like-charge attraction of colloidal macro-ions [117, 118], and hydrodynamically induced surface accumulation of microparticles [55, 119, 120] have been only shortly addressed, with an occasional reference to related literature, or even not mentioned at all. The present notes can serve as a good preparation for an improved understanding of research papers on these additional phenomena.

Acknowledgments These notes include and complement material which was presented in August 2014 at the SOMATAI summer school in Berlin, and in March 2014 at the SOMATAI workshop in Jülich. It is our pleasure to thank the organizers of these events, and here in particular Peter Lang (ICS-3, FZ Jülich), for having invited us to present a lecture on the colloidal hydrodynamics of microparticles and associated interfacial effects. We thank Maria Ekiel-Jeżewska (Polish Academy of Sciences, Warsaw) for having provided us with the streamlines figures of a point force near a planar interface, and Jonas Riest and Rafael Roa (ICS-3, FZ Jülich) and Roland Winkler (IAS-2, FZ Jülich) for helpful discussions. Moreover, we are grateful to Ulrike Nägele (FZ Jülich, ICS-3) for her help with travel and accommodation.

References

1. J.K.G. Dhont, *An Introduction to Dynamics of Colloids.* Elsevier, Amsterdam (1996)
2. P. Szymczak, M. Cieplak, J. Phys. Condens. Matter. **23**, 033102 (2011)
3. R.G. Larson, J.J. Magda, Macromolecules **22**, 3004 (1989)
4. H. Tanaka, J. Phys.: Condens. Matter **13**, 4637 (2001)
5. P. Wojtaszczyk, J.B. Avalos, Phys. Rev. Lett. **80**, 754 (1998)
6. C. Korn, U.S. Schwarz, Phys. Rev. Lett. **97**, 138103 (2006)
7. O.B. Usta, J.E. Butler, A.J.C. Ladd, Phys. Fluids **18**, 031703 (2006)
8. E. Guazzelli, J.F. Morris, *A Physical Introduction to Suspension Dynamics* (Cambridge University Press, Cambridge, 2012)
9. G. Nägele, Colloidal Hydrodynamics, in *Physics of Complex Colloids*, ed. by C. Bechinger, F. Sciortino, P. Ziherl. Proceedings of the International School of Physics "Enrico Fermi", vol. 184 (IOS Press, Amsterdam; SIF, Bologna, 2012), p. 451
10. G. Nägele, Dynamics of charged-particles dispersions, in *Proceedings of the 5th Warsaw School of Statistical Physics* (Warsaw University Press, 2014), p. 83
11. E.J. Hinch, Hydrodynamics at low Reynolds numbers: A brief and elementary introduction, in *Disorder and Mixing*, vol. 2, ed. by E. Guyon, J.-P. Nadal, Y. Pomeau (Springer, Dordrecht, 1988), pp. 43–55
12. P.N. Pusey, Colloidal suspensions, in *Liquids, Freezing and Glass Transition*, ed. by J. P. Hansen, D. Levesque, J. Zinn-Justin (Elsevier, Amsterdam, 1991), p. 763
13. R.B. Jones, P.N. Pusey, Annu. Rev. Phys. Chem. **42**, 137 (1991)
14. S. Kim, S.J. Karrila, *Microhydrodynamics: Principles and Selected Applications* (Butterworth-Heinemann, Boston, 1991)
15. J. Happel, H. Brenner, *Low Reynolds Numbers Hydrodynamics* (Kluwer, Dordrecht, 1991)
16. Z. Zapryanov, S. Tabakova, *Dynamics of Bubbles, Drops and Rigid Particles, Fluid Mechanics and Its Applications* (Springer, Dordrecht, 2011)
17. G.K. Batchelor, *An Introduction to Fluid Dynamics.* Cambridge Mathematical Library (Cambridge University Press, Cambridge, 2000)
18. L.D. Landau, E.M. Lifshitz, *Fluid Mechanics, Course of Theoretical Physics* (Pergamon Press, London, 1987)
19. E. Guyon, J.P. Hulin, L. Petit, *Physical Hydrodynamics* (Oxford University Press, Oxford, 2001)
20. G.I. Taylor, *Low Reynolds Number Flows.* National Committee For Fluid Mechanics Films (1996), http://web.mit.edu/hml/ncfmf.html
21. W.B. Russel, D.A. Saville, W.R. Schowalter, *Colloidal Dispersions, Cambridge Monographs on Mechanics* (Cambridge University Press, Cambridge, 1989)
22. E. Lauga, M.P. Brenner, H.A. Stone, The no-slip boundary condition, in *Springer Handbook of Experimental Fluid Mechanics*, ed. by C. Tropea, A. Yarin, J.F. Foss (Springer, Berlin, 2007
23. L.G. Leal, *Laminar Flow and Convective Transport Processes* (Butterworth-Heinemann, Boston, 1992
24. N. Lecoq, R. Anthore, B. Cichocki, P. Szymczak, F. Feuillebois, J. Fluid Mech. **513**, 247 (2004)
25. R. Tuinier, T. Taniguchi, J. Phys.: Condens. Matter **17**, L9 (2005)
26. D. Lopez, E. Lauga, Phys. Fluids **26**, 071902 (2014)
27. K.H. Lan, N. Ostrowsky, D. Sornette, Phys. Rev. Lett. **57**, 17 (1986)
28. B. Cichocki, M.L. Ekiel-Jezewska, E. Wajnryb, J. Chem. Phys. **136**, 071102 (2012)
29. J.F. Brady, G. Bossis, Annu. Rev. Fluid Mech. **20**, 111 (1988)
30. I.M. Jánosi, T. Tél, D.E. Wolf, J.A.C. Gallas, Phys. Rev. E **56**, 2858 (1997)
31. M.L. Ekiel-Jeżewska, E. Wajnryb, Phys. Rev. E **83**, 067301 (2011)
32. D.L. Ermak, J.A. McCammon, J. Chem. Phys. **69**, 1352 (1978)

33. G. Nägele, Brownian Dynamics simulations, in *Computational Condensed Matter Physics*, vol. 32 (Forschungszentrum Jülich Publishing, 37th IFF Spring School edition, 2006)
34. A.T. Chwang, T. Wu, J. Fluid Mech. **67**, 787 (1975)
35. C. Pozrikidis, *Boundary Integral and Singularity Methods for Linearized Viscous Flow* (Cambridge University Press, Cambridge, 1992)
36. J.D. Jackson, *Classical Electrodynamics*, 3rd edn. (Wiley, New York, 1998
37. B. Cichocki, B.U. Felderhof, K. Hinsen, J. Chem. Phys. **100**, 3780 (1994)
38. M.L. Ekiel-Jeżewska, E. Wajnryb, Precise multipole method for calculating hydrodynamic interactions between spherical particles in the stokes flow, in *Theoretical Methods for Micro Scale Viscous Flows*, ed. by F. Feuillebois, A. Sellier (2009), pp. 127–172
39. M. Lisicki, arXiv:1312.6231 [physics.flu-dyn] (2013)
40. L.G. Leal, Annu. Rev. Fluid Mech. **12**, 435 (1980)
41. S.E. Spagnolie, E. Lauga, J. Fluid Mech. **700**, 105 (2012)
42. H. Luo, C. Pozrikidis, J. Eng. Math. **62**, 1 (2007)
43. G.K. Batchelor, J. Fluid Mech. **44**, 419 (1970)
44. M.M. Tirado, C.L. Martinez, J.G. de la Torre, J. Chem. Phys. **81**, 2047 (1984)
45. R.G. Cox, J. Fluid Mech. **44**, 791 (1970)
46. J.B. Keller, S.I. Rubinow, J. Fluid Mech. **75**, 705 (1976)
47. G.G. Stokes, Trans. Camb. Philos. Soc. **9**, 8 (1851)
48. J.L. Anderson, Annu. Rev. Fluid Mech. **21**, 61 (1989)
49. J.H. Masliyah, S. Bhattacharjee, *Electrokinetic and Colloid Transport Phenomena* (Wiley, New York, 2006
50. H.J. Keh, J.L. Anderson, J. Fluid Mech. **153**, 417 (1985)
51. A.S. Khair, T.M. Squires, Phys. Fluids **21**, 042001 (2009)
52. R. Golestanian, T.B. Liverpool, A. Ajdari, New J. Phys. **9**, 126 (2007)
53. H.A. Stone, A.D. Samuel, Phys. Rev. Lett. **77**, 4102 (1996)
54. C. Contreras-Aburto, G. Nägele, J. Chem. Phys. **139**, 134110 (2013)
55. F. Weinert, D. Braun, Phys. Rev. Lett. **101**, 168301 (2008)
56. J. Rotne, S. Prager, J. Chem. Phys. **50**, 4831 (1969)
57. H. Yamakawa, J. Chem. Phys. **53**, 436 (1970)
58. E. Wajnryb, K.A. Mizerski, P.J. Zuk, P. Szymczak, J. Fluid Mech. **731**, R3 (2013)
59. R. Courant, D. Hilbert, *Methods of Mathematical Physics II* (Interscience, New York, 1962)
60. F.M. Weinert, D. Braun, Phys. Rev. Lett. **101**, 168301 (2008)
61. P.J. Zuk, E. Wajnryb, K.A. Mizerski, P. Szymczak, J. Fluid Mech. **741**, R5 (2014)
62. B. Carrasco, J. Garcia de la Torre, Biophys. J. **76**, 3044 (1999)
63. B. Cichocki, R.B. Jones, R. Kutteh, E. Wajnryb, J. Chem. Phys. **112**, 2548 (2000)
64. S. Bhattacharya, J. Blawzdziewicz, E. Wajnryb, Physica A **356**, 294 (2005)
65. M. Kedzierski, E. Wajnryb, J. Chem. Phys. **133**, 154105 (2010)
66. D.J. Acheson, *Elementary Fluid Dynamics* (Oxford University Press, Oxford, 1990)
67. D.J. Jeffrey, Y. Onishi, J. Fluid Mech. **139**, 261 (1984)
68. R. Tadmor, J. Phys.: Condens. Matter **13**, L195 (2001)
69. M.L. Ekiel-Jeżewska, R. Boniecki, Stokes Flow generated by a point force in various geometries II. Velocity field, Technical report, IFTR. Polish Acad. Sci. (2010)
70. B. Cichocki, M.L. Ekiel-Jeżewska, G. Nägele, E. Wajnryb, J. Chem. Phys. **121**, 2305 (2004)
71. J. Blake, Proc. Camb. Philos. Soc. **70**, 303 (1971)
72. K. Aderogba, J.R. Blake, Bull. Aust. Math. Soc. **18**, 345 (1978)
73. J. Blake, A. Chwang, J. Eng. Math. **8**, 23 (1974)
74. E. Lauga, T.M. Squires, Phys. Fluids **17**, 103102 (2005)
75. M. Lisicki, B. Cichocki, S.A. Rogers, J.K.G. Dhont, P.R. Lang, Soft Matter **10**, 4312 (2014)
76. B. Cichocki, R.B. Jones, Phys. A **258**, 273 (1998)
77. H.A. Lorentz, *Abhandlung über Theoretische Physik* (B. G. Teubner, Leipzig, 1907)
78. H. Faxén, Ark. Mat. Astron. Fys. **17**, 1 (1923)
79. H. Brenner, Chem. Eng. Sci. **16**, 242 (1961)
80. A.J. Goldman, R.G. Cox, H. Brenner, Chem. Eng. Sci. **22**, 637 (1967)

81. A.J. Goldman, R.G. Cox, H. Brenner, Chem. Eng. Sci. **22**, 653 (1967)
82. W. Dean, M. O'Neill, Mathematika **10**, 13 (1963)
83. W. Dean, M. O'Neill, Mathematika **11**, 67 (1964)
84. R.B. Jones, J. Chem. Phys. **123**, 164705 (2005)
85. B. Lin, J. Yu, S. Rice, Phys. Rev. E **62**, 3909 (2000)
86. R. Sadr, C. Hohenegger, H. Li, P.J. Mucha, M. Yoda, J. Fluid Mech. **577**, 443 (2007)
87. P. Huang, K. Breuer, Phys. Rev. E **76**, 046307 (2007)
88. L. Lobry, N. Ostrowsky, Phys. Rev. B **53**, 12050 (1996)
89. K. Ishii, T. Iwai, H. Xia, Opt. Express **18**, 7390 (2010)
90. M.A. Plum, W. Steffen, G. Fytas, W. Knoll, B. Menges, Opt. Express **17**, 10364 (2009)
91. M.A. Plum, J. Řička, H.-J. Butt, W. Steffen, New J. Phys. **12**, 103022 (2010)
92. P. Holmqvist, J.K.G. Dhont, P.R. Lang, Phys. Rev. E **74**, 021402 (2006)
93. P. Holmqvist, J.K.G. Dhont, P.R. Lang, J. Chem. Phys. **126**, 044707 (2007)
94. M. Hosoda, K. Sakai, K. Takagi, Phys. Rev. E **58**, 6275 (1998)
95. M. Lisicki, B. Cichocki, J.K.G. Dhont, P.R. Lang, J. Chem. Phys. **136**, 204704 (2012)
96. S.A. Rogers, M. Lisicki, B. Cichocki, J.K.G. Dhont, P.R. Lang, Phys. Rev. Lett. **109**, 098305 (2012)
97. B.J. Berne, R. Pecora, *Dynamic Light Scattering: With Applications to Chemistry, Biology, and Physics*. Dover Books on Physics Series (Dover Publications, Mineola, 2000
98. R. Sigel, Curr. Opin. Colloid Interface Sci. **14**, 426 (2009)
99. G. Nägele, Phys. Rep. **272**, 215 (1996)
100. B. Cichocki, M.L. Ekiel-Jeżewska, E. Wajnryb, J. Chem. Phys. **140**, 164902 (2014)
101. W.B. Russel, E.J. Hinch, L.G. Leal, G. Tieffenbruck, J. Fluid Mech. **83**, 273 (1977)
102. S.-M. Yang, L.G. Leal, J. Fluid Mech. **136**, 393 (1983)
103. E.P. Ascoli, D.S. Dandy, L.G. Leal, J. Fluid Mech. **213**, 287 (1990)
104. I. Cantat, C. Misbah, Phys. Rev. Lett. **83**, 880 (1999)
105. U.S. Agarwal, A. Dutta, R.A. Mashelkar, Chem. Eng. Sci. **49**, 1693 (1994)
106. C. Berdan, L.G. Leal, J. Colloid Interface Sci. **87**, 62 (1982)
107. R. Trouilloud, T. Yu, A. Hosoi, E. Lauga, Phys. Rev. Lett. **101**, 048102 (2008)
108. E. Lauga, T.R. Powers, Rep. Prog. Phys. **72**, 096601 (2009)
109. J. Elgeti, R.G. Winkler, G. Gompper, Rep. Prog. Phys. **78**, 056601 (2015)
110. J. Lighthill, *Mathematical Biofluiddynamics* (SIAM, Philadelphia, 1975)
111. E.M. Purcell, Am. J. Phys. **45**, 3 (1977)
112. J.W. Swan, J.F. Brady, R.S. Moore, Phys. Fluids **23**, 071901 (2011)
113. A.P. Berke, L. Turner, H.C. Berg, E. Lauga, Phys. Rev. Lett. **101**, 038102 (2008)
114. E. Lauga, W.R. DiLuzio, G.M. Whitesides, H.A. Stone, Biophys. J. **90**, 400 (2006)
115. R. Di Leonardo, D. Dell'Arciprete, L. Angelani, V. Iebba, Phys. Rev. Lett. 038101 (2011)
116. P.P. Lele, J.W. Swan, J.F. Brady, N.J. Wagner, E.M. Furst, Soft Matter **7**, 6844 (2011)
117. T. Squires, M. Brenner, Phys. Rev. Lett. **85**, 4976 (2000)
118. T.M. Squires, J. Fluid Mech. **443**, 403 (2001)
119. R. Di Leonardo, F. Ianni, G. Ruocco, Langmuir **25**, 4247 (2009)
120. J. Morthomas, A. Würger, Phys. Rev. E **81**, 051405 (2010)

Part IV
Experimental Techniques

Chapter 11
Advanced Light Scattering Techniques

Andreas Charles Völker, Andreas Vaccaro and Frédéric Cardinaux

Abstract Three popular light scattering techniques used for characterization of soft matter samples such as colloidal suspensions or emulsions are presented. Static and dynamic light scattering, as well as diffusing wave spectroscopy will be discussed in detail. These techniques allow determination of a wide range of important soft matter properties, including particle size, size distribution, microrheology and molecular weight, to mention only a few examples. These can significantly influence the properties and dynamics of an interface directly or indirectly. They are thus vital tools to study soft matter at aqueous interfaces. Since the theoretical fundamentals of the presented light scattering techniques are related, we will discuss them in a comprehensible order while presenting some selected applications for each technique, with a focus on the most recent improvements. We will also discuss the most important technical considerations for each technique.

11.1 Introduction

The light scattering techniques presented in this chapter are typically used for characterization of soft matter samples such as colloidal suspensions or emulsions. Static and dynamic light scattering, as well as diffusing wave spectroscopy, are discussed. These techniques allow determination of a wide range of important soft matter properties such as particle size and size distribution, microrheology or molecular weight, to mention only a few examples. While none of these properties characterize an interface on their own, they can significantly influence the properties and dynamics of an interface directly or indirectly. They are thus vital tools to study soft matter at aqueous interfaces. Since the theoretical fundamentals of the presented light scattering techniques are related, we will discuss them in a

A.C. Völker (✉) · A. Vaccaro · F. Cardinaux
LS Instruments AG, Fribourg, Switzerland
e-mail: info@lsinstruments.ch

© Springer International Publishing Switzerland 2016
P.R. Lang and Y. Liu (eds.), *Soft Matter at Aqueous Interfaces*,
Lecture Notes in Physics 917, DOI 10.1007/978-3-319-24502-7_11

comprehensible order while presenting some selected applications for each technique, with a focus on the most recent advanced developments. We will also give the most important technical considerations for each technique.

11.2 Static Light Scattering Fundamentals

In a Static Light Scattering (SLS) experiment, a collimated monochromatic beam of light illuminates a colloidal sample and the intensity I of the scattered light is measured as a function of the detection angle θ. This allows determination of the molecular mass M, the radius of gyration R_g which is a measure of the size of the colloidal particles, and the (osmotic) second virial coefficient B_2 which describes the magnitude of the interaction among the particles.

A typical SLS experimental setup consists of a monochromatic light beam, be it an Hg-arc lamp or a monochromatic laser, and a detector whose position defines the scattering angle θ. The incident light and the scattering angle define the scattering plane (Fig. 11.1). Customarily used detectors are photomultiplier tubes (PMTs), avalanche photodiodes (APDs), or standard photodiodes. The first two types are normally mounted on a goniometer arm that rotates concentrically around the sample and thus allows measurements at different scattering angles. An advantage of using APDs and PMTs, despite their relatively high cost compared to standard photodiodes, is that dynamic light scattering measurements can be simultaneously performed.

To discuss the physical properties of an SLS experiment, we employ standard electromagnetic wave theory [1]. Upon interaction of the incident light—an electromagnetic wave—with the electronic cloud of the molecules constituting the colloidal particles, the electronic cloud oscillates. This oscillation corresponds to an oscillating current, which results in the emission of light. It is scattering. If there is no loss of energy in the scattering process, the latter is said to be elastic and the scattered light has the same wavelength λ as that of the incident light. The corresponding physics is well described by the classic equation of the electric field radiated by an oscillating dipole found in standard electromagnetics textbooks.

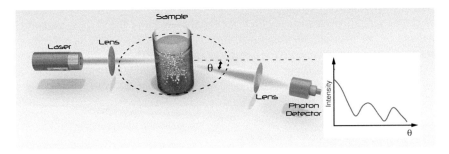

Fig. 11.1 SLS experiment including a photon detector that can rotate concentrically around the sample [2] (Figure reprinted with permission. Copyright by LS Instruments AG, Switzerland)

For a colloidal particle of volume dV suspended in a medium of refractive index n_s, much smaller than λ, the expression for the radiated electric field reads [1]

$$d\mathbf{E}_s = \frac{k^2}{n_s} \frac{\exp(ikR)}{4\pi R} \hat{\mathbf{e}} \times \mathbf{p}(t) \times \hat{\mathbf{r}}, \tag{11.1}$$

where $k = 2\pi/\lambda$ is the module of the wave-vector $\mathbf{k}_i$ of the incident light, $\hat{\mathbf{e}}$ is the unit vector indicating the scattering direction, and R is the distance of the detector from the particle. By means of the polarizability $d\alpha$, the oscillating dipole moment $\mathbf{p}$ (t) can be expressed in terms of the incident field $\mathbf{E}_i(t) = \mathbf{E}_0 \exp[i(\mathbf{k}_i \cdot \mathbf{r} - \omega t)]$ at the position $\mathbf{r}$ of the particle as follows:

$$\mathbf{p}(t) = n_s^2 d\alpha \mathbf{E}_i(t). \tag{11.2}$$

For objects smaller than λ and with refractive index n, the polarizability can be calculated by means of the Lorentz-Lorenz equation as

$$d\alpha = 3dV \frac{m^2 - 1}{m^2 + 2} \tag{11.3}$$

where $m \equiv \frac{n}{n_s}$ is the relative refractive index. Piecing everything together, the modulus of the scattered field can now be written as

$$dE_s = dV \frac{k^2}{4\pi n_s} \frac{m^2 - 1}{m^2 + 2} \frac{\exp[i(kR - \omega t)]}{R} E_0 \exp[i\mathbf{k}_i \cdot \mathbf{r}], \tag{11.4}$$

where E_0 is the module of $\mathbf{E}_0$. Referencing spatial positions with respect to a reference point $\mathbf{r}_0$, which corresponds to the variable substitutions $\mathbf{r} = \mathbf{r}' + \mathbf{r}_0$ and $R = R_0 - \mathbf{r}' \cdot \hat{\mathbf{r}}$, this yields

$$dE_s = E_0 dV \Delta\rho \frac{\exp[i(kR_0 + \mathbf{k}_i \cdot \mathbf{r}_0 - \omega t)]}{R_0} \exp[-i(\mathbf{k}_s - \mathbf{k}_i) \cdot \mathbf{r}'], \tag{11.5}$$

where we introduced the scattering length density, $\Delta\rho$ expressed as

$$\Delta\rho \equiv \frac{k^2}{4\pi n_s} \frac{m^2 - 1}{m^2 + 2}. \tag{11.6}$$

To obtain Eq. (11.5), we exploited the fact that in case of elastic scattering, the modules of the incident and scattered wave vectors are the same ($\mathbf{k}_s = k\hat{\mathbf{e}}$)

Equation (11.5) shows that the scattered field is proportional to the incident field module, the particle volume and the scattering length density. Furthermore, the phase shift due to the displacement of a particle from a reference position $\mathbf{r}_0$ to the position $\mathbf{r}' + \mathbf{r}_0$ is $-(\mathbf{k}_s - \mathbf{k}_i) \cdot \mathbf{r}'$. Hence the quantity $\mathbf{q} \equiv \mathbf{k}_s - \mathbf{k}_i$, known as scattering vector, takes a crucial role in the scattering theory as it represents an experimental quantity that determines the length scale at which interferences in the system take place. By means of simple geometrical considerations, its module is calculated as

$$q = \frac{4\pi n_s}{\lambda} \sin(\theta/2). \tag{11.7}$$

If we now consider a homogeneous colloidal particle of any size made up of many infinitesimally small sub regions dV, the total scattered field can be obtained by integrating Eq. (11.1) over the volume V of the particle:

$$E_s = E_0 \Delta\rho \frac{\exp[i(kR_0 + \mathbf{k}_i \cdot \mathbf{r}_0 - \omega t)]}{R_0} \int_V dV \exp[-i\mathbf{q} \cdot \mathbf{r}']. \tag{11.8}$$

In Eq. (11.8), we implicitly assumed that the incident field exciting each sub-element dV is unaffected by the presence of the surrounding elements. We thus assumed that we can neglect reflection of the incident light at the particle/solvent interface and that the scattering is so weak that the attenuation of the incident field is negligible. This assumption is known as the Rayleigh-Gans-Debye (RGD) approximation and can be mathematically formulated as

$$|1 - m| \ll 1 \quad a \ll \frac{\lambda}{2\pi n_s |1 - m|}. \tag{11.9}$$

Equation (11.8) is easily extended to a system of N identical particles by means of a simple summation

$$E_s = E_0 \Delta\rho \frac{\exp[i(kR_0 + \mathbf{k}_i \cdot \mathbf{r}_0 - \omega t)]}{R_0} \sum_j \int_{V_j} dV_j \exp\left[-i\mathbf{q} \cdot \mathbf{r}'_j\right]. \tag{11.10}$$

Using each individual particles' position $\mathbf{R}_j$ in a straightforward variable substitution $\mathbf{r}'_j = \mathbf{R}_j + \mathbf{r}_j$ simplifies the equation. This amounts to expressing the internal positions of each particle sub-element with respect to the particle position $\mathbf{R}_j$. Performing the variable substitution we obtain

$$E_s(t) = E_0 \frac{\exp[i(kR_0 + \mathbf{k}_i \cdot \mathbf{r}_0 - \omega t)]}{R_0} \sum_j b_j \exp\left[-i\mathbf{q} \cdot \mathbf{R}_j(t)\right] \tag{11.11}$$

where we defined the so-called scattering length of particle j as

$$b_j \equiv \Delta\rho \int_{V_j} dV_j \exp\left[-i\mathbf{q} \cdot \mathbf{r}'_j\right] \tag{11.12}$$

Equation (11.11) clearly illustrates how the internal interference, embodied in the particle's scattering length, and the interparticle interference factor, are accounted for by the scattering vector.

Equation (11.11) is, however, not yet useful for the interpretation of experimental data, since we do not measure the electric field but the temporal average of the corresponding scattered intensity. If we assume the sample to be ergodic, time averages can be replaced by ensemble averages indicated by the notation $\langle \cdot \rangle$ and we can write

$$I_s(\mathbf{q}) = \left\langle |E_s|^2 \right\rangle = \frac{I_0}{R_0^2} \left\langle \sum_{j=1}^{N} \sum_{k=1}^{N} b_j b_k^* \exp\left[-i\mathbf{q} \cdot (\mathbf{R}_j - \mathbf{R}_k) \right] \right\rangle \tag{11.13}$$

For a system composed by identical particles considerable math leads to the important result that

$$I_s(\mathbf{q}) = \frac{I_0}{R_0^2} N V^2 \Delta\rho^2 P(\mathbf{q}) S(\mathbf{q}) \tag{11.14}$$

where we introduced the particle form factor, $P(\mathbf{q})$ defined as

$$P(\mathbf{q}) = \frac{\langle |b(\mathbf{q})| \rangle}{\Delta\rho^2} \tag{11.15}$$

and the structure factor

$$S(\mathbf{q}) = \frac{1}{N} \left\langle \sum_{j=1}^{N} \sum_{k=1}^{N} \exp\left[-i\mathbf{q} \cdot (\mathbf{R}_j - \mathbf{R}_k) \right] \right\rangle \tag{11.16}$$

$P(\mathbf{q})$ accounts for the internal particle interference while $S(\mathbf{q})$ accounts for the interparticle interference. Note that Eq. (11.14) marks a crucial result of the RGD theory, because inter- and intra- particle interferences appear as independent multiplicative terms. The magnitude of the signal is thus proportional to the incident intensity, the number of scatterers, the square of the particle volume, and the square of a contrast term. Finally, it is important to note that the interference is fully described by a single experimental quantity, namely the scattering vector.

For small angles, and hence for small scattering vector modules ($R_g q \ll 1$), the following general expansion holds

$$P(\mathbf{q}) = 1 - \frac{R_g^2 q^2}{3} + O\left(R_g^4 q^4 \right) \tag{11.17}$$

where R_g is the optical radius of gyration of the colloidal particles under investigation. The radius of gyration effectively represents a measure of the size of the colloidal particles present in the suspension. Therefore, Eq. (11.17) allows estimation of particle size by means of a fit of the intensities collected as a function of the scattering angle (and hence the scattering vector module) at small angles.

Note that the structure factor depends on the sample concentration, as it accounts for the interparticle interference. In dilute samples, no interparticle correlation exists and the structure factor is equal to 1 for all angles. For more concentrated systems, however, depending on whether the interparticle interactions are attractive or repulsive, the zero angle value will be either larger or smaller than 1, respectively. This behavior is described by the following particle concentration expansion of $S(0)$

$$S(0) = \frac{1}{1 + 2\bar{B}_2 w + O(w^2)} \qquad (11.18)$$

Here w is the sample mass concentration and $\bar{B}_2$ the mass-based second virial coefficient. This equation therefore allows estimation of the second virial coefficient from a series of measurements performed at low angles and different particle concentrations.

11.2.1 Characterization of Macromolecules in Solution: Zimm Plot

In practice, we will have to consider scattering caused not only by particles, but also by the solvent. To account for this, we introduce the Rayleigh ratio $\mathcal{R}$, defined as the time-averaged scattering intensity per unit solid angle, unit scattering volume, and unit illumination intensity [3]. The Rayleigh ratio is a measure of the absolute scattering of the sample, and includes contributions from both the solvent and the macromolecules. Additional measurement of the pure solvent scattering ($\mathcal{R}_{sol}$) is necessary to isolate the macromolecules' contribution: $\mathcal{R}_{ex} = \mathcal{R} - \mathcal{R}_{sol}$. The excess Rayleigh ratio $\mathcal{R}_{ex}$ contains information about the macromolecules, and can be used to characterize systems such as polymers, surfactant aggregates, or proteins.

In the following, we present a method called Zimm-plot, from which the molecular weight M, the radius of gyration R_g, and the second virial coefficient B_2 of macromolecules in solution can be obtained [3, 4].

Based on the RGD theory, and considering the influence of the experimental setup, the excess Rayleigh Ration of a solution of macromolecules is [3–5]:

$$\mathcal{R}_{ex}(q, w) = KwM\, P(q)\, S(q, w), \qquad (11.19)$$

where K is a constant and w is the concentration of macromolecules $P(q)$ is the form factor and $S(q)$ the structure factor, as defined previously. The constant K requires prior knowledge about the solution's refractive index increment $\left(\frac{dn}{dw}\right)$ and is defined as:

$$K = \frac{\pi^2 n_s^2}{N_A \lambda^4} \left(\frac{dn}{dw}\right)^2, \tag{11.20}$$

where N_A is the Avogadro number. For $qR_g \ll 1$, the form factor can be approximated by the Guinier approximation [7]:

$$P(q) \approx 1 - \frac{R_g^2}{3} q^2, \tag{11.21}$$

In the limit of small q, the structure factor is related to the osmotic compressibility of the fluid. Moreover, an expansion for small concentrations of the compressibility allows to link the structure factor to a single interaction parameter called the second virial coefficient B_2 [5, 6]. The structure factor reads:

$$\lim_{w,q \to 0} S(q, w) = 1 - 2\bar{B}_2 w, \tag{11.22}$$

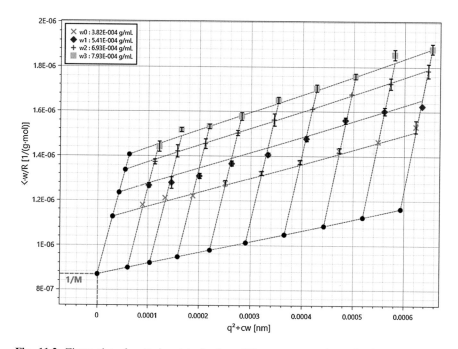

Fig. 11.2 Zimm plot of scattering data for four different concentrations of polystyrene chains dispersed in toluene and measured at a temperature of 25 °C. The data points are presented in color, while the extrapolations of the global fit to q = 0 and w = 0 are shown as black dots. We obtained a radius of gyration of 43 nm, a molecular weight of 1.15 10^6 g/mol and a second virial coefficient of 390 cm^3/g, consistent with the chain conformation expected for this well-documented model system [8] (Figure reprinted with permission. Copyright by LS Instruments AG, Switzerland)

Using the results of Eqs. (11.21) and (11.22), and rearranging the terms in Eq. (11.19) yield the following expression:

$$\frac{Kw}{\mathcal{R}_{ex}(q,w)} \approx \frac{1}{M}\left[1 + \frac{R_g^2}{3}q^2\right](1 + 2\bar{B}_2 w), \qquad (11.23)$$

where we used $(1-x)^{-1} \approx 1+x$ for small x.

A convenient way to obtain R_g, M, and B_2 using Eq. (11.23) is the so-called Zimm plot [3–6]. It is a schematic representation based on a series of SLS measurements at different scattering angles and different sample concentrations. The plot linearizes the q^2-dependence of $\mathcal{R}_{ex}$ while separating the dependence on q and w by plotting $Kw/\mathcal{R}_{ex}(q,w)$ versus $q^2 + cw$, where c is a constant to be chosen arbitrarily.

Figure 11.2 shows an example of Zimm plot for linear polystyrene chains dispersed in toluene. The intercept of the linear interpolation at fixed q yields the inverse of the molecular mass M, while the slope is proportional to the second virial coefficient. Similarly, the radius of gyration R_g is obtained by linear interpolation of the Zimm plot at a fixed concentration.

11.3 Fundamentals of Dynamic Light Scattering

Dynamic Light Scattering (DLS)—also known as Photon Correlation Spectroscopy (PCS) or Quasi-Elastic Light Scattering (QELS)—is an experimental technique that studies the temporal fluctuations of light scattered by particles, droplets, or bubbles [3, 4].

The technique is typically used to study the dynamics of a wide variety of soft matter systems such as micro- and nano-emulsions, nanoparticles, colloids, micelles, polymers or proteins [5, 9]. One of the most popular applications of DLS consists in the measurement of particle size from few nanometers up to few microns, and size distribution in dilute systems [10]. DLS could also be applied to more concentrated systems, where dynamics contains additional information about the structure of the dispersion [11–15].

In a DLS experiment, a coherent laser light source illuminates a volume containing N particles while a detector, positioned at angle θ from the incident light and placed at a distance R from the sample, collects the scattered intensity $I(\theta, t)$. The spatial distribution of the scattered intensity is that of a speckle pattern (grains of light), resulting from the interferences of the scattered fields from individual particles. Since particles are moving (i.e. they undergo Brownian motion), the scattered fields are time-dependent, which leads to a temporal fluctuation of the intensity of a speckle ("boiling" speckle). The intensity fluctuations therefore entail information about particles' motion and can be characterized using the intensity auto-correlation function, defined as:

$$g_2(\boldsymbol{q}, \tau) = \frac{\langle I(\boldsymbol{q}, 0) I(\boldsymbol{q}, \tau) \rangle}{\langle I(\boldsymbol{q})^2 \rangle}, \qquad (11.24)$$

where the brackets ... denote an average over the ensemble of realizations of the fluctuations, and $\boldsymbol{q}$ is the scattering vector. Auto-correlation simply compares the intensity at time t with the intensity at time $t + \tau$ for a range of lag times $\boldsymbol{\tau}$ covering the decay of g_2. The intensity auto-correlation function can be related to the field auto-correlation function $g_1(\boldsymbol{q}, \tau)$ using the Siegert relation [9]:

$$g_2(\boldsymbol{q}, \tau) = 1 + \beta [g_1(\boldsymbol{q}, \tau)]^2, \qquad (11.25)$$

where β is a pre-factor related to the number of speckles detected within the available surface of the detector.

While intensity is the accessible quantity in a DLS experiment, the amplitude and phase of the scattered field is necessary to link the intensity fluctuations to the particles' motion. The electric field scattered by a collection of N particles of arbitrary shape and size reads (cf. 1.2.1):

$$E(\boldsymbol{q}, t) = \sum_{j=1}^{N} b_j(\boldsymbol{q}, t) e^{-i\boldsymbol{q}\boldsymbol{R}_j(t)}, \qquad (11.26)$$

where b_j is the amplitude of the field scattered by particle j and $\boldsymbol{R}_j$ is the position of particle j. Note that terms that do not fluctuate over time have been omitted in the expression of the scattered field. The field auto-correlation function can be written as:

$$g_1(\boldsymbol{q}, \tau) = \frac{\langle E(\boldsymbol{q}, 0) E^*(\boldsymbol{q}, \tau) \rangle}{\langle I(\boldsymbol{q}) \rangle} = \frac{\sum_j \sum_k \left\langle b_j(\boldsymbol{q}, 0) b_k^*(\boldsymbol{q}, \tau) e^{-i\boldsymbol{q}[\boldsymbol{R}_j(0) - \boldsymbol{R}_k(\tau)]} \right\rangle}{\sum_j \sum_k \left\langle b_j(\boldsymbol{q}, 0) b_k^*(\boldsymbol{q}, 0) e^{-i\boldsymbol{q}[\boldsymbol{R}_j(0) - \boldsymbol{R}_k(0)]} \right\rangle} \qquad (11.27)$$

where the star indicates the complex conjugate of the field. The auto-correlation function can be simplified by considering a system of N identical spheres. In this case, the field amplitudes are the same for all particles and do not depend on time, i.e. $b_i(\boldsymbol{q}, \tau) = b(q)$. Furthermore, for centro-symmetrical particles, the amplitude of the scattered field only depends on the module of the scattering vector $q = 4\pi n \sin \theta / \lambda$, where n is the suspension's refractive index and λ the laser wavelength. Applying these simplifications to the Eq. (11.27) yields:

$$g_1(q, \tau) = \frac{\sum_j \sum_k \left\langle e^{-i\boldsymbol{q}[\boldsymbol{R}_j(0) - \boldsymbol{R}_k(\tau)]} \right\rangle}{\sum_j \sum_k \left\langle e^{-i\boldsymbol{q}[\boldsymbol{R}_j(0) - \boldsymbol{R}_k(0)]} \right\rangle} = \frac{F(q, \tau)}{F(q, 0)}, \qquad (11.28)$$

where $F(q, \tau)$ is called the intermediate scattering function and $F(q, 0) = S(q)$ is the static structure factor introduced in the section on static light scattering. It now becomes apparent how the dynamics of identical interacting spheres is influenced by the fluid structure.

For diluted systems, one can further simplify the results for the field auto-correlation function by considering that the positions $R_{i,j}$ of the particles are uncorrelated. As a consequence, the cross terms in the summation are on average zero, and only self-terms persist:

$$g_1(q, \tau) = N^{-1} \sum_j \left\langle e^{-i\boldsymbol{q}[\boldsymbol{R}_j(0) - \boldsymbol{R}_j(\tau)]} \right\rangle$$
$$= \left\langle e^{-i\boldsymbol{q}[\boldsymbol{R}(0) - \boldsymbol{R}(\tau)]} \right\rangle \qquad (11.29)$$
$$= e^{-i\boldsymbol{q}\cdot\Delta R(\tau)}$$

The displacement $\Delta \boldsymbol{R}(\tau)$ originates from particles' Brownian motion. The field auto-correlation function can now be evaluated:

$$g_1(q, \tau) = e^{-\frac{q^2}{6}\langle \Delta R^2(\tau)\rangle} = e^{-q^2 D_0 \tau}, \qquad (11.30)$$

where we used $\langle \Delta R^2(\tau) \rangle = 6D_0\tau$. The field auto-correlation function is therefore an exponential decay having a decay rate modulated by the free diffusion coefficient D_0 and the square of the scattering vector.

In practice, the diffusion coefficient can be obtained from a fit of the intensity auto-correlation function. The Stokes-Einstein relation allows to calculate the hydrodynamic radius a of the particles, for a given temperature T and a solvent viscosity η:

$$D_0 = \frac{k_B T}{6\pi\eta a}. \qquad (11.31)$$

Note that the theory of Dynamic Light Scattering is only valid for single scattered light. As for most scattering methods interpretation becomes exceedingly difficult for systems with non-negligible contributions from multiple scattering, small contributions of multiple scattering can already result in large analysis errors. In the case of larger particles with a high scattering contrast, this limits the technique to very low particle concentrations. A large variety of systems are therefore excluded from investigations with conventional dynamic light scattering. However, it is possible to suppress multiple scattering in DLS via the cross-correlation approach. Different implementations of cross-correlation light scattering have been developed and applied. The same method can also be used to correct Static Light Scattering (SLS) data for multiple scattering contributions. Alternatively, in the limit of strong multiple scattering, a variant of dynamic light scattering called Diffusing Wave Spectroscopy (DWS) can be applied (cf. section about DWS).

11.3.1 Sizing Using the Method of Cumulants

An important application of dynamic light scattering is the determination of particle size. In the introductory chapter about DLS, we focused on results for idealized systems (i.e. perfectly identical sphere). In the following, we extend the theory to dispersions of spherical objects having a small degree of polydispersity, and present a reliable method to analyze DLS data [3, 5, 9].

The field auto-correlation function (Eq. 11.27) can be simplified assuming that the system under study is a diluted solution of polydisperse spheres. In this case, the field amplitude $b_j(q,t)$ is independent of time and only depends on the module of the scattering vector. For diluted dispersions, particles' positions are uncorrelated and the field auto-correlation is given by [5]:

$$g_1(q,\tau) = \frac{\sum_j b_j^2(q) \left\langle e^{-i\boldsymbol{q}[\boldsymbol{R}_j(0)-\boldsymbol{R}_j(\tau)]} \right\rangle}{\sum_j b_j^2(q)}$$
$$= \frac{\sum_j b_j^2(q) e^{-q^2 D_{0j}\tau}}{\sum_j b_j^2(q)},$$

(11.32)

where $D_{0j} = k_B T/(6\pi\eta a_j)$ is the free diffusion coefficient of particles j with radius a_j.

We see that $g_1(q,\tau)$ is a sum of exponential decays, each of which is weighted by the scattered intensity b_j^2 for particles with radius a_j. If only a small degree of polydispersity is allowed, the correlation function will be close to a single exponential, and it is therefore natural to try to parameterize the D_{0j} term using an average diffusion coefficient $\bar{D}$. Here we need to use the (intensity-) weighted average diffusion coefficient $\bar{D}$ defined as:

$$\bar{D} = \frac{\sum_j b_j^2(q) D_j}{\sum_j b_j^2(q)},$$

(11.33)

and similarly,

$$\overline{D^2} = \frac{\sum_j b_j^2(q) D_j^2}{\sum_j b_j^2(q)},$$

(11.34)

A substitution of $D_{0j} = \bar{D} + (D_{0j} - \bar{D})$ in Eq. (11.32) yields:

$$g_1(q,\tau) = e^{-q^2 \bar{D}\tau} \frac{\sum_j b_j^2(q) e^{-q^2[D_{0j}-\bar{D}]\tau}}{\sum_j b_j^2(q)},$$

(11.35)

where the exponential inside the sum can be expanded,

$$g_1(q,\tau) = e^{-q^2\bar{D}\tau} \frac{\sum_j b_j^2(q)\left[1 - (D_{0j} - \bar{D})q^2\tau + \frac{1}{2}(D_{0j} - \bar{D})^2 q^4\tau^2 + \cdots\right]}{\sum_j b_j^2(q)}$$

$$= e^{-q^2\bar{D}\tau}\left[1 + \frac{1}{2}\left(\frac{\overline{D^2} - \bar{D}^2}{\bar{D}^2}\right)(q^2\bar{D}\tau)^2 + \cdots\right] \tag{11.36}$$

The terms into brackets in Eq. (11.36) represent the expansion of an exponential, which can be finally substituted,

$$g_1(q,\tau) = e^{-q^2\bar{D}\tau + \frac{1}{2}\left(\frac{\overline{D^2} - \bar{D}^2}{\bar{D}^2}\right)(q^2\bar{D}\tau)^2 + \cdots}. \tag{11.37}$$

The next step is to find a relation between the variance of the distribution of diffusion coefficients $\left(\overline{D^2} - \bar{D}^2\right)/\bar{D}^2$ and the variance of the size distribution itself. Here we take advantage of the fact that a relation between the average diffusion coefficient and the particles' radius exists in the limit $q\bar{a} \leq 1$. Indeed, under these conditions, the scattered intensity b_j^2 is proportional to the second power of the particles' volume (i.e. $b_j^2 \propto a_j^6$). Using $\bar{a^n} = \left(\sum_j^N a_j^n\right)/N$, we can therefore write:

$$\frac{\overline{D^2} - \bar{D}^2}{\bar{D}^2} = \frac{\overline{a^4} \cdot \overline{a^6}}{\left(\overline{a^5}\right)^2} - 1, \tag{11.38}$$

which can be approximated by,

$$\frac{\overline{D^2} - \bar{D}^2}{\bar{D}^2} = \frac{(1 + 6\sigma^2)(1 + 15\sigma^2)}{(1 + 10\sigma^2)} - 1 \approx \sigma^2, \tag{11.39}$$

for a distribution of radii $P(a)$ having a variance $\sigma^2 = \left(\overline{a^2} - \bar{a}^2\right)/\bar{a}^2$. The approximation used in Eq. (11.39) is valid for narrow size distributions and reads [5]:

$$\overline{a^n} = \bar{a}^n\left(1 + \frac{n(n-1)}{2}\sigma^2 + \cdots\right), \tag{11.40}$$

The result of Eq. (11.39) demonstrates that the auto-correlation function measured in a DLS experiment not only contains information about the average radius of the particles, sometimes also called the apparent radius $\bar{a} = k_B T/(6\pi\eta\bar{D})$, but also about the variance of the size distribution σ^2 [11, 16]. This development holds for diluted suspensions of spherical particles having a narrow size distribution and for $q\bar{a} \leq 1$.

In practice, the Siegert relation can be used to relate the measured intensity correlation function to the field correlation function of Eq. (11.37) [9], and can be further linearized to accommodate a simple fit having the form of an expansion of cumulants of the distribution [11, 16]:

$$\ln[g_2(q, \tau) - 1] = \ln[cst] - 2\Gamma\tau + \mu\tau^2 + \cdots \qquad (11.41)$$

where the average size is obtained from the first cumulant $\Gamma = q^2\bar{D}$, and the second cumulant μ, which is related to the standard deviation σ of the distribution as $\sigma = \sqrt{\mu/\Gamma^2}$, gives an estimate of the polydispersity. A detailed discussion about the type of the minimization procedure and the accuracy of the resulting fits can be found in [11].

11.3.2   Technical Considerations for a DLS Experiment

Laser intensity: The required laser intensity depends on the detection efficiency of the employed detector. While PMTs often have a quantum efficiency QE of less than 10 %, APDs can reach more than 60 % for detection of red light (wavelength λ, 600–700 nm). In the case of a PMT, a red laser would thus require approximately 6 times as much intensity to obtain the same result as an APD. As a rule of thumb one can use:

$$I \times \text{QE}\,(\lambda) > 10\ \text{mW}$$

More advanced DLS techniques or very weakly scattering particles might require more intensity.

Laser stability: While the long term stability (>1 min) of the laser intensity $I(t)$ is not critical in a DLS experiment, fluctuations on the time scale of the decay time will result in significant errors.

Laser coherence length: This should be larger than the scattering volume, which is typically <1 cm.

The scattering angle: The correlation function in a DLS experiment depends on the wave vector and thus the scattering angle. Moreover, the intensity scattered at a given scattering angle depends on particle size. There is thus an optimum angle of detection for each particle size. A high quality analysis should always be performed at several scattering angles (multi angle DLS). This becomes even more important in case of polydisperse samples with unknown particle size distribution since at certain angles, the scattering intensity of some particles will completely overwhelm the weak scattering signal of other particles, thus making them invisible to the data analysis at this angle.

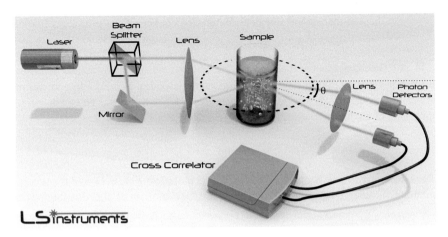

Fig. 11.3 Suppression of multiple scattering using the 3D cross correlation technology [21] (Figure reprinted with permission. Copyright by LS Instruments AG, Switzerland)

DLS instruments working exclusively at a fixed angle can only deliver good results for some particles. Therefore, special attention should be paid when considering the precision of a DLS instrument.

Detectors: The timescale of the detected intensity fluctuations $I(t)$ in a DLS experiment is on the order of ms and less, so should be the temporal resolution of the detectors used in DLS. APDs and PMTs offer such a temporal resolution. Another important detector parameter is the dead time. This is the time after detection of a single photon in which no further detection is possible. The dead time will influence the linearity of the intensity detection. Furthermore, all PMT- and APD-based photon detectors have a certain probability to produce a second electronic pulse after they detect a photon (the so called "After-Pulsing Effect"). An after-pulsing probability as little as 2 % can result in significant errors for intensity fluctuations on timescales below 1 μs. By cross correlating two detectors (pseudo cross-correlation), the after-pulsing effect can be eliminated, thus allowing measurements down to the full-time resolution of the detector, only limited by the temporal resolution of the correlator (Fig. 11.3).

11.3.3 Suppressing Multiple Scattering in DLS and SLS

Accurate dynamic and static light scattering measurements require the detection and analysis of single scattering events; even relatively limited occurrences of multiple scattering in the sample can dramatically compromise measurement results in DLS and SLS (Fig. 11.4b). Therefore, highly scattering, turbid samples must usually be diluted and tedious verification measurements must be executed to ensure data validity. Furthermore, in many cases sample processes such as aggregation and

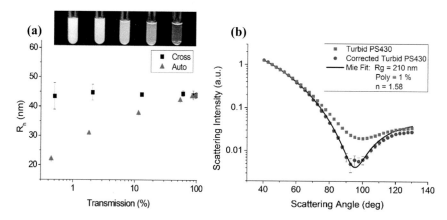

Fig. 11.4 a Results of DLS experiments on a dilution series of latex spheres in water and the corresponding picture of the samples. The results obtained using the 3D cross-correlation scheme (cross) are unaffected by multiple scattering, while the single beam experiment records (auto) show a significant decrease of the particle radius with increasing turbidity [22]. **b** Corrected and uncorrected SLS data along with a Mie fit for a turbid sample of 430 nm polystyrene particles in water (5×10^{-5} w/w, 46 % transmission) [23] (Figure reprinted with permission. Copyright by LS Instruments AG, Switzerland)

gelation depend on high particle concentrations and so dilution is unacceptable as it modifies the very samples properties to be measured.

The onset of multiple scattering can be assessed by comparing the suspension mean free path l with the total length of the light path L within the sample. The mean free path is the average distance travelled by photons between two consecutive scattering events. Prediction for l can be obtained using Mie theory. Single scattering samples have $L/l \ll 1$. For $L \sim l$, contributions from multiple scattering can already affect measurements. Finally, when $L/l \gg 1$ samples are opaque and DLS results can be wrong by orders of magnitude.

An elegant solution to these problems was proposed by Schätzel in 1991 [17] and applied later using different experimental realizations [18]. The key concept common to these experiments is that only singly scattered light will produce correlated intensity fluctuations when the intensity of two identical but independent experiments are cross-correlated.

There are different experimental ways to do this, but in practice using two laser beams at the same scattering angle, but displaced symmetrically above and below the scattering plane has proven to be the most reliable. This technique is often referred to as "3D cross correlation" or simply "3D". In a 3D light scattering experiment, two parallel laser beams are focused onto the scattering volume using a lens (Fig. 11.3). An identical detection lens placed at the same distance from the scattering volume guides scattered light to the detection optics. The two signals are then cross correlated and since only the signal produced by single scattering is identical on both detectors, the multiple scattering is suppressed. Moreover, the intercept of the cross-correlation function is proportional to the multiple scattering and can be used to correct the scattering intensity (Fig. 11.4b). The 3D technique can thus be applied to both DLS and SLS to correct for multiple scattering [19].

One drawback of the 3D cross-correlation technique is that one photon detector measures the scattered light intensity at the desired scattering vector, but also receives a contribution from the second illumination beam operating at the same wavelength. A four-fold reduction in the cross-correlation intercept arises from cross-talk between the two simultaneous scattering experiments executed in this way. The correlation intercept strongly influences measurement accuracy due to its pivotal role in accurately fitting models to the measured data. For strongly scattering samples where only a small component of the detected light is singly-scattered, the signal-to-noise ratio of the measurement becomes unacceptably low as the magnitude of the cross-correlation intercept falls into the noise of the baseline fluctuations.

To avoid this problem, Block and Scheffold [20] developed a new measurement technique in which the two scattering experiments are temporally separated by modulating the incident laser beams and gating the detector outputs at frequencies exceeding the timescale of the system dynamics. This robust modulation scheme eliminates cross-talk between the two beam-detector pairs and leads to a four-fold improvement in the cross-correlation intercept, while fully suppressing the negative effects of multiple scattering (Fig. 11.5).

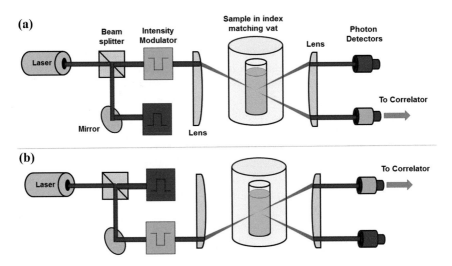

Fig. 11.5 Schematic of the 3D cross-correlation scheme using beam modulation to avoid simultaneous illumination of the scattering volume with the two beams. Acousto optical modulators allow for modulation rates >1 MHz, which is sufficient for most samples [20] (Figure reprinted with permission. Copyright by LS Instruments AG, Switzerland)

11.4   Fundamentals of Diffusing Wave Spectroscopy

Diffusing Wave Spectroscopy (DWS) extends the application of DLS to concentrated and highly scattering (optically white) media in which light is multiply scattered. As in the case of DLS, scattering could be caused by either solid particles, liquid droplets or gaseous bubbles within the medium. For simplicity, we will refer to any of these as scattering particles.

Observing coherent light that is transmitted through such turbid media onto a screen, one notes a fluctuating pattern of bright and dark spots (Fig. 11.6) similar to the case of DLS. In DWS, however, the pattern typically fluctuates much faster, since light is scattered several times. Nevertheless, the rate at which it fluctuates relates to the dynamics of the particles within the sample; namely, the faster the particle's motion, the faster the speckle pattern's fluctuations. DWS theory establishes an analytical expression connecting the intensity fluctuations of scattered light to the motion of the particle to obtain their mean square displacement, similar to DLS but in a different scattering regime. The different scattering regimes can be distinguished by the mean free path l, which stands for the average distance travelled by a photon between two consecutive scattering events (Fig. 11.7). In dilute dispersions, l is given by:

$$l = \frac{1}{\rho\sigma}, \tag{11.42}$$

Fig. 11.6 Basic DWS setup in which a laser illuminates a suspension of particles in Brownian motion in water. Observation of the transmitted light on a screen reveals a boiling speckle pattern [37] (Figure reprinted with permission. Copyright by LS Instruments AG, Switzerland)

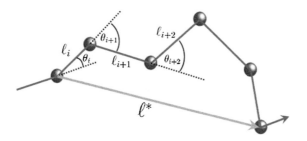

Fig. 11.7 Transport mean free path l^* and mean free path l_i (Figure reprinted with permission. Copyright by LS Instruments AG, Switzerland)

where ρ is the number density of particles and σ the total scattering cross-section of a single particle. Comparison of l and the thickness L of the sample defines the scattering regime. Three cases are typically identified:

- $l \gg L$: single scattering limit where DLS applies;
- $l \sim L$: intermediate regime involving a few scattering events (a regime can only be accessed with 3D light scattering);
- $l \ll L$: multiple scattering regime where DWS applies.

In the first two cases, the ratio T of transmitted light intensity I over the input laser intensity I_0 decays according to Beer's law as $T = I/I_0 = exp\ (-L/l)$. This expression no longer holds in the regime of multiple scattered light.

11.4.1 The Diffusion Approximation

In the multiple scattering regime, where photons are scattered many times, the diffusion approximation can be applied, hence the appellation DWS [24]. In this case, photon propagation can be modeled by a random walk. In the case of very small particles, light is scattered in an isotropic fashion; the direction of the scattered light is thus randomized after each scattering event. This is a true random walk with step lengths equal to the mean free path l. For bigger particles, however, light scattering becomes anisotropic, and is peaked in the forward direction. The scattering direction of several consecutive scattering events is correlated (biased random walk). Additional scattering events are required to ensure randomization of the direction of light propagation. As a consequence, another photon random walk step length scale that takes into account scattering anisotropy, is defined; it is the transport mean free path l^*, which relates to l as follows:

$$l^* = \frac{l}{1 - \langle cos\theta \rangle},$$

(11.43)

where $< cos\theta > (= g$, the scattering anisotropy parameter) is an ensemble average of the scattering angle θ over many scattering events (Fig. 11.7). For isotropic scattering, $< cos\ \theta > = 0$, and $l^* = l$. For anisotropic scattering, $l^* > l$. Transmitted photons will, therefore, have undergone $\sim (L/l^*)^2$ random steps with l^*/l scattering events per step, for a total average number of scattering events of $N \sim (L/l^*)^2\ (l^*/l)$. This number must be large for the diffusion approximation to be valid. Kaplan et al. experimentally showed that this approximation holds for $L > 4\ l^*$ for anisotropic (Mie) scattering and $L > 8\ l^*$ for isotropic scattering [25]. In this limit the transmission T of light scattered through a slab of infinite transverse extent and finite thickness L can be calculated:

$$T = \frac{\frac{5l^*}{3L}}{1 + \frac{4l^*}{3L}}. \tag{11.44}$$

Using Eq. (11.44), it is possible to determine l^* of a sample by comparing its transmitted intensity with that of a reference sample with known l^*, provided that both samples are in cells with the same thickness L, and they are illuminated with the same intensity I_0. In practice, this means that one will calibrate a DWS transmission experiment with an aqueous dispersion of particles of given size for which calculation of l^* is straightforward.

11.4.2 DWS Autocorrelation Functions

For each random walk with path length $s = N\,l$, the autocorrelation function g_1^s [26]:

$$g_1^s(t) = e^{-\frac{1}{3}k_0^2 \langle \Delta_r^2(\tau) \rangle \frac{s}{l^*}}, \tag{11.45}$$

for the case of uncorrelated scatterers, such as tracer beads at a volume concentration below 5 %. Here k_0 denotes the modulus of the photon wave vector in the medium of propagation, and τ the correlation lag time. One notices that the term $k_0^2 < \Delta r^2(\tau) >/3$ reflects the decay of the correlation function due to a single scattering event but averaged over all scattering vectors q, weighted by the form factor of the particle. The average number of random steps s/l^* in each photon path of length s represents the contribution of multiple scattering. Thus each individual particle only needs to move, on average, a very small fraction of the wavelength λ_0 to produce a significant decay in $g_1^s(\tau)$. This is in stark contrast with DLS where a particle has to move a distance of $\sim \lambda_0$ to lead to a substantial decay of the autocorrelation function since there is only one scattering event. The ability to probe particle dynamics down to sub-nanometer length scales is one of the most outstanding features of DWS.

To obtain the full autocorrelation function of the field, we need to weight Eq. (11.45) with the probability density function of the path lengths $P(s)$:

$$g_1(\tau) = \int_0^\infty P(s)g_1^s(\tau)ds. \tag{11.46}$$

$P(s)$ is determined through the use of the diffusion equation for light. This can be solved analytically for some simple experimental geometries. The most commonly used geometries are transmission through and backscattering from a slab of finite thickness L and of infinite transverse extent (Fig. 11.8). In the case of a plane wave illumination, in transmission, Eq. (11.46) becomes:

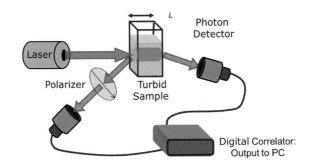

Fig. 11.8 Typical experimental geometries used in DWS. Transmission: scattered light is detected from the side opposite to the illumination; backscattering: scattered light detected from the same side (Figure reprinted with permission. Copyright by LS Instruments AG, Switzerland)

$$g_1(\tau) = \frac{\left(\frac{L}{l^*} + \frac{4}{3}\right)\sqrt{k_0^2\langle\Delta_r^2(\tau)\rangle}}{\left[1 + \frac{4}{9}k_0^2\langle\Delta_r^2(\tau)\rangle\right]sinh\left[\frac{L}{l^*}\sqrt{k_0^2\langle\Delta_r^2(\tau)\rangle}\right] + \frac{4}{3}\sqrt{k_0^2\langle\Delta_r^2(\tau)\rangle}cosh\left[\frac{L}{l^*}\sqrt{k_0^2\langle\Delta_r^2(\tau)\rangle}\right]}.$$

$$(11.47)$$

In the case of backscattering, Eq. (11.46) becomes:

$$g_1(\tau) = exp\left(-\gamma\sqrt{k_0^2\langle\Delta_r^2(\tau)\rangle}\right),$$ (11.48)

where $\gamma = z_0/l^* + 2/3$. z_0 is the distance that separates the source of diffusing intensity in the medium from the face of incident illumination ($z_0 \sim l^*$).

While the only fitting parameter in Eq. (11.48) is $< \Delta r^2(\tau) >$, γ being a constant, it seems simpler to conduct measurements in backscattering. However, γ was shown to depend on both the polarization state of the backscattered light and particle size. Typically, γ varies between 1.5 and 2.7 as particle size and polarization states are varied. This is related to the fact that backscattering will always have contributions of light for which the diffusion approximation does not hold, thus one finds that, in practice, the results obtained in transmission are far more precise than those obtained in backscattering. Nevertheless, DWS measurements in backscattering can be very useful, whenever measurements in transmission are not possible. This is for example the case when the sample is too dense (too high particle concentration) to allow transmission of light.

Finally, one needs to note that, just as in a DLS experiment, one cannot measure the field $E(t)$, but only the intensity $I(t)$ from which one computes the intensity autocorrelation function $g_2(\tau)$. The field autocorrelation function $g_1(\tau)$ is directly related to $g_2(\tau)$ through the Siegert relation (Eq. 11.25).

11.4.3   Technical Considerations for DWS Setups

Laser: In DWS experiments, samples must be illuminated by a laser with sufficient coherence length, which should correspond to the longest relevant paths travelled by the photons in the sample. For a turbid sample of thickness L = 1 cm, this can easily be more than 10 cm. Simple diode lasers, however, often have a coherence length of less than 1 mm. While it is mathematically possible to consider a cut-off path length in the path length distribution $P(s)$, this introduces significant errors, since the distribution at very short paths have a minor contribution compared to the incoherent light of the longer paths that will result in noise. The coherence length of the laser is thus an important quality factor in DWS experiments and should always be on the order of 1 m.

Detector: The intensity fluctuations of backscattered or transmitted light can be collected by point-like detectors such as single-mode fibers coupled to photomultipliers (PMT) or avalanche photodiodes (APD), or by standard Charged Coupled Device (CCD) or Complementarity Metal-Oxide-Semiconductor (CMOS) cameras. In the case of fibers, a single speckle of light is detected, whereas there are as many detected speckles as pixels in the camera, if the detection optics is adjusted in such a way that the speckle size matches the pixel size of the camera (typically in the order of 10 μm). Doing DWS with a camera is referred to as Multi-Speckle DWS (MSDWS) [27]. In MSDWS, a number n (in the order of 10^5–10^6) of independent speckles are simultaneously sampled, and n correlation functions are computed. The ensemble-averaged intensity autocorrelation function $g_2(\tau)$ is subsequently obtained by averaging over all pixels. Thus the acquisition time is reduced by a factor equal to n, compared to a single detector. The highest acquisition rate, however, is typically limited to 30 Hz for standard CCD and CMOS cameras, which limits the smallest measurable characteristic relaxation time of $g_2(\tau)$ to $\tau \sim 10$ ms. By contrast, the time resolution of a single-mode fiber coupled to an APD or a PMT and a hardware correlator (photon counting device) can be as fast as 12.5 ns. Fast relaxation dynamics can be therefore probed in the latter case. To give an idea, for 220 nm diameter polystyrene particles suspended in water at a volume fraction of 7 %, the amplitude of $g_2(\tau)$ already decreases by 50 % after a correlation time τ of 0.85 μs, as measured in transmission using a single-mode detection scheme. Such very fast relaxation dynamics would not be observed by MSDWS. The single-mode detection scheme is therefore the most suitable one to detect very fast relaxation dynamics. On the other hand longer measurement times are required to measure slow relaxation times with a single detector. It could take several hours to well measure a very viscous dispersion, for example.

DWS Echo technology: One way to solve this problem is to introduce a ground glass in the light path, mounted on a motor with adjustable angular speed, as shown in Fig. 11.9 [28]. Here the ground glass scrambles the light illuminating the sample. After a full revolution of the ground glass, the speckle pattern reappears and generates an echo peak in the measured correlation function. The sample intensity

Fig. 11.9 DWS setup
integrating a rotating ground
glass [38]

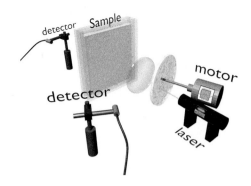

autocorrelation function is now determined for long correlation times that are separated by the motor rotation period. The shape of the echo peaks is defined by the rotation speed while the height follows the correlation function of the sample internal dynamics. The echo mode allows drastic reduction of the measurement time, from several hours to a few minutes. By merging the acquired short and long time parts of the recorded sample correlation function, one can therefore access an extremely wide range of correlation times in a matter of minutes for all kinds of samples with a single fast detector.

11.4.4 Applications of DWS

DWS is an optical technique based on multiple light scattering in concentrated colloidal materials, and hence naturally applies to everyday life soft matter systems such as suspensions (varnishes, paints, inks, etc.), emulsions (mayonnaise, cosmetic creams, milk, etc.), foams (shaving cream, beer or coffee foam, etc.) or gels (yogurt, gelatin, etc.). These materials typically consist of colloidal particles, droplets, or bubbles dispersed at high volume fractions in a continuous liquid phase.

Maybe the most popular application is the combination of DWS with microrheology, which consists in computing the frequency-dependent storage $G'(\omega)$ and loss $G''(\omega)$ moduli of a colloidal system by applying the generalized Stokes-Einstein relation to the mean square displacement of the scattering colloids [29]. Since, moreover, DWS probes particle motion at short timescales, such a combination allows determination of $G'(\omega)$ and $G''(\omega)$ over a very large frequency range, up to 10^6 rad/s, while most of mechanical rheometers are limited to angular frequencies of $\sim 10^3$ rad/s. As such, DWS microrheology is complementary to mechanical rheology, and, in some cases, can even replace it. We note, in addition, that such high frequencies are only obtainable with the use of the single-mode detection scheme, since the acquisition rate of CCD and CMOS cameras is limited to ~ 30 Hz.

Another application, which uses DWS in backscattering, is particle sizing [30]. Although γ depends on both particle size and polarization detection, the mean value $< \gamma >$ obtained by averaging the values of γ as measured using two different polarization detection modes (i.e. either vertically or horizontally polarized light is detected), was experimentally shown to be independent of particle size. As a consequence, using $< \gamma >$ and running two consecutive measurements by tuning the polarization detection was demonstrated to give reliable and very accurate estimates of the hydrodynamic radius of particles dispersed at high volume fractions in liquid solvents. Also static measurements of the transmitted intensity as measured in DWS can provide the diameter of bubbles in foams like shaving cream [31]. Particle size distribution, however, cannot be assessed by DWS.

Another field of application of DWS, in addition, is the monitoring of processes like gelation (e.g. transformation of milk into yogurt, drying of paint films, etc.) [27, 32], aging (e.g. Ostwald ripening in foams, emulsions, coarsening of foams, etc.) [33, 34], etc., which are of industrial relevance.

DWS can also detect fluctuations of surfaces of low-viscosity droplets dispersed in a solvent [35], or detect formation of a skin layer at the interface between air and a drying suspension [36].

Overall, DWS covers a broad range of applications including both fundamental and industrial aspects.

References

1. J.D. Jackson, *Classical Electrodynamics*, 3rd edn. (Wiley, New York, 1999)
2. LS Instruments AG, Static Light Scattering. (LS Instruments AG, 2014). http://www.lsinstruments.ch/technology/static_light_scattering_sls/. Accessed 27 October 2014
3. B.J. Berne, R. Pecora, in *Dynamic Light Scattering*, ed. by S.B. Ross-Murphy (Wiley, New York, 1976), p. 376
4. W. Brown, *Dynamic Light Scattering* (Oxford University, Oxford, 1993)
5. P. Pusey, in *Neutrons, X-rays and Light: Scattering Methods Applied to Soft Matter*, ed. by P. Linder, Th. Zemb (Elsevier, Amsterdam, 2002), p. 3
6. B. J. Zimm, J. Chem. Phys. **16**, 1093 (1948)
7. A. Guinier, G. Fournet, in *Small Angle Scattering of X-rays*, ed. by M.G. Mayer (New York, Wiley Interscience, 1955)
8. J.R.R. Lugo, N. Brauer, F. Cardinaux, Characterization of Macromolecules in Solutions: Zimm Plot. (LS Instruments AG, 2014), http://www.lsinstruments.ch/applications/application_notes_dls_sls/characterization_of_macromolecules_in_solutions_zimm_plot/. Accessed August 2014
9. K. Schätzel, in *Dynamic Light Scattering*, ed. by W. Brown (Oxford University, Oxford, 1993), p. 76
10. M. Muschol, F. Rosenberger, J. Chem. Phys. **107**, 1953 (1995)
11. B. Frisken, Applied Optics **40**(24), 4087 (2001)
12. P.N. Segrè, S.P. Meeker, P.N. Pusey, C.K. Poon, Phys. Rev. Lett. **75**, 958 (1995)
13. G. Brambilla, D.El. Masri, M. Pierno, L. Berthier, L. Cipelletti, G. Petekidis, A. B. Schofield, Phys. Rev. Lett. **102**, 085703 (2009)
14. C. Haro-Pérez, G.J. Ojeda-Mendoza, L.F.J. Rojas-Ochoa, J. Chem. Phys. **134**, 244902 (2011)
15. T. Eckert, E. Bartsch, Phys. Rev. Lett. **89**, 125701 (2002)
16. D.E. Koppel, J. Chem. Phys. **57**, 4814 (1972)
17. K. Schätzel, J. Mod. Opt. **38**, 9 (1991)
18. P.N. Pusey, Curr. Opin. Colloid Interface Sci. **4**, 177 (1999)

19. C. Urban and P. Schurtenberger, J. Colloid Interface Sci., 207, 150 (1998)
20. I. Block, F. Scheffold, Rev. Sci. Instrum. **81**, 123107 (2010)
21. LS Instruments AG, 3D Cross-Correlation Light Scattering. (LS Instruments AG, 2014). http://www.lsinstruments.ch/technology/dynamic_light_scattering_dls/3d_cross-correlation_light_scattering/. Accessed 27 October 2014
22. I.D. Block, Particle Sizing in Highly Scattering Samples. (LS Instruments AG, 2010). http://www.lsinstruments.ch/applications/application_notes_dls_sls/particle_sizing_in_highly_scattering_samples/. Accessed July 2010
23. I.D. Block, Characterizing Concentrated Colloidal Suspensions. (LS Instruments AG, 2010). http://www.lsinstruments.ch/applications/application_notes_dls_sls/characterizing_concentrated_colloidal_suspensions/. Accessed June 2010
24. D.A. Weitz, D.J. Pine, Diffusion wave spectroscopy. in *Dynamic Light Scattering* ed. by W. Brown (Oxford University Press, New York, 1993), p. 652
25. P.D. Kaplan, M.H. Kao, A.G. Yodh, D.J. Pine, Appl. Opt. **32**, 3828 (1993)
26. D.A. Weitz, J.X. Zhu, D.J. Durian, H. Gang, D.J. Pine, Phys. Scripta **49**, 610 (1993)
27. L. Brunel, A. Brun, P. Snabre, L. Cipelletti, Opt. Express **15**, 15250 (2007)
28. L. Brunel, A. Brun, P. Snabre, L. Cipelletti, Opt. Express **15**, 15250 (2007)
29. T.G. Mason, D.A. Weitz, Phys. Rev. Lett. **74**, 1250 (1995)
30. F. Scheffold, J. Dispersion Sci. Technol. **23**, 591 (2002)
31. D.J. Durian, D.A. Weitz, D.J. Pine, Phys.Rev. A **44**, R7902 (1991)
32. M. Alexander, D.G. Dalgleish, Curr. Opin. Colloid Interface Sci. **12**, 179 (2007)
33. Y. Hemar, D. S. Horne, Colloids Surf., B **12**, 239 (1999)
34. N. Isert, G. Maret, C.M. Aegerter, Eur. Phys. J. E **36**, 116 (2013)
35. H. Gang, A.H. Krall, D.A. Weitz, Phys. Rev. Lett. **73**, 3435 (1994)
36. A.M. König, T.G. Weerakkody, J.L. Keddie, D. Johannsmann, Langmuir **24**, 7580 (2008)
37. LS Instruments AG, Diffusing Wave Spectroscopy. (LS Instruments AG, 2014). http://www.lsinstruments.ch/technology/diffusing_wave_spectroscopy_dws/. Accessed 27 Oct 2014
38. P. Zakharov F. Scheffold, *Advances in Dynamic Light Scattering, Light Scattering Reviews 4.* ed. by A.A. Kokhanovsky (Springer-Praxis Publishing Ltd., Chichester, 2009)

Chapter 12
Scattering Techniques Applied to Soft Matter Interfaces

Jean Daillant

Abstract An introduction to scattering techniques for the investigation of interfaces and thin films is given in this lecture. A unified formalism for reflectivity, scattering and diffraction is used in order to underline the common concepts. The different methods are illustrated with several examples of liquid surfaces, lipid membranes, polymer films, nanoparticles, using either x-rays or neutrons.

12.1 Introduction

Surfaces and interfaces play an important role in a large number of systems, in particular in soft-condensed matter where many systems have a very large specific area. From the 1980s, x-ray and neutron scattering techniques (diffraction, reflectivity, surface scattering) have been widely used to determine the structure of soft surfaces and interfaces in complement with, or combined to other methods, sometimes developed at the same time like scanning probe microscopies. With the development of more powerful sources, detectors and methods, they are now also increasingly used to understand interfacial processes.

Different surface scattering techniques have been developed to answer specific questions. Grazing incidence diffraction (GID) allows the determination of the in-plane structure of crystalline samples. It was first applied to Langmuir films at the air/water interface were published in 1987 [1, 2]. The density profile normal to an interface is obtained in reflectivity experiments, developed at the same time, with the first reflectivity experiments on the bare water surface being published in 1985 [3]. In reflectivity studies, only density profiles averaged over the surface can be measured and from the beginning of the 90s, the analysis of diffuse scattering was developed to access in-plane inhomogeneities or fluctuations [4, 5]. Finally, grazing

J. Daillant (✉)
Synchrotron SOLEIL, L'orme des Merisiers, Saint-Aubin BP48,
91192 Gif-sur-Yvette Cedex, France
e-mail: jean.daillant@synchrotron-soleil.fr

© Springer International Publishing Switzerland 2016
P.R. Lang and Y. Liu (eds.), *Soft Matter at Aqueous Interfaces*,
Lecture Notes in Physics 917, DOI 10.1007/978-3-319-24502-7_12

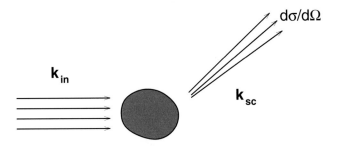

Fig. 12.1 Definition of differential scattering cross-section

Incidence X-ray Scattering (GISAXS) and Grazing Incidence Neutron Scattering (GISANS) are the surface couterpart of small angle scattering and are used to characterize surface structures in the range ≈ 1–100 nm.

Many soft systems have been studied using scattering techniques: simple liquids, liquid crystals, surfactants, polymers, membranes, ionic liquids, nanoparticles. The investigation of liquid–liquid and solid–liquid interfaces has also become possible, opening wider fields of research.

This chapter is organised as follows. After a short introduction to the interaction of x-rays with matter, scattering cross-sections for the different methods will be derived within an unified frame. Experimental aspects will be illustrated with several examples.

12.2 The Scattering Cross-Section

12.2.1 Differential Scattering Cross-Section and Intensity

The full interpretation of a scattering experiment requires the detailed comparison of an experimentally determined scattered intensity with a model calculation. This is most conveniently achieved by calculating the differential scattering cross-section $d\sigma/d\Omega$ which is defined as the intensity scattered per unit solid angle in the direction $\mathbf{k}_{sc}$ (polarisation $\mathbf{e}_{sc}$) for a unit incident flux in the direction $\mathbf{k}_{in}$ (polarisation $\mathbf{e}_{in}$), see Fig. 12.1. The scattered intensity is then obtained by convoluting $d\sigma/d\Omega$ with the experimental resolution function.

$$I = \Phi_0 \int_{\Omega} \frac{d\sigma}{d\Omega} d\Omega, \qquad (12.1)$$

where Φ_0 is the incident flux and Ω the solid angle subtended by the detector. In some cases (reflectivity, diffraction), intensity will be peaked in a given direction and Eq. (12.1) will amount to a convolution with instrumental resolution. In other cases (diffuse scattering), intensity varies slowly with angle and as a result of integration over $d\Omega$ in Eq. (12.1), intensity will be roughly proportional to Ω. Having a proper estimate of the resolution is then as critical as the scattering cross-section calculation to evaluate the scattered intensity.

12.2.2   The Born Approximation

In the most simple Born approximation (kinematic approximation) which neglects multiple scattering, a given scatterer only "sees" the incident wave. Considering first that all scatterers in the medium are identical, the waves scattered by two scatterers separated by a distance $\mathbf{r}$ only differ by a phase factor $e^{i\mathbf{q}\cdot\mathbf{r}}$, where $\mathbf{q} = \mathbf{k}_{sc} - \mathbf{k}_{in}$ is the wave-vector transfer (Fig. 12.2). Summing over all scatterers in the medium, the total amplitude scattered by the medium will be proportional to $\left|\sum_j e^{i\mathbf{q}\cdot\mathbf{r}_j}\right|^2$ and the scattering cross-section which is proportional to the intensity can be written:

$$\frac{d\sigma}{d\Omega} = b^2 \left|\sum_j e^{i\mathbf{q}\cdot\mathbf{r}_j}\right|^2 = b^2 \left|\int d\mathbf{r}\rho(\mathbf{r})e^{i\mathbf{q}\cdot\mathbf{r}}\right|^2, \tag{12.2}$$

where the proportionality factor b has the dimension of a length and is called the scattering length. If there are different kinds of scatterers, on can write

$$\frac{d\sigma}{d\Omega} = \left|\sum_j b_j e^{i\mathbf{q}\cdot\mathbf{r}_j}\right|^2 = \left|\int d\mathbf{r}\rho_b(\mathbf{r})e^{i\mathbf{q}\cdot\mathbf{r}}\right|^2, \tag{12.3}$$

where ρ_b is called the scattering length density.

As only intensities or scattering cross-sections are measured in a scattering experiment and not amplitudes, a scattering experiment will generally not lead to a

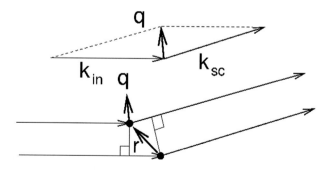

Fig. 12.2 Kinematics of scattering. The phase shift between the waves scattered in direction $\mathbf{k}_{sc}$ by two scatterers separated by $\mathbf{r}$ is $\mathbf{q} \cdot \mathbf{r}$

direct determination of scattering length densities. Starting from Eq. (12.2), we have:

$$
\begin{aligned}
\frac{d\sigma}{d\Omega} &= b^2 \left| \int d\mathbf{r} \rho(\mathbf{r}) e^{i\mathbf{q}\cdot\mathbf{r}} \right|^2 \\
&= b^2 \int d\mathbf{r} d\mathbf{r}' \rho(\mathbf{r}) \rho(\mathbf{r}') e^{i\mathbf{q}\cdot(\mathbf{r}-\mathbf{r}')} \\
&= b^2 \int d\mathbf{r} d\mathbf{R} \rho(\mathbf{0}) \rho(\mathbf{R}) e^{i\mathbf{q}\cdot\mathbf{R}} \\
&= b^2 V \int d\mathbf{R} \langle \rho(\mathbf{0})\rho(\mathbf{R}) \rangle e^{i\mathbf{q}\cdot\mathbf{R}},
\end{aligned}
\tag{12.4}
$$

where the step between the last two lines is just the definition of the average of $\rho(\mathbf{0})\rho(\mathbf{R})$ over the scattering volume, with $\mathbf{R} = \mathbf{r} - \mathbf{r}'$. $\langle \rho(\mathbf{0})\rho(\mathbf{R}) \rangle$ is called the density-density correlation function. It is the average of the densities at two points separated by $\mathbf{R}$ and plays a particularly important role in scattering, in particular in disordered systems. A scattering experiment will determine correlation functions and give a statistical average description of the sample structure.

12.2.3 Scattering Length and Refractive Index

X-rays interact with matter through different mechanisms. They can be absorbed by an atom with the subsequent emission of another photon (fluorescence) or electron (photoelectric effect). Anomalous scattering will happen close to an absorption edge where the interaction will become resonant. Raman scattering, Compton scattering or magnetic scattering might also happen depending on the system or radiation energy [6, 7]. In this chapter, we shall mainly consider Thomson (charge) scattering: A charged particle is accelerated by the oscillating electric field and re-radiates.

The velocity $\mathbf{v}$ of free electrons of mass m in an electric field $\mathbf{E}$ is such as $md\mathbf{v}/dt = -e\mathbf{E}$. For a $e^{i\omega t}$ time dependence $\mathbf{v} = (ie/m(\omega))\mathbf{E}$, the displacement is $\mathbf{x} = -i\mathbf{v}/\omega$ and we have an oscillating dipole

$$
\mathbf{p} = -e\mathbf{x} = -\frac{e^2}{m_e \omega^2} \mathbf{E} e^{i\omega t}
\tag{12.5}
$$

associated with each electron. The field radiated by a dipole $\mathbf{p}$ at a distance r in the far field is:

$$
\mathbf{E}_{sc} = \frac{k_0^2 e^{-ik_0 r}}{4\pi\varepsilon_0 r} (\mathbf{r} \times \mathbf{p}) \times \mathbf{r}.
\tag{12.6}
$$

Denoting $E_{in} = \hat{e}_{in}$ and $E_{sc} = \hat{e}_{sc}$, $(\mathbf{r} \times \mathbf{p}) \times \mathbf{r} = r\mathbf{p} - (\mathbf{p} \cdot \mathbf{r})\mathbf{r}$. As $\hat{e}_{sc}$ is normal to $\mathbf{r}$, $(r\mathbf{p} - (\mathbf{p} \cdot \mathbf{r})\mathbf{r}) \cdot \hat{e}_{sc} = pr(\hat{e}_{in} \cdot \hat{e}_{sc})$. The intensity scattered in a unit solid angle is $|E|^2/r^2$, which, using $k_0 = 2\pi/\lambda = \omega/c$, leads to

$$b = \frac{e^2}{4\pi\varepsilon_0 mc^2}(\hat{e}_{in} \cdot \hat{e}_{sc}) = r_e(\hat{e}_{in} \cdot \hat{e}_{sc}). \tag{12.7}$$

$$r_e = \frac{e^2}{4\pi\varepsilon_0 mc^2} = 2.818 \times 10^{-15} \text{ m}, \tag{12.8}$$

with $\varepsilon_0 = 8.85 \times 10^{-12}$ the electric permittivity of vacuum. r_e is called classical electron radius or Thomson radius.

The scattering is more efficient for the light electrons than for the heavy nuclei (mass m in Eq. (12.7)). As long as the frequency of the electromagnetic field is much larger than the characteristic atomic frequencies, which is the case considered above for the generally light atoms considered in soft-condensed matter, the electrons can be considered as free electrons [8, 9], and a material can be simply characterised by its electron density ρ_e.

For heavier atoms, the classical model described above can be pushed further by adding a restoring force $m\omega_0^2$ describing the interaction with the nucleus and a damping force due to the emission of light and transfer of energy to other electrons. Then, $md^2\mathbf{x}/dt^2 + \gamma d\mathbf{x}/dt + m\omega_0^2\mathbf{x} = -e\mathbf{E}$, and

$$b = r_e\frac{\omega^2}{\omega^2 - \omega_0^2 - i\gamma\omega}(\hat{e}_{in} \cdot \hat{e}_{sc}) = r_e(f + f' + if'')(\hat{e}_{in} \cdot \hat{e}_{sc}) \tag{12.9}$$

which describes anomalous scattering, i.e. the variation of the scattering cross-section close to the absorption edge at ω_0. Note that Eq. (12.9) is equivalent to the full quantum mechanical description.

For neutrons, the two main interactions are the strong interaction with the nuclei and, as the neutron is a spin 1/2 particle, the magnetic interaction with the existing magnetic moments (nuclear and electronic) [10]. Neutrons obey the Schrödinger equation and interact with nuclei via the short-range Fermi pseudo-potential:

$$V_F(r) = b\left(\frac{2\pi\hbar^2}{m_n}\right)\delta(\mathbf{r}), \tag{12.10}$$

where $\delta(\mathbf{r})$ is the Dirac delta-function. We have

$$\left(-\frac{\hbar^2}{2m_n}\nabla^2 + \frac{2\pi\hbar^2}{m_n}\sum_i \rho_i b_i\right)\psi(r) = \mathscr{E}\psi(r), \tag{12.11}$$

with m_n the neutron mass and b_i the tabulated scattering length of nuclei i.

An important consideration for neutrons is incoherent scattering. A material will indeed consist of a random distribution of isotopes and spin states, implying a random distribution of scattering lengths. For a given atom i in the system, one can write $b_i = \langle b \rangle + \delta b_i$, where $\langle \delta b \rangle = 0$. Coming back to Eq. (12.3) and developing,

$$
\frac{d\sigma}{d\Omega} = \left| \sum_j b_j e^{i\mathbf{q}\cdot\mathbf{r}_j} \right|^2 = \sum_i \sum_j (\langle b \rangle + \delta b_i)(\langle b \rangle + \delta b_j) e^{i\mathbf{q}\cdot(\mathbf{r}_i - \mathbf{r}_j)}
$$

$$
= \langle b \rangle^2 \left| \sum_j b_j e^{i\mathbf{q}\cdot\mathbf{r}_j} \right|^2 + N(\langle b^2 \rangle - \langle b \rangle^2),
$$

(12.12)

with N the number of nuclei in the system. Indeed $\langle \delta b_i \rangle = \langle \delta b_j \rangle \langle \delta b_i \rangle_{i \neq j} = 0$ and $\langle \delta b_i \delta b_i \rangle = \langle b^2 \rangle - \langle b \rangle^2$. The first term in Eq. (12.12) called coherent scattering is the only interesting term for structure determination and the second called incoherent scattering will only give rise to an isotropic background.

A complementary, often most useful description is based on the use of a refractive index like in optics. For x-rays in a homogeneous medium, we have the Maxwell equations and the constitutive relation,

$$
\nabla \times \mathbf{E} = -\frac{\partial \mathbf{B}}{\partial t}
$$

$$
\nabla \times \mathbf{B} = \mu_0 \mathbf{j} + \mu_0 \frac{\partial \mathbf{D}}{\partial t},
$$

(12.13)

$$
\mathbf{D} = \varepsilon_0 n^2 \mathbf{E},
$$

where ∇ is the $(\partial/\partial x, \partial/\partial y, \partial/\partial z)$ operator. $\mathbf{E}$ is the electric field and $\mathbf{D}$ the electric induction. $\mathbf{B}$ is the magnetic field. $\mu_0 = 4\pi \times 10^{-7} \mathrm{N/A^2}$ is the magnetic permeability of vacuum with $\varepsilon_0 \mu_0 c^2 = 1$. These equations can be combined to give the (Helmholtz) propagation equation:

$$
\nabla \times \nabla \times \mathbf{E} - n^2(\mathbf{r}) \frac{\omega^2}{c^2} \mathbf{E} = 0
$$

$$
= -\nabla^2 \mathbf{E} - n^2(\mathbf{r}) \frac{\omega^2}{c^2} \mathbf{E},
$$

(12.14)

where n is the refractive index with $\varepsilon = n^2$. We have $k = n\omega/c = nk_0$. Note that in Eq. (12.14), all the complexity of the system is carried by $n^2(\mathbf{r})$.

Using Eq. (12.5), the local polarization of the medium is $\mathbf{P}(\mathbf{r}) = \rho_{el}(\mathbf{r})\mathbf{p}(\mathbf{r})$ with $\rho_{el}(\mathbf{r})$ the local electron density. Using $\mathbf{D} = \varepsilon_0 n^2 \mathbf{E} = \varepsilon_0 \mathbf{E} + \mathbf{P}$, we can now define the refractive index:

$$n = 1 - \frac{\rho_{el} e^2}{2 m_e \varepsilon_0 \omega^2} = 1 - \frac{\lambda^2}{2\pi} \rho_{el} r_e \approx 1 - 10^{-6}, \qquad (12.15)$$

for x-rays with $\lambda \approx 1$ Å.

As the Schrödinger equation Eq. (12.11) has exactly the same structure as the propagation equation Eq. (12.14) for x-rays with b_i is the scattering length of nucleus i of density ρ_i, we have

$$n = 1 - \frac{\lambda^2}{2\pi} \sum_i \rho_i b_i.$$

12.3 Reflectivity

12.3.1 Reflectivity in the Born Approximation

Integrating Eq. (12.2) for a perfect dioptre of electron density ρ_{sub}, one obtains

$$\frac{d\sigma}{d\Omega} = b^2 \rho_{sub}^2 \int dz dz' \int d\mathbf{r} d\mathbf{r}' e^{i\mathbf{q}_\| \cdot (\mathbf{r}_\| - \mathbf{r}'_\|)} e^{iq_z z} e^{iq_z z'}$$
$$= \frac{4\pi^2 A b^2 \rho_{sub}^2 \delta(\mathbf{q}_\|)}{q_z^2}, \qquad (12.16)$$

where we have used $\int d\mathbf{r}_\| e^{i\mathbf{q}_\| \mathbf{r}_\|} = 4\pi^2 \delta(\mathbf{q}_\|)$. In Eq. (12.16), A is the illuminated area, $\mathbf{q}_\|$ is the wave-vector transfer component in the plane of the surface and q_z the wave-vector transfer component normal to the surface. The condition $\delta(\mathbf{q}_\|)$ yields specular condition $\theta_{sc} = \theta_{in} = \theta$, $\psi = 0$ and $q_z = 4\pi \sin \theta / \lambda$. The $\delta(\mathbf{q}_\|)$ condition ensures that for an infinite surface, intensity will be reflected only in the $\theta_{sc} = \theta_{in}$ direction (Fig. 12.3).

Integrating over the angular acceptance of the detector $\delta\Omega = d\theta_{sc} d\psi = (2/k_0 q_z) d\mathbf{q}_\|$, and normalising to the incident intensity (leading to a factor $A \sin \theta$ since $d\sigma/d\Omega$ is normalised to a unit incident flux), one obtains for the reflectivity coefficient within the Born approximation

$$R = 16\pi^2 b^2 \rho_{sub}^2 / q_z^4. \qquad (12.17)$$

Due to the inner product $(\hat{\mathbf{e}}_{in} \cdot \hat{\mathbf{e}}_{sc})$ in Eq. (12.7) for b, the reflectivity is 0 if $\hat{\mathbf{e}}_{in}$ and $\hat{\mathbf{e}}_{sc}$ are perpendicular to each other. In other words, the Brewster angle is exactly 45° for x-rays in the Born approximation.

Another important point is that in a binary system or a colloidal system, averaging Eq. (12.16) over all possible orientations of local surfaces (or wave-vectors),

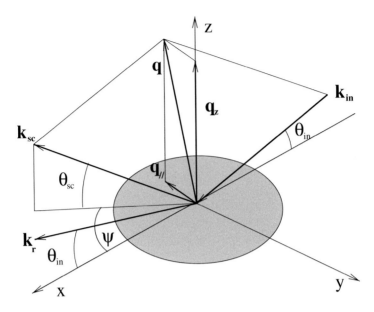

Fig. 12.3 Geometry and notations used in this chapter. θ_{in} is the angle of incidence (which is defined with respect to the interface, differently from the conventions in optics) and θ_{sc} the scattering angle. $\mathbf{k}_{in}$ is the incident wave vector and $\mathbf{k}_r$ the wave vector of the reflected wave. $\mathbf{q}$ is the wave-vector transfer, and $\mathbf{q}_{\parallel}$ its projection onto the sample surface. In a reflectivity experiment, $\theta_{sc} - \theta_{in} - 0$, $\psi - 0$ and the wave vector transfer is $q_z - 4\pi \sin\theta/\lambda$

again leads to a Q^{-4} dependence, the so-called Porod law of small angle scattering [11].

In fact, the Born approximation is not always accurate enough as can be seen from the divergence at $q_z = 0$ in Eq. (12.17), and this can be in particular the case for surfaces.

12.3.2   The Optical Index and Total External Reflection

To solve this problem, the alternative description of matter through a refractive index introduced in the previous section is advantageous. In the limit discussed above where the electromagnetic frequencies are much larger than the characteristic atomic frequencies, this index is local [8, 9], i.e. only averaged over resolution volumes, which explains why atomic resolution can be achieved using x-rays. The generalization of Eq. (12.15) is:

$$n = 1 - \delta - i\beta; \quad \text{with} \quad \delta = \frac{\lambda^2}{2\pi}\rho_b, \tag{12.18}$$

with λ the wavelength, and ρ_b the scattering length density.

From Eq. (12.18), it is easy to see that surface scattering may be large. Indeed, total external reflection will occur whenever going from a medium with a larger index to a medium with a lower index. For x-rays, this happens when going from air or a vacuum to a dense medium since the refractive index of matter is (slightly) less than 1. Indeed, using the Snell-Descartes law of refraction with medium 1 a vacuum and medium 2 a dense medium,

$$\cos \theta_1 = n \cos \theta_2, \qquad (12.19)$$

with $n = 1 - \delta$, we have $\cos \theta_2 = 1$ for $\cos \theta_1 = 1 - \delta$, and total external reflection occurs for grazing angles of incidence $\theta_{in} \leq \theta_c = \sqrt{2\delta} \approx 10^{-3}$ [12]. This phenomenon is of great help for the study of surfaces since for $\theta_{in} < \theta_c$ only an evanescent wave propagates below the surface (with a penetration depth equal to a few nm), and hence surface sensitivity is considerably enhanced. On the other hand, scattering cross-sections are large in the total external reflection region, multiple scattering cannot be neglected, and the simple kinematical approach is no longer good enough.

For an $\exp i(\omega t - k_z z)$ wave, the penetration depth is $1/2\mathscr{I}m(k_z)$ (Fig. 12.4). Using Eq. (12.19), it is easy to show that the imaginary part of the normal component of the wave vector in the medium is:

$$\mathscr{I}m(k_z) = \frac{1}{\sqrt{2}} k_0 \sqrt{[(\theta^2 - 2\delta_i)^2 + 4\beta^2]^{1/2} - (\theta^2 - 2\delta)}. \qquad (12.20)$$

The penetration depth is a few nms below the critical angle, increases around θ_c and is limited by absorption above θ_c.

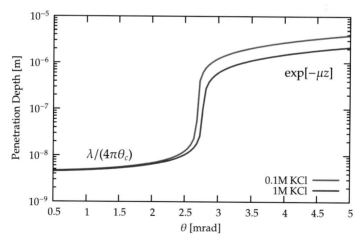

Fig. 12.4 Penetration depth of a 8 keV radiation in a KCl solution as a function of the grazing angle of incidence for two concentrations (0.1 and 1 M)

12.3.3 Fresnel Reflectivity and Optical Formalism

Using boundary conditions (continuity of the tangential component of $\mathbf{E}$ and of the normal component of $\mathbf{D} = n^2\mathbf{E}$, one finds the transmission $r_{0,1}$ and reflection $t_{0,1}$ coefficients for the electric field for a perfect dioptre [12],

$$r = r_{0,1} = \frac{k_{z,0} - k_{z,1}}{k_{z,0} + k_{z,1}}, \tag{12.21}$$

$$t = t_{0,1} = \frac{2k_{z,0}}{k_{z,0} + k_{z,1}} \tag{12.22}$$

The Fresnel reflection coefficient for the intensity is $R = R_F = |rr^*|$.

$$R_F = \frac{I}{I_0} = \frac{|E_{\text{ref}}|^2}{|E_{\text{in}}|^2} = \left| \frac{q_z - \sqrt{q_z^2 - q_c^2}}{q_z + \sqrt{q_z^2 - q_c^2}} \right|^2, \tag{12.23}$$

where E_{in} is the incident field and E_{ref} the reflected field, and we have noted $q_c = (2\pi/\lambda)\theta_c$. Fresnel reflectivity is plotted on Fig. (12.5) where it can be seen that Eq. (12.23) converges to Eq. (12.17) for large q_z values far from total reflection.

Equation (12.23) can be generalized for stratified media using either recursive methods [13] or a matrix formalism [12, 14, 15]. In the latter case, propagation in each medium is characterized by a transfer matrix and refraction at interfaces is also taken care of by a refraction matrix. Reflectivity can be obtained by the multiplication of the matrices characterizing the Fresnel reflectivities of the stratified medium. Interestingly, it can be shown [12] that a very good approximation to the reflectivity of a complex system is given by:

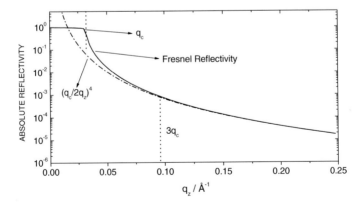

Fig. 12.5 Calculated reflectivity of a flat silicon wafer and asymptotic law after [12]

$$R = R_F \left| \frac{1}{\rho_{\text{sub}}} \int \frac{\partial \rho}{\partial z} e^{iq_z z} dz \right|^2, \tag{12.24}$$

where R_F is the Fresnel reflectivity of the system. Integrating by parts, assuming absorption in the substrate for convergence, one also has

$$R = R_F \left| \frac{q_z}{\rho_{\text{sub}}} \int \rho(z) e^{iq_z z} dz \right|^2, \tag{12.25}$$

close to the Born expression Eq. (12.17), but with a R_F factor instead of the diverging $1/q_z^4$ factor. Equation (12.25) shows that reflectivity will give access to the scattering length density profile normal to the surface.

For a single film of scattering length density ρ_f and thickness ℓ on a substrate, Eq. (12.25) becomes:

$$R = R_F \left[\rho_f^2 + (\rho_f - \rho_{\text{sub}})^2 + 2\rho_f(\rho_f - \rho_{\text{sub}}) \cos(q_z \ell) \right]^2 / \rho_{\text{sub}}^2. \tag{12.26}$$

the reflectivity curve will exhibit fringes resulting from the interference between x-rays or neutrons reflected at the film free surface and the film/substrate interface. The interfringe distance θ is such that $q_z \ell = 4\pi \sin \theta / \lambda \ell = 2\pi$, leading to $2\ell \sin \theta = \lambda$, similar to Bragg's law.

12.3.4 Example: Neutron Reflectivity Investigation of the Swelling of a Polyacrylamide Film

X-ray reflectivity can be easily measured using laboratory sources (x-ray tubes or rotating anodes). Synchrotron radiation allows time-resolved studies, possibly down to ms using a dispersive setup [16]. Radiation damage can be an issue with intense x-ray sources which is avoided with neutrons. Another advantage of neutrons is the possibility to do contrast variation, for example using hydrogenated and deuterated samples with different scattering lengths in order to determine the precise distribution of a given molecule by comparing the two experiments (Fig. 12.6).

As an example, the swelling dynamics of the ultrathin polyacrylamide spin-coated films in saturated vapor of D_2O and H_2O was studied using neutron and X-ray reflectivity in Ref. [17]. A uniform scattering length density (SLD) profile represents the dry films (Fig. 12.7), whereas the SLD profiles corresponding to the swollen films were characterized with a decreasing solvent concentration along the film thickness from the top surface to the film/substrate interface. As the scattering length densities of D_2O and polyacrylamide are known, the overall scattering length density profile of the composite film ρ_{comp} can be converted into concentration profiles using $\rho_{\text{comp}} = c_{D_2O}\rho_{D_2O} + (1 - c_{D_2O})\rho_{\text{PAM}}$. This was used in Ref. [17] to understand the anomalous swelling dynamics of those samples.

Fig. 12.6 Swelling of a polyacrylamide film after [17]. Swelling of the film (*top*) and schematics of the neutron reflectometer at Laboratoire Léon Brillouin (*bottom*)

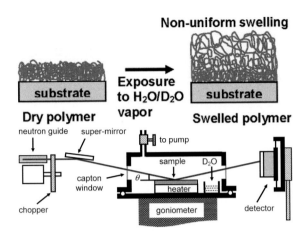

12.4 Perturbation Methods: The Distorted-Wave Born Approximation

As we have seen in Sect. 12.3.1, the Born approximation is not accurate when $\theta < \theta_c$ which is a very interesting limit due to surface sensitivity. More accurate approximations like the distorted-wave Born approximation have therefore been developed to treat this case. A distorted-wave Born approximation (DWBA) is a perturbation theory using as unperturbed state a system as close as possible to the system of interest in which the electric field can be calculated exactly using the Born approximation [18]. For example, if one is interested in the diffraction by a monolayer on a substrate, an appropriate reference system could be composed of the substrate and an homogeneous layer having the average refractive index of the monolayer. If one is interested in the surface roughness of a dioptre, an appropriate reference system would be the same dioptre without roughness. This means that reflection and refraction would be exactly (dynamically) taken into account, but not diffraction from the monolayer, or diffuse scattering from the roughness. More complicated reference states can be considered. An example when dealing with gratings is given in Ref. [19]. Two different formalisms have been developed for the DWBA. The first one [20–22], initiated by Croce and Vineyard is a full optical formalism. The second one, initiated in particular by Sinha [11] is a quantum formalism, which has the advantage of being more simple to handle.

In general the scattering is weak enough that only the first order DWBA can be used, i.e. only single diffraction or diffuse scattering events are considered. The second-order theory [23] is only important to treat reflectivity, i.e. scattering with $q_\parallel = 0$ which cannot be accounted for by single scattering events. Of course, the closest the reference state to the actual experimental system, the better the approximation. For this reason, it has been proposed to use graded reference states to give a better account of scattering by surface roughness [24, 25]. This is however

Fig. 12.7 Swelling of a polyacrylamide film after [17]. **a** Neutron reflectivity data (*symbols*) with fitted profiles (*lines*) of the swelling polymer films when exposed to D$_2$O, from *top* to the *bottom* as a function of increasing exposure time. **b** Scattering length density profiles corresponding to the reflectivity data

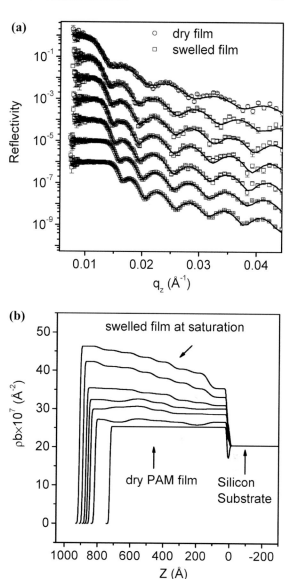

not necessary for soft-condensed matter systems which generally contain only weak scatterers, and for which the only strong contrast is at the surface. For this reason, we use a simpler approximation in the following, in which reflection and refraction are only taken into account at the top or substrate surface. This approximation is good enough for soft-condensed matter systems, and allows a unified treatment of diffraction and diffuse scattering. The interested reader will find the details of the electromagnetic calculations in Ref. [26].

As mentioned in Sect. 12.2.3, all the complexity of the system can now be described using the refractive index in the propagation equation Eq. (12.14):

$$\nabla \times \nabla \times \mathbf{E}(\mathbf{r}) - n_{\mathrm{ref}}^2(\mathbf{r})k_0^2\mathbf{E}(\mathbf{r}) = \delta n^2(\mathbf{r})k_0^2\mathbf{E}(\mathbf{r}). \tag{12.27}$$

Using $\delta\mathbf{P}(\mathbf{r}') = \varepsilon_0\delta n^2(\mathbf{r}')\mathbf{E}(\mathbf{r}')$, we can write

$$\mathbf{E} = \mathbf{E}_{\mathrm{ref}} + \delta\mathbf{E}, \tag{12.28}$$

where $\delta\mathbf{E}$ is the field created by the fictitious dipolar source $\delta\mathbf{P}$ in the reference case. This is an exact equation and the approximations will consist in evaluating this field (Fig. 12.8). Formally, as we are considering linear electromagnetism, this can be done using Green functions $\mathscr{G}(\mathbf{R},\mathbf{r})$ formally defined as:

$$\nabla \times \nabla \times \mathscr{G}(\mathbf{R},\mathbf{r}) - n_{\mathrm{ref}}^2(\mathbf{R})k_0^2\mathscr{G}(\mathbf{R},\mathbf{r}) = k_0^2\varepsilon_0\delta(\mathbf{R}-\mathbf{r}), \tag{12.29}$$

i.e. for a unit dipole instead of $\delta\mathbf{P}$. The perturbation $\delta\mathbf{E}$ is then obtained by summing over all dipoles $\delta\mathbf{P}$:

$$\begin{aligned}
\mathbf{E}(\mathbf{R}) &= \mathbf{E}_{\mathrm{ref}}(\mathbf{R}) + \delta\mathbf{E}(\mathbf{R}) \\
&= \mathbf{E}_{\mathrm{ref}}(\mathbf{R}) + \int d\mathbf{r}\delta\mathbf{P}(\mathbf{r})\mathscr{G}(\mathbf{R},\mathbf{r}) \\
&= \mathbf{E}_{\mathrm{ref}}(\mathbf{R}) + \varepsilon_0 \int d\mathbf{r}\delta n^2(\mathbf{r})\mathbf{E}(\mathbf{r})\mathscr{G}(\mathbf{R},\mathbf{r}).
\end{aligned} \tag{12.30}$$

Using the reciprocity theorem, it can be shown that the Green function can be simply calculated as the electric field in $\mathbf{r}$ at the surface due to a unit dipole in $\mathbf{R}$ (detector) [18, 20]. In a vacuum, it is simply Eq. (12.6) which ensures

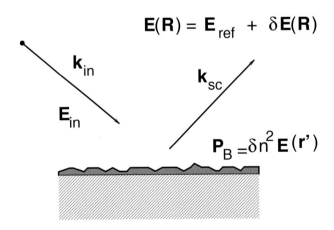

Fig. 12.8 Schematics of the perturbation theory

consistency with the Born approximation. At a surface, it can be calculated like in Sect. 12.3.2. Indeed, the surface electric field is the superposition $(1 + r)E_0$ of the incident and reflected fields above the surface and the transmitted field below the surface tE_0. As the field is continuous, either of them can be used for small heights $q_z\ell \ll 1$, and we simply have in the far field:

$$\mathscr{G}(\mathbf{R}, \mathbf{r}) = t_{0,1}^{sc} k_0^2 \frac{e^{-ik_0|R-r|}}{4\pi\varepsilon_0 |R - r|}. \tag{12.31}$$

Developing $|R - r|$ as $R - \mathbf{r} \cdot \mathbf{R}$ to lowest order, with $\mathbf{R}$ the distance between the detector in the far field, i.e. much farther than the sample size and an arbitrary origin on the sample, one finally gets a plane wave development for the Green function:

$$\mathscr{G}(\mathbf{R}, \mathbf{r}) = t_{0,1}^{sc} k_0^2 \frac{e^{-ik_0R}}{4\pi\varepsilon_0 R} e^{i\mathbf{k}_{sc} \cdot \mathbf{r}} \widehat{\mathbf{e}}_{sc}, \tag{12.32}$$

with $\widehat{\mathbf{e}}_{sc}$ the unit vector giving the polarisation of the scattered field. Equation (12.32) differs from the Born case Eq. (12.6) only by the transmission coefficient $t_{0,1}^{sc}$ calculated in the scattering direction, giving an estimate of the surface electric field. $t_{0,1}$ exhibits a peak at the critical angle (Fig. 12.9) called Vineyard peak in the surface diffraction literature and Yoneda peak in the surface scattering or GISAXS literature. That tE is a good approximation of the surface field is demonstrated in Fig. 12.9 where it is compared to the fluorescence of a

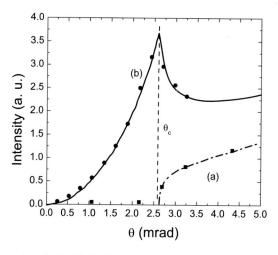

Fig. 12.9 a Penetration depth (*dashed line*) and **b** square of the modulus of the Fresnel Transmission coefficient for polarization (*s*) as a function of the grazing angle of incidence θ (*solid line*). Points are experimental data for Mn K_α fluorescence of respectively, a 10^{-3}M MnCl$_2$ solution (**a**) and a monolayer of Mn^{2+} ions adsorbed at the water surface below a behenic acid monolayer (**b**). The fluorescence intensity emitted by an atom is proportional to E^2 and is therefore a direct measure of the local surface field with the monolayer and the penetration with the solution

monolayer of ions at the water surface. As the fluorescence emitted by an atom is proportional to E^2, it is a direct probe of the electric field. More, generally, the element sensitivity of fluorescence is widely used in surface scattering techniques like grazing incidence fluorescence, like in Fig. 12.9 or standing waves. In the latter, the standing wave field created by the interference between the incident field and the field diffracted by a periodic structure like a crystal or monolayer with a resolution given by a fraction of the period is used to precisely locate atoms.

12.4.1 General Scattering Cross-Section for Scattering in Thin Films

Inserting Eq. (12.32) into Eq. (12.30) with $E = t^{in}_{0,1} e^{-i\mathbf{k}_{in}\cdot\mathbf{r}} \widehat{\mathbf{e}}_{in}$, one obtains:

$$E(\mathbf{R}) = E_{\text{ref}}(\mathbf{R}) + k_0^2 E_0 \frac{e^{-ik_0 R}}{4\pi R} t^{in}_{0,1} t^{sc}_{0,1} \int d\mathbf{r}\,\delta n^2(\mathbf{r}) e^{i\mathbf{k}_{sc}\cdot\mathbf{r}} e^{-i\mathbf{k}_{in}\cdot\mathbf{r}} (\widehat{\mathbf{e}}_{in}\cdot\widehat{\mathbf{e}}_{sc}). \quad (12.33)$$

Using Eq. (12.18) for the refractive index and calculating the flux in a unit solid angle for a unit incident flux, one obtains for the scattering cross-section:

$$d\sigma/d\Omega - d\sigma/d\Omega_{\text{ref}} + \left| t^{in}_{0,1} \right|^2 \left| t^{sc}_{0,1} \right|^2 \left| \int d\mathbf{r}\,\delta\rho_b(\mathbf{r}) e^{i\mathbf{q}\cdot\mathbf{r}} \right|^2, \quad (12.34)$$

with the wave vector transfer $\mathbf{q} = \mathbf{k}_{sc} - \mathbf{k}_{in}$. $t^{in}_{0,1}$ and $t^{sc}_{0,1}$ are the Fresnel transmission coefficients between the upper (0) and lower (1) media, for respectively the angle of incidence θ_{in} and the scattering angle in the scattering plane θ_{sc}. The coefficient $t^{in}_{0,1}$ is an approximation of the actual field scattered by the electron density fluctuations $\delta\rho$, and $t^{sc}_{0,1}$ describes how this field propagates to the detector. $\delta\rho_b$ is the difference between the actual scattering length density and that of the perfect dioptre.

12.4.2 Grazing Incidence Diffraction

We consider here diffraction in a flat thin film, for example a Langmuir film. Denoting $\rho_{uc}(\mathbf{r})$ the electron density in the unit cell, $\mathbf{a}_1$ and $\mathbf{a}_2$ the lattice vectors, and N_1 and N_2 the number of repeat units along $\mathbf{a}_1$ and $\mathbf{a}_2$ respectively (Fig. 12.10),

$$\delta\rho(\mathbf{r}) = \sum_{i_1=1}^{N_1} \sum_{i_2=1}^{N_2} \rho_{uc}(\mathbf{r} - i_1\mathbf{a}_1 - i_2\mathbf{a}_2). \quad (12.35)$$

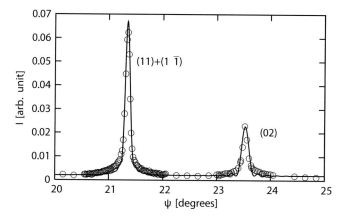

Fig. 12.10 Grazing incidence diffraction from a Langmuir film of behenic acid $C_{22}H_{44}O_2$ in the S phase at temperature $T = 16.1\,^{\circ}C$ and surface pressure $\Pi = 32.1$ mN/m (*circles*). Calculation was made using Eq. (12.37) with $a_1 = 0.5027$ nm and $a_2 = 0.7265$ nm. $N_1 = 140$ and $N_2 = 80$

Then,

$$S(\mathbf{q}) = \left\langle \left| \int d\mathbf{r} \delta\rho(\mathbf{r}) e^{i\mathbf{q}\cdot\mathbf{r}} \right|^2 \right\rangle$$

$$= \left| \int d\mathbf{r} \sum_{i_1=1}^{N_1} \sum_{i_2=1}^{N_2} \rho_{uc}(\mathbf{r}) e^{i(\mathbf{q}_\parallel\cdot\mathbf{a_1})i_1 + i(\mathbf{q}_\parallel\cdot\mathbf{a_2})i_2} e^{i\mathbf{q}\cdot\mathbf{r}} \right|^2. \tag{12.36}$$

Performing first the summation and integrating, one obtains:

$$d\sigma/d\Omega = b^2 \left| t_{0,1}^{in} \right|^2 \left| t_{0,1}^{sc} \right|^2 |\tilde{\rho}_{uc}(\mathbf{q})|^2 \frac{\sin^2\left(\frac{N_1 \mathbf{q}\cdot\mathbf{a_1}}{2}\right)}{\sin^2\left(\frac{\mathbf{q}\cdot\mathbf{a_1}}{2}\right)} \frac{\sin^2\left(\frac{N_2 \mathbf{q}\cdot\mathbf{a_2}}{2}\right)}{\sin^2\left(\frac{\mathbf{q}\cdot\mathbf{a_2}}{2}\right)}, \tag{12.37}$$

where $\tilde{\rho}_{uc}(\mathbf{q})$ is the Fourier transform of $\rho_{uc}(\mathbf{r})$. If N_1 and N_2 are large,

$$d\sigma/d\Omega \approx 4\pi^2 N_1 N_2 b^2 \left| t_{0,1}^{in} \right|^2 \left| t_{0,1}^{sc} \right|^2 |\tilde{\rho}_{uc}(\mathbf{q})|^2 \sum_n \sum_p \delta(\mathbf{q}\cdot\mathbf{a_1} - 2n\pi)\delta(\mathbf{q}\cdot\mathbf{a_2} - 2p\pi)$$

$$\approx \sum_{\mathbf{G}} 4\pi^2 A b^2 \left| t_{0,1}^{in} \right|^2 \left| t_{0,1}^{sc} \right|^2 \frac{|\tilde{\rho}_{uc}(\mathbf{q})|^2}{|\mathbf{a_1} \times \mathbf{a_2}|^2} \delta(\mathbf{q}_\parallel - \mathbf{G}), \tag{12.38}$$

where $\mathbf{G} = (2n\pi/a_1, 2p\pi/a_2)$ is a reciprocal lattice wave vector and one recovers Bragg peaks at reciprocal lattice positions and a scattered intensity proportional to the crystal size. In the z direction $\tilde{\rho}_{uc}(\mathbf{q})$ gives access to the Fourier transform of the unit cell electron density at reciprocal space position $\mathbf{G}$. Note that $A/|\mathbf{a_1} \times \mathbf{a_2}|$ is the number of unit cells.

12.4.3 Example: Structure Determination of Amphiphilic Films at the Air-Water Interface

Langmuir films usually form two-dimensional powders at the water surface. With fatty acids or alcohols, the indexation can be made by choosing a rectangular centered cell with two atoms per cell with $a_1 < a_2$ (choosing a hexagonal cell is also possible) (Fig. 12.11).

If $a_2 = a_1\sqrt{3}$, the cell is hexagonal, and if all molecules in the cell are equivalent, the three first-order peaks (02), (11) and $(1\bar{1})$ are degenerate. This is the case in the so-called LS phase. If $a_2 \neq a_1\sqrt{3}$, the cell is said to be distorted-hexagonal. This is generally the case if the molecules are tilted. If the chain molecules are tilted along one of the cell axes, $[10]$ ($\mathbf{a_1}$) towards their nearest neighbours (NN) like in the L_2 phase, or $[01]$ ($\mathbf{a_2}$) towards their next-nearest neighbours (NNN) like in the L_2' phase, the reflections (11) and $(1\bar{1})$ are degenerate, and distinct from (02). The unit cell parameters a_1 and a_2 can be easily determined from simple geometrical considerations: $a_1 = 4\pi/\sqrt{4q_{(11)}^2 - q_{(02)}^2}$ and $a_2 = 4\pi/q_{(02)}$.

If the two molecules in the unit cell have the same form factor (implying in particular that there is no preferential orientation of the backbone planes, or this

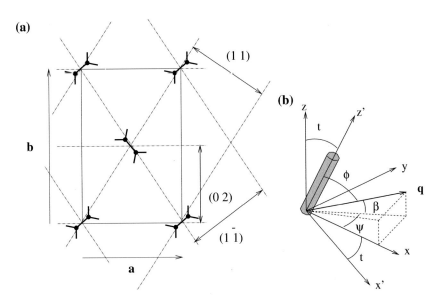

Fig. 12.11 a Indexation of the first-order grazing incidence diffraction peaks for a *centered rectangular cell*. The angle between the backbone planes of the central molecule and the molecules at the *corners* is 90° in the herringbone structure and 40° in the pseudo-herringbone structure. **b** Geometry for the calculation of the form factor of a cylindrical molecule

preferential orientation is the same for all the molecules, contrary to what is represented in Fig. 12.11a). The Fourier transform of the electron density in the unit cell in Eq. (12.38) can be written as:

$$\tilde{\rho}_{uc}(\mathbf{q}) = \tilde{\rho}_{mol}(\mathbf{q}) \left[1 + e^{i\mathbf{q}\cdot[1/2(\mathbf{a}_1 + \mathbf{a}_2)]} \right],$$

where $\tilde{\rho}_{mol}(\mathbf{q})$ is the molecular form factor. It immediately follows that (hk) reflections with odd values of $h+k$ will be forbidden since they will include a factor $1 + cos(\pi(h+k)) = 0$.

The vertical dependence of the scattered intensity depends on the form factor of the tilted cylindrical molecules which can be easily calculated (full calculation using atomic positions is given for example in Ref. [27]) using cylindrical coordinates with an axis parallel to the long axis of the molecule (Fig. 12.11b),

$$\tilde{\rho}_{mol}(\mathbf{q}) = \int_0^{a_{mol}} r dr \int_0^{l_{mol}} dz' \int_0^{2\pi} d\omega \; \rho \; e^{i\mathbf{q}\cdot\mathbf{r}},$$

where z' is the length along the l_{mol} long cylinder of radius a_{mol}. A point $\mathbf{r}$ within the cylinder has for coordinates $\mathbf{r} = (r\cos\omega, r\sin\omega, z')$. The integration is most easily performed using the angle ϕ between the cylinder and the wave-vector transfer as variable:

$$\tilde{\rho}_{mol}(\mathbf{q}) = \rho \frac{e^{iql_{mol}\cos\phi} - 1}{q\cos\phi} \int_0^{a_{mol}} 2\pi r dr J_0(qr\sin\phi).$$

One obtains,

$$\tilde{\rho}_{mol}(\mathbf{q}) = N_{e^-} \left(\frac{e^{iql_{mol}\cos\phi} - 1}{ql_{mol}\cos\phi} \right) \left(\frac{2J_1(qa_{mol}\sin\phi)}{qa_{mol}\sin\phi} \right), \tag{12.39}$$

where N_{e^-} is the total number of electrons in the molecule. From Eq. (12.39) it appears that the molecular form factor is maximum near $\phi = \pi/2$, i.e. is peaked in a direction perpendicular to the tilt direction. We deduce from this that for NN tilt the intensity of the (11) peak is above the horizon by an angle equal to the tilt angle, and that the (02) peak has its maximum intensity on the horizon (Fig. 12.12). For NNN tilt, both peaks have their maximum above the horizon. In any case, the tilt angle t and azimuth can be simply deduced from the peak location. For NN tilt, $\tan t = q_z^{(11)} / \sqrt{q_{\parallel}^{(11)^2} - q_{\parallel}^{(02)^2}/4}$, and for NNN tilt, $\tan t = q_z^{(11)}/q_{\parallel}^{(11)}$. In that case, $q_z^{(11)} = 2q_z^{(02)}$.

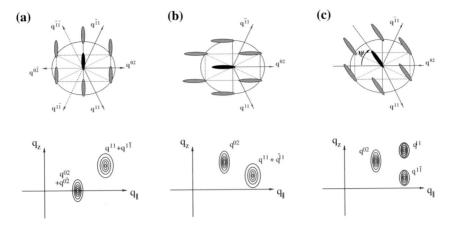

Fig. 12.12 Relation between the molecular tilt azimuth (*top*) and the diffraction pattern (*bottom*) in a distorted-hexagonal cell. The azimuth is indicated by the long axis of the ellipses. **a** Tilt in the nearest-neighbour direction, **b** tilt in the next nearest-neighbour direction, **c** tilt in an intermediate direction

12.5 Surface, Interface and Thin Films Fluctuations

Fluctuations are the rule in soft-condensed matter systems, and they are not restricted to fluctuations in the positional order in ordered systems: surfaces and interfaces may strongly fluctuate. Because these fluctuations are not related to any discontinuous symmetry, they do not give rise to scattering close to a singularity, and scattering is centered at the origin of reciprocal space.

12.5.1 Random Surfaces

A fluctuating surface can be described as a random surface via probability functions. The one-point distribution function $p_1(\mathbf{r}_{\parallel}, z)$ will be the probability to find the surface with height z in $\mathbf{r}_{\parallel}$. The mean height of the surface is then

$$\langle z \rangle (\mathbf{r}_{\parallel}) = \int_{-\infty}^{\infty} z(\mathbf{r}_{\parallel}) p_1(\mathbf{r}_{\parallel}, z) dz. \tag{12.40}$$

Higher order probability functions can also be defined like the two-point distribution function $p_2(\mathbf{r}_{1\parallel}, z_1; \mathbf{r}_{2\parallel}, z_2)$ which gives the probability of finding the surface both at z_1 in $\mathbf{r}_{1\|}$ and z_2 in $\mathbf{r}_{2\|}$. A very important quantity is the height-height correlation function

$$\langle z_1 z_2 \rangle = \int_{-\infty}^{\infty} z_1 z_2 p_2(\mathbf{r}_{1\|}, z_1, \mathbf{r}_{2\|}, z_2) dz_1 dz_2. \tag{12.41}$$

Very often, one will also use the power spectrum:

$$\langle \tilde{z}(\mathbf{q}_\|) \tilde{z}(-\mathbf{q}_\|) \rangle = \int d\mathbf{r}_\| e^{i\mathbf{q}_\| \cdot \mathbf{r}_\|} \langle z(\mathbf{0}_\|) z(\mathbf{r}_\|) \rangle, \tag{12.42}$$

with:

$$\tilde{z}(\mathbf{q}_\|) = \int z(\mathbf{r}_\|) e^{i\mathbf{q}_\| \cdot \mathbf{r}_\|} d\mathbf{r}_\|. \tag{12.43}$$

12.5.2 Scattering Cross-Section

The simplest case is that of a single rough surface separating two homogeneous media (the upper medium being for example a vacuum). Height fluctuations are density fluctuations as the density of the upper and lower phase are different at an interface. Then, Eq. (12.34) becomes:

$$\begin{aligned}
\frac{d\sigma}{d\Omega} &= b^2 \left| t_{0,1}^{in} \right|^2 \left| t_{0,1}^{sc} \right|^2 \left| \int d\mathbf{r} \delta \rho^2 e^{i\mathbf{q} \cdot \mathbf{r}} \right|^2 \\
&= b^2 \left| t_{0,1}^{in} \right|^2 \left| t_{0,1}^{sc} \right|^2 \Delta \rho^2 \left| \int d\mathbf{r}_\| \int_0^{z(\mathbf{r}_\|)} dz e^{i\mathbf{q} \cdot \mathbf{r}} \right|^2.
\end{aligned} \tag{12.44}$$

Integrating over z:

$$\frac{d\sigma}{d\Omega} = b^2 \left| t_{0,1}^{in} \right|^2 \left| t_{0,1}^{sc} \right|^2 \frac{\Delta \rho^2}{|q_z|^2} \left| \int d\mathbf{r}_\| [e^{iq_z z(\mathbf{r}_\|)} - 1] e^{i\mathbf{q}_\| \cdot \mathbf{r}_\|} \right|^2. \tag{12.45}$$

Then,

$$\frac{d\sigma}{d\Omega} = b^2 \left| t_{0,1}^{in} \right|^2 \left| t_{0,1}^{sc} \right|^2 \frac{\Delta \rho^2}{|q_z|^2} \int d\mathbf{r}_\| \int d\mathbf{r}'_\| [e^{iq_z z(\mathbf{r}_\|)}) - 1] [e^{-iq_z z(\mathbf{r}'_\|)} - 1] e^{i\mathbf{q}_\| \cdot (\mathbf{r}_\| - \mathbf{r}'_\|)}. \tag{12.46}$$

Making the change of variables $\mathbf{R}_\parallel = \mathbf{r}_\parallel - \mathbf{r}'_\parallel$ and integrating over $\mathbf{R}_\parallel$:

$$\frac{d\sigma}{d\Omega} = Ab^2 \left|t_{0,1}^{in}\right|^2 \left|t_{0,1}^{sc}\right|^2 \frac{\Delta\rho^2}{|q_z|^2} \int d\mathbf{R}_\parallel \left[\langle e^{iq_z z(\mathbf{R}_\parallel) - iq_z^*(0)}\rangle - \langle e^{iq_z z(\mathbf{R}_\parallel)}\rangle \langle e^{-iq_z^*(\mathbf{R}_\parallel)}\rangle\right] e^{i\mathbf{q}_\parallel \cdot \mathbf{R}_\parallel}.$$

(12.47)

Assuming Gaussian statistics, and, in any case, at second order,

$$\langle e^{iq_z z(\mathbf{R}_\parallel)}\rangle = e^{-\frac{1}{2}q_z^2 \langle z(\mathbf{R}_\parallel)\rangle^2},$$

(12.48)

one finally obtains:

$$\frac{d\sigma}{d\Omega} = Ab^2 \left|t_{0,1}^{in}\right|^2 \left|t_{0,1}^{sc}\right|^2 \Delta\rho^2 \frac{e^{-q_z^2 \langle z^2\rangle}}{|q_z|^2} \int d\mathbf{R}_\parallel \left[e^{|q_z|^2 \langle z(\mathbf{R}_\parallel)z(0)\rangle} - 1\right] e^{i\mathbf{q}_\parallel \cdot \mathbf{R}_\parallel},$$

(12.49)

where $\langle z^2\rangle$ is the surface r.m.s. roughness, and $\langle z(0)z(\mathbf{r}_\parallel)\rangle$ is the height-height correlation function.

Note that in the limit of small q_z's:

$$d\sigma d\Omega = Ab^2 \left|t_{0,1}^{in}\right|^2 \left|t_{0,1}^{sc}\right|^2 \Delta\rho^2 \langle z(\mathbf{q}_\parallel)z(-\mathbf{q}_\parallel)\rangle,$$

(12.50)

i.e. the scattering cross-section is proportional to the power spectrum. An important point to note is the absence of $\delta(\mathbf{q})$ functions in Eqs. (12.49), (12.50), implying in particular that the scattered intensity will be proportional to the solid angle subtended by the detector.

12.5.3   Example: The Liquid Vapor Interface

An interesting case is that of liquid interfaces for which the surface roughness is due to capillary waves.

A first approach to the structure of liquid surfaces was initiated in 1893 by van der Waals who described the liquid-vapour interface as a region of smooth transition from the density of the liquid to that of the gas [28, 29]. Conversely, in the 1965 capillary wave model of Buff et al. [30], a step-like local profile was assumed for the liquid-vapour interface whose large-scale width results from the wandering of the interface due to the propagation of thermally excited capillary waves. Light scattering experiments have shown that this model gives an accurate description of the liquid surface for in-plane length scales larger than one micron [31–33].

We give below as an example the calculation of the scattering cross-section starting from the free energy of fluctuations. The free energy of a deformed liquid surface can be written:

$$\mathscr{H} = \int \frac{1}{2}\Delta\rho g\zeta^2 + \gamma\left(\sqrt{1 + \left(\frac{\partial\zeta}{\partial x}\right)^2 + \left(\frac{\partial\zeta}{\partial x}\right)^2} - 1\right)$$

$$= \frac{1}{2}A\sum\zeta(\mathbf{q}_\parallel)\zeta^*(\mathbf{q}_\parallel)\left[\Delta\rho g\zeta^2 + \gamma q_\parallel^2\right]. \tag{12.51}$$

$\Delta\rho$ is the density difference between the lower and upper liquid, γ is the surface tension, g the acceleration of gravity and A the area. The first term gives the gravitational energy difference when replacing vapor by liquid and the second term the increase in surface energy, proportional to the surface area increase. Applying equipartition of energy, the capillary wave spectrum can be written:

$$\langle\zeta(\mathbf{q}_\parallel)\zeta(-\mathbf{q}_\parallel)\rangle = \frac{1}{A}\frac{k_B T}{\Delta\rho g + \gamma q_\parallel^2}. \tag{12.52}$$

with the thermal energy $k_B T$. Equation (12.52) describes thermally excited capillary waves limited by gravity at large scales and by surface tension at distances smaller than the so-called capillary length ($l_c = \sqrt{\gamma/\Delta\rho g} \approx 2.7\,\text{mm}$ for water). More precisely, the $\gamma q_\parallel^2$ term stems from the increase in interfacial area due to the deformation. Fourier transforming, we obtain the height-height correlation function:

$$\langle\zeta(0)\zeta(\mathbf{r}_\parallel)\rangle = k_B T/(2\pi\gamma)K_0(r_\parallel\sqrt{\Delta\rho g/\gamma}). \tag{12.53}$$

K_0 is the modified second kind Bessel function of order 0. $K_0(x)_{x\to 0} \approx Log 2 - \gamma_E Log x$ with γ_E Euler's constant, and $\lim_{x\to\infty} K_0(x) = 0$. Then, the scattering cross-section can be written to a good approximation [5, 11, 34] as:

$$\frac{d\sigma}{d\Omega} = Ab^2\rho_\text{sub}^2\left|t_{0,1}^\text{in}\right|^2\left|t_{0,1}^\text{sc}\right|^2\frac{2k_B T}{\gamma q_\parallel^2}\left(\frac{q_\parallel}{q_\text{max}}\right)^\tau, \tag{12.54}$$

where $\tau = (k_B T/2\pi\gamma)q_z^2$, $q_\text{min} = \sqrt{\Delta\rho g/\gamma}$ is the minimum wave vector in the capillary wave spectrum, and q_max is the largest one, on the order $2\pi/\text{molecular}$ size.

In fact, this is not enough to fully describe the liquid surface as acoustic waves in the liquid must also be included. In that case, the scattering cross-section is [18]:

$$d\sigma/d\Omega = b^2\left|t_{0,1}^\text{in}\right|^2\left|t_{0,1}^\text{sc}\right|^2\int_{-\infty}^{0}dz\int_{-\infty}^{0}dz' e^{iq_{z,\text{sub}}z}e^{-iq_{z,\text{sub}}^*z'}$$

$$\int d\mathbf{r}_\parallel\langle\delta\rho(0,z')\delta\rho(\mathbf{r}_\parallel,z)\rangle e^{i\mathbf{q}_\parallel\cdot\mathbf{r}_\parallel}, \tag{12.55}$$

where $\mathbf{q}_{z,\text{sub}}$ is the normal component of the wave-vector transfer in the substrate, and the dependence of the bulk correlation function on $\mathbf{r}_\parallel$ and z have been explicitly shown. In the important case of a liquid, $\langle \delta\rho(\mathbf{r})\delta\rho(\mathbf{r}')\rangle = \rho_{\text{sub}}^2 k_B T \kappa_T \delta(\mathbf{r} - \mathbf{r}')$, where κ_T is the liquid isothermal compressibility, and the integration yields:

$$d\sigma/d\Omega = Ab^2 \left|t_{0,1}^{\text{in}}\right|^2 \left|t_{0,1}^{\text{sc}}\right|^2 \rho_{\text{sub}}^2 \frac{k_B T \kappa_T}{2\mathscr{I}m(q_{z,\text{sub}})}. \tag{12.56}$$

$2\mathscr{I}m(q_{z,\text{sub}})$ is the effective penetration length in the liquid. The total scattering due to surface and bulk fluctuations is therefore:

$$d\sigma/d\Omega = A\rho_{\text{sub}}^2 b^2 \left|t_{0,1}^{\text{in}}\right|^2 \left|t_{0,1}^{\text{sc}}\right|^2 \left[\frac{k_B T}{\gamma q_{\parallel}^2}\left(\frac{q_{\parallel}}{q_{\text{max}}}\right)^{\tau} + \frac{k_B T \kappa_T}{2\mathscr{I}m(q_{z,\text{sub}})}\right]. \tag{12.57}$$

Depending on whether the grazing angle of incidence is below or above the critical angle for total external reflection, the effective penetration length varies from less than 6 nm whatever the scattering angle (evanescent wave) to more than 10 μm, allowing the control of the relative weight of the surface and bulk terms (Fig. 12.13). Figure 12.13a gives a striking picture of the very nature of a liquid surface. Starting from high $q_{\parallel}$ values, i.e. at local scales, one observes the static peak due to the short-range order of the nearest neighbours. The same peak is observed in the bulk if the penetration length is increased. Then, going to larger length scales (smaller $q_{\parallel}$ values) for the surface, we observe the characteristic divergence in the spectrum. On the contrary, there is no divergence in the low-$q_{\parallel}$ end of the bulk spectrum which is well described by the white spectrum of acoustic waves. If we now look for a more precise understanding of the surface experimental data, they are correctly described by the simple capillary wave spectrum only up to $q_{\parallel} \approx 5 \times 10^8$ m^{-1}. A much better agreement can be obtained by including the bulk fluctuations contribution within the penetration depth, but the experimental scattering is still significantly larger than calculated for 5×10^8 m$^{-1} \lesssim q_{\parallel} \lesssim 10^{10}$ m^{-1}. This additional scattering does not come from the bulk and is due to the surface which is thus found rougher than expected from the simple model. The corresponding larger thermally excited surface fluctuations can be attributed to a lower surface energy, and the scale-dependent surface energy $\gamma(q_{\parallel})$ could be estimated from the measurements. An explanation for this lowering of the surface energy can be found in Ref. [35] taking into account the long-range power law decay of the dispersion (van der Waals) forces always existing between molecules. Indeed, the origin of the surface energy lies in the non-compensation of molecular interactions at the interface. A simple interpretation of the observed effect (smaller surface tension for small wavelength fluctuations) is that this non-compensation of molecular interactions at the surface is reduced for a corrugated interfacial configuration having a wavelength shorter than the molecular interaction range, resulting in a lowering of the surface energy or surface tension.

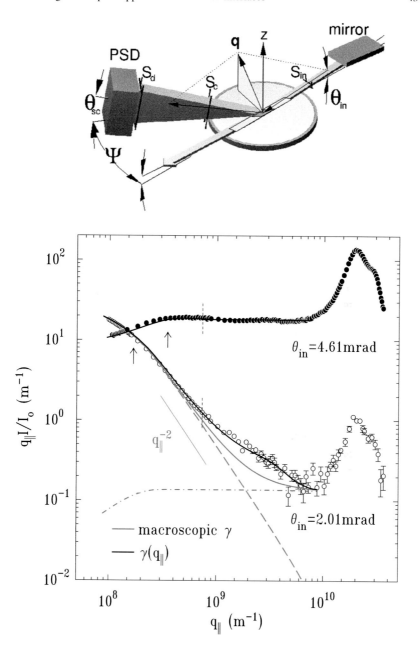

◀ **Fig. 12.13** Scattering at the air-water interface. Geometry of the experiment (*top*). Scans in the *horizontal plane* using a vertically mounted PSD (*bottom*). Signal integrated between $\theta_{sc} = 10$ mrad and 0.1 rad for two values of the grazing angle of incidence: $\theta_{in} = 4.61$ mrad above the critical angle for total external reflection θ_c (*black circles*, bulk scattering dominant) and $\theta_{in} = 2.01$ mrad $< \theta_c$ (*empty circles*, capillary wave contribution dominant for all but high $q_\parallel$ values). The *continuos grey lines* are the result of calculations using Eq. (12.57), split into capillary-wave contribution using the macroscopic surface tension γ (*grey long-dashed lines*), and acoustic-wave contribution (*grey short-dashed lines*). The *continuous black lines* have been calculated using $\gamma(q_\parallel)$ given by the theory [35]. There is no adjustable parameter in any of those calculations. The experimental signal has been multiplied by $q_\parallel$ in order to compensate approximately for resolution effects and to obtain curves proportional to the scattering cross-section or to the fluctuation spectra (at least above a resolution cut-off indicated by *arrows*). Note the peak at $q_\parallel \approx 2 \times 10^{10}$ m^{-1} giving the short-range structure of nearest neighbours in water. The two-peaks for $\theta_{in} = 4.61$ and 2.01 mrad superpose precisely when accounting for the penetration depth difference

12.5.4 Thin Films and Membranes

More generally, it is important to note that for the equilibrium structures considered in soft-condensed matter, the equipartition of energy will always yield $\langle \zeta(\mathbf{q}_\parallel)\zeta(-\mathbf{q}_\parallel)\rangle = k_B T / \mathcal{H}(\mathbf{q})$, where $\mathcal{H}(\mathbf{q})$ is the energy necessary to deform the interface as a sine wave of amplitude $\zeta(\mathbf{q})$. This means that the scattered intensity will always be inversely proportional to the energy necessary to deform the interface, and that we will be able to determine this important quantity at all length scales.

Particularly interesting is the case of thin films. Considering for simplicity only the case of an homogeneous film of thickness d, Eq. (12.34) leads to:

$$
\frac{d\sigma}{d\Omega} = \frac{A}{q_z^2} b^2 \left|t_{0,1}^{in}\right|^2 \left|t_{0,1}^{sc}\right|^2 \left[\rho_{film}^2 e^{-q_z^2 \langle \zeta_f^2 \rangle} \int d\mathbf{r}_\parallel \left(e^{q_z^2 \langle \zeta_f(0)\zeta_f(\mathbf{r}_\parallel)\rangle} - 1\right)e^{i\mathbf{q}_\parallel \cdot \mathbf{r}_\parallel}\right]
$$
$$
+ 2\rho_{film}^2(\rho_{sub}^2 - \rho_{film}^2)e^{-\frac{1}{2}q_z^2(\langle \zeta_f^2 \rangle + \langle \zeta_s^2 \rangle)}\cos(q_z d)\int d\mathbf{r}_\parallel \left(e^{q_z^2 \langle \zeta_f(0)\zeta_s(\mathbf{r}_\parallel)\rangle} - 1\right)e^{i\mathbf{q}_\parallel \cdot \mathbf{r}_\parallel}
$$
$$
+ \rho_{sub}^2 e^{-q_z^2 \langle \zeta_s^2 \rangle} \int d\mathbf{r}_\parallel \left(e^{q_z^2 \langle \zeta_s(0)\zeta_s(\mathbf{r}_\parallel)\rangle} - 1\right)e^{i\mathbf{q}_\parallel \cdot \mathbf{r}_\parallel},
$$

$$(12.58)$$

where ρ_{film} is the film electron density, and ζ_f and ζ_s denote the film-vacuum and film-substrate interface positions respectively. Equation (12.58) shows that the cross-correlation $\langle \zeta_f(0)\zeta_f(\mathbf{r}_\parallel)\rangle$ between the interfaces can be determined since the contrast of the interference pattern directly depends of this correlation, and the different contributions may be separated. In a system at equilibrium, this correlation will result from interactions which can therefore be determined. An interesting example is that of wetting films on rough surfaces [36].

Similar phenomena will happen in membranes. Interactions between two lipid bilayers supported on a solid substrate in water have been investigated in [37]. Two bilayers of DSPC, ($L - \alpha$ 1,2-distearoyl-sn-glycero-3-phosphocholine) were used in

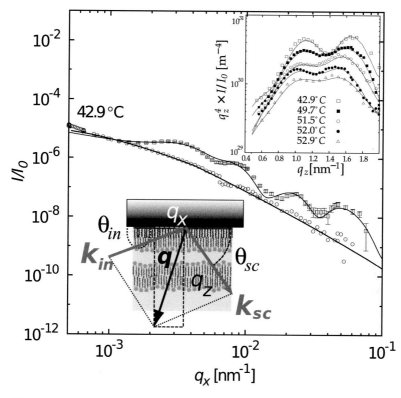

Fig. 12.14 Off-specular reflectivity from silicon substrate (○) and a DSPC bilayer at T = 42.9 °C (□) as a function of q_x. *Continuous lines* represent best fits. *Top inset* off-specular reflectivity as a function of q_z zoomed into the region where it is most sensitive to the potential at T = 42.9 °C (□), 49.7 °C (■), 51.5 °C (○), 52.0 °C (●), 52.9 °C (△). Note the shift in minimum and decrease in contrast with increasing temperature. *Bottom inset* Schematics of the experiment

Ref. [37] (Fig. 12.14). It was shown in Ref. [38] that the second bilayer was able to freely fluctuate in the potential of the first bilayer and of the substrate.

$$\mathscr{H} = \sum_q \frac{1}{2} \sum_{i=1}^2 (U_i'' + \gamma_i q^2 + \kappa_i q^4 + U_{12}'')|u_i(q)|^2 - U_{12}''u_1(q)u_2(q), \qquad (12.59)$$

with γ the membrane tension ≤1 mN/m, κ its bending rigidity, U_i'' the second derivative of the interaction potential of membrane i with the substrate, and U_{12}'', the second derivative of the inter-membrane potential [39]. $u_1(q)$ is the deformation of the first membrane close to the surface and $u_2(q)$ that of the floating membrane. Using the same method as above, not only the membrane structure, but also γ, κ and the potentials can be accurately determined (Fig. 12.14). Indeed, according to Eq. (12.58), the interference term depends on the cross-correlation between the two surfaces. A zoom of the off-specular reflectivity in the region where it is most sensitive to the interaction potential is shown in the inset of Fig. 12.14 for different temperatures. It is important to note here that the second derivative of the inter-bilayer interaction potential is directly linked to the depth of the minimum in the diffuse scattering curve around $q_z \approx 1.0 - 1.5$ nm^{-1} without much coupling to the other parameters. Similarly, the interbilayer water thickness is strongly correlated to the q_z position of that minimum. Hence, it can be seen directly in the inset of Fig. 12.14 that the interaction potential becomes weaker (the minimum is less pronounced) when the interlayer water thickness increases (left shift of the minimum) at higher temperatures. Careful modeling of the scattering in Ref. [37] allowed the authors to evidence the very weak electrostatic repulsion between almost neutral bilayers and to discriminate between different entropic potentials.

12.5.5 Grazing Incidence Scattering

Let us finally consider the case of domains or particles at a surface or an interface. We need to determine $\delta\rho(\mathbf{r})$ and insert it into Eq. (12.34). The particles are located at $\mathbf{r}_{i\|}$, have a scattering length density $\Delta\rho$ in excess of the reference, and their surface density is ρ_{part}. We have:

$$\delta\rho(\mathbf{r}) = \Delta\rho \sum_i \delta(\mathbf{r}_\| - \mathbf{r}_{i\|}) \otimes F(\mathbf{r}), \tag{12.60}$$

where $F(\mathbf{r})$ is a function describing the shape of the particles, taking a value of 1 inside the particle and 0 outside. The term $\left| \int d\mathbf{r}\delta\rho(\mathbf{r})e^{i\mathbf{q}\cdot\mathbf{r}} \right|^2$ in Eq. (12.34) becomes:

$$\left| \int d\mathbf{r}\delta\rho(\mathbf{r})e^{i\mathbf{q}\cdot\mathbf{r}} \right|^2 = \Delta\rho^2 \left| \tilde{F}(\mathbf{q}) \right|^2 \left| \int d\mathbf{r}_\| \sum_i \delta(\mathbf{r}_\| - \mathbf{r}_{i\|})e^{i\mathbf{q}_\|\cdot\mathbf{r}_\|} \right|^2. \tag{12.61}$$

One can then write:

$$\left| \int d\mathbf{r}_\| \sum_i \delta(\mathbf{r}_\| - \mathbf{r}_{i\|})e^{i\mathbf{q}_\|\cdot\mathbf{r}_\|} \right|^2 = \int d\mathbf{r}_\| d\mathbf{r}'_\| \sum_{i,j} \delta(\mathbf{r}_\| - \mathbf{r}_{i\|})\delta(\mathbf{r}'_\| - \mathbf{r}_{j\|})e^{i\mathbf{q}_\|\cdot\mathbf{r}_\|}e^{-i\mathbf{q}_\|\cdot\mathbf{r}'_\|}$$

$$= N + \sum_{i \neq j} e^{i\mathbf{q}_\|\cdot(\mathbf{r}_{i\|} - \mathbf{r}_{j\|})}$$

$$= N + \int d\mathbf{r}_\| \sum_{i \neq j} \delta(\mathbf{r}_\| - \mathbf{r}_{i\|} + \mathbf{r}_{j\|})e^{i\mathbf{q}_\|\cdot\mathbf{r}_\|}$$

$$= N + N\rho_{\text{part}} \int d\mathbf{r}_\| g(\mathbf{r}_\|)e^{i\mathbf{q}_\|\cdot\mathbf{r}_\|},$$

$$\tag{12.62}$$

with N the number of particles and we have defined the pair correlation function :

$$g(\mathbf{r}_{\parallel}) = \frac{1}{\rho_{part}} \langle \sum_{i \neq 0} \delta(\mathbf{r}_{\parallel} - \mathbf{r}_{i\parallel}) \rangle. \tag{12.63}$$

In Eq. (12.63, ρ_{part} is the surface density of particles. $\rho_{part} g(\mathbf{r}_{\parallel})$ is the probability to find a particle at $\mathbf{r}_{\parallel}$, knowing that there is a particle at $\mathbf{0}$ (Fig. 12.15).

We then have for the scattering cross-section:

$$\frac{d\sigma}{d\Omega} = A\rho_{part} \left| t_{0,1}^{in} \right|^2 \left| t_{0,1}^{sc} \right|^2 \Delta\rho^2 \left| \tilde{F}(\mathbf{q}) \right|^2 \left[1 + \rho_{part} \int d\mathbf{r}_{\parallel} g(\mathbf{r}_{\parallel}) e^{i\mathbf{q}_{\parallel} \cdot \mathbf{r}_{\parallel}} \right], \tag{12.64}$$

which is the equivalent of small angle scattering for a two-dimensional array of particles. $A\rho_{part}$ is the total number of particles. For a sphere of radius R,

$$F(\mathbf{q}) = \int_V e^{i\mathbf{q}\cdot\mathbf{r}} d\mathbf{r} = \int_0^R 4\pi r^2 \frac{\sin(qr)}{qr} dr$$
$$= 4\pi \frac{\sin(qR) - qR\cos(qR)}{q^3} \tag{12.65}$$

The square of $F(\mathbf{q})$ divided by the volume of the particle is usually called the form factor $P(\mathbf{q})$.

$$P_{sphere}(\mathbf{q}) = 9 \frac{(\sin(qR) - qR\cos(qR))^2}{(qR)^6}. \tag{12.66}$$

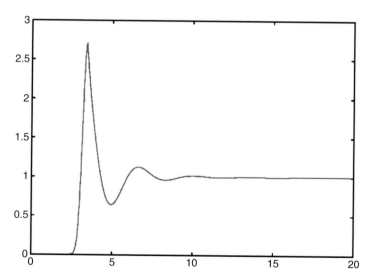

Fig. 12.15 Example of pair correlation function

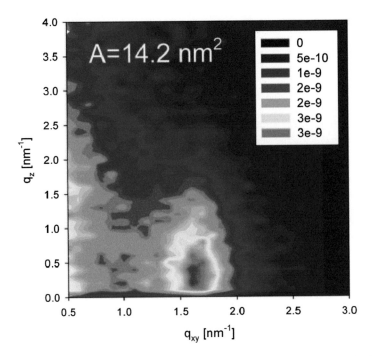

Fig. 12.16 Grazing incidence x-ray scattering of gold nanoparticles stabilized by a mixed ligand shell of 1-hexanethiol and 11-mercapto-1-undecanol at the n-tetradecane/water interface after [40]

An example of grazing incidence x-ray scattering of gold nanoparticles stabilized by a mixed ligand shell of 1-hexanethiol and 11-mercapto-1-undecanol at the n-tetradecane/water interface after [40] is given in Fig. 12.16. The particles have a radius of 1.1 ± 0.1 nm. The in-plane position of the peak reflects the first maximum in $g(\mathbf{r}_\parallel)$ with an interparticle distance of $\approx 2\pi/1.6 = 3.9$ nm. The z-dependence of the intensity results from the particle shape.

Scattering methods (reflectivity and GISAXS) where also used in Ref. [41] to investigate the interfacial synthesis of gold nanoparticles.

References

1. P. Dutta, J.B. Peng, B. Lin, J.B. Ketterson, M. Prakash, P. Georgopoulos, S. Ehrlich, X-ray diffraction studies of organic monolayers on the surface of water. Phys. Rev. Lett. **58**(21), 2228–2231 (1987)
2. K. Kjaer, J. Als-Nielsen, C.A. Helm, L.A. Laxhuber, H. Möhwald, Ordering in lipid monolayers studied by synchrotron x-ray diffraction and fluorescence microscopy. Phys. Rev. Lett. **58**(21), 2224–2228 (1987)
3. A. Braslau, M. Deutsch, P.S. Pershan, A.H. Weiss, J. Als-Nielsen, J. Bohr, Surface roughness of water measured by x-ray reflectivity. Phys. Rev. Lett. **54**(2), 114–117 (1985)

4. D.K. Schwartz, M.L. Schlossman, E.H. Kawamoto, G.J. Kellogg, P.S. Pershan, B.M. Ocko, Thermal diffuse x-ray scattering studies of the water-vapor interface. Phys. Rev. A **41**(10), 5687–5690 (1990)
5. M.K. Sanyal, S.K. Sinha, K.G. Huang, B.M. Ocko, X-ray scattering study of capillary-wave fluctuations at a liquid surface. Phys. Rev. Lett. **66**(5), 628–631 (1991)
6. R.W. James, *The Optical Principles of the Diffraction of X-Rays* (Bell and sons, London, 1948)
7. F. de Bergevin, Interaction of x-rays (and neutrons) with matter, in *X-Ray and Neutron Reflectivity: Principles and Applications*, vol. 770, Lecture Notes in Physics, ed. by J. Daillant, A. Gibaud (Springer, Berlin, 2009), pp. 1–60
8. L.D. Landau, E.M. Lifshitz, *Electrodynamics of Continuous Media*, vol. 8, Course of Theoretical Physics (Pergamon Press, Oxford, 1960)
9. D.W. Oxtoby, F. Novack, S.A. Rice, The Ewald-Oseen theorem in the x-ray frequency region: a microscopic analysis. J. Chem. Phys. **76**, 5278–5284 (1982)
10. C. Fermon, F. Ott, A. Menelle, Neutron reflectometry, in *X-Ray and Neutron Reflectivity: Principles and Applications*, vol. 770, Lecture Notes in Physics, ed. by J. Daillant, A. Gibaud (Springer, Berlin, 2009), pp. 194–247
11. S.K. Sinha, E.B. Sirota, S. Garoff, X-ray and neutron scattering from rough surfaces. Phys. Rev. B **38**(4), 2297–2311 (1988)
12. A. Gibaud, G. Vignaud, Specular reflectivity from smooth and rough surfaces, in *X-ray and Neutron Reflectivity: Principles and Applications*, vol. 770, Lecture Notes in Physics, ed. by J. Daillant, A. Gibaud (Springer, Berlin, 2009), pp. 87–139
13. L.G. Parratt, Surface studies of solids by total reflection of x-rays. Phys. Rev. **95**, 359 (1954)
14. A. Herpin, Calcul du pouvoir réflecteur d'un systême stratifié quelconque. C. R. Acad. Sci. Paris **225**, 182 (1948)
15. M. Born, E. Wolf, *Principles of Optics* (Pergamon, Oxford, 1993)
16. T. Matsushita, E. Arakawa, Y. Niwa, Y. Inada, T. Hatano, T. Harada, Y. Higashi, K. Hirano, K. Sakurai, M. Ishii, M. Nomura, A simultaneous multiwavelength dispersive x-ray reflectometer for time-resolved reflectometry. Eur. Phys. J. Special Topics, **167**, 113–119 (2009) (An original scheme with strong potential for dispersive reflectivity)
17. M. Mukherjee, A. Singh, J. Daillant, A. Menelle, F. Cousin, Effect of solvent-polymer interaction in swelling dynamics of ultrathin polyacrylamide films: a neutron and x-ray reflectivity study. Macromolecules **40**(4), 1073–1080 (2007)
18. J. Daillant, S. Mora, A. Sentenac, Diffuse scattering, in *X-Ray and Neutron Reflectivity: Principles and Applications*, vol. 770, Lecture Notes in Physics, ed. by J. Daillant, A. Gibaud (Springer, Berlin, 2009), pp. 121–162
19. T. Baumbach, P. Mikulik, X-ray reflection from multilayer gratings, in *X-Ray and Neutron Reflectivity: Principles and Applications*, vol. 770, Lecture Notes in Physics, ed. by J. Daillant, A. Gibaud (Springer, Berlin, 2009), pp. 266–274
20. P. Croce, Sur la propagation des ondes électromagnétiques dans les milieux stratifiés et diffusants traitée par la méthode de Green. J. Optics (Paris) **14**(4), 213–220 (1983)
21. L. Névot, P. Croce, Caractérisation des surfaces par réflexion rasante de rayons x. Application à l'étude du polissage de quelques verres silicates. Revue Phys. Appl. **15**, 761–779 (1980)
22. G.H. Vineyard, Grazing-incidence diffraction and the distorted-wave approximation for the study of surfaces. Phys. Rev. B **26**(8), 4146–4159 (1982)
23. D.K.G. de Boer, Influence of the roughness profile on the specular reflectivity of x-rays and neutrons. Phys. Rev. B **49**, 5817–5820 (1994)
24. I.A. Artyukov, A.Y. Karabekov, I.V. Kozhevnikov, B.M. Alaudinov, V.E. Asadchikov, Experimental observation of the near surface layer effects on x-ray reflection and scattering. Phys. B **198**, 9–12 (1994)
25. S. Dietrich, A. Haase, Scattering of x-rays and neutrons at interfaces. Phys. Rep. **260**, 1–138 (1995)
26. J. Daillant, A. Gibaud (eds.), *X-Ray and Neutron Reflectivity: Principles and Applications*, vol. 770, Lecture Notes in Physics (Springer, Berlin, 2009)

27. J. Pignat, J. Daillant, S. Cantin, F. Perrot, O. Konovalov, Grazing incidence x-ray diffraction study of the tilted phases of Langmuir films: determination of molecular conformations using simulated annealing. Thin Solid Films **515**(14), 5691–5695 (2007)

28. J.D. van der Waals, The thermodynamic theory of capillarity under the hypothesis of a continuous variation of density. Verhandel. Konink. Akad. Weten. Amsterdam (Sect. 1), **1**(8) (1893)

29. J.W. Cahn, J.E. Hilliard, Free energy of a nonuniform system. I. interfacial free energy. J. Chem. Phys. **28**, 258–267 (1958)

30. F.P. Buff, R.A. Lovett, R.H. Stillinger, Interfacial density profile for fluids in the critical region. Phys. Rev. Lett. **15**(15), 621–623 (1965)

31. R. Loudon, Theory of thermally induced surface fluctuations on simple liquids. Proc. R. Soc. Lond. **372**, 275–295 (1980)

32. R. Loudon, Ripples on liquid interfaces, in *Surface Excitations*, vol. 9, Modern Problems in Condensed Matter Sciences, ed. by V.M. Agranovich, R. Loudon (North-Holland Physics Publishing, Amsterdam, 1984), pp. 589–638

33. D. Beysens, M. Robert, Thickness of fluid interfaces near the critical point from optical reflectivity measurements. J. Chem. Phys. **87**(5), 3056–3061 (1987)

34. M. Fukuto, R.K. Heilmann, P.S. Pershan, J.A. Griffiths, S.M. Yu, D.A. Tirrel, X-ray measurements of non-capillary spatial fluctuations from a liquid surface. Phys. Rev. Lett. **81** (16), 3455–3458 (1998)

35. K.R. Mecke, S. Dietrich, Effective hamiltonian for liquid-vapor interfaces. Phys. Rev. E **59**, 6766–6784 (1999)

36. I.M. Tidswell, T.A. Rabedeau, P.S. Pershan, J.P. Folkers, M.P. Baker, G.M. Whitesides, Wetting films on chemically modified surfaces: an x-ray study. Phys. Rev. B **44**, 10869–10879 (1991)

37. A. Hemmerle, L. Malaquin, T. Charitat, S. Lecuyer, G. Fragneto, J. Daillant, Controlling interactions in supported bilayers from weak electrostatic repulsion to high osmotic pressure, in *Proceedings of the National Academy of Sciences* (2012)

38. J. Daillant, E. Bellet-Amalric, A. Braslau, T. Charitat, G. Fragneto, F. Graner, S. Mora, F. Rieutord, B. Stidder, Structure and fluctuations of a single floating lipid bilayer. Proc. Nat. Acad. Sci. U S A **102**, 11639 11644 (2005)

39. L. Malaquin, T. Charitat, S. Lecuyer, G. Fragneto, J. Daillant, Controlling interactions in supported bilayers from weak electrostatic repulsion to high osmotic pressure. Eur. Phys. J. E **31**(3), 285–301 (2010)

40. S. Kubowicz, M.A. Hartmann, J. Daillant, M.K. Sanyal, V.V. Agrawal, C. Blot, O. Konovalov, H. Mohwald, Gold nanoparticles at the liquid-liquid interface: x-ray study and Monte Carlo simulation. Langmuir **25**(2), 952–958 (2009)

41. M.K. Sanyal, V.V. Agrawal, M.K. Bera, K.P. Kalyanikutty, J. Daillant, C. Blot, S. Kubowicz, O. Konovalov, C.N.R. Rao, Formation and ordering of gold nanoparticles at the toluene-water interface. J. Phys. Chem. C **112**(6), 1739–1743 (2008)

Chapter 13
Characterization of Soft Matter at Interfaces by Optical Means

Benoit Loppinet

Abstract In this chapter, I give a brief overview, which is biased by personal experience, on how various optical techniques can be used for characterization of soft matter at interfaces, including ellipsometry, light scattering, and total internal reflection geometries. Without discussing the technical details and theoretical foundations of the methods, I focus on what can be learned by applying the individual techniques.

13.1 Introduction

In this chapter, I will try to give a brief introduction to various optical techniques that can be used to gain quantitative characterization of surfaces and interfaces of soft matter systems such as polymer or colloids deposited at interfaces or surfaces but also the surface of soft matter samples. Typical examples of such systems would be adsorbed colloids at air-water interfaces, or polymer chains adsorbed or grafted on a solid flat substrate (as in polymer brushes). The variety of samples and the ease of access of optical techniques imply that a large number of techniques will be available. Indeed recent developments of optical instrumentation have seeded the development of numerous techniques and have been the objects of monograph and reviews. In view of this variety the newcomer may feel some confusion. Typically experimental techniques will aim to quantify structure and morphology, composition, and properties. Optical techniques are well suited for structure and morphology. Other specific properties can be measure through dedicated like Composition can be addressed through spectroscopic techniques (see chapter D.14 by A. Erbe et al.). More than often, one technique can not bring all the required information and complementary analysis using different techniques are often needed. Most

B. Loppinet (✉)
IESL-FORTH, Heraklion, Greece
e-mail: benoit@iesl.forth.gr

© Springer International Publishing Switzerland 2016
P.R. Lang and Y. Liu (eds.), *Soft Matter at Aqueous Interfaces*,
Lecture Notes in Physics 917, DOI 10.1007/978-3-319-24502-7_13

optical techniques, have by construction a resolution which is limited by the light wavelength. That may seem a bit limited for the precise investigation of interfaces. But the ease of implementation, as compared to e.g. x-ray techniques, and the availability makes them popular. Moreover use of interference offers possibilities to increase the resolution. The aim of this chapter is not so much to give a detailed description of specific techniques as to provide an introduction of the basic principles. It is intended to be complementary to the chapters D.11–D.15.

Optical fields have long been used to characterize interfaces. The historical example of Benjamin Franklin noticing the fast spread of oil droplet on water surface (at the end of 18th century) and Lord Rayleigh estimating molecular size by dividing the volume of spread oil to the area of the spread (in the nid 19th century) provides a good example how qualitative and quantitative measurement can sometimes simply be obtained.

13.2 Optical Field Propagation and Reflection [1]

Optical fields are described as electromagnetic plane waves propagating in vacuum with velocity c. The direction of the electric field E defines the polarization of the wave. The wave vector k is situated along the propagation axis and $|k| = k = 2\pi/\lambda$ where λ is the wavelength. Further the radial frequency is ω. As the wave propagates of one wave length during one period: $\omega c - 2\pi/\lambda$

In material the propagation of the electromagnetic wave is described through the use of a refractive index n, so that $\omega c = 2\pi n/\lambda$. Dielectric (transparent) media have a real refractive index. If absorption is present, the refractive index can become complex. A better description is done in term of real and imaginary part of the complex electric permittivity ε and $\varepsilon = n^2$. As I will consider simple dielectric cases, I will use refractive index in what follows.

When light encounters an optical interface, characterized by a variation of refractive index in space, this local variation of refractive index affects the light propagation, leading to the well-known refraction and reflection. The reflected/transmitted light therefore carries information about the refractive index profile of the interface. That serves as base of a number of reflectivity techniques used for surface or interface characterization.

The simplest model for an interface is an abrupt, discontinuous change of refractive index from medium 1 with refractive index n_1 to medium 2 with refractive index n_2 at position z_0, where the z-direction is defined along the interface normal. This is known as Fresnel interface. In this case the reflection and transmission coefficient have the well known forms derived from continuity equation for the electric field. Two reflection coefficients can be derived depending on the polarization of the incident light.

If the electric field oscillates parallel to the plane of incidence (which is span by the wave vector k and the interface normal, see also Fig. 13.1) it is referred to as a p-wave, the polarization normal to this plane is referred to as s-wave. The

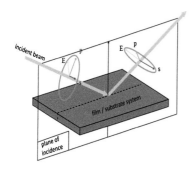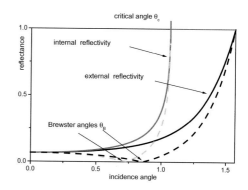

Fig. 13.1 *Left* Reflectivity geometry, *Right* angular dependence of the reflection coefficient rp and rs for external reflection n1 > n2 (*black lines*) and internal reflection n2 > n1 (*grey lines*) as a function of incidence angle for n2/n1 = 1.52/1.33. broken: rp;full: rs. At incidence greater than critical, the internal reflection introduces a phase shift

reflectivity coefficients are expressed in various ways using either refractive index n and incidence angle θ or wave vectors $\boldsymbol{k} = \boldsymbol{q} + \boldsymbol{Q}$ where $\boldsymbol{q}$ and $\boldsymbol{Q}$ are the normal and in plane components of the wave vector $\boldsymbol{k}$.

$$r_s = \frac{q_1 - q_2}{q_1 + q_2} = \frac{n_1 \cos\theta_1 - n_2 \cos\theta_2}{n_1 \cos\theta_1 + n_2 \cos\theta_2} = -\frac{\sin(\theta_1 - \theta_2)}{\sin(\theta_1 + \theta_2)} \qquad (13.1)$$

$$r_p = \frac{\frac{q_1}{n_1^2} - \frac{q_2}{n_2^2}}{\frac{q_1}{n_1^2} + \frac{q_2}{n_2^2}} = \frac{n_2 \cos\theta_1 - n_1 \cos\theta_2}{n_2 \cos\theta_1 + n_1 \cos\theta_2} = \frac{\tan(\theta_2 - \theta_1)}{\tan(\theta_2 + \theta_1)} \qquad (13.2)$$

Moreover, Snell law accounts for the continuity at the interface $\boldsymbol{Q}_1 = \boldsymbol{Q}_2$ or $n_1 \sin\theta_1 = n_2 \sin\theta_2$

The reflection coefficients can be complex numbers, with a modulus and a phase. Reflectance is the square of the modulus. Complex reflectivity coefficients are encountered in the total internal reflection regime, where the reflectance is equal to one, but a phase shift appears.

Two specific angles can be noted on the graph on the right of Fig. 13.1. At the Brewster angle θ_B, the p-reflectivity is zero so that all the p-polarized light is transmitted. The second angle is the critical angle of internal reflection θ_c where both s and p reflectivity reach one. At incidence angles large than the critical angle total internal reflection occurs, where both reflectivity coefficients have modulus one, but have a different phase shift.

In the presence of a thin layer of thickness d and refractive index n_f between semi infinite medium 1 and 2, multiple reflection gives rise to interference. The total reflection coefficient is a function of the reflection coefficients at each interface and the phase shift depends on the thickness of the layer. For both s and p polarization the total reflection coefficient is of the form [2]:

$$r = \frac{r_1 + r_2 \exp\left(-2iq_f d\right)}{1 + r_1 r_2 \exp\left(-2iq_f d\right)} \tag{13.3}$$

where $q_f = n_f \cos\theta_f$ and $n_1 \sin\theta_1 = n_f \sin\theta_f$. r_1 is the reflection coefficient between the medium 1 and the film, r_2 is the reflection coefficient between the film on the medium 2 with the appropriate incidence angles. The expression is valid for both s and p polarization.

The reflectivities of multilayer systems can be computed through the matrix method where every layer can be represented by a matrix I and every interface by a matrix M. The total system is then expressed by a matrix [3]

$$S = \begin{pmatrix} s_{11} & s_{12} \\ s_{21} & s_{22} \end{pmatrix} = I_1 M_1 I_2 M_2 \ldots I_n M_n I_{n+1}$$

with $I_j = \dfrac{1}{t_j}\begin{pmatrix} 1 & r_j \\ r_j & 1 \end{pmatrix}$ and $M_j = \begin{pmatrix} \exp(-2iq_j) & 0 \\ 0 & \exp(-2iq_j) \end{pmatrix}$ where t_j and r_j are the transmission and reflection coefficients between medium j and j + 1 and q_j is the phase shift induced by the jth layer of thickness d_j.

The total reflectivity r is then $r = s_{12}/s_{11}$.

This matrix method allows the computation of reflectivities for layered systems for both s and p polarization. This approach can also be used to calculate reflectivities arising from continuous refractive index profile which have to be approximated for this purpose by a layered system, as schematically shown in th eleft part of Fig. 13.2.

Of course the Fresnel model is at best an approximation, but as molecular sizes that will control the change of material are small compared to the typical wavelength of light, it is a good approximation for optically flat surfaces.

Measuring reflectivity as a function of angle or wavelength can therefore bring information on the refractive index profile. Practically optical reflectivity techniques

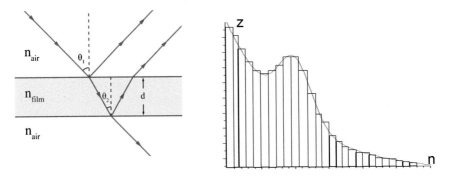

Fig. 13.2 *Left* thin film interference and *right* discretized refractive index profile

are not able to characterize thin layers. This can be achieved through the use of radiation with much shorter wavelength like X-ray or neutrons (see chapter D.12 by J. Daillant)

Optical reflectivities are nonetheless used to probe the formation of thin layers, especially at Brewster angle incidence (Brewster angle reflectivity). There, a non-zero reflectivity can be attributed to the formation of an interfacial layer. This can be used to monitor adsorption for example.

Other, somewhat more sophisticated reflectivity techniques use resonant layer substrates (optical wave guide, surface Plasmon resonance). The reflected intensities typically show a minimum when the light is exciting resonantly the structure. This excitation can be achieved through internal reflection. The exact onset of this resonananance is very sensitive to the refractive index on the other side of the wave guide (optical wave guide spectroscopy). The reflectivity is then very sensitive to the refractive index, and small changes lead to a shift in the reflectivity spectrum, that can easily be measured. But probably the most used and versatile reflectivity techniques is ellipsometry that relies on measuring the ratio r_p/r_s.

13.3 Ellipsometry

Ellipsometry is probably the oldest, most established and best known reflectivity technique. Ellipsometry is the subject of a number of dedicated text books [3–6] and its applications to various fields of science and technology have been and are regularly reviewed [7–9]. It measures the ratio of the p and s polarized reflectivity with the aim to characterize the refractive index profile that gives rise to the reflectivity. The ellipsometric ρ ratio is defined as:

$$\rho = \frac{r_p}{r_s} = \tan \psi \, e^{i\Delta} = \frac{|r_p|}{|r_s|} \exp i(\delta_p - \delta_s) \tag{13.4}$$

$\tan\Psi$ describes the ratio of the p and s reflectivities amplitudes, and Δ the phase shift between the p and s polarised reflected light induced by the reflecting interface. Practically it is implemented in many different ways that measure Ψ and Δ. Ellipsometers come in a variety of geometries. A broad range of instrumentation exists, ranging from sophisticated commercial setups including variable angle and spectroscopic (variable wavelength) possibilities to simpler fixed angle manual nulling ellipsometer. The scheme of nulling ellipsometry is the simplest way and often the most precise.

The relation between the measured quantities (angular, wavelength dependence of the ellipsometric ratio) and what the user would like to know (refractive index profile) is neither "intuitive" nor straightforward. It is therefore difficult to provide simple, meaningful introduction to ellipsometry. Despite this apparent limitation, ellipsometry remains a very useful and powerful technique, due to its high sensitivity and its very wide range of applicability. It can provide information that is difficult to obtain with other (possibly more intuitive) techniques. The analysis and interpretation of the measured data in terms of the looked-for quantities (composition, distribution, homogeneity,...) often relies on the use of analysis software which is installed on most commercial instruments. This type of analysis is an inverse problem that relies on the use of models to produce refractive index profiles. These models often need more parameters than can be safely deduced from the data, so that the obtained results are often model dependent and independent checks of the validity of the models are not easy.

13.3.1 Thick Film Limit

In case of thick layer or multilayers, the refractive index profile can be modelled by a homogeneous (box) multi-layer. Then the ellipsometric ratio can be calculated using the matrix formalism exposed above. For accurate measurements, variable angle and possibly variable wavelength will be used. Minimization software will be used to return the best solution of the profile. However, cautious users will remember that, as mentioned above, the obtained profile will most likely not be the unique one that provide a good fit of the data.

A good example for the application of ellipsometry to thick soft matter layers is the case of concentration profiles in extended polymer brushes. If made thick enough such systems can extend to hundreds of nanometres and therefore fall into the thick film category. There, multiple angle incidence coupled with internal reflection can be used (see Ref. [8] and reference therein). The resolution is not good enough to allow a direct quantification of the refractive index profile, but using a profile function, good fits were obtained. They quantify the change of thickness and concentration profiles following swelling and deswelling induced by changes of temperature or pH.

Similarly, it is also possible to follow and evaluate the change of thicknesses for polymer brush following external triggering of the brush (by change of temperture or pH for example) at a single angle of detection, using spectroscopic ellipsometry, or even single wavelength ellipsometry. Those changes will not affect the adsorbed amount, but only the distribution of the polymer in the direction normal to the surface. For a recent example of application to polyelectrolyte brushes see [10].

13.3.2 Thin Film Limit

The use of refractive index profile loses its sense when the layer thickness is reduced to small dimension compared to the light wavelength. This thin film limit is relevant in the case of adsorption of molecules and macromolecules on solid substrate or liquid/air or liquid/liquid interfaces. Ellipsometry is widely used to study such systems. In those cases the use of refractive index profile is not recommended, as refractive index and thickness may not be measured independently. The ellipsometric signal will be dominated by the substrate-ambient signal, and the contribution of the film will be small. A perturbation expansion of the reflection coefficient can be obtained in terms of the ellipsometric invariants introduced by Lekner [2]. Those invariants are named this way as they measure moments of the refractive index function and are independent of the precise profile of refractive index. In the case of adsorption of molecules/macromolecules on a solid substrate from solutions, [8, 11] the first invariant can be expressed in terms of the refractive index increment dn/dc of the solution and of the adsorbed amount (the enrichment layer) Γ (typically in mg/m^2). The ellipsometric ratio expansion can be put into the following forms:

$$\rho - \rho_0 = iK(\theta, n_1, n_2)\, \Gamma \frac{dn}{dc} \tag{13.5}$$

$\Gamma \frac{dn}{dc}$ has the dimension of an optical thickness ($n.d$). It is independent of the precise distribution of the adsorbat's refractive index as long as the thickness remains small compared to the light wavelength.

The constant K depends on the measurement condition (refractive index of substrate and solution and incidence angle) as:

$$K = \left(\frac{2\pi}{\lambda}\right)^2 \frac{4(n_1^2 - n_2^2)n_2^2 q_x^2 q_1}{(q_1 - q_2)^2 (n_2^2 q_1 + n_1^2 q_2)^2} \frac{n_1^2 - n_2^2}{n_1} \tag{13.6}$$

Adsorption measurements are often done at the Brewster angle, where the sensitivity is expected to be largest. Note that the first expansion term (for dielectric layers) is an imaginary number. A special approach to measure directly this term has been introduced by Beaglehole [12].

It should be clear that in this first order expansion case, ellipsometric measurements will not be able to provide independently the overall extend (thickness) of the adsorbed layer and its refractive index. Only the adsorbed amount will be measurable in a micro-balance type a measurement. If the adsorbed amount is already known, then ellipsometry can be expected to be redundant. For example, the use of ellipsometry in the thin film limit to measure conformation changes at air water interface can be expected to be limited, as the signal will mostly arise from the molecules surface concentration, independently of the extension of the formed layer.

As films grow thicker, a second term of the perturbation development may become measureable. When the two terms can be measured independently, then it becomes possible to deduce independently both the thickness and the refractive index of the formed layer. This is best done in kinetics experiments (signal evolution) as the noise/parasitic signal is expected to remain constant [12]. The method has been well demonstrated by M. Tirrell and co-workers for the case of diblock copolymer adsorption, measured at Brewster angle [12].

13.3.3 Effective Medium Approximation
for Non-Homogeneous Layers

As a specular reflectivity technique, ellipsometry can in principle only address the refractive index profile normal to the interface. Possible in plane variation of the refractive index can arise from roughness, mixed composition. The standard approach to treat such inhomogeneities in ellipsometry, is to use an effective refractive index for the medium. This type of approximation, illustrated in Fig. 13.3 is known as effective medium approximation.

It originates from the need of describing a complex medium with a single refractive index number. Various models exist to calculate the refractive index of a multi-constituent layer. Such models are usually available in the ellipsometry software [5, 6]. Such an approach is commonly used to evaluate porosity in thin film material for example through ellipsometric porosimetry. Therefore ellipsometry analysis software often offers the possibility to build such models. The number of parameters for the model can become quite large.

In the field of soft matter systems, such an approach has recently been proposed for the study of colloidal particles deposited at fluid-fluid interfaces. In particular several groups have tried to relate the ellipsometric signal to the precise position of colloidal particles at interfaces, in an attempt to measure contact angle of small colloidal particles. It relies on the use of an effective medium theory that allows to transform the position and concentration of the particles at the interface in a refractive index profile $n(z)$ with refractive index and thickness. The thickness of the effective layer is expected to relate with the position of the colloid at the interface. $n(z)$ is parameterized with quantities like the particles size, their surface concentration and their position on the interface (the contact angle). The parameter values can be found using a minimization (fitting) procedure of the predicted ellipsometric ratio to the measured ones. Measurements were done at the Brewster angle. Knowledge of the refractive index of the colloids and both fluids is required. The found contact angles were in reasonable agreement with values measured with macroscopic samples [13].

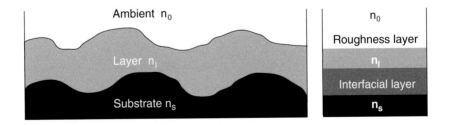

Fig. 13.3 Typical example of effective medium approximation for a rough film deposited on a rough surface. *Left* is the real structure, *right* is the effective refractive index used for fitting ellipsometric results. Please provid a Figure of higher quality

13.4 Other Optical Techniques

Apart from ellipsometry, very many other optical techniques are able to provide some level of characterization of surfaces and interfaces. Imaging techniques, that attempt to image the morphology, composition of surfaces and interfaces exist in various forms. Surface light scattering techniques can be used, especially dynamic light scattering techniques are able to probe temporal fluctuation at surface or interface.

13.4.1 Imaging of Surfaces or Interfaces Morphology

Specular reflectivity techniques probe only the direction normal to the interface. It is therefore not well suited for the characterization of in plane morphology. A number of techniques exist that can provide an image, or a mapping of the interfaces refractive index, either as a direct image or through scanning methods.

An important example is the Brewster angle microscopy [14] (BAM). Introduced about 25 years ago, it provides a way to image (on a grey scale) the refractive index inhomogeneities of surface layers. The contrast arises from the strong dependence of the p-reflectivity at the Brewster angle on the presence or absence of a thin layer. As in dark field microscopy, the bare Fresnel interface will appear black and a layer will appear grey, the thicker/denser the brighter. BAM requires the use of oblique incidence which reduces the field of view. Therefore it is often associated with scanning abilities, and large size image can be obtained as a superposition of images. BAM is particularly well adapted for the air water interface, and in particular for application in combination with a Langmuir through where it allows for controlling the homogeneity of the monolayers.

Over the years other types of reflectivity microscopy have been used to determine surface structure. For example the standard white light EPI microscope is well adapted to characterize film thicknesses deposited on reflective surfaces. In the presence of a thin optical layer, the multiple reflections may give rise to interference, which are wavelength dependent and produce colourful pictures which provide a simple way to measure refractive index homogeneities of films and layers. Being an interferometric technique it has a high sensitivity in the z-direction. It has the resolution of the microscope detection in the plane.

A range of instruments, known as optical surface profilometers aim at providing a quantitative refractive index profile mapping, with good in plane resolution. Different approaches can be used, based on white light interferometry or confocal scanning reflectivity. Typical examples of application range from measuring the roughness of polymer films to contact angles of colloidal particles at air-water interface. Buried interfaces have also been imaged this way.

Progresses in microscopy, like confocal fluorescence microscopy with high resolution, using fluorescence of single molecules, have also benefited the study of interfaces structures and dynamics. One can take advantage of the ability to image single molecules and to follow them in time. The fluorescent tracers required for fluorescence microscopy are nowadays easily available, being quantum dots, simple fluorescent molecules, or modified polymers or colloids. Two examples of use of advanced microscopy technique for interfacial soft matter phenomena are the observation of capillary waves in phase separated colloid-polymer mixtures using confocal microscopy [15] or the adsorption and diffusion of single fluorescently labelled molecules on a solid substrate [16, 17]. Interfacial/surface systems present the advantage that there extend in the normal direction is reduced, making the microscopic observation simpler than in 3D systems.

Total internal reflection (TIR) can be used in conjunction with microscopy and imaging. TIR microscopy and TIR fluorescent microscopy provide a neat way to measure time-dependent positions of particles and relate to relate the position fluctuation to the interaction potential between the particles and the interface [18].

13.4.2 Light Scattering and Dynamics

Light scattering is well adapted to the study of soft matter (see chapter D.11 by C. A. Völker). A number of established techniques exist. Distinction is often made between static light scattering that measures time averaged intensities to obtain structural information and dynamic light scattering that analyses fluctuating intensities in the time/frequency domains, and relates it to the dynamics of refractive index fluctuation that are thermally excited at a given temperature, (which is often related to Brownian motion).

The angular dependence of the scattered light is often used as it allows for the variation of the scattering vector (defined as the difference of the scattered wave vector and the incident wave vector), which itself provides a spatial frequency or a wavelength of the probed fluctuation.

The case of scattering by surfaces and interfaces has long been considered. (see chapter D. 12 by J. Daillant). Due to the large wavelength, light is not the most appropriate probe to resolve the structure of nanometre length scale. X-rays or neutrons are more appropriate for that purpose. Dynamic light scattering provides the ability to access dynamical information on interfaces and their fluctuations. The information does not only lie in the value of the scattered intensity and its angular dependence, but rather in the time fluctuation of the scattered intensity. The origin of such light intensity fluctuation can be found in the presence of fluctuation of the refractive index close to the interface. The temporal analysis of the scattered light intensity provides the characteristic life time of the fluctuation of given length scale (defined by the scattering vector). Different type of analysis are available depending on the type of fluctuation one is looking at and looking forTwo specific techniques are presented below with some details.

13.4.3 Surface Quasi Elastic Light Scattering

Surface Quasi Elastic Light Scattering designates a technique used to probe the dynamics of surface wave present at fluid-fluid interfaces [19, 20]. Fluid surfaces and interfaces at finite temperature present a spectrum of thermally excited surface waves, with in particular capillary waves. Scattering by propagating surface waves leads to a Doppler shift in the scattered wave compare to the incident one. The related frequency ω can be found in the time domain as well as in the frequency domain. The frequency provides a measure of the propagation of the specific surface waves at the specific scattering vector q.

At a given scattering angle, one would then detect the light scattered by the mode of wavelength $2\pi/q$ and measure their frequency, as well as their damping (how far the thermally excited waves do propagate)

Varying the scattering wave vector through the scattering angle provides a way to obtain the dispersion curves (ω vs q) for the specific surface waves.

The technique is used to obtain information on the spectrum of capillary waves and especially their damping and relates them to mechanical properties of the interfacial layer. The analysis is realized either in the time domain or in the frequency domain (providing by the way a nice illustration of the equivalence of the two). The results are then used to deduce the mechanical modulus characteristic of the interface, very much like the sound velocity is used to measure the compression modulus of solids. However capillary waves are more complicated than sound waves, and the interpretation is not always straightforward.

13.4.4 Evanescent Wave Dynamic Light Scattering

Another light scattering technique well adapted to interfacial studies is the so-called evanescent wave dynamic light scattering (EWDLS). Evanescent wave denotes a type of near field wave, i.e. waves localized near an interface. They are therefore ideal candidates to probe interfacial phenomenon. Near field radiations are present in a number of situations, and in particular in the case of total internal reflection (see Appendix 1). Such radiation is excited through total internal reflection but also through other related approaches such as optical wave guides or surface plasmons.

The evanescent wave is used as the incoming beam of a scattering experiment, providing a light scattering experiment with a very reduced, anisotropic scattering volume. This is set by the penetration depth of the evanescent field and the beam dimension in the other directions. The applications of EW-DLS have been reviewed by Sigel in [21].It has so far been used to probe near wall diffusion of polymers and colloids. Interfacial effects can be measured with resolutions that are of the order of tens of nanometres ($\sim\lambda/50$).

13.5 Conclusion

Optical techniques have long proven very useful for qualitative and quantitative characterization of surfaces and interfaces. The continuous development of photonic devices, and the relative ease to manipulate optical signals makes optical fields very attractive to characterize surfaces and interfaces despite the limitation due to the long wavelength of visible light (~ 0.5 μm).The development of new approaches and instrumentation are bound to lead to progress in our understanding of the structure and dynamics of those surfaces and interfaces.

Acknowledgments Part of the work was presented as as a lecture at the SOMATAI summer school 2014 in Berlin. The support of the Greek ESPA programme Areistea RINGS is acknowledged.

Appendix 1: Total Internal Reflection and Evanescent Field

Electromagnetic near fields designate non-radiative fields that are localized near an object, so to say at the surface of the object. Optical near fields offer a convenient way to probe interfaces. They extend over one wavelength or so. In particular they can be used to excite scattering or fluorescence.

One such near field is the evanescent waves present in the medium of lower refractive index at total internal reflection. When reflectivity coefficients are one, there is nonetheless a near field penetrating the medium of lower refractive index. It is well described by the reflectivity coefficients, with r_p and r_s being complex numbers of module 1 (Fig. 13.4).

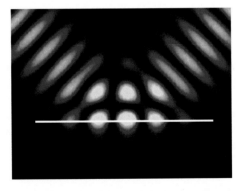

Fig. 13.4 Optical field at TIR the bottom correspond to the low refractive index medium (wiki-commons). The white line is added to materialize the interface. Below is the medium of low refractive index and above the one with high refractive index. The brightness relates to the amplitude of the electric field (the brighter the higher) The standing waves due to the interference between the incoming and reflected beam is clearly seen. The evanescent field is also clearly visible below the white line. (image wiki-commons)

In particular it is possible to compute the penetration depth.

The evanescent field writes [2] as $E_{ev} = E_0\, e^{-\kappa z} e^{i(kx - \omega t)}$ with the "penetration wave vector" $\kappa = 1/d_p = \frac{2\pi n_1}{\lambda} \sqrt{\sin^2 \theta - \frac{n_2^2}{n_1^2}}$ and the "propagation wave vector" along the interface $k = \frac{2\pi n_1}{\lambda} \sin \theta$.

The other may be less appreciated length scale associated with TIR is the Goss Haenchen shift, that describes the lateral shift of the beam [2]. Looking into the reflectivity coefficient, a phase shift appears under TIR, which also depends on the polarization and the incidence angle. This phase shift is the sign of a time (length) associated with the TIR.

The EW produced at TIR can be associated with many different detection schemes [22], ellipsometry, light scattering, fluorescence, Raman, IR, and recently optical rotation.

The critical angle is independent of the polarization for non-birefringent materials, and so is the penetration depth. However the s and p polarized light undergo different phase shift under TIR. The difference of phase shift is what ellipsometry under total internal reflection will measure. The penetration depth can be varied by changing the incidence angle. A larger optical contrast between the two materials will provide a shorter penetration depth, as will larger incidence angles. As an example, the field penetration depth d_p at interface between high refractive index incidence medium ($n_1 = 2$) with water ($n_2 = 1.33$) can become as low as 60 nm for a wavelength of 532 nm.

References

1. M. Born, E. Wolf, *Principles of Optics* (Pergamon Press, Oxford, 1975)
2. J. Lekner, *Theory of Reflection:of Electromagnetic and Particle Waves* (Springer, Dordrecht, 1987)
3. R.M.A. Azzam, N.M. Bashara, *Ellipsometry and Polarized Light* (Elsevier, Amsterdam, 1987)
4. H.G. Tompkins, *A User's Guide to Ellipsometry* (Dover Publication Inc, New. York, 2006)
5. H. Fujiwara, *Spectroscopic Ellipsometry: Principles and Applications* (Wiley, New York, 2007)
6. Handbook of Ellipsometry, ed. by H.G. Tompkins, E.A. Irene (Springer, Heidelberg, 2004)
7. J.L. Keddie, Curr. Opin. Colloid Interface Sci. **6**, 102 (2001)
8. D. Johannsmann, Investigations of soft organic films with ellipsometry. in *Functional Polymer Films*, ed. by W. Knoll, R.C. Advincula (Wiley-VCH Verlag GmbH & Co. KGaA, 2011)
9. W. Ogieglo, H. Wormeester, K.-J. Eichhorn, M. Wessling, N.E. Benes, In situ ellipsometry studies on swelling of thin polymer films: a review. *Progress in Polymer Science*, 2014, in ress. (http://www.sciencedirect.com/science/article/pii/S0079670014001063)
10. J.D. Willott, T.J. Murdoch, B.A. Humphreys, S. Edmondson, G.B. Webber, E.J. Wanless, Critical salt effects in the swelling behavior of a weak polybasic brush. Langmuir **30**(7), 1827–1836 (2014)
11. J.C. Charmet, P.G. Degennes, J. Opt. Soc. Am. **73**, 1777 (1983)
12. R. Toomey, M. Tirrell, In situ investigation of adsorbed amphiphilic block copolymers by ellipsometry and neutron reflectometry in soft matter characterization. ed. by R. Borsali, R. Pecora (Springer, Netherlands, 2008)

13. A. Stocco, G. Su, M. Nobili, M. Inab, D. Wang, In situ assessment of the contact angles of nanoparticles adsorbed at fluid interfaces by multiple angle of incidence ellipsometry. Soft Matter **10**, 6999–7007 (2014)
14. B. Desbat, S. Castano, Brewster angle microscopy and imaging ellipsometry. in *Encyclopedia of Biophysics* ed. by G.C.K. Roberts (Springer, 2013), pp. 196–200
15. D.G.A.L. Aarts, M. Schmidt, H.N.W. Lekkerkerker, Direct visual observation of thermal capillary waves. Science **304**(5672), 847–850 (2004)
16. S. Granick, S.C. Bae, Molecular motion at soft and hard interfaces: from phospholipid bilayers to polymers and lubricants. Annu. Rev. Phys. Chem. **58**, 353 (2007)
17. C. Yu, J. Guan, K. Chen, S.C. Bae, Steve G., Single-molecule observation of long jumps in polymer adsorption. ACS Nano **7**, 9735 (2013)
18. D. Prieve, Measurement of colloidal forces with TIRM. Adv Coll. Interface Sci. **82**, 93 (1999)
19. P. Cicuta, I. Hopkinson, Recent developments of surface light scattering as a tool for optical-rheology of polymer monolayers. Colloids Surf A: Physicochem Eng. Aspects **233**, 97–107 (2004)
20. A.R. Esker, C. Kim, H. Yu, Polymer monolayer dynamics. Adv. Polym. Sci. **209**(1), 59–110 (2007)
21. R. Sigel, Light scattering near and from interfaces using evanescent wave and ellipsometric light scattering. Curr. Opin. Colloid Interface Sci. **14**(6), 426–437 (2009)
22. F. de Fornel, *Evanescent Waves, From Newtonian Optics to Atomic Optics,* vol 73 (Springer Series in Optical Sciences, 2010)

Chapter 14
Optical Absorption Spectroscopy at Interfaces

Andreas Erbe, Adnan Sarfraz, Cigdem Toparli, Kai Schwenzfeier and Fang Niu

Abstract This chapter summarises the physical principles of optical absorption spectroscopy and its use for the characterisation of surfaces and interfaces. After a brief discussion of the fundamentals of absorption spectroscopy and its relation to quantum mechanics, the chapter discusses the basics of optics at interfaces, focusing on the absorption of light by molecules in the interfacial region. Because of fundamental similarities, the chapter will touch on spectroscopy of both electronic and vibrational transitions, with a strong focus on infrared absorption experiments. There is a brief discussion, with reference to examples, of experiments in internal and external reflection geometry, including a brief discussion of the measurement of spectra on different classes of substrates (metallic vs. transparent).

14.1 Motivation

This chapter shall illustrate the use of optical absorption spectroscopy in the characterisation of interfaces, with a particular emphasis on the detection of molecules at interfaces between condensed media. While this chapter's main focus is on reflection absorption techniques, physical principles of related techniques, such as ellipsometry (see Chap. 13 by B. Loppinet), X-ray and neutron reflectivity and scattering techniques (see Chap. 12 by J. Daillant), and non-linear optical spectroscopy (second harmonic and sum frequency generation spectroscopy; see Chap. 15 by M. Hoffmann et al.) are discussed elsewhere. In this chapter, reflection absorption techniques of both electronic and vibrational transitions shall be discussed, though the former are typically probed in the ultraviolet (UV) and visible (VIS) spectral range, and the latter in the mid-infrared (IR). From the spectroscopic

A. Erbe (✉) · A. Sarfraz · C. Toparli · K. Schwenzfeier · F. Niu
Max-Planck-Institut für Eisenforschung GmbH, Max-Planck-Str. 1,
40237 Düsseldorf, Germany
e-mail: a.erbe@mpie.de

© Springer International Publishing Switzerland 2016
P.R. Lang and Y. Liu (eds.), *Soft Matter at Aqueous Interfaces*,
Lecture Notes in Physics 917, DOI 10.1007/978-3-319-24502-7_14

side, the physical basics of the different transitions are closely related, so that it appears justified to treat them in a single text.

The chapter shall briefly summarise the quantum mechanical basics of optical absorption spectroscopy, further it shall point out what type of information on molecular systems can be obtained using absorption spectroscopy at interfaces. The heart of the chapter is the analysis of the peculiarities when working at interfaces.

The main motivation for using optical spectroscopy is its versatility. Many of the characterisation techniques that yield the most detailed information, such as electron-probe based techniques, including microscopy, spectroscopy, scattering and diffraction techniques, are limited to vacuum conditions, or single crystalline surfaces. Their extension to ambient conditions is a field of very active current research. Currently, many techniques cannot be applied routinely to soft matter at aqueous (or other solution-based) interfaces. Optical techniques, especially spectroscopy, have proven to be valuable tools for the characterisation of the composition, conformation, molecular orientation and kinetics of soft interfacial systems. Optical spectroscopy is suitable for application in complex environments, e.g. aqueous media.

This chapter is not another textbook, and not another review article. Rather, it is meant to briefly summarise textbook knowledge, enrich it with current examples, and to refer to modern and classical reviews of applications, especially from the field of soft matter, for further reading. A good general introduction into the physical basics of optical spectroscopy, especially its foundation in quantum mechanics, is given by Hollas [1] and Levine [2]. The effect of molecular symmetry on vibrational and electronic spectra is discussed by Harris and Bertolucci [3]. (Almost) all analytical aspects of electronic and vibrational spectroscopy in the bulk are summarised in a book edited by Gauglitz and Vo-Dinh [4]. An introduction specifically into the vibrational spectroscopy of molecular solids is given by Sherwood [5]—while its description of computations is largely outdated, the book is well-suited for understanding the principles of waves in crystalline solids. A further introduction into the optical properties of crystalline (and amorphous) solids is given by Fox [6]. The physical basics for investigations of interfaces is described by Tolstoy, Chernyshova, and Skryshevsky, with the major focus on IR spectroscopy [7]. Further, interface specific applications of optical spectroscopy are discussed in a number of reviews [8–15]. In this chapter, we will also refer to a number of original research articles as examples of application of one or another technique, with some bias to work from our own laboratory.

Before proceeding, a word on notation. In this chapter, vectorial quantities are represented bold underlined as e.g. $\underline{\mathbf{r}}$. The same letter as a normal symbol, e.g. r, may have a different meaning than the vectorial quantity. The modulus of vectorial quantity is denoted as $|\underline{\mathbf{r}}|$. In a subscript or superscript, an italic symbol, e.g. r in N_r, is a symbol which is a placeholder for a number, while a non-italic symbol, e.g. the r in N_r, is an abbreviation. Operators are designated by a hat, as $\hat{r}$.

14.2   Principles

14.2.1   Type of Information Obtained

The central quantity measured experimentally in an absorption spectroscopy experiment, both in a volume phase as well as at an interface, is the absorbance

$$A = -\log_{10}\left(\frac{I_{\text{sample}}}{I_{\text{reference}}}\right) = -\log_{10} M. \tag{14.1}$$

Here, I is the irradiance/intensity of light recorded at the detector (See, e.g. [16] for a discussion of the background of the irradiance for experimentalists). While the index "sample" refers to the actual system under investigation (e.g. a polymer adsorbed to an interface), the index "reference" refers to a reference state, which is ideally the same interface without the presence of the sample. In an experiment in a bulk phase, "reference" is simply a blank measurement, i.e. a measurement without any sample in the beam path. In experiments involving interfaces, there is no such unique choice of a reference system, and all practical reference measurements involve reflection at a certain reference interface, so that the reflectance absorbance A_{r} (which will be mainly discussed in this chapter) is given as

$$A_{\text{r}} = -\log_{10}\left(\frac{R_{\text{sample}}}{R_{\text{reference}}}\right), \tag{14.2}$$

where the two reflectivities R can be used synonymously with the respective intensities from Eq. 14.1, and will be discussed briefly in Sect. 14.3. A_{r} describes the difference of an interface to a reference state, and needs to be carefully interpreted as such.

Equation 14.1 also introduces the "transmittance" M, which is frequently used in the chemical literature discussing bulk spectra, but is less useful for the quantitative analysis of reflection spectra.

In a bulk phase, the absorbance recorded after light is passing through a sample of length l is proportional to the concentration c of an absorbing species via Lambert Beer's law,

$$A = \alpha_{\text{c}} l c, \tag{14.3}$$

where the absorption coefficient α_{c} is an intrinsic quantity of the transition studied, and shall be discussed below. The strict linearity in concentration c assumes non-interacting species, which absorb light. At higher concentrations, deviations from the linear behaviour $A \propto c$ are found. Therefore, α_{c} is a well-defined quantity in dilute systems (at infinite dilution), or in a pure phase. In a pure phase, $\alpha_{\text{c}} c = \alpha_{\text{p}}$, as concentration is fixed in a pure phase at constant pressure (see also Sect. 14.2.3).

Also in a reflection experiment, the absorbance increases with increasing amount of species present. In addition, especially for thin and ordered systems, the absorbance is affected by the orientation of the molecules. In some cases, effects of concentration and orientation can be separated (see Sect. 14.3). Absorption of light at a characteristic wavelength is characteristic for the presence of a certain bond or group —though certain bonds are easier to detect than others. Electronic transitions (involving valence electrons) are furthermore sensitive to the environment the absorbing group is immersed in, e.g. the solvation state. Some vibrational modes also "feel" this presence of the surrounding medium, while others are non-sensitive to it. Many vibrational modes vary with the conformation of a molecule. Overall, optical absorption spectroscopy at interfaces is able to yield (a) qualitative information on the presence of certain groups, (b) (semi)quantitative information on the amount of certain species present, (c) quantitative information about the orientation of certain groups, (d) qualitative information about the environment (e.g. polar vs. non-polar) a certain group is in and (e) information on the conformation of molecules.

Amongst the optical spectroscopic techniques, vibrational spectroscopy is probably the most versatile. The large number of vibrational modes of organic molecules—especially those comprising "soft matter"—ensure that almost all molecular substances, and many solids, can in principle be detected [17–19]. Mid-IR light used in IR absorption experiments, with its photon energy of about 0.1 eV is one of the softest probes widely available, so the effect on the sample is minimal, which is not always true for alternative techniques.

14.2.2 The Spectroscopic Process of Dipole Transitions

While significantly more complicated in solvated large molecules as encountered in soft matter at aqueous interfaces, the fundamentals of the spectroscopic process can still be understood on the basics of quantum mechanics of simple systems. The treatment here closely follows [1]. A quantum system is characterised by its wave function $\Psi(\mathbf{r}, t)$, which in general depends on the positional coordinates $\mathbf{r}$, and for simplicity we will disregard the dependence on time t. In this picture, energy eigenvalues E_i are obtained for the ith eigenstate as a solution of the Schrödinger equation,

$$\hat{H}\Psi = E_i\Psi, \tag{14.4}$$

with the quantum mechanical Hamiltonian

$$\hat{H} = \hat{T} + \hat{V}, \tag{14.5}$$

consisting of the sum over all contributions to kinetic energy $\hat{T}$ and potential energy $\hat{V}$. For a molecule, $\hat{H}$ has contributions from the kinetic energy of both electrons (index e) and nuclei (index n), and from the interaction potentials of electrons and nuclei with each other,

$$\hat{H} = \hat{T}_e + \hat{T}_n + \hat{V}_{en} + \hat{V}_{ee} + \hat{V}_{nn}. \tag{14.6}$$

Because motion of electrons is significantly faster than motion of nuclei, e.g. because of the different masses, this Hamiltonian can be separated into an electronic contribution for fixed, static nuclei,

$$\hat{H}_e = \hat{T}_e + \hat{V}_{en} + \hat{V}_{ee}, \tag{14.7}$$

and a contribution from the motion of the nuclei itself,

$$\hat{H}_n = \hat{T}_n + \hat{V}_{nn} + E_e. \tag{14.8}$$

This Born-Oppenheimer approximation enables us to solve the Schrödinger equation independently for electronic states, and states associated with a motion of the nuclei. It implies that the total wave function can be factorised into contributions from nuclei and electrons, $\Psi = \Psi_e \Psi_n$, while the total energy is the sum of electronic and nuclear contributions, $E_{tot} = E_e + E_n$.

Spectroscopically, we can observe transitions between the energy levels of the electrons ("electronic transitions"). The nuclear part contains contributions from vibrations and rotations, from which we will discuss only the transitions between vibrational levels ("vibrational transitions"). In other words, different Hamiltonians are used, resulting in different wave functions, and different energy eigenvalues.

Light shall be described as an electromagnetic plane wave with angular frequency ω, with space and time dependence of the electric field $\mathfrak{E}$

$$\underline{\mathfrak{E}}(\underline{r},t) = \underline{\mathfrak{E}}^{(0)} e^{\pm i(\mathbf{k} \cdot \mathbf{r} - \omega t)}, \tag{14.9}$$

with wave vector $\mathbf{k}$, amplitude $\underline{\mathfrak{E}}^{(0)}$ and . Light can be absorbed by a system if its photon energy $E_{photon} = \hbar \omega$ equals the difference between (e.g. electronic or vibrational) energy levels (Here, $\hbar$ is the Planck constant h divided by 2π). The rate of change in occupancy N_2 of an excited state 2, which ultimately determines the total absorbed light intensity is given as

$$\frac{dN_2}{dt} = (N_1 - N_2) B_{21} \rho(\omega), \tag{14.10}$$

where index 1 indicates the ground state, B_{21} the Einstein coefficient and $\rho(\omega)$ the spectral radiation density. The Einstein coefficient is directly related to the wave functions of the involved states,

$$B_{21} = \frac{2\pi^3}{3\pi\varepsilon_0 h^2} \left|\underline{\mathbf{M_{21}}}\right|^2, \text{i.e. } B_{21} \propto \left|\underline{\mathbf{M_{21}}}\right|^2, \tag{14.11}$$

with transition dipole moment

$$\underline{\mathbf{M_{21}}} = \int \underline{\Psi_2}^* \, \hat{\underline{\mu}} \, \underline{\Psi_1} \, \mathrm{d}\underline{\mathbf{r}} = \langle \Psi_2 | \hat{\underline{\mu}} | \Psi_1 \rangle, \tag{14.12}$$

where we use the convenient bra-ket notation (ε_0 denotes the vacuum permittivity). Here, the * denotes the complex conjugate and $\hat{\underline{\mu}} = \sum q_i \underline{\mathbf{r_i}}$ is the dipole moment operator, summing over the product of all partial charges q_i and positions, in analogy to the definition of the dipole moment. The dipole moment operator is a vectorial quantity, and hence the transition dipole moment is a vectorial quantity too. While the Einstein coefficient, which ultimately determines the strength of an absorption, is essentially the squared modulus of the transition dipole moment, in interfacial systems the vectorial nature of the transition dipole moment will become important, as essentially $\underline{\mathbf{M_{21}}} \cdot \underline{\mathfrak{E}}^{(0)}$ determines the strength of an absorption (In isotropic solution, the orientation of the transition dipole moment with respect to the electric field takes an average value, while at interfaces, a preferential orientation is typically imposed).

The Einstein coefficient B_{21} (and hence the modulus of the transition dipole moment) is related to the integral over the absorption band with maximum at wavenumber $\tilde{\nu}_{21}$ from its beginning at wavenumber $\tilde{\nu}_1$ till its end at wavenumber $\tilde{\nu}_2$

$$\int_{\tilde{\nu}_1}^{\tilde{\nu}_2} \alpha_c \mathrm{d}\tilde{\nu} = \frac{N_A h \tilde{\nu}_{21} B_{21}}{\ln 10}, \tag{14.13}$$

where N_A represents the Avogadro constant. It is sometimes convenient to use the dimensionless oscillator strength

$$f_{21} = \frac{4\varepsilon_0 m_e c^2 h \tilde{\nu}_{21}}{e^2} B_{21} \tag{14.14}$$

to express the same information (Here, m_e and e represent mass and charge of the electron, respectively). The fact that we can relate a quantum mechanical quantity to the integral over the absorption band is convenient in some cases, but complicates interpretation of spectra in condensed phase. For a full quantitative description here, a model including the band shape is needed, and a description is needed which yields details of the interaction of light with matter. This purpose is served by the dielectric function, which will be introduced as a macroscopic concept below.

Before proceeding to the next section, we should comment on the fundamental selection rule, which follows from Eqs. 14.11 and 14.12. A transition can only be directly excited, if its transition dipole moment is non-zero. We need a change in dipole moment during the transition for the transition to be excited. The simplest way to understand the transition rule is based on the analysis of a vibrational transition of a two-atomic molecule, where the dipole moment is oriented along the molecular axis, and its change in the course of the absorption with elongation along the molecular axis x can be written as as a Taylor expansion

$$\mu = \mu_{eq} + \frac{d\mu}{dx}x + \dots \tag{14.15}$$

Here, μ is the magnitude of the dipole moment as a function of difference $x = x' - x_0$ from the equilibrium atomic separation x_0. Inserting Eq. 14.15 as the dipole moment operator into Eq. 14.12 yields

$$\underline{\mathbf{M}_{21}} = \mu_{eq}\langle \Psi_2 | \Psi_1 \rangle + \frac{d\mu}{dx}\langle \Psi_2 | x | \Psi_1 \rangle + \dots \tag{14.16}$$

The first term in this sum vanishes as $\langle \Psi_2 | \Psi_1 \rangle = 0$ when using orthogonal, normalised wave functions, while the second exists for certain combinations of wavefunctions and if $d\mu/dx \neq 0$ (For vibrations of more complex molecules, mass–weighted normal coordinates need to be introduced and take the role of x, see [20]).

In a simple absorption spectroscopy experiment as outlined here, a transition between two energy levels can hence only be excited if the dipole moment changes in the course of the transition. Another kind of spectroscopy, Raman spectroscopy, yields signals in a different experiment, if the polarisability of a system changes in the course of the transition. While Raman spectroscopy is also popular in the characterisation of interfacial systems [7, 21–23], it shall not be discussed here in more detail.

14.2.3 From the Molecule to the Dielectric Function

Optical phenomena within matter involve interactions between the electromagnetic radiation and atoms, ions and/or electrons [6]. On a macroscopic level, the complex dielectric function ε_r describes the electrical (at low frequencies) and optical (at high frequencies) properties of a material. ε_r is strongly dependent on frequency/photon energy, and contains the entire information about optical transitions, mainly about the dipole transitions discussed in this chapter. The dielectric function can be determined directly by ellipsometry experiments or derived from absorption, reflection or transmission experiments [24]. In general, the dielectric function is related to the complex refractive index m as

$$m^2(\omega) = \varepsilon_r(\omega)\mu_r(\omega). \tag{14.17}$$

Though exception exist (e.g. [25]), this chapter shall consider only non-magnetic systems with a relative permeability $\mu_r = 1$.

If an optical medium is excited by the periodic transverse electric field of a light beam, it has a dielectric response due to its electrons. The medium is polarised by the applied field, and an induced dipole moment is the result. The dipole moment per unit volume V defines the dielectric polarisation $\underline{\mathbf{P}} = \frac{N}{V}\underline{\underline{\alpha}}^* \cdot \underline{\mathfrak{E}}$.[1] There are three types of polarisation: (1) electronic polarisation, a displacement of electrons with respect to nucleus, (2) atomic polarisation, a distortion of atomic position in a molecule or lattice, and (3) orientational polarisation, an alignment of polar molecules by the electric field [26]. The polarisability tensor $\underline{\underline{\alpha}}^*$ describes on the molecular level how "strong" the reaction of the system is to an electric field in a certain direction.

In a classical description of dielectric media, electrons are assumed to be bound by harmonic forces to positively charged ions [27]. In classical physics, solving the equation of a damped harmonic oscillator leads to the Lorentzian-type dielectric function (e.g. [16] and Chap. 12 by J. Daillant). From the spectroscopy point of view, a quantum mechanical approach is desired. For this purpose, we need to solve the stationary Schrödinger equation, Eq. 14.4, and obtain the wave functions of a system. Their time dependence can be treated e.g. in a perturbation approach, for details see [27].

The polarisation can be written as the expectation value of the dipole operator, in close analogy to Eq. 14.12

$$\underline{\mathbf{P}}(t) = \mathfrak{N}_0 \langle \Psi_2(t) | \underline{\hat{\mu}} | \Psi_1(t) \rangle. \tag{14.18}$$

Here $\mathfrak{N}_0$ is the number density of the mutually independent atoms in the system. Provided they are known, one can insert the wave functions into Eq. 14.18, and solve the integral to obtain the response to an electric field from the material. Transformation into the frequency domain gives for each excitation frequency a dielectric susceptibility $\underline{\underline{\chi}}$, which relates the polarisation at this frequency to the incident electric field,

$$\underline{\mathbf{P}}(\omega) = \underline{\underline{\chi}}(\omega)\underline{\mathfrak{E}}(\omega). \tag{14.19}$$

In this quantum mechanical approach, atoms are represented as a collection of oscillators with different transition frequencies ω_{pq}. The resulting equation

$$\chi(\omega) = \frac{\mathfrak{N}_0 e^2}{2m_e} \sum_p \frac{f_{pq}}{\omega_{pq}} \left(\frac{1}{\omega + \omega_{pq} + i\gamma_{pq}} - \frac{1}{\omega - \omega_{pq} + i\gamma_{pq}} \right) \tag{14.20}$$

closely resembles the shape of the response of a classical harmonic oscillator [16],

---

[1]The term "polarisation" is ambiguous in this text, because we follow general literature usage. Polarisation can stand for the polarisation of light, as will be extensively used from Sect. 14.3 onward. In this paragraph, dielectric "polarisation" means the induction of an electric field, opposing an external field, in matter. A third meaning of polarisation, which is not used in this chapter, however, in Chap. 2 by C. D. Fenández-Solis et al. is the application of a controlled electrode potential other than the open circuit potential.

with oscillator strength $f_{pq} \propto B_{pq}$, see Eq. 14.14. Adding the strength of all oscillators by summing over all final states q leads to the oscillator strength sum rule,

$$\sum_q f_{pq} = 1. \tag{14.21}$$

This rule shows that the total transition strength in an atom can be represented as one oscillator which includes many partial oscillators. In Eq. 14.20 we have ignored the tensorial nature of the susceptibility, however, from Eq. 14.19 it is clear that the components of the electric field in the different directions of space can induce a polarisation in the three different directions of space.

The dielectric function is essentially $\varepsilon_r(\omega) = 1 + \chi(\omega)$,[2] and when we include effects beyond this picture in $\varepsilon_r(\infty)$ (which should $= 1$, but in this formulation contains all contributions not accounted for, but present at $\omega \gg \omega_T$), we obtain for a single excitation (see also Chap. 12 by J. Daillant)

$$\varepsilon_r(\omega) = \varepsilon_r(\infty) + \frac{\omega^2_P}{\omega_T^2 - \omega^2 - i\omega\gamma} \tag{14.22}$$

where ω_T is the transition frequency of the excitation. Our resulting complex dielectric function has a real and imaginary part, as has the complex refractive index . The shape of Eqs. 14.20 and 14.22 shows that real and imaginary part of the dielectric function are not independent. When there is no light absorption, $k = 0$ and $\varepsilon_1 = n^2$. Further, from k one obtains the absorption coefficient α_P when dividing by the wavelength λ

$$\alpha_P = \frac{4\pi k}{\lambda}. \tag{14.23}$$

We should finally note that the susceptibility $\chi(\omega)$ is defined for a collection of atoms or molecules. On the other hand, the polarisability α^* is the analogous quantity on the level of an individual atom or molecule. Nevertheless, their role is closely related: both describe the response of a system to an electric field the system is exposed to.

14.2.4 Electronic Transitions

The energy differences between electronic ground states and electronic excited states of many molecular systems are such that they can be excited by light with photon energies in the UV, with some systems extending into the VIS. The absorption of UV or VIS corresponds to the excitation of valence electrons.

---

[2] As a side remark, we note that in general, both χ and ε_r are tensorial quantities, which we ignore in these equations for simplicity.

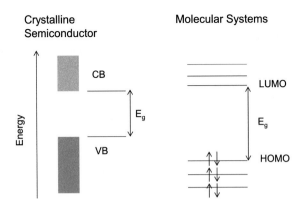

Fig. 14.1 Schematic view of a transition by light absorption in a solid from valence band to conduction band (*left*) and in a molecular system from HOMO to LUMO (*right*). The vibrational fine structure of the different electronic energy levels is not shown in this figure

In molecular systems, which are common in soft matter, the lowest energy electronic transition is a transition from electrons in the highest occupied molecular orbital (HOMO) to the lowest unoccupied molecular orbital (LUMO), as illustrated in Fig. 14.1. After excitation, the system may relax back to its original, energetically more stable state, by releasing the energy difference as photons. If the molecule is exposed to light photon energy equal to E_g, the HOMO-LUMO energy gap, this wavelength will be absorbed [1].

While in this picture, we need essentially to consider the HOMO and LUMO (and all other molecular orbitals) of the complete molecule, often characteristic groups of the molecule are dominating light absorption at a certain wavelength. These groups are called chromophores, and their presence can easily be detected in a UV/VIS spectroscopic experiment [4].

While soft matter is typically composed of molecular systems, at interfaces it may be in touch with a crystalline solid, e.g. a semiconductor or a metal. In this case, there are no well-defined energy levels, but there are bands. Optical properties of crystalline solids are connected with transitions between the bands, and the lowest energy transitions, the analogue of the HOMO-LUMO transition, is the transition between valance band and conduction band, as illustrated in Fig. 14.1 [28, 29]. As we have a continuum of transitions above the band gap E_g, the spectra in this case do not consist of discrete bands, but show strong absorption over an extended wavelength range.

14.2.5 Vibrational Transitions

Radiation in the mid-IR wavelength range excites vibrational modes. One can think of these in two different ways: (a) as vibrations of molecules or (b) as vibration of lattices. In molecular soft matter systems, the molecular picture is more useful to understand spectra, however, at solid/liquid interfaces, lattice vibrations of the solid may become important.

The picture of the vibration of molecules (a) is the more easy to comprehend for (small) molecules at an interfaces. If we look at the structure of a molecule, say water, H_2O, feel free to regard the bonds between the atoms as "quantum mechanical springs" [20], for which we can obtain solutions of the Schrödinger equation with harmonic (or anharmonic) oscillators representing the potential. Every molecule is thus a collection of springs, and at certain frequencies these springs will oscillate—these are the eigenfrequencies of a certain system. The light will provide the excitation energy of these systems, so the light intensity will decrease after passing through a sample.

The second picture, the picture of lattice modes in a crystal, is more popular in the world of crystalline solids [30]. Here, lattice vibrations are known as phonons, and are the analogue to vibrations of molecules (See [5] for a clear statement on the distinction between lattice vibrations and phonons).

14.2.5.1   Molecular Vibrations

In order to understand the vibrations of a molecule, we want to think of water as an example. The molecule consists of three atoms. Every molecule with X atoms has $3X - 6$ vibrational degrees of freedom, except if the molecule is linear, which has only $3X - 5$. What oscillations these modes come down to can be determined "from scratch" by a normal coordinate analysis [1, 20]. In the simple picture presented above, we need to replace the actual coordinates by mass-weighted normal coordinates, in which the respective analysis can be performed.

For water with its three atoms, we should look for 3 oscillatory modes. These three turn out to be (a) the antisymmetric stretching mode, (b) the symmetric stretching mode, and (c) the bending mode. The movement of bonds in these different modes is depicted in Fig. 14.2. Please understand that when looking at the motion of the atoms the picture is a bit different, because in a vibrational mode, the centre of mass of the molecule needs to be fixed.

In the case of water, all three modes can be observed in the IR spectrum (Fig. 14.3). The absorbance band at around 1645 cm^{-1} is assigned to water bending mode [δ (OH$_2$)], while the band at 3000–3700 cm^{-1} contains the stretching modes [ν (OH)].

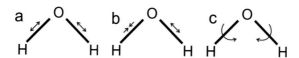

Fig. 14.2 Motion of the bonds in the different normal modes of a water molecule: **a** symmetric stretching mode, **b** antisymmetric stretching and **c** bending mode

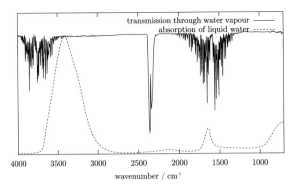

Fig. 14.3 IR absorption spectrum (*A* from Eq. 14.1) of liquid water (*bottom*) and water vapour (*top*, transmittance, *M* from Eq. 14.1). The peak around 2300 cm^{-1} is the absorption of atmospheric carbon dioxide, which was present in the system during measurement, but won't be discussed here

Figure 14.3 illustrates the strong differences between a spectrum from the gas phase and a spectrum of a liquid. In the gas phase, rotational excitation show up as fine structure around the vibrational transitions. While the bending mode is centred at the same wavenumber in gas phase and in liquid phase, the stretching modes are shifted to lower wavenumber in liquid phase. In water, this shift is caused by the strong hydrogen bonding. Hydrogen bonding also leads to the occurrence of the restricted rotational and translational modes at lower wavenumbers ("librations") [31–34]. At this point, it is worth to direct the interested reader to the website of Chaplin [35], which discusses in detail the absorption spectrum of water.

For complicated molecules, the possible modes are also getting more complicated. Even a relatively simple molecule like hexane, C_6H_{14}, has 20 Atoms, and therefore 54 vibrational modes. Some of these modes will be very close in their frequencies and some even degenerate, however for chemically more diverse systems, as typically encountered in soft matter, this does not need to be the case. In the case of complex molecules as present in soft matter systems, frequently almost all modes can be visible in both the IR absorption as well as the Raman spectrum, though with different intensities.

In general, some modes will be localized in certain parts of a molecules and therefore be highly specific for the presence of a certain group. Examples are the CH_2 stretching modes, to whom only the methylene groups contribute. These modes hint to the presence of certain groups in a molecules [4, 19]. Also, small difference in a vibrational frequency of a certain group (e.g. of the CH_2 stretching modes) can be an indication of different conformations of these groups. A very systematic collection of the vibrational modes typically observed in organic compounds with a great variety of substitution patterns was compiled by Nyquist [19]. Nowadays, also software is available to analyse vibrational spectra for the presence of certain characteristic groups based on empirically collected knowledge [36].

Other modes will be delocalised over the whole molecules, meaning that all bonds are collectively oscillating. These are the ones responsible for the "fingerprint" of the molecule, making IR spectroscopy useful as a proof of identity [4, 19].

At interfaces, the bond between a certain molecule and the interface may give rise to additional vibrational modes not present in the free molecules. Likewise, coupling may occur between vibrational modes of species on both sides of the interface, or between electronic transitions of species on one and vibrational transitions of species on another side of an interface.

14.2.5.2 Lattice Vibrations (Phonons)

Similar to atoms in a molecule, the atoms in a crystal oscillate around their average position. In a crystalline solid, these oscillations happen in a synchronised way between different unit cells, and are called lattice vibrations, also known as phonons [30]. A crystal will also have phonons that are highly localized and that resemble e.g. the stretching of a bond. As an example, the O–Si–O stretching modes occur at similar frequencies in a quartz crystal and in a silicon-organic compound. Other phonons are delocalized over the complete unit cell of a crystal. Details are presented in textbooks of solid state physics [5, 37].

In a crystal containing x atoms in y molecules per unit cell, containing in total X atoms, there will also be $3X$ normal modes, each with energy $\hbar\omega$. Of these modes, $3(y-1)$ will be translational modes (see [5] for more details).

The phonons may couple to electronic transitions, giving rise to interesting physics, e.g. for the electrical conductivity [38]. Also in solids, typical phonon energies are comparable with IR photon energies. When adsorbing molecules to interfaces, the lattice vibrations of the crystalline material may couple with the vibrational modes of the molecules, giving rise to completely new phenomena due to the presence at the interface.

14.3 What's Special at Interfaces? Principles and Application

At an interface, reflection and refraction of light will occur. If light-absorbing species are present at the interface, absorption will occur in addition. Most information about the state of the interface can be obtained by analysing the reflected light, which is why most of this section will discuss the different options to perform reflection experiments (We will, however, also briefly discuss the experiments in a transmission geometry). Within the scope of this book, we will focus on systems containing soft matter at interfaces where at least one side is an aqueous solution. We will exclusively discuss specular reflection experiments (angle of incidence = angle of reflection), and not discuss diffuse reflection and scattering.

The central quantities in reflection spectroscopy are the amplitude reflection coefficients r, which give the ratio of reflected time-averaged electric field amplitude $\mathfrak{E}^{(0),\text{refl}}$ to incident time-averaged electric field amplitude $\mathfrak{E}^{(0),\text{inc}}$, $r = \mathfrak{E}^{(0),\text{refl}}/\mathfrak{E}^{(0),\text{inc}}$. As the electric field has been introduced as vectorial quantity in Sect. 14.2 (Eq. 14.9), r has a strong directional dependence. To take this directional dependence into account, the polarisation of light will be introduced below. A more detailed discussion of the amplitude reflection coefficient is presented in Chap. 13 by B. Loppinet. For this chapter, we still have to note that we will discuss intensities rather than field amplitudes. The reflectivity R describes the analogous intensity ratio between reflected I^{refl} and incident I^{inc} intensity of light, $R = I^{\text{refl}}/I^{\text{inc}}$, and is the quantity already used in Eq. 14.2 to define the reflectance absorbance.

Reflection experiments at interfaces can be divided into two broad categories, internal and external reflection and are schematically shown in Fig. 14.4.

The external reflection geometry is the simplest geometry of a reflection experiment. Light is incident through a sample-containing solution, the medium with the lower refractive index, and reflected from the interface under investigation. Because in this experiment, there is a uniform surface, therefore molecules in the entire illuminated part of the surface contribute to the final spectrum in the same fashion. This type of experiment can also be used to study the surface of bulk metals, including single crystals, or rough surfaces as used in industrial processes. Using flat surfaces opens up the possibility to determine the orientation of molecules with respect to a well-defined interface, or any other well-defined reference direction. Big drawback of this geometry is the interaction of the light with the bulk sample medium in a similar fashion as in transmission geometry, hence tricks are needed to extract interface-specific information.

Fig. 14.4 Schematic representation of internal and external reflection experiments performed at an interface. For the sake of clarity, refracted beams in transmission are not depicted

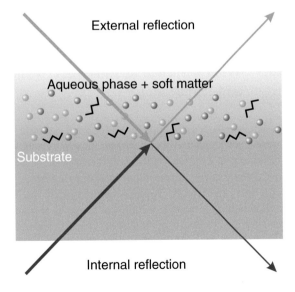

In internal reflection, light is impinging on the interface under study from a medium with the higher refractive index above the critical angle of total reflection. Hence, an evanescent wave is generated at the interface, probing the sample medium. In this fashion, there is limited interaction of the light with the bulk sample medium. However, use of internal reflection is limited to few incidence media.

Before proceeding to different cases of application, it is worth to point out that at interfaces, an analysis of the polarisation of light is critical. Because the linear polarisations of light are eigenpolarisations of a planar interface, i.e. impinging linearly polarised light result in reflected linearly polarised light, we shall limit ourselves to linearly polarised light in this chapter. For a thorough introduction to polarisation of light, see [39]. In Eq. 14.9, for a transverse wave as discussed here, $\underline{\mathfrak{E}} \perp \mathbf{k}$, hence $\underline{\mathfrak{E}}^{(0)} \perp \mathbf{k}$, which corresponds to amplitude components perpendicular to the propagation direction of the electromagnetic wave. The conventional coordinate system in the optics of interfaces is shown in Fig. 14.5.

When the oscillation of the electric field vector occurs perpendicular to the paper plane, this is called perpendicular polarisation (or vertical, or transverse electric, or simply s-polarisation; labeled "s"; referring to the German word "senkrecht" for "perpendicular", in Fig. 14.5). The electric field therefore has only components in y-direction, and no component in the x- and z-direction. An oscillation of the electric field vector within the x-z-plane, but perpendicular to the direction of polarisation is

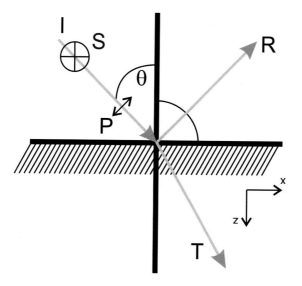

Fig. 14.5 Coordinate system used (x/z axes) to describe light impinging on a surface under an angle of incidence θ. Reflected (R) and transmitted (T) beams are observed. The two modes of linear polarisation are shown: in s-polarisation, the electric field vector oscillates perpendicular to the x-z-plane, the plane of incidence, while for p-polarisation it oscillates within that plane and perpendicular to the direction of propagation of the incident beam. The medium from which the light impinges on the surface is medium 1 with the refractive index n_1, and the exit medium is medium 2 with a refractive index n_2

shown as an arrow and called parallel polarisation (or horizontal, or transverse magnetic, or simply p-polarisation; labeled "p" in Fig. 14.5). The electric field of the p-polarised wave has therefore components in the x and z direction, but is 0 in y-direction.[3]

Because the oscillation of the electric field vectors are in different planes for the different polarizations, they can excite different components of the vectorial transition dipole moment to a different extend. One gets maximum signal if the transition dipole moment is parallel to the direction of polarisation. That is the basis for obtaining information about the orientation of the molecules at an interface.

Quantification of the reflectivities from the interfaces are discussed in Chap. 13 by B. Loppinet. Here, we want to point out that based on the complex refractive index of a well-defined layer systems, reflectivities can be calculated, hence a reflectance absorbance can be defined. Calculation from the complex refractive indices of all ingredients may be performed using matrix methods [40, 41]. For such calculations, the author's lab has developed the open source computer program reflcalc, which is flexible but not too user-friendly [42]. For geometries with in-plane structure, sophisticated methods of solving the Maxwell equations are needed, e.g. finite difference time domain [43] or finite element methods [44].

14.3.1 Transmission Spectroscopy for Interface Characterisation

The simplest spectroscopic experiment to characterise an interface is a transmission experiment, as sketched in Fig. 14.6. Here, light is simply going through the structure under investigation, the transmitted light is collected and analysed. This experiment is convenient if all parts of a structure are sufficiently optically transparent. For the investigation of metals or other strongly absorbing substrates, a wire grind can be manufactured and introduced into the beam [45]. The solution, including the interface under study, must be placed between two windows transparent for the wavelength range in use.

In this type of experiment a spectrum contains information averaged over all molecules with the respective absorption in the beam. This includes molecules adsorbed to the surface, but also those present in the bulk solution, or a thin film. This problem is simplified if only adsorbed molecules are present in the system, e.g. if unbound molecules can be rinsed away.

Metals, however, are not transparent to light—their investigation is, however, highly desired e.g. when discussing electrochemical phenomena (see Chap. 2 by C. D. Fenández-Solis et al.). Near the surface of an ideal metal ($n = 0$, $k \rightarrow$ infinity) [46],

---

[3] Actually, the magnetic field vector is perpendicular to the electric field vector. Therefore, for the p-wave, the magnetic field vector has only a y-component that is different from 0, which is where the name transverse-magnetic originates from.

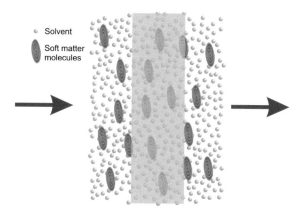

Fig. 14.6 Sketch of the propagation of light (*arrows*) in a transmission experiment. The *central darker* structure symbolises a metal grid, or an IR-transparent semiconductor. The *dots* symbolise the solvent, while the *curved* structure depicts the soft matter molecules under study

the boundary conditions show that the electric field vanishes directly at the interface for normal incidence [40, 46]. In real metals, which have values $n \sim 5–20$ and $k \sim 5–50$ [24], the boundary conditions are somewhat softened in this respect. Therefore, some intensity of light will be present at the interface. The intensity will be maximised at large angles of incidence, near grazing incidence and depends on the polarisation [40, 47].

A further complication in this approach arises from the strong IR absorption of most solvents, including water. This problem is known in solution transmission IR spectroscopy, and is usually solved by using only short paths through the electrolyte, down to <10 µm, if vibrational modes near the solvent absorption should be investigated. In such thin cells, adsorption experiments are, however, hard to perform, because of diffusion limited transport to the grid.

Because of its appealing simplicity, however, the transmission experiment has been very popular (see e.g. [45, 48]). It is advantageous for qualitative analysis of finely dispersed structures, where species at an interface make up a significant fraction of the total amount of molecules to be analysed. The transmission approach is particularly suitable to investigate differences between different states, and if bound molecules have a significantly different spectroscopic signature than those in free solution.

14.3.2   External Reflection Spectroscopy on Metals

In the visible, and even more so in the infrared, metals are good mirrors. At room temperature, they posses a continuum of electronic transitions below the plasma edge. Since the photon energy $\hbar\omega$ of infrared radiation is lower than that, metals absorb strongly in the infrared wavelength range. Consequently, electromagnetic waves, and therefore light as well, cannot be transmitted through macroscopic metal samples above a thickness of several tens of nm. Therefore, bulk metal surfaces cannot be studied by transmission (see Sect. 14.3.1) or internal reflection (see Sect. 14.3.4) techniques. Thus, external reflection experiments provide the most significant amount of information [7, 49].

Naively, one would expect this to be a very straightforward technique, since metals do have a high reflectivity. However, due to the boundary conditions for electromagnetic waves at a metallic interface, the field is almost zero directly at the interface. This is because, as already mentioned, the electromagnetic field does not penetrate the metal due to its material properties, so close to the metal surface, electromagnetic waves already "prepare" for the presence of the metal. Figure 14.7 depicts the situation for a reflected electromagnetic wave at normal incidence. It shows that at normal incidence, there will be a maximum of intensity at about a quarter of the wavelength away from the surface. One can therefore highly sensitively probe materials somewhere about this distance away from the surface, but not in the first nanometres away from the surface.

Because of electromagnetic reasons, only transition dipole moment components perpendicular to the metal surface can be excited near a metal surface [7, 49, 50]. This so-called "surface selection rule" is governed by the interaction of the transition dipole moment with its image dipole, as illustrated in Fig. 14.8. As light has a wavelength that is much larger than the transition dipoles of molecular dimension,

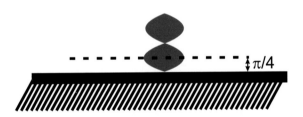

Fig. 14.7 Standing wave pattern formed by a reflected electromagnetic wave that is impinging on a metal surface at normal incidence. The pattern formed by interference between incident and reflected wave has a knot at or near the surface. The maximum of intensity is somewhere around a quarter of the wavelength inside the incidence medium

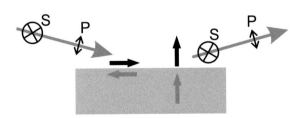

Fig. 14.8 A transition dipole moment on the surface of a conductor with free charge carriers, e.g. a metal, interacts with its image dipole. Hence, components parallel to the metal surface cancel, while components perpendicular to the interface are amplified

light will probe the sum over the transition dipole and its image dipole. Transition dipole moment components parallel to the interface, however, cancel with their image dipole. On the other hand, components perpendicular to the surface are enhanced. S-polarised light would excite only components which are parallel to the surface, hence no absorption of light is measured in s-polarisation. The components which are active on the surface can be excited in p-polarisation at large angles of incidence.

One example, where external reflection spectroscopy can be conveniently used on metal substrates are monomolecular layers on metals in air or vacuum. Here, we shall briefly discuss the case of self-assembled monolayers (SAMs), however, similar lines of reasoning apply to Langmuir-Blodgett films, or other adsorbate structures. SAMs are essentially spontaneous assemblies of organic molecules chemisorbed onto the metal substrate through a head group. A representation of a typical SAM on a metal substrate is shown in Fig. 14.9. The interesting properties of such a system to be studied pertains to the interface between the organic molecule and the metal surface. This means in principle the angle, order/disorder, morphology, and coverage of the deposited SAMs. IR spectroscopy is a common method to study such systems—though most of the time not in contact with aqueous solutions. The most commonly studied system of SAMs are molecules with thiol headgroups assembled onto a gold surface, by the strong S-Au bond [51–55].

IR spectra can be recorded, with spectra from the bare substrate serving as a reference (background in Eq. 14.2). An example of a spectrum is shown in Fig. 14.10. This figure shows the CH stretching modes of dodecanethiol adsorbed to gold. The spectrum differs from a spectrum in homogeneous solution in the absorbance ratios between the different peaks (surface selection rule applied to an oriented layer!), and in the exact peak position (conformationally ordered chains with all-trans conformation in SAM, disordered in solution). See [56, 57] for a detailed discussion of the spectra of alkyl chains in different conformations.

Because of the fact that only the transition dipole moments perpendicular to the surface can be probed at all, the determination of the orientation of molecules requires a reference system. This reference system can either be a quantitative

Fig. 14.9 Representation of a self assembled monolayer on a metal substrate, as an example for organic thin films on metals

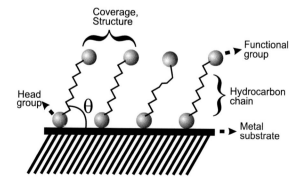

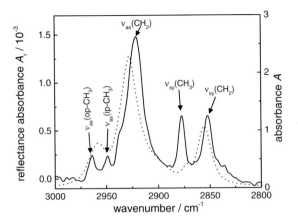

Fig. 14.10 IR reflection absorption spectrum of a decanethiol SAM on Au(111) (*full line*, scale on the *left*), compared to spectrum of decanethiol in CCl_4 (*dotted line*, scale on the *right*). Different CH stretching modes are shown, with the respective assignment. The reflection spectrum has been recorded at grazing incidence (incidence angle 80°) with p-polarised light

comparison to a transmission spectrum [58]. Alternatively, an internal reference, i.e. a comparison of groups whose transition dipole moment directions are not equal, is suitable, which we shall briefly illustrate here. We start from the approach that

$$\int_{band} A_r d\tilde{v} \propto B_{21} \beta_z, \tag{14.24}$$

where β_z gives the projection of the transition dipole moment components on the z-axis, according to the surface selection rule, and $\int_{band} A_r d\tilde{v}$ represents the integrated reflectance absorbance. For β_z, one can use a model developed for the determination of the orientation of protein fibres from its linear dichroism [59], which yields for a uniaxial system

$$\beta_z = S_2 \cos^2 \gamma + 1/3(1 - S_2), \tag{14.25}$$

where γ is the angle between transition dipole moment and molecular axis, and S_2 the orientational order parameter,

$$S_2 = \frac{\langle \cos^2 \delta \rangle_N - 1}{2}, \tag{14.26}$$

with tilt angle δ of the molecular structure with respect to the z-axis, and $\langle . \rangle_N$ indicating an ensemble average. Let's consider now the ratio of reflectance absorbance between two modes, one with $\gamma^{(1)} = 0°$, and the other with $\gamma^{(2)} = 90°$. Examples may be SAMs terminated by —NO_2 or —COO^- groups. These groups posses both symmetric and antisymmetric stretching modes, with transition dipole moments which are in first approximation perpendicular to each other. Hence absorption peaks from these modes can be analysed quantitatively for determination of the orientation of these groups. Calculating the ratio of the reflectance absorbances by putting Eq. 14.26 into Eq. 14.25, substituting the result for β_z into Eq. 14.24, and simplifying, yields

$$\frac{\int_{\text{band}} A_r^{(1)} d\tilde{v}}{\int_{\text{band}} A_r^{(2)} d\tilde{v}} = \frac{C^{(1)} B_{21}^{(1)} (2S_2 + 1)}{C^{(2)} B_{21}^{(2)} (1 - S_2)}, \tag{14.27}$$

where $C^{(2)}$ and $C^{(2)}$ are the constants of proportionality from Eq. 14.24. The ratio $D^{(\text{refl})} = \int_{\text{band}} A_r^{(1)} d\tilde{v} / \int_{\text{band}} A_r^{(2)} d\tilde{v}$ can then be obtained from the integrated bands from the reflectance absorbance spectrum, while the ratio $D^{(\text{trans})} = C^{(1)} B_{21}^{(1)} / C^{(2)} B_{21}^{(2)}$ is obtained from the integrated absorbance of the same bands in a spectrum of the same species in solution, where it is randomly distributed. Substituting the ratios into Eq. 14.27 yields $D^{(\text{refl})} = D^{(\text{trans})} \cdot (2S_2 + 1)/(1 - S_2)$, which can be solved directly for the squared cosine of the tilt angle, using Eq. 14.26, to yield

$$\langle \cos^2 \delta \rangle_N = \frac{D^{(\text{refl})}}{D^{(\text{refl})} + 2D^{(\text{trans})}}. \tag{14.28}$$

With this result, the orientation of a molecular group can be determined by comparing the absorbance ratio of two modes in reflection with those in a random orientation, e.g. from a transmission spectrum in free solution.

In modern works, a calculated spectrum (e.g. by density functional theory, DFT) can be directly compared to a measured spectrum. DFT is also quite useful to obtain the directions of transition dipole moments. The ratio $D^{(\text{trans})}$ may hence also be obtained from a calculation. When comparing measured spectra to experimentally obtained, most of the time only the calculated frequencies are compared, not calculated spectra (see e.g. [60] for one example). Calculations of the full dielectric function and hence the full spectral information probed by an experiment are also possible, though challenging [61]. It should be pointed out that calculated vibrational frequencies apply not only to the IR absorption experiments, but also to transitions in Raman or sum frequency generation spectroscopy experiments (see also Chap. 15 by M. Hoffmann et al.) [60, 62].

Because of the surface selection rule, quantification of the exact amount of transition dipoles near an interface is rather challenging. Even a surface which is fully covered may not show any contribution in the spectrum, if the molecular transition dipoles are oriented parallel to the surface. As one textbook puts it [63], the surface of a metal is the perfect place to hide a thin coating of organic material.

Thicker soft matter structures can also be studied, e.g. [64]. In experiments where thick layers of soft matter are investigated, it has to be pointed out that because of the distance-dependence of the field strength (Fig. 14.7), molecular groups which are located at a different distance from the surface contribute to the spectrum differently, which again makes quantification of concentrations as in transmission spectroscopy of homogeneous solutions difficult.

The problem arising from the fact that in the case of metals, species in solution contribute to the spectra to a larger extend than species on the surface is obvious for this geometry. Further, solvent absorption is a problem, which is why optical path lengths through the sample solution need to be minimised. Due to the large angle of incidence needed, the solvent layer thickness needs to be even smaller than for transmission experiments.

The lack of interface specificity of the direct external reflection experiment can be overcome by modulating the polarisation of the incoming light. The physical principle of the technique uses the fact that during propagation through an isotropic medium, all possible linear polarisation states of light are affected equally. However, p-polarised light interacts differently with an interface than s-polarised light (see also Chap. 13 by B. Loppinet). Because the linear polarisations are the eigenpolarisations of a planar interface, the differences in the linear polarisation states are the quantities that contain the information about the interface [40, 47, 65]. The measured reflection spectrum is obtained by signal processing as the difference of the intensity measured with parallel polarisation and the intensity measured with perpendicular polarisation [66–68]. The signal is then usually normalised with respect to the sum of signals from both polarisations. Because of the surface selection rule, the analysis is considerably simplified in the case of metal substrates compared to transparent substrates [66]. For non-metallic substrates, e.g. semiconductors, or H_2O, the use of the difference signal can lead to cases where contributions to the absorption from light with parallel and perpendicular polarisations cancel each other, and no absorption is seen in the spectrum despite the presence of molecules on the surface.

The length scale of the interface specificity of this technique is on the length scale of the differences in the intensity reflection coefficients of the two linear polarisations, i.e. in the subwavelength range, usually a few 10 nm in the mid-IR [47]. The constraint that only a thin layer of electrolyte may be present above the surface, here between the entrance window and the sample surface, is still present when using polarisation modulation.

14.3.3 External Reflection Experiments on Transparent Substrates

The surface selection rule, caused by the free electrons in the metal, makes metals special in reflection spectroscopy experiments. External reflection experiments can, however, also be performed with substrates which are transparent, such as MgF_2 or CaF_2, which are transparent from UV to IR. Substrates containing polarisable molecules, such as H_2O, also fall in this group. Though H_2O is strongly absorbing in the IR, its absorption from excitation of vibrational modes is still considerable weaker than the absorption caused by electronic transitions in metals.

Typically, the reflectivity of these materials is much lower than the reflectivity of metals, therefore, the total intensity at the detector is typically rather low, making experiments on non-metallic substrates much more challenging compared to metals discussed in Sect. 14.3.2. On the other hand, there is no mechanism to decrease the intensity of certain species, as the surface selection rule for metal surfaces. A rough determination of orientation is possible from measurements with two polarizations at one angle of incidence, however, a better method are angular-resolved measurements, which need to be fit to a multilayer model, similar to the analysis of ellipsometric data (see Chap. 13 by B. Loppinet).

The angular and polarisation dependence of the absorbance is not trivial. The complex relation results from the presence of a transmitted and reflected beam. In s-polarisation, absorbance bands are typically negative, while in p-polarisation they change the sign around the Brewster angle. Bands are most intense in p-polarisation near the Brewster angle. For more details, see e.g. [7]. A careful optimisation of the angle of incidence is needed if optimum sensitivity should be achieved. For examples, the reader is also referred to the substrate-specific original literature, e.g. [69–73].

14.3.4 Internal Reflection Spectroscopy—Attenuated Total Reflection (ATR)

In internal reflection experiments, measurements are performed at an incidence angle above the critical angle θ_{cr} of total internal reflection, which is given as $\sin(\theta_{cr}) = n_2/n_1$ from the ratio of the refractive indexes n_1 and n_2 of incidence and exit material, respectively.[4] The larger the refractive index n_1, the smaller θ_{cr}, and therefore the larger the range of incidence angles one can use [74, 75]. The incident wave generates an evanescent wave with an exponentially decaying electric field into the medium with lower refractive index,

$$\mathfrak{E}(z) = \mathfrak{E}_0 \, e^{-z/d_p}. \tag{14.29}$$

Here, $\mathfrak{E}_0$ is the electric field at the interface, z is the distance from the surface, and

$$d_p = \lambda_0/2\pi \sqrt{(n_1 \sin \theta)^2 - n_2^2}, \tag{14.30}$$

---

[4]The same phenomenon is referred to in Chap. 12 by J. Daillant as "total external reflection", because at X-ray wavelengths, the refractive index situation is reversed compared to optical wavelengths, i.e. the medium with lower refractive index here may be the medium with higher refractive index when using X-rays.

with λ_0 the vacuum wavelength of the incident radiation, is the characteristic length of the exponential, the penetration depth. The nature of the evanescent wave is e.g. discussed in [76]. Visualisations can be found e.g. at [77].

Placing an absorbing medium in contact with the evanescent wave leads to an attenuation of the total internal reflection, which is why this method is also known as attenuated total reflection (ATR) spectroscopy. An overview over physical basics and general applications is given elsewhere [7, 15, 49, 74, 78]. This method has also been quite popular for the characterisation of lipid membranes, small peptides, and membrane proteins [12, 78]. Polymers at interfaces have also been investigated extensively [79].

Internal reflection spectroscopy requires a transparent, high-index medium of incidence. Typical examples for materials serving as "internal reflection elements", i.e. the high refractive index incident media, are silicon, germanium, and zinc selenide in the IR, and ZrO_2 or high-index glasses in the UV/VIS. Especially semiconductors with a low band gap are very suitable incidence media in the IR, though they can be used in transmission and external reflection experiments as well. The transparency of the substrate is usually only limited by the absorption from lattice vibrations (see Sect. 14.2.5.2). Metals are unsuitable media of incidence, due to their strong light absorption. However, thin metal films can be deposited onto the surface of a transparent medium of incidence, and a tunnelling of the light through the metal films used for spectroscopic applications. Such a method has been used to characterise peptides [80, 81] and lipid bilayers on gold surfaces [82, 83]. Interlayer systems can be applied to enhance the transparency of the metal layers [84, 85].

Using internal reflection spectroscopy, there is a certain surface specificity given by the penetration depth d_p. The penetration depth of the evanescent wave depends on the refractive index difference (or rather, depends on how "far" away one is from the critical angle), and a higher refractive index difference with a higher angle of incidence gives a lower penetration depth, therefore increasing the "surface-specificity", i.e. the amount the contributions near the interface are amplified over the contributions from the bulk medium 2. It is worth noting that absorbance is governed by $|\underline{\mathfrak{E}}(z)|^2$, implying that the characteristic length of the decay of the absorbance contribution is $d_p/2$. In typical situations, with silicon or germanium as incidence media and aqueous electrolytes as rarer medium, $d_p \sim 200\text{--}400$ nm. Even if thin metal (or other) films are present on the surface of the medium of incidence, the decay length and hence the interface specificity are determined by the medium of incidence and the exit medium. The intermediate films modify the amplitudes $\underline{\mathfrak{E}}_0$, but d_p is left unchanged. There is a significant decrease of $\underline{\mathfrak{E}}_0$ on thin metal films, leading to a low overall sensitivity. It must be noticed, however, that the exponentially decaying intensity profile means that solution species are still detected in the spectrum, but species nearer to the interface are detected with a larger weighting factor.

The big advantage of this method is that no propagation of light through the medium under investigation is needed. Therefore, there are no geometrical constraints to the solution side of the experiment and no need for thin layer cells.

The orientation of a molecule on a surface can easily be estimated in internal reflection spectroscopy from the dichroic ratio $D = (\int_{\text{band}} A_{r,p} d\tilde{v})/(\int_{\text{band}} A_{r,s} d\tilde{v})$ via S_2 (Eq. 14.26). D is the ratio of the integrated reflectance absorbance with s- $(A_{r,s})$ and p-polarisation $(A_{r,p})$. Details can be found elsewhere [15, 78]. The orientation can also be determined if thin solid layers of known thickness are present between medium of incidence and solution [82, 83]. As there is no orientation of the transition dipole moment in which absorption is fully extinct, the concentration of species on the surface can be determined as well [86–88]. Furthermore, the evanescent wave is one of the most repeatable waveforms that can be created for spectroscopic application, which is why its use for illumination has big advantages for highly quantitative work [32, 74, 75].

14.3.5 Evanescent-Wave Illuminated Spectroscopy in External Reflection

An evanescent wave generated on one interface can also be used to illuminate a second interface in the vicinity of the one where the evanescent wave is generated. This is illustrated in Fig. 14.11. Frequently, this geometry is also referred to as ATR geometry [10, 11], but because the mechanism in which the light interacts with the sample surface is different from the one described in Sect. 14.3.4, we decided to devote an extra subsection to it. Here, from the viewpoint of the sample interface, the optics is similar to the optics described in Sect. 14.3.2, except that the illumination is now happening through an evanescent wave instead of a freely propagating wave.

In this geometry, bulk metal electrodes can be illuminated, but a large price has to be paid in terms of the interface specificity. Because in this geometry, the evanescent wave is generated several 100 nm away from the sample surface, there

Fig. 14.11 Sketch of an optical configuration in which an evanescent wave generated at a semiconductor/solvent interface is used to probe a sample surface close-by through a solution

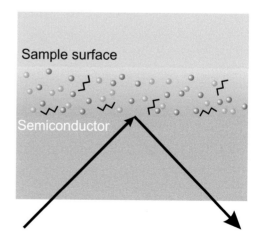

is an exponentially larger sensitivity to species near the incidence medium than to species near the sample surface. Furthermore, the solution layer between incidence medium and sample surface must be extremely thin, leading again to a very demanding cell design.

On the other hand, the interface to the internal reflection geometry offers an increase in overall sensitivity when compared to external reflection or transmission experiments, which can be achieved by using multiple reflections in the medium of incidence [12]. Applications of this geometry are described elsewhere [10, 11].

14.3.6 Surface Enhancement from Rough Interfaces

Using rough metal surfaces, it is possible to obtain special "surface enhancement" of the absorption in the spectra, which leads e.g. in the IR to "surface enhanced infrared absorption" (SEIRA) spectroscopy. The enhancement originates mainly from near-field effects: incident light is scattered with a high intensity into e.g. radial fields near particles or other small structures [7, 65, 89]. Therefore, there is an increase in interface specificity, as well as an increase in the overall sensitivity to species. A number of applications in this field have been reviewed recently [13]. There are certain analogies in the mechanisms to the mechanisms of surface-enhanced Raman spectroscopy, though the details of the mechanisms must take into account the different mechanisms of Raman and IR absorption spectroscopy [90, 91]. While sketched in principle in Fig. 14.12 for the more popular internal reflection geometry [13], surface enhancement is obtained in both internal and external reflection geometry [92].

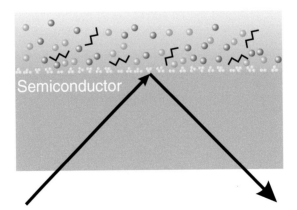

Fig. 14.12 Optical configuration of creating surface enhancement. A metal-island structure (*dots on the surface*) is deposited on the surface of an IR-transparent medium (*bottom*). IR light (*arrows*) propagates inside the semiconductor medium probing solvent and sample near the metal nanostructures

Due to the near-field nature of the experiment, this is the method with the largest interface specificity, which can reach only a few nm. Both the large sensitivity and the large interface do, however, imply the need for using rough surfaces. On these surfaces, quantitative work is again difficult, because different parts of the surface contribute to the overall spectrum to a different extend. Furthermore, some regions of the surface are not illuminated at all [90]. In addition, the option to determine the orientation of molecules has disappeared, mainly because of the ill-defined surface. Adsorption studies to such surfaces may also be rather specific to the actual surface morphology and structure.

One of the current sticking points of SEIRA in the applications described here is the need for rather ill-defined surfaces. The use of well-defined surfaces may bring further advantages to the technique [92].

14.4   Conclusion

Interfaces can be probed by optical spectroscopy using a variety of reflection techniques, but under some circumstances also by transmission spectroscopy. The physical principles of dipole transitions are very similar for probing electronic and vibrational transitions. However, the latter offer more transitions, therefore they are more frequently probed in soft matter science. Depending on the type of sample and the available sampling techniques, optical absorption spectroscopy may give (a) qualitative information on the presence of certain groups, (b) (semi)quantitative information on the amount of certain species present, (c) quantitative information about the orientation of certain groups, (d) qualitative information about the environment (e.g. polar vs. non-polar) a certain group is in and (e) information of the conformation of molecules.

From the spectroscopic techniques discussed here, SEIRA spectroscopy has the largest interface specificity, of only a few nanometres, but it depends on the exact geometry of the surface. The interface specificity of polarisation modulated external reflection experiments is less pronounced, followed by direct internal reflection spectroscopy. All other methods discussed here are not interface specific. However, interface specificity itself is only important if species near the surface are to be investigated in a background of species in a bulk solution or film. A better speci-ficity to interface contributions is obtained by non-linear optical techniques, see Chap. 15 by M. Hoffmann et al.

For optical reasons, the sensitivity to the amount of adsorbed species is high for SEIRA spectroscopy, but due to the use of multiple internal reflection can be increased almost arbitrarily using internal reflection spectroscopy in a direct way or using the generated evanescent wave to illuminate a different surface.

Overall, the internal reflection geometry (ATR) is the best suited for quantitative work, though it is rather limited in the number of interfaces which can be inves-tigated. Both orientation of species and surface concentrations can be determined in ATR geometry. On the other hand, spectroscopy in external reflection shows largely different spectra comparing metal and transparent substrates. On metals, the "surface selection rule" implies that transition dipoles in a certain direction cannot be probed, making quantification of the surface coverage difficult. Overall, all methods discussed here are useful for qualitative in situ and operando analysis of the composition of interfaces, which is likely the most significant contribution of optical spectroscopy to interface science.

Acknowledgments This work is supported by the Cluster of Excellence RESOLV (EXC 1069) funded by the Deutsche Forschungsgemeinschaft. C. T. thanks the International Max Planck Research School for Surface and Interface Engineering in Advanced Materials (IMPRS-SurMat) for a scholarship. The authors thank Prof. M. Stratmann for continuous support.

References

1. J.M. Hollas, *Modern Spectroscopy*, 4th edn. (Wiley, Chichester, 2004)
2. I.N. Levine, *Molecular Spectroscopy* (Wiley, New York, 1975)
3. D.C. Harris, M.D. Bertolucci, *Symmetry and Spectroscopy—An Introduction to Vibrational and Electronic Spectroscopy* (Dover, New York, 1978)
4. G. Gauglitz, T. Vo-Dinh (eds.), *Handbook of Spectroscopy* (Wiley, Weinheim, 2003)
5. P.M.A. Sherwood, *Vibrational Spectroscopy of Solids* (Cambridge University Press, Cambridge, 1972)
6. M. Fox, *Optical Properties of Solids* (Oxford University Press, Oxford, 2001)
7. V.P. Tolstoy, I.V. Chernyshova, V.A. Skryshevsky, *Handbook of Infrared Spectroscopy of Ultrathin Films* (Wiley, Hoboken, 2003)
8. W. Plieth, G.S. Wilson, C. Gutierrez de la Fe, Spectroelectrochemistry: a survey of *in situ* spectroscopic techniques. Pure Appl. Chem. **70**, 1395–1414 (1998)
9. W. Plieth, G.S. Wilson, C. Gutierrez de la Fe, Erratum. Pure Appl. Chem. **70**, 2409–2412 (1998)
10. D. Marshall, P.R. Rich, Mitochondrial function, in *Methods in Enzymology*, vol. 456, Studies of Complex I by Fourier Transform Infrared Spectroscopy (Elsevier, Amsterdam, 2009), pp. 53–74
11. P.R. Rich, M. Iwaki, Methods to probe protein transitions with ATR infrared spectroscopy. Mol. BioSyst. **3**, 398–407 (2007)
12. C. Vigano, J.-M. Ruysschaert, E. Goormaghtigh, Sensor applications of attenuated total reflection infrared spectroscopy. Talanta, **65**, 1132–1142 (2005)
13. K. Ataka, J. Heberle, Biochemical applications of surface-enhanced infrared absorption spectroscopy. Anal. Bioanal. Chem. **388**, 47–54 (2007)
14. F. Zaera, Probing liquid/solid interfaces at the molecular level. Chem. Rev. **112**, 2920–2986 (2012)
15. D.A. Woods, C.D. Bain, Total internal reflection spectroscopy for studying soft matter. Soft Matter **10**, 1071–1096 (2014)
16. C.F. Bohren, D.R. Huffman, *Absorption and Scattering of Light by Small Particles* (Wiley, New York, 1983)
17. C.D. Bain, P.R. Greene, Spectroscopy of adsorbed layers. Curr. Opin. Colloid Interface Sci. **6**, 313–320 (2001)
18. A. Barth, Infrared spectroscopy of proteins. Biochim. Biophys. Acta (Bioenergetics) **1767**, 1073–1101 (2007)
19. R.A. Nyquist, *Interpreting Infrared, Raman, and Nuclear Magnetic Resonance Spectra* (Academic Press, San Diego, 2001)

20. E.B. Wilson, J.C. Decius, P.C. Cross, *Molecular Vibrations* (Dover Publications, New York, 1955)
21. Z.-Q. Tian, B. Ren, Adsorption and reaction at electrochemical interfaces as probed by surface-enhanced raman spectroscopy. Annu. Rev. Phys. Chem. **55**, 197–229 (2004)
22. D.H. Murgida, P. Hildebrandt, Disentangling interfacial redox processes of proteins by SERR spectroscopy. Chem. Soc. Rev. **37**, 937–945 (2008)
23. S.L. Kleinman, R.R. Frontiera, A.-I. Henry, J.A. Dieringer, R.P. Van Duyne, Creating, characterizing, and controlling chemistry with SERS hot spots. Phys. Chem. Chem. Phys. **15**, 21–36 (2013)
24. E.D. Palik, *Handbook of Optical Constants of Solids*, vols. 1–4 (Academic Press, San Diego, 1985–1997)
25. T.N. Stanislavchuk, T.D. Kang, P.D. Rogers, E.C. Standard, R. Basistyy, A.M. Kotelyanskii, G. Nita, T. Zhou, G.L. Carr, M. Kotelyanskii, A.A. Sirenko, Synchrotron radiation-based far-infrared spectroscopic ellipsometer with full mueller-matrix capability. Rev. Sci. Instr. **84**, 023901 (2013)
26. H. Fujiwara, *Spectroscopic Ellipsometry: Principles and Applications* (Wiley, Chichester, 2007)
27. H. Haug, S.W. Koch, *Quantum Theory of Optical and Electronic Properties of Semiconductors*, 5th edn. (World Scientific, Singapore, 2009)
28. F. Wooten, *Optical Properties of Solids* (Academic Press, San Diego, 1972)
29. W. Vogel, D.-G. Welsch, *Quantum Optics* (Wiley, Weinheim, 2006)
30. C. Kittel, *Introduction to Solid State Physics*, 7th edn. (Wiley, New York, 1996)
31. M. Olschewski, S. Knop, J.R.G. Lindner, P. Vöhringer, From single hydrogen bonds to extended hydrogen-bond wires: low-dimensional model systems for vibrational spectroscopy of associated liquids. Angew. Chem. Int. Ed. **52**, 9634–9654 (2013)
32. J.E. Bertie, Z. Lan, Infrared intensities of liquids XX: The intensity of the OH stretching band of liquid water revisited, and the best current values of the optical constants of H_2O (l) at 25 °C between 15,000 and 1 cm^{-1}. Appl. Spectrosc. **50**, 1047–1057 (1996)
33. D.M. Leitner, M. Havenith, M. Gruebele, Biomolecule large-amplitude motion and solvation dynamics: modelling and probes from THz to x-rays. Int. Rev. Phys. Chem. **25**, 553–582 (2006)
34. G. Niehues, M. Heyden, D.A. Schmidt, M. Havenith, Exploring hydrophobicity by THz absorption spectroscopy of solvated amino acids. Faraday Discuss. **150**, 193–207 (2011)
35. M. Chaplin, Water absorption spectrum. http://www1.lsbu.ac.uk/water/water_vibrational_spectrum.html. Accessed Oct 2014
36. Labcognition Analytical Software. irAnalyze/RAMalyze. http://www.labcognition.com. Accessed Oct 2014
37. N.W. Ashcroft, D. Mermin, *Solid State Physics* (Holt, Rinehart and Winston, Dumfries, 1976)
38. R.A. Jishi, M.S. Dresselhaus, G. Dresselhaus, Electron-phonon coupling and the electrical conductivity of fullerene nanotubes. Phys. Rev. B **48**, 11385–11389 (1993)
39. R.M.A. Azzam, N.M. Bashara, *Ellipsometry and Polarized Light* (Elsevier Science, Amsterdam, 1999)
40. J. Lekner, *Theory of Reflection of Electromagnetic and Particle Waves* (Martinus Nijhoff, Dordrecht, 1987)
41. M. Schubert, Polarization-dependent optical parameters of arbitrarily anisotropic homogeneous layered systems. Phys. Rev. B **53**, 4265–4274 (1996)
42. A. Erbe, Reflcalc. http://home.arcor.de/aerbe/en/prog/a/reflcalc.html. Accessed June 2014
43. A. Taflove, S.C. Hagness, *Computational Electrodynamics: The Finite-Difference Time-Domain Method*, 3rd edn. (Artech House, Boston, 2005)
44. J.L. Volakis, A. Chatterjee, L.C. Kempel, Finite element method for electromagnetics, in *IEEE/OUP Series on Electromagnetic Wave Theory* (IEEE Press/Oxford University Press, New York/Oxford, 1998)
45. D. Moss, E. Nabedryk, J. Breton, W. Mäntele, Redox-linked conformational changes in proteins detected by a combination of infrared spectroscopy and protein electrochemistry. Eur. J. Biochem. **187**, 565–572 (1990)

46. J.D. Jackson, *Classical Electrodynamics* (Wiley, Hoboken, 1998)
47. W.N. Hansen, Electric fields produced by the propagation of plane coherent electromagnetic radiation in a stratified medium. J. Opt. Soc. Am. **58**, 380–390 (1968)
48. D.D. Schlereth, W. Mäntele, Redox-induced conformational changes in myoglobin and hemoglobin: electrochemistry and ultraviolet-visible and fourier transform infrared difference spectroscopy at surface-modified gold electrodes in an ultra-thin-layer spectroelectrochemical cell. Biochemistry **31**, 7494–7502 (1992)
49. M.K. Debe, Optical probes of organic thin films: photons-in and photons-out. Prog. Surf. Sci. **24**, 1–282 (1987)
50. M. Moskovits, Surface selection rules. J. Chem. Phys. **77**, 4408–4416 (1982)
51. C. Vericat, M.E. Vela, G. Benitez, P. Carro, R.C. Salvarezza, Self-assembled monolayers of thiols and dithiols on gold: new challenges for a well-known system. Chem. Soc. Rev. **39**, 1805–1834 (2010)
52. E. Pensa, E. Cortés, G. Corthey, P. Carro, C. Vericat, M.H. Fonticelli, G. Benítez, A.A. Rubert, R.C. Salvarezza, The chemistry of the sulfur gold interface: in search of a unified model. Acc. Chem. Res. **45**, 1183–1192 (2012)
53. F. Schreiber, Self-assembled monolayers: from 'simple' model systems to biofunctionalized interfaces. J. Phys.: Condens. Matter **16**, R881–R900 (2004)
54. J.C. Love, L.A. Estroff, J.K. Kriebel, R.G. Nuzzo, G.M. Whitesides, Self-assembled monolayers of thiolates on metals as a form of nanotechnology. Chem. Rev. **105**, 1103–1170 (2005)
55. M. Kind, C. Wöll, Organic surfaces exposed by self-assembled organothiol monolayers: preparation, characterization, and application. Prog. Surf. Sci. **84**, 230–278 (2009)
56. R.G. Snyder, H.L. Strauss, C.A. Elliger, Carbon-hydrogen stretching modes and the structure of n-alkyl chains. 1. Long, disordered chains. J. Phys. Chem. **86**, 5145–5150 (1982)
57. R.A. Macphail, H.L. Strauss, R.G. Snyder, C.A. Elliger, C-H stretching modes and the structure of normal alkyl chains. 2. Long, all-trans chains. J. Phys. Chem. **88**, 334–341 (1984)
58. D.L. Allara, R.G. Nuzzo, Spontaneously organized molecular assemblies. 2. Quantitative infrared spectroscopic determination of equilibrium structures of solution-adsorbed n-alkanoic acids on an oxidized aluminum surface. Langmuir **1**, 52–66 (1985)
59. R.D.B. Fraser, T.P. MacRae, *Conformation in Fibrous Proteins and Related Synthetic Polypeptides* (Academic Press, New York, 1973)
60. J. Liu, B. Schupbach, A. Bashir, O. Shekhah, A. Nefedov, M. Kind, A. Terfort, C. Wöll, Structural characterization of self-assembled monolayers of pyridine-terminated thiolates on gold. Phys. Chem. Chem. Phys. **12**, 4459–4472 (2010)
61. C. Hogan, R. Del Sole, G. Onida, Optical properties of real surfaces from microscopic calculations of the dielectric function of finite atomic slabs. Phys. Rev. B **68**, 035405 (2003)
62. M.I. Muglali, A. Erbe, Y. Chen, C. Barth, P. Koelsch, M. Rohwerder, Modulation of electrochemical hydrogen evolution rate by araliphatic thiol monolayers on gold. Electrochim. Acta **90**, 17–26 (2013)
63. B.E. Hayden, Vibrational spectroscopy of molecules on surfaces, in *Reflection Absorption Infrared Spectroscopy* (Plenum Press, New York, 1987)
64. C. Zafiu, G. Trettenhahn, D. Pum, U.B. Sleytr, W. Kautek, Structural control of surface layer proteins at electrified interfaces investigated by in situ Fourier transform infrared spectroscopy. *Phys. Chem. Chem. Phys.* **13**, 13232–13237 (2011)
65. M. Born, E. Wolf, *Principles of Optics: Electromagnetic Theory of Propagation, Interference and Diffraction of Light*, 7th edn. (Cambridge University Press, Cambridge, 1999)
66. T. Buffeteau, B. Desbat, J.M. Turlet, Polarization modulation FT-IR spectroscopy of surfaces and ultra-thin films: experimental procedure and quantitative analysis. Appl. Spectrosc. **45**, 380–389 (1991)
67. L.A. Nafie, D.W. Vidrine, in *Fourier transform infrared spectroscopy.* Double Modulation Fourier Transform Spectroscopy, vol. 3 (Academic Press, San Diego, 1982), pp. 83–123

68. K.W. Hipps, G.A. Crosby, Applications of the photoelastic modulator to polarization spectroscopy. J. Phys. Chem. **83**, 555–562 (1979)
69. A. Kerth, A. Erbe, M. Dathe, A. Blume, Infrared reflection absorption spectroscopy of amphipathic model peptides at the air/water interface. Biophys. J. **86**, 3750–3758 (2004)
70. M. Schiek, K. Al-Shamery, M. Kunat, F. Traeger, C. Wöll, Water adsorption on the hydroxylated H-(1 × 1) O-ZnO(0001̄) surface. Phys. Chem. Chem. Phys. **8**, 1505–1512 (2006)
71. E. Amado, A. Kerth, A. Blume, J.R.G. Kressler, Infrared reflection absorption spectroscopy coupled with brewster angle microscopy for studying interactions of amphiphilic triblock copolymers with phospholipid monolayers. Langmuir **24**, 10041–10053 (2008)
72. M. Buchholz, P.G. Weidler, F. Bebensee, A. Nefedov, C. Wöll, Carbon dioxide adsorption on a ZnO(101̄0) substrate studied by infrared reflection absorption spectroscopy. Phys. Chem. Chem. Phys. **16**, 1672–1678 (2014)
73. A. Erbe, A. Kerth, M. Dathe, A. Blume, Interactions of KLA amphipathic model peptides with lipid monolayers. ChemBioChem **10**, 2884–2892 (2009)
74. N.J. Harrick, *Internal Reflection Spectroscopy* (Harrick Scientific, Ossining, 1987)
75. M. Milosevic, *Internal Reflection and Spectroscopy. ATR* (Wiley, New York, 2012)
76. M. Milosevic, On the nature of the evanescent wave. Appl. Spectrosc. **67**, 126–131 (2013)
77. M.W. Davidson, Florida State University, Molecular expressions. http://micro.magnet.fsu.edu/. Accessed Oct 2014
78. E. Goormaghtigh, V. Raussens, J.-M. Ruysschaert, Attenuated total reflection infrared spectroscopy of proteins and lipids in biological membranes. Biochim. Biophys. Acta (Biomembranes) **1422**, 105–185 (1999)
79. M. Müller, B. Torger, E. Bittrich, E. Kaul, L. Ionov, P. Uhlmann, M. Stamm, In-situ ATR-FTIR for characterization of thin biorelated polymer films. Thin Solid Films **556**, 1–8 (2014)
80. M. Boncheva, H. Vogel, Formation of stable polypeptide monolayers at interfaces: controlling molecular conformation and orientation. Biophys. J. **73**, 1056–1072 (1997)
81. M. Liley, T.A. Keller, C. Duschl, H. Vogel, Direct observation of self-assembled monolayers, ion complexation, and protein conformation at the gold/water interface: an FTIR spectroscopic approach. Langmuir **13**, 4190–4192 (1997)
82. Y. Cheng, N. Boden, R.J. Bushby, S. Clarkson, S.D. Evans, P.F. Knowles, A. Marsh, R.E. Miles, Attenuated total reflection fourier transform infrared spectroscopic characterization of fluid lipid bilayers tethered to solid supports. Langmuir **14**, 839–844 (1998)
83. A. Erbe, R.J. Bushby, S.D. Evans, L.J.C. Jeuken, Tethered bilayer lipid membranes studied by simultaneous attenuated total reflectance infrared spectroscopy and electrochemical impedance spectroscopy. J. Phys. Chem. B **111**, 3515–3524 (2007)
84. M. Reithmeier, A. Erbe, Dielectric interlayers for increasing the transparency of metal films for mid-infrared attenuated total reflection spectroscopy. Phys. Chem. Chem. Phys. **12**, 14798–14803 (2010)
85. M. Reithmeier, A. Erbe, Application of thin-film interference coatings in infrared reflection spectroscopy of organic samples in contact with thin metal films. Appl. Opt. **50**, C301–C308 (2011)
86. P. Wenzl, M. Fringeli, J. Goette, U.P. Fringeli, Langmuir **10**, 4253–4264 (1994)
87. S. Nayak, P. Ulrich Biedermann, M. Stratmann, A. Erbe, A mechanistic study of the electrochemical oxygen reduction on the model semiconductor n-Ge(100) by ATR-IR and DFT. Phys. Chem. Chem. Phys. **15**, 5771–5781 (2013)
88. S. Nayak, P.U. Biedermann, M. Stratmann, A. Erbe, In situ infrared spectroscopic investigation of intermediates in the electrochemical oxygen reduction on n-Ge (100) in alkaline perchlorate and chloride electrolyte. Electrochim. Acta **106**, 472–482 (2013)
89. B.J. Messinger, K. Ulrich von Raben, R.K. Chang, P.W. Barber, Local fields at the surface of noble-metal microspheres. Phys. Rev. B **24**, 649–657 (1981)
90. G. Vasan, A. Erbe, Incidence angle dependence of the enhancement factor in attenuated total reflection surface enhanced infrared absorption spectroscopy studied by numerical solution of the vectorial Maxwell equations. Phys. Chem. Chem. Phys. **14**, 14702–14709 (2012)

91. G. Vasan, Y. Chen, A. Erbe, Computation of surface-enhanced infrared absorption spectra of particles at a surface through the finite element method. J. Phys. Chem. C **115**, 3025–3033 (2011)
92. M. Fan, A.G. Brolo, Self-assembled Au nanoparticles as substrates for surface-enhanced vibrational spectroscopy: optimization and electrochemical stability. ChemPhysChem **9**, 1899–1907 (2008)

Chapter 15
Introduction to Quantitative Data Analysis in Vibrational Sum-Frequency Generation Spectroscopy

Matthias Josef Hofmann and Patrick Koelsch

Abstract Analyzing molecules at aqueous interfaces in situ, in vitro, or even in vivo without the need for labels and/or disruptive sample preparation is crucial for the understanding and optimization of material's interactions with its surrounding. In this context, a central theme is the ability to differentiate between molecules in the respective bulk phases and those that are located at the interface. Here we introduce vibrational sum-frequency generation (SFG) spectroscopy, a nonlinear optical technique that is capable to selectively probe molecules at interfaces. SFG spectroscopy can be applied under *ex vacuo* conditions and allows to record vibrational spectra from molecules at interfaces. Though this technique holds great potential in research themes involving aqueous interfaces, the data analysis of SFG spectra can get quite complex and often requires a comprehensive understanding of the underlying nonlinear optical processes. This chapter introduces experimental and theoretical aspects of SFG spectroscopy with a strong focus on data analysis. It is meant for scientists new to the field of SFG spectroscopy who like to explore its applicability and theoretical background or are starting to apply SFG spectroscopy in their own research.

15.1 Introduction

Vibrational sum-frequency generation (SFG) spectroscopy has become an established nonlinear optical method for surface analysis of aqueous interfaces. It allows to record vibrational spectra of molecules with superior surface specificity and

M.J. Hofmann
Universität Regensburg, Universitätsstrasse, 31, 93053 Regensburg, Germany

P. Koelsch (✉)
National ESCA and Surface Analysis Center for Biomedical Problems, Department of Bioengineering, University of Washington, Seattle, Washington 98195, USA
e-mail: koelsch@uw.edu

© Springer International Publishing Switzerland 2016
P.R. Lang and Y. Liu (eds.), *Soft Matter at Aqueous Interfaces*,
Lecture Notes in Physics 917, DOI 10.1007/978-3-319-24502-7_15

491

sensitivity down to submonolayer coverages. Its working principle is based on the spatial and temporal overlap between visible and IR laser pulses at interfaces to generate a signal at the sum-frequency of the two incident beams ($\omega_{SFG} = \omega_{IR} + \omega_{VIS}$). This process can be understood as the excitation of a vibration with the IR beam followed by an upconversion through the visible beam in a coherent anti-Stokes Raman process to generate SFG signals as illustrated in Fig. 15.1. In a typical experimental setting, the visible wavelength will be kept constant while the IR frequency is either tuned or broadband in nature. In an experiment utilizing picosecond (ps) laser pulses, the frequency is narrowband (typically < 5 cm^{-1}) and an SFG spectrum is acquired by tuning the IR frequency. Repetition rates for ps lasers are on the order of 10 and acquiring a spectrum over a range of 100 cm^{-1} takes several tens of minutes. In an experiment utilizing femtosecond (fs) laser pulses, the IR frequency is broadband (up to several 100 cm^{-1}) so that every SFG pulse contains information within this spectral range. This together with repetition rates for fs laser sources being 1000 Hz and higher results in spectral acquisition times in the sub-second regime. Independent of the laser source is the principle that once the incident IR energy is in concert with a resonant mode that is both IR and Raman active, an SFG signal can be produced. This SFG beam is spatially separated from the reflected visible and IR beams and can be isolated by a proper set of mirrors and filters. Detection schemes for narrowband experiments involve a monochromator and a photomultiplier to measure the SFG photon flux at each IR wavelength. For broadband experiments, SFG spectra are recorded by a spectrograph that is dispersing the signal onto a CCD camera.

The differences to linear *first-order* vibrational spectroscopies, as discussed in Chap. 14 by A. Erbe et al. are in the selection rules: *second-order*—in fact all even-order—nonlinear optical processes are forbidden in media that are isotropic and in molecular arrangements that possess inversion symmetry. Molecules that are self-assembling at interfaces do have an asymmetric arrangement, for example the isotropic surrounding of molecules typical for all gases and most liquid phases is disrupted at the interface. Applying SFG selection rules to such a scenario yields that molecules in the isotropic bulk phase cannot generate second-order signals—

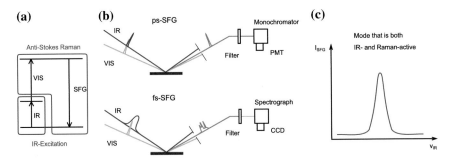

Fig. 15.1 **a** Energy diagram of the SFG process. **b** Probing schemes for ps- and fs-SFG including a frequency representation of the involved beams. **c** Typical presentation for an SFG spectrum

only molecules that are both residing in an ordered fashion within the asymmetric interphase and do not possess inversion symmetry in its molecular structure can produce a signal (Fig. 15.2). Also consider that many crystal structures are centrosymmetric and cannot generate SFG, while the centrosymmetry is necessarily broken at interfaces. This makes vibrational SFG spectroscopy a powerful technique to probe molecules at solid/liquid, liquid/liquid, solid/air, and air/liquid interfaces [1–17].

Let us have a closer look on light-matter interactions and the difference between linear and nonlinear optics. Light can be expressed as an electromagnetic wave that is traveling in time and space. The respective electric field **E** may induce a polarization **P** in the medium, where **P** is a macroscopic quantity denoting the induced dipole moment per unit volume. In linear optics, a direct proportionality between the electric field and the polarization can be found. This *linear* relation is the result of energy considerations within molecules that follow the principles of a harmonic oscillator, i.e., the restoring force is proportional to the elongation. In analogy, the electric field is proportional to the polarization, which is generating a wave that has the same frequency as the incoming wave. The corresponding proportionality factor is a material constant that is called the *susceptibility* χ. If we only consider completely elastic light-matter interactions, this scenario—on an abstract level—is describing linear optics: the intensity of light may be reduced due to absorption processes, the speed of light might differ as a result of a change in refractive index, but no new frequencies can be generated—i.e., the polarization wave in the medium has the same frequency as the incoming light. The linear relation between electric field and polarization holds true unless the incident fields are strong enough so that potential energy levels within the molecule are reached that cannot be approximated by an harmonic oscillator anymore, but follows the real potential—i.e., a Morse potential. In this case, the polarization vector **P** can be represented as a power series of the incident fields **E** and the linear relation between the electric field and the polarization no longer holds true, this is in fact where the term *nonlinear* optics stems from. The respective material constants $\chi^{(i)}$ are referred to as susceptibility tensors of i-th order. The i-th component of the **P**-vector can be obtained by the following summation of contributions arising from expanding **P** into a Taylor series:

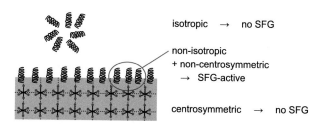

Isotropic → no SFG

non-isotropic
+ non-centrosymmetric
→ SFG-active

centrosymmetric → no SFG

Fig. 15.2 Illustration of the selection rules: media that are isotropic or molecular arrangements that possess inversion symmetry cannot produce SFG signals. These symmetries are typically broken at interfaces making SFG spectroscopy inherently surface specific

$$P_i = \epsilon_0 \left(\chi_{ij}^{(1)} \cdot E_j + \frac{1}{2} \chi_{ijk}^{(2)} \cdot E_j E_k + \frac{1}{6} \chi_{ijkl}^{(3)} \cdot E_j E_k E_l + \dots \right), \qquad (15.1)$$

with ε_0 being the vacuum permittivity. In case of second-order nonlinear spectroscopy, the third rank tensor $\chi^{(2)}$ is the relevant material constant. Even though the absolute values of its elements are several orders of magnitude lower compared to the corresponding first order $\chi^{(1)}$ tensor known from linear spectroscopies, such as conventional IR-spectroscopy, $\chi^{(2)}$-effects can become significant when using higher input fields, such as those generated by ultrashort laser pulses. In this case, the whole endeavor of recording SFG signals comes back to measuring tensor elements of $\chi^{(2)}$. The goal of this chapter is to outline strategies for measuring these tensor elements and to relate the measured data to molecular arrangements such as order and orientation of molecular moieties at interfaces. None of the presented strategies and results can be viewed as new insights into SFG spectroscopy. This chapter is more intended to provide an overview of analysis routines for those researchers who are new to the field. It is arranged in the following way: we first define appropriate laboratory and molecular coordinate systems. Then we will outline the structure of the $\chi^{(2)}$ tensor and discuss symmetry considerations to identify non-vanishing tensor elements. These tensor elements are required to be understood in the context of the experimental geometry to finally relate the measured SFG signal to a $\chi^{(2)}$ tensor element.

15.1.1 Definition of Laboratory Coordinate System

The first step to an appropriate interpretation of SFG data is to define coordinate systems. There are three types to be considered: the laboratory fixed coordinate system xyz, the surface fixed coordinate system XYZ and the molecular fixed coordinate system abc. These orthonormal coordinate systems can be transferred into each other by the application of an appropriate Euler-transformation, which is discussed later in this chapter. Here we first define the laboratory fixed coordinate system xyz with respect to the direction of propagation for the participating IR-, VIS- and SFG light (Fig. 15.3). The surface plane is typically the xy-plane, whereas the surface normal is referred to as z-axis. With the direction of propagation for the laser beams projected onto the surface being the x-axis, the plane of incidence coincides with the xz-plane. In a typical experimental setting, the polarization of incident and emitted laser beams is set to p and s where p denotes an electric field vector oscillating parallel to the plane of incidence and s denotes an electric field vector perpendicular to the plane of incidence (Fig. 15.3). The terms s-polarized and p-polarized are derived from the German expressions for "*senkrecht*"—meaning perpendicular—and "*parallel*".

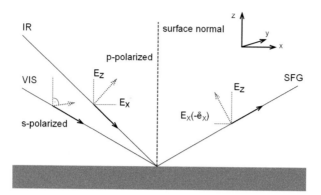

Fig. 15.3 Definition of the directions of propagation and the electric field vectors; the plane of incidence is given by the xz-plane, the surface plane is given by the xy-plane; the *solid-line arrows* denote the unit vectors of propagation of the IR-, VIS- and SFG light, the *dotted-line arrows* indicate the p-polarized components of the electric fields. The decomposition of the p-polarized component along the laboratory x-axis and z-axis of the electric field for the IR beam is shown as grey dashed lines. Note that s-polarized light has only a contribution along the y-axis and that the unit vector along the x-axis is reversed for the E_X component of the SFG field

15.2 Structure of the $\chi^{(2)}$ Tensor and Symmetry Considerations

The linear $\chi^{(1)}$ tensor is a second-rank tensor and has [3×3] elements. The second-order $\chi^{(2)}$ is a third-rank tensor with a total of [3×3×3] elements. These can be probed by polarization dependent measurements, for example by utilizing p- and s-polarization. The polarization combination is typically noted in order of decreasing energy of the involved beams (or increasing wavelength: SFG, Vis, and IR). The use of *sps*-polarization means that the emission at the sum-frequency is detected under s-polarization, whereas the incident visible beam is p-polarized and the incident IR-beam is set to be s-polarized. Taking into account that s-polarization has only a contribution along the y-axis whereas the p-polarization can be decomposed into components along the x-axis and z-axis, it is evident that various tensor elements of the nonlinear susceptibility of second order $\chi^{(2)}$ can be probed using a given set of polarization combinations.

Table 15.1 explicitly shows the tensor components that are probed for all sets of eigen-polarization combinations. Studies at interfaces often involve *azimuthal isotropy* meaning that the properties of the surface and its covering layer are identical in both x-direction and y-direction. These types of surfaces are also referred to as rotationally symmetric about the surface normal. There is no possibility to distinguish between the x-axis and y-axis in terms of optical properties of the surface—rotating the sample by any angle will lead to the same SFG spectrum. For this reason, we can use the equality for x-direction and y-direction to simplify the nonlinear susceptibility of second order $\chi^{(2)}$. For example, a symmetry

Table 15.1 Probed tensor elements under the given polarization combinations. The underlined elements are non-vanishing in an azimuthally isotropic system

Polarization combination	Number	Tensor elements
ppp-polarization	2.2.2	$\underline{\chi^{(2)}_{zzz}}$, $\underline{\chi^{(2)}_{zzx}}$, $\underline{\chi^{(2)}_{zxz}}$, $\underline{\chi^{(2)}_{xzz}}$, $\underline{\chi^{(2)}_{zxx}}$, $\underline{\chi^{(2)}_{xzx}}$, $\underline{\chi^{(2)}_{xxz}}$, $\chi^{(2)}_{xxx}$
pps-polarization	2.2.1	$\underline{\chi^{(2)}_{zzy}}$, $\chi^{(2)}_{zxy}$, $\chi^{(2)}_{xzy}$, $\chi^{(2)}_{xxy}$
psp-polarization	2.1.2	$\chi^{(2)}_{zyz}$, $\chi^{(2)}_{zyx}$, $\chi^{(2)}_{xyz}$, $\chi^{(2)}_{xyx}$
spp-polarization	1.2.2	$\chi^{(2)}_{yzz}$, $\chi^{(2)}_{yzx}$, $\chi^{(2)}_{yxz}$, $\chi^{(2)}_{yxx}$
pss-polarization	2.1.1	$\underline{\chi^{(2)}_{zyy}}$, $\chi^{(2)}_{xyy}$
sps-polarization	1.2.1	$\underline{\chi^{(2)}_{yzy}}$, $\chi^{(2)}_{yxy}$
ssp-polarization	1.1.2	$\underline{\chi^{(2)}_{yyz}}$, $\chi^{(2)}_{yyx}$
sss-polarization	1.1.1	$\chi^{(2)}_{yyy}$

operation that is rotating the sample by $180°$ changes the signs of the unity vectors along the x and y-direction for the laboratory fixed coordinate system xyz and can be expressed as $x \rightarrow -x$ and $y \rightarrow -y$. Let us now consider all 27 tensor elements of the nonlinear susceptibility of second order $\chi^{(2)}$ and their transformation behavior along the x- and y-axis upon the application of the symmetry operation. For an azimuthally isotropic system, the transformations $x \rightarrow -x$ and $y \rightarrow -y$ need to have the same results. For example the elements xxz and $(-x)(-x)z$ must be the same, which is the case since $(-x)(-x)z = xxz$. Let us consider the element xzz: changing the sign of x will change the sign of xzz. Therefore, xzz and $-xzz$ can only be the same if xzz is zero. Exercising the same formalism to all 27 $\chi^{(2)}$ elements reveals that only 13 elements are non-zero for a system having a net orientation in z-direction and being azimuthally isotropic otherwise (underlined $\chi^{(2)}$ elements in Table 15.1).

For a completely isotropic system, the transformation $x \rightarrow -x$, $y \rightarrow -y$ and $z \rightarrow -z$ have to result in the same response. Performing this transition for any of the 27 tensor elements yields the same tensor element with a negative sign (e.g. $xxx = (-x)(-x)(-x) = -xxx$ and so on). Again, the same element being positive and negative can only mean that all 27 tensor elements are zero. That is the reason why all even-order processes are forbidden in isotropic media. In these cases all $\chi^{(2n)}$ elements with $n = 1, 2, 3, \ldots$ are zero.

15.2.1 From the Molecular (Hyper)Polarizability β to the Macroscopic Susceptibility $\chi^{(2)}$

Quantitative orientational analysis is more demanding from a theoretical point of view. In practice it relies on the comparison of experimentally assessed SFG intensity ratios with expressions gained from theoretical considerations. The first exercise in this context is to find a relation between the molecular fixed coordinate system *abc* and the laboratory fixed coordinate system *xyz*. These transformations can be expressed by means of Euler angles and corresponding rotation matrices. The molecular fixed coordinate system *abc* is the appropriate system for describing the molecular properties under investigation. Theoretical calculations concerning transition dipole moments, polarizabilities, etc. are carried out at this level. One order above is the so-called surface fixed coordinate system *XYZ*. This frame serves to describe, e.g., the orientation of domains at surfaces or inclination of the probed surface, for example when investigating solid surfaces. The laboratory fixed coordinate system *xyz* denotes the coordinate system in which the polarization of the incident and emitted beams are defined. Typically, the surface fixed coordinate system *XYZ* and laboratory fixed coordinate system *xyz* coincide due to the fact of azimuthal isotropy and appropriate leveling of the sample.

In this case, only a transition between the molecular fixed coordinate system *abc* and the laboratory fixed coordinate system *xyz* has to be considered. This can be achieved by means of an Euler-transformation [18] about the Euler angles θ, Φ and Ψ shown in Fig. 15.4. Here we define the geometrical relations between Euler angles and coordinate systems as follows:

- the tilt angle θ denotes the angle between the molecular *c*-axis and the laboratory *z*-axis.
- the azimuthal angle Φ describes the angle between the molecular *a*-axis and the laboratory *x*-axis for a rotation about the laboratory *z*-axis.
- the twist angle Ψ denotes the rotation of the molecular *ab*-plane about the molecular *c*-axis. It is only the application of this step, that the molecular *a*-axis becomes displaced from the laboratory *xy*-plane.

In the trivial case that all Euler angles are zero, the laboratory fixed coordinate system *xyz* and the molecular fixed coordinate system *abc* coincide. Otherwise

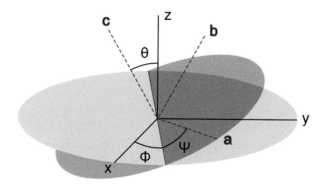

Fig. 15.4 Definition of the Euler angles θ, Φ and Ψ describing the relative orientation between molecular fixed coordinate system *abc* and laboratory fixed coordinate system *xyz*

every possible relative orientation between the molecular fixed coordinate system *abc* and the laboratory fixed coordinate system *xyz* can be established allowing the Euler angles θ, Φ and Ψ to vary in ranges from $0°$ to $180°$ for the tilt angle θ and from $0°$ to $360°$ for both the azimuthal angle Φ and the twist angle Ψ.

The transformation from the molecular fixed coordinate system *abc* to the laboratory fixed coordinate system *xyz* is equivalent to the application of three successive transformations: a rotation about the *c*-axis followed by a rotation about the *a*-axis followed by a rotation about the *c*-axis. Therefore, the Euler-transformation matrix $\hat{R}$ can be obtained by multiplication of three rotation matrices. The subsequent rotation about the initial *z*-axis, followed by rotations about the rotated *x*-axis and the doubly rotated *z*-axis is referred to as the *zxz*-convention [19]. The corresponding matrix describing the overall rotation is given by

$$\hat{R} = \begin{pmatrix} \cos\Psi\cos\Phi - \cos\theta\sin\Phi\sin\Psi & -\sin\Psi\cos\Phi - \cos\theta\sin\Phi\cos\Psi & \sin\theta\sin\Phi \\ \cos\Psi\sin\Phi + \cos\theta\cos\Phi\sin\Psi & -\sin\Psi\sin\Phi + \cos\theta\cos\Phi\cos\Psi & -\sin\theta\cos\Phi \\ \sin\theta\sin\Psi & \sin\theta\cos\Psi & \cos\theta \end{pmatrix}. \tag{15.2}$$

The transformation of the molecular hyper-polarizability β_{lmn} given in the molecular fixed coordinate system *abc* to $\chi_{ijk}^{(2)}$ in the laboratory fixed coordinate system *xyz* is achieved by two steps: first transforming the molecular hyper-polarizability β_{lmn} from the molecular fixed coordinate system *abc* to the laboratory fixed coordinate system *xyz* by means of an Euler-transformation and second by an orientational averaging:

$$\chi_{ijk}^{(2)} = \frac{N_s}{\epsilon_0}\langle\beta_{ijk}\rangle = \frac{N_s}{\epsilon_0}\langle\hat{R}\beta_{lmn}(\Phi, \theta, \Psi)\rangle. \tag{15.3}$$

These steps can be summarized in the following equation that relates each tensor element of $\chi_{ijk}^{(2)}$ to a summation of rotations and averaging over the molecular hyper-polarizability β_{lmn} [20]:

$$\chi_{ijk}^{(2)} = \frac{N_s}{\epsilon_0}\sum_{lmn}\langle[\hat{i}\cdot(\hat{R}\hat{l})][\hat{j}\cdot(\hat{R}\hat{m})][\hat{k}\cdot(\hat{R}\hat{n})]\rangle\beta_{lmn}, \tag{15.4}$$

with N_s being the number density. In this equation $\chi_{ijk}^{(2)}$ denotes the observable tensor element given in the laboratory frame and β_{lmn} the corresponding molecular property. The pointed brackets denote orientational averaging that is equivalent to the integration over the Euler angles θ, φ and Ψ. The quantities $\hat{i}, \hat{j}, \hat{k}$ and $\hat{l}, \hat{m}, \hat{n}$ represent the unit vectors of the coordinate systems of the laboratory and molecular frame, respectively. The expressions within the rectangular delimiters represent the projection of the rotated molecular unit vector of the molecular fixed coordinate system *abc* onto the laboratory unit vector of interest.

It is obviously important to get insights into the tensor elements of the molecular hyper-polarizability β_{lmn}. In this context, symmetry considerations are of great importance when it comes to simplifying the molecular hyper-polarizability β_{lmn} without knowing the exact value of each element. In analogy to symmetry considerations for the macroscopic quantity $\chi^{(2)}$, we will extend this discussion to the microscopic scale by analyzing the symmetry of individual molecular moieties. A common example for molecules under investigation in SFG is a methylene group (Fig. 15.5). The corresponding symmetry operations can be found from character tables given in literature, e.g., in almost every standard textbook of physical chemistry. Every symmetry operation within the point group needs to be accounted for in order to attain a full description of all molecular hyper-polarizability β_{lmn} elements. Assuming a perfect C_{2v}-symmetry for the methylene group without any kind of distortion, we find two mirror planes σ_1 and σ_2 as well as a rotation about the main axis C_2 and the identity operation E. Table 15.2 lists the corresponding transformation of basis vectors from the character tables.

Let us consider in the following the transformation behavior for the various symmetry operations within the point group C_{2v}. It is important to include all symmetry operations of the point group in order to identify non-vanishing contributions. For the tensor element β_{bba} the relation $\beta_{bba} = -\beta_{bba}$ is a result from the symmetry operations σ_1 and C_2 and can only be fulfilled when this tensor element assumes a value of zero (Table 15.3). Consequently, the tensor element β_{bba} is zero. Let us now consider the tensor element β_{bbc}. Table 15.4 shows that for the β_{bbc} element all transformations lead to a similar result as demanded by the point group. Therefore, the element β_{bbc} *can* lead to a non-vanishing contribution to the overall SFG signal intensity. Performing the same analysis for all other tensor elements leads to the following: non-vanishing contributions can only be found in case of an odd number of contributions of the c-components, i.e., only for one or

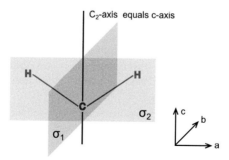

Fig. 15.5 Definition of the CH$_2$-molecule in the laboratory coordinate system; the c-axis is along the C_2-main axis of the methylene group; the plane of the CHC-group coincides with the ac-plane

Table 15.2 Transformation of basis vectors upon application of the symmetry operations within the C_{2v}-point group

Symmetry operation	Transformation of components
E	$a \rightarrow a$
	$b \rightarrow b$
	$c \rightarrow c$
σ_1	$a \rightarrow -a$
	$b \rightarrow b$
	$c \rightarrow c$
σ_2	$a \rightarrow a$
	$b \rightarrow -b$
	$c \rightarrow c$
C_2	$a \rightarrow -a$
	$b \rightarrow -b$
	$c \rightarrow c$

Table 15.3 Transformation of the tensor element β_{bba} upon application of the symmetry operations within the C_{2v}-point group

Symmetry operation	Transformation of components
E	$bba \rightarrow bba$
σ_1	$bba \rightarrow bb(-a) = -bba$
σ_2	$bba \rightarrow (-b)(-b)a = bba$
C_2	$bba \rightarrow (-b)(-b)(-a) = -bba$

Table 15.4 Transformation of the tensor element β_{bbc} upon application of the symmetry operations within the C_{2v}-point group

Symmetry operation	Transformation of components
E	$bbc \rightarrow bbc$
σ_1	$bbc \rightarrow bbc$
σ_2	$bbc \rightarrow (-b)(-b)c = bbc$
C_2	$bbc \rightarrow (-b)(-b)c = bbc$

three c-components. Also excluded are combinations that include all three components, e.g., β_{abc}. Therefore, in the case of C_{2v}-symmetry, the molecular hyper-polarizability β_{lmn} has only 7 non-vanishing tensor elements: β_{ccc}, β_{caa}, β_{aca}, β_{aac}, β_{bbc}, β_{bcb} and β_{cbb}.

In summary, only by symmetry considerations, we reduced the number of the molecular hyper-polarizability β_{lmn} tensor elements from 27 to 7. This procedure exemplifies the power of considering symmetry arguments for the prediction of non-vanishing tensor elements of molecular hyper-polarizability β_{lmn}.

In the following a practical example is discussed: the orientation of a terminal methyl group in an aliphatic chain that is isotropic in the azimuthal and twist angles. Here we are interested in the vibrations that involve a transition dipole moment along the molecular axis, i.e., the c-axis. This scenario is equivalent to a molecular arrangement that is cylindrically symmetric with symmetry along the c-axis. The

Table 15.5 Probed tensor elements of the $C_{\infty v}$-symmetry group under all eigen-polarization combinations

Polarization combination	Contributing tensor elements
ppp-polarization	$\chi^{(2)}_{zzz}$, $\chi^{(2)}_{xxz}$, $\chi^{(2)}_{xzx}$, $\chi^{(2)}_{zxx}$
sss-polarization	
ssp-polarization	$\chi^{(2)}_{yyz}$
sps-polarization	$\chi^{(2)}_{yzy}$
pss-polarization	$\chi^{(2)}_{zyy}$
psp-polarization	
spp-polarization	
pps-polarization	

corresponding symmetry group is $C_{\infty v}$.[1] Given that the visible and SFG frequencies are far from a resonance, the tensor components that include a and b in the third place are negligible, i.e., at an IR frequency with a vanishing IR transition dipole moment perpendicular to the c-axis. In this case only three non-zero tensor elements can be found within the molecular hyper-polarizability β_{lmn}. These are β_{aac}, β_{bbc} and β_{ccc}. If we consider additionally azimuthal isotropy, these elements are linearly related to each other such that

$$\beta_{aac} = \beta_{bbc} \quad \text{and} \quad \frac{\beta_{aac}}{\beta_{ccc}} = \frac{\beta_{bbc}}{\beta_{ccc}} = r. \tag{15.5}$$

Therefore the molecular hyper-polarizability β_{lmn} of a single CH_3 group can be described by the two variables β_{ccc} and the ratio r.

On a macroscopic level, the $\chi^{(2)}$ tensor can be reduced to 13 elements due to azimuthal isotropy as mentioned earlier. Additional reductions are related to the molecular symmetry that excludes elements with all indices, such as xyz, yxz etc. Therefore, only 7 elements are left to consider for $\chi^{(2)}$. These are listed in Table 15.5. Also note that in azimuthal isotropy, only five tensor elements are different: $\chi^{(2)}_{zzz}$, $\chi^{(2)}_{zzx}$, $\chi^{(2)}_{xxz} = \chi^{(2)}_{yyz}$, $\chi^{(2)}_{xzx} = \chi^{(2)}_{yzy}$ and $\chi^{(2)}_{zxx} = \chi^{(2)}_{zyy}$. Furthermore, if the visible and SFG frequencies are away from electronic resonances, the tensor elements $\chi^{(2)}_{xzx}$ and $\chi^{(2)}_{zxx}$ as well as $\chi^{(2)}_{yzy}$ and $\chi^{(2)}_{zyy}$ are very similar, so that only four different elements remain: $\chi^{(2)}_{zzz}$, $\chi^{(2)}_{zzx}$, $\chi^{(2)}_{xxz} = \chi^{(2)}_{yyz}$ and $\chi^{(2)}_{xzx} = \chi^{(2)}_{yzy} = \chi^{(2)}_{zxx} = \chi^{(2)}_{zyy}$. Finally, in case the methyl groups would be oriented primarily upright, the laboratory fixed coordinate system xyz and molecular fixed coordinate system abc are similar. As mentioned above, we are only considering transition dipole moments along the z-axis, or c-axis for that matter, meaning that tensor elements $\chi^{(2)}_{xzx}$ and $\chi^{(2)}_{zxx}$—with an x-direction in the IR polarization—would be very small for the same reasons stated above for the

---

[1] The molecular symmetry of a methyl group is C_{3v}, but its macroscopic arrangement as defined here is $C_{\infty v}$.

molecular hyper-polarizability β_{lmn}. In such a scenario, only $\chi_{zzz}^{(2)}$ and $\chi_{xxz}^{(2)} = \chi_{yyz}^{(2)}$ are non-vanishing. However, here we are assuming the most general case for which we consider the three different elements of $\chi^{(2)}$.

In order to obtain the susceptibility tensor elements $\chi_{zzz}^{(2)}$, $\chi_{xxz}^{(2)}$, and $\chi_{xzx}^{(2)}$ in the laboratory frame, all three tensor elements β_{aac}, β_{bbc} and β_{ccc} arising from the molecular fixed coordinate system abc have to be taken into account in accordance to Eq. 15.4. In a first step, we multiply the Euler-transformation matrix $\hat{R}$ and the molecular hyper-polarizability β_{lmn} which results in

$$\chi_{zzz}^{(2)} = \frac{N_s}{\in_0} \left\langle \sin^2 \theta \ \sin^2 \Phi \ \cos \theta \ \beta_{xxz}^{(2)} + \sin^2 \theta \ \cos^2 \Phi \ \cos \theta \ \beta_{yyz}^{(2)} + \cos^3 \theta \ \beta_{zzz}^{(2)} \right\rangle. \tag{15.6}$$

An equivalent derivation for the macroscopically observable tensor elements $\chi_{xxz}^{(2)}$ and $\chi_{xzx}^{(2)}$ yields

$$\begin{aligned}\chi_{xxz}^{(2)} = \chi_{yyz}^{(2)} = \frac{N_s}{\in_0} \Big\langle & (\cos \Psi \ \cos \Phi - \cos \theta \ \sin \Phi \ \sin \Psi)^2 \cos \theta \ \cdot \ \beta_{xxz}^{(2)} \\ & + (\cos \Psi \ \sin \Phi + \cos \theta \ \cos \Phi \ \sin \Psi)^2 \cos \theta \ \cdot \ \beta_{yyz}^{(2)} \\ & + \sin^2 \theta \ \sin^2 \Psi \ \cos \theta \ \cdot \ \beta_{zzz}^{(2)} \Big\rangle \end{aligned} \tag{15.7}$$

and

$$\begin{aligned}\chi_{xzx}^{(2)} = \chi_{yzy}^{(2)} = \chi_{zyy}^{(2)} = \chi_{zxx}^{(2)} = \frac{N_s}{\in_0} \Big\langle & (\cos \Psi \ \cos \Phi - \cos \theta \ \sin \Phi \ \sin \Psi) \sin^2 \theta \ \sin \Phi \ \sin \Psi \ \cdot \ \beta_{xxz}^{(2)} \\ & - (\cos \Psi \ \sin \Phi + \cos \theta \ \cos \Phi \ \sin \Psi) \sin^2 \theta \ \cos \Phi \ \sin \Psi \ \cdot \ \beta_{yyz}^{(2)} \\ & + \sin^2 \theta \ \sin^2 \Psi \ \cos \theta \ \cdot \ \beta_{zzz}^{(2)} \Big\rangle. \end{aligned} \tag{15.8}$$

Considering an isotropic distribution of the azimuthal angle Φ and the twist angle Ψ, i.e., no dependency of molecular hyper-polarizability β_{lmn} for these angles, we can integrate Eqs. 15.6, 15.7 and 15.8 from 0° to 360° for Φ and Ψ which leads to [20]

$$\chi_{zzz}^{(2)} = \frac{N_s}{\in_0} \beta_{ccc} \left[\langle r \cos \theta \rangle + \langle \cos^3 \theta \rangle (1 - r) \right],$$

$$\chi_{xxz}^{(2)} = \chi_{yyz}^{(2)} = \frac{1}{2} \frac{N_s}{\in_0} \beta_{ccc} \left[\langle \cos \theta \rangle (1 + r) - \langle \cos^3 \theta \rangle (1 - r) \right] \quad \text{and}$$

$$\chi_{xzx}^{(2)} = \chi_{yzy}^{(2)} = \chi_{zyy}^{(2)} = \chi_{zxx}^{(2)} = \frac{1}{2} \frac{N_s}{\in_0} \beta_{ccc} \left[\langle \cos \theta \rangle - \langle \cos^3 \theta \rangle \right] (1 - r).$$

$$\tag{15.9}$$

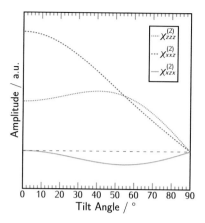

Fig. 15.6 Plot of the tensor elements $\chi^{(2)}_{zzz}$, $\chi^{(2)}_{xxz}$ and $\chi^{(2)}_{xzx}$ as a function of the tilt angle θ. The plots show the angular dependence assuming a value of 2.4 for r and 1.2 for the polarizability tensor element along the main molecular axis β_{ccc}

A plot of the tensor elements $\chi^{(2)}_{zzz}$, $\chi^{(2)}_{xxz}$, and $\chi^{(2)}_{xzx}$ as a function of the tilt angle θ is given in Fig. 15.6 for $r = 2.4$ and $\beta_{ccc} = 1.2$ [20]. An orientational analysis is typically performed using *ratios* of measured (and subsequently fitted) intensities recorded under specific polarization combinations, which will be discussed in the next chapter. In such a case, the set of three equations for the non-vanishing $\chi^{(2)}$ tensor elements will allow determining the parameters $\beta^{(2)}_{ccc}$, r, and tilt angle θ. Note that considering ratios does eliminate the dependency of the number density N_s. This is possible if we, for example, use a delta-function for the distribution, i.e., the integral over tilt angle θ yields its identity. Such a procedure determines the *average* tilt angle θ, but not its distribution. Using SFG to quantify orientations as presented here would require an explicit distribution function which cannot be determined *a priori* from conventional SFG measurements—so SFG spectroscopy serves as a technique that allows to deduce average tilt angles of molecular fragments rather than an explicit distribution function if not combined with simulations [21] or other techniques [22].

15.2.2   From $\chi^{(2)}$ to $\chi^{(2)}_{\mathit{eff}}$

The three tensor elements $\chi^{(2)}_{zzz}$, $\chi^{(2)}_{zzx}$, $\chi^{(2)}_{xxz} = \chi^{(2)}_{yyz}$, $\chi^{(2)}_{xzx} = \chi^{(2)}_{yzy} = \chi^{(2)}_{zxx} = \chi^{(2)}_{zyy}$ can be deduced from SFG measurement with three different input and output polarization combinations, for example *ssp*, *sps*, and *ppp*. In order to relate experimental data to theoretical considerations as introduced here, we need to take into account local field corrections L that include experimental parameters such as angle of incidences and refractive indices of the involved bulk phases. This is achieved by a transition from the nonlinear susceptibility of second order $\chi^{(2)}$ to the closely related effective nonlinear susceptibility of second order $\chi^{(2)}_{\text{eff}}$:

$$\chi^{(2)}_{\text{eff},ijk} = [L_{ii}(\omega_i) \cdot \hat{e}_i(\omega_i)] \cdot \chi^{(2)}_{ijk} : [L_{jj}(\omega_j) \cdot \hat{e}_j(\omega_j)][L_{kk}(\omega_k) \cdot \hat{e}_k(\omega_k)]\chi^{(2)}_{ijk}, \quad (15.10)$$

where $\hat{e}$ stands for the unit polarization vectors. The local field corrections along the orthogonal axes of the laboratory fixed coordinate system xyz are given by [1, 20, 23, 24]

$$L_{xx}(\omega_i) = \frac{2n_1(\omega_i \cos \gamma_i)}{n_1(\omega_i) \cos \gamma_i + n_2(\omega_i) \cos \alpha_i}$$

$$L_{yy}(\omega_i) = \frac{2n_1(\omega_i \cos \alpha i)}{n_1(\omega_i) \cos \alpha i + n_2(\omega_i) \cos \alpha_i} \quad \text{and} \quad (15.11)$$

$$L_{zz}(\omega_i) = \frac{2n_2(\omega_i \cos \alpha i)}{n_1(\omega_i) \cos \gamma_i + n_2(\omega_i) \cos \alpha_i} \left(\frac{n_1(\omega_i)}{n'(\omega_i)}\right)^2.$$

In these equations, the parameters $n_i(\omega_k)$ denote the refractive indices in medium i (1 for the phase that carry the beams) as a function of the incident or emitted frequency for the respective light source. The relation between the angles of incidence α_i and the corresponding refracted angles γ_i is given by Snell's law of refraction.

The parameter $n'(\omega_i)$ denotes the effective refractive index within the interfacial layer as introduced by Shen and coworkers. It is further discussed in references [1, 20, 23, 25]. The interfacial refractive index is the only unknown parameter within the calculation of the local field factors. The dielectric function associated with the effective interfacial refractive index can be associated to the ratio between microscopic local field factors parallel and perpendicular to the interface $(n'^2_{\text{eff}} = \epsilon'_{\text{eff}} = L_{\parallel}/L_{\perp})$. Therefore, the effective interfacial refractive index only influences intensities when p-polarization is used for one of the beams. Its value is usually in the range between 1 and values for the dielectric constant of the adjacent bulk phases.

$\chi^{(2)}_{\text{eff}}$ depends on the polarization combinations used in the experiments. It is a linear combination of the tensor elements that are probed in a certain polarization combination. Table 15.5 lists all tensor elements of the $C_{\infty v}$-symmetry for which the respective polarization combinations include non-zero $\chi^{(2)}$ elements, namely ppp, ssp, sps, and pss. For this symmetry, $\chi^{(2)}_{\text{eff}}$ is a linear combination of:

$$\chi^{(2)}_{\text{eff, }ppp} = \sin(\alpha_{SFG})\sin(\alpha_{Vis})\sin(\alpha_{IR}) \cdot L_{zz}(\omega_{SFG})L_{zz}(\omega_{Vis})L_{zz}(\omega_{IR}) \cdot \chi_{zzz}$$
$$- \cos(\alpha_{SFG})\cos(\alpha_{VIS})\sin(\alpha_{IR}) \cdot L_{xx}(\omega_{SFG})L_{xx}(\omega_{Vis})L_{zz}(\omega_{IR}) \cdot \chi_{xxz}$$
$$- \cos(\alpha_{SFG})\sin(\alpha_{Vis})\cos(\alpha_{IR}) \cdot L_{xx}(\omega_{SFG})L_{zz}(\omega_{Vis})L_{xx}(\omega_{IR}) \cdot \chi_{xzx}$$
$$+ \sin(\alpha_{SFG})\cos(\alpha_{Vis})\cos(\alpha_{IR}) \cdot L_{zz}(\omega_{SFG})L_{xx}(\omega_{Vis})L_{xx}(\omega_{IR}) \cdot \chi_{zxx}$$

$$\tag{15.12}$$

$$\chi^{(2)}_{\text{eff, }ssp} = \sin(\alpha_{IR}) \cdot L_{yy}(\omega_{SFG})L_{yy}(\omega_{Vis})L_{zz}(\omega_{IR}) \cdot \chi_{yyz} \tag{15.13}$$

$$\chi^{(2)}_{\text{eff, }sps} = \sin(\alpha_{Vis}) \cdot L_{yy}(\omega_{SFG})L_{zz}(\omega_{Vis})L_{yy}(\omega_{IR}) \cdot \chi_{yzy} \tag{15.14}$$

$$\chi^{(2)}_{\text{eff, }pss} = \sin(\alpha_{SFG}) \cdot L_{zz}(\omega_{SFG})L_{yy}(\omega_{Vis})L_{yy}(\omega_{IR}) \cdot \chi_{zyy} \tag{15.15}$$

The sign of the individual contribution from a $\chi^{(2)}$ tensor element that has an x-component in SFG, i.e., xzx or xxz, is negative, because of the direction for the SFG polarization unit vector being in $(-x)$-direction (see Fig. 15.3). Interestingly, in the off-resonant case for visible and SFG, the tensor elements $\chi^{(2)}_{xzx}$ and $\chi^{(2)}_{zxx}$ as well as the prefactors that include trigonometric functions and local field factors are very similar. Therefore, if the angle of incidence for vis and IR are close to one another, the tensor elements $\chi^{(2)}_{xzx}$ and $\chi^{(2)}_{zxx}$ almost cancel each other given that one is positive and the other is negative. In this case, $\chi^{(2)}_{\text{eff, }ppp}$ is dominated by the $\chi^{(2)}_{zzz}$ and $\chi^{(2)}_{xxz}$ components.

In all optical experiments, the frequencies for the electric field are too high to be measured directly. Intensities scale with the absolute value squared of the electric field and consequently, the SFG intensity is proportional to $\left|\mathbf{P}^{(2)}\right|^2$. In this context, relating the SFG Intensity I_{SFG} to $\chi^{(2)}_{\text{eff}}$ involves again Fresnel factors that account for the experimental geometry as well as for the refractive indices of the respective bulk phases (sometimes also referred to as macroscopic field corrections). Considering these macroscopic field factors, one can define the SFG intensity as:

$$I_{SFG} = \frac{8\pi^3\omega^2 \sec^2\alpha_{SFG}}{c_0^3 n_1(\omega_{SFG})n_1(\omega_{Vis})n_1(\omega_{IR})} \cdot \left|\chi^{(2)}_{\text{eff}}\right|^2 \cdot I(\omega_{Vis})I(\omega_{IR}), \tag{15.16}$$

with c_0 being the speed of light in vacuum. In the following we used Eqs. 15.9, 15.10 and 15.12–15.16 to calculate the SFG intensity for 60° angle of incidence for the visible beam and 55° for the IR beam. The corresponding intensity plot is shown in Fig. 15.7 for $\beta_{aac} = 2.88$, $\beta_{bbc} = 2.88$, and $\beta_{ccc} = 1, 2$ [20].

Additionally, we considered similar symmetry arguments for C_{2v}-symmetry and C_{3v}-symmetry and performed the same calculations as for the $C_{\infty v}$-symmetry. For these calculations, we also assumed azimuthal and twist angle isotropy. Note that the directions of the transition dipole moments for symmetric and asymmetric

Fig. 15.7 Calculated SFG
intensities for a molecular
group assuming $C_{\infty v}$-
symmetry. Note that *pss*-
polarization and *sps*-
polarization nearly overlap
and that *ppp*-polarization is
multiplied by a factor of 5

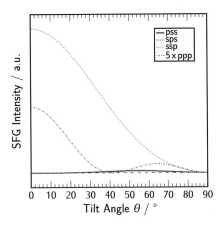

vibrations differ and therefore other elements of the molecular hyper-polarizability β_{lmn} are non-zero. All values used in our calculations are referring to CH_2 and CH_3 vibrations; if not otherwise noted, a value for $\beta_{ccc} = 1$ was used for normalization purposes: for $C_{2v,\text{symmetric}}$: $\beta_{aca}/\beta_{ccc} = 2$ [26]; for $C_{2v,\text{asymmetric}}$: $\beta_{aca}/\beta_{ccc} = 0.27$ [27]; for $C_{3v,\text{symmetric}}$: $\beta_{aca}/\beta_{ccc} = 2.4$, $\beta_{ccc} = 1.2$ [20]; and for $C_{3v,\text{asymmetric}}$: $\beta_{aca}/\beta_{ccc} = 3.4$ [27]. The corresponding plots for the calculated SFG Intensity versus the tilt angle θ of the transition dipole moment for the symmetric and asymmetric vibration are shown in Fig. 15.8 for C_{2v}-symmetry—note that the difference to the $C_{\infty v}$-symmetry for the symmetric vibration is related to different values within the molecular hyper-polarizability β_{lmn}. Figure 15.9 display the corresponding plots for molecular fragments that have C_{3v}-symmetry. Here the plot

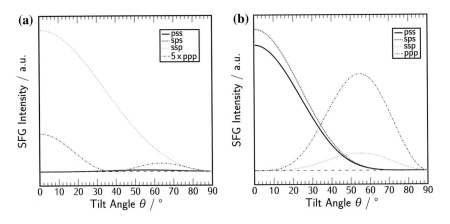

Fig. 15.8 Calculated intensities of molecular fragments assuming C_{2v}-symmetry, **a** for the symmetric stretch (*ppp*-polarization is multiplied by a factor of 5) and **b** for the asymmetric stretch. Note that *pss*-polarization and *sps*-polarization nearly overlap

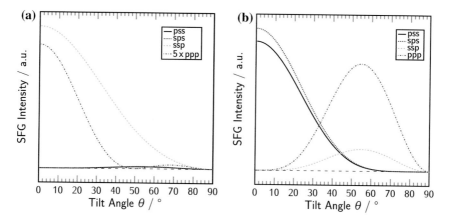

Fig. 15.9 Calculated intensities of molecular fragments assuming C_{3v}-symmetry, **a** for the symmetric stretch (*ppp*-polarization is multiplied by a factor of 5) and **b** for the asymmetric stretch. Note that *pss*-polarization and *sps*-polarization nearly overlap

for the symmetric vibration is the same as for $C_{\infty v}$-symmetry, but differs for the asymmetric vibration due to an orientation of the transition dipole moment in the *a*-axis direction.

These plots are very helpful when it comes to identifying SFG spectral contributions in various polarization combinations. For example, the symmetric CH_3 stretching vibration would be almost absent in *ppp*-polarization at a tilt angle θ around 45°, but at maximum intensity in *ssp*-polarization. Also, for the asymmetric stretching vibration, *ppp*-polarization dominates over *ssp*-polarization. In literature, discussions about spectral contributions without taking into account the used polarization combination have led to confusion in the assignments of vibrational bands. For example, the Fermi resonance of the CH_3 symmetric stretch can be misinterpreted as the asymmetric stretching vibration, which in fact was only absent at a certain tilt angle θ and chosen polarization combination.

15.3 Comparison of Various Fitting Methods

The previous sections introduced the theoretical background needed to relate the molecular hyper-polarizability β_{lmn} in the molecular fixed coordinate system *abc* to elements of $\chi_{eff}^{(2)}$ in the laboratory fixed coordinate system *xyz*. These relations finally ended up in equations that allowed us to deduce average orientations. Here we finally outline how to measure elements of $\chi_{eff}^{(2)}$ using Eq. 15.16. The first important fact to recognize is the proportionality between the SFG intensity and the absolute square of the non-linear susceptibility $\left|\chi^{(2)}\right|^2$:

$$I_{SFG} \propto \left|\mathbf{P}^{(2)}\right|^2 \propto \left|\chi_{\text{eff}}^{(2)}\right|^2 I_{IR} I_{VIS}. \tag{15.17}$$

$\chi_{\text{eff}}^{(2)}$ itself includes possible contributions from a non-resonant background as a result of electronic transitions and from resonant vibrational modes. The corresponding decomposition of the effective nonlinear susceptibility of second order $\chi_{\text{eff}}^{(2)}$ in the frequency domain can be written as

$$\chi_{\text{eff}}^{(2)} = \left|\chi_{NR}^{(2)}\right| \exp^{i\phi} + \sum_q \frac{\chi_q A_{R,q}}{\omega_{R,q} - \omega_{IR} - i\Gamma_{R,q}}, \tag{15.18}$$

in which ϕ is the phase of the non-resonant background. Other equivalent forms of this equation can be found in the literature and the form presented here is consistent with the time-domain description discussed by Wang and co-workers [28, 29].

The indices R and NR denote the corresponding contributions due to resonant and non-resonant effects, respectively. $\Gamma_{R,q}$ represents the line width of the q-th oscillating mode, $\omega_{R,q}$ the resonant frequency of the q-th mode χ_q, and $A_{R,q}$ the associated amplitude value:

$$A_{R,q} = -\frac{1}{2\,\epsilon_0\,\omega_q} \frac{\partial \zeta_{i,j,k}^{(1)}}{\partial Q_q} \frac{\partial \mu_k}{\partial Q_q} \tag{15.19}$$

in which $\partial \zeta_{i,j,k}^{(1)}/\partial Q_q$ and $\partial \mu_k/\partial Q_q$ are the partial derivatives of the Raman polarizability tensor and the IR transition dipole moment, Q_q is the normal coordinate of the q-th mode. Fitting the spectra with Eq. 15.18 yields the Amplitude $A_{R,q}$, which represents the tensor element of $\chi_{\text{eff}}^{(2)}$ for the polarization combination used in the experiment, e.g., for ppp the amplitude $A_{R,q}$ obtained by fitting a spectrum is equal to $\chi_{\text{eff},\,ppp}^{(2)}$.

Given that Eq. 15.17 includes $\left|\chi_{\text{eff}}^{(2)}\right|^2$ to obtain I_{SFG}, it is important to realize that cross terms can occur. This is a major difference to linear IR spectroscopy, where peaks add up linearly in the spectrum that can be fitted for example with Lorentzian peak shapes as given by

$$\frac{1}{2\pi} \frac{A_{R,q}}{\left(\omega_{IR} - \omega_{R,q}\right)^2 + \left(\frac{1}{2}\Gamma_{R,q}\right)^2}. \tag{15.20}$$

SFG spectra, in particular those with overlapping peaks, strongly depend on the phase between the peaks. For example, two peaks that do overlap and have opposing phases may cancel each other. It is therefore of great importance to use the correct fitting function to obtain quantitative results to be used for data analysis. The phase in Eq. 15.18 adds another parameter to the fitting routine for SFG spectra, which may over-parametrize the fitting routine itself. Hence it is often necessary to predefine certain parameters, such as the position of peaks that are known from other experiments or theoretical modelling approaches. Also the peak width can be limited as many molecular moieties do have a similar surrounding leading to the same damping constants. Other theoretical methods and algorithms for spectral analysis have been introduced recently, including the maximum entropy method (MEM) [30–33] and iterative phase matching between MEM and intensity fits (iMEMfit) [34]. It is also possible to experimentally access the real and imaginary part of $\chi_{\text{eff}}^{(2)}$ by performing so-called phase-sensitive SFG, in which the SFG signal is interfered with a reference signal. Such a discussion is beyond the scope of this chapter.

Some reports on SFG data use Lorentzian line shapes in their analysis. Again, for single peaks the results for fitting with Eq. 15.18 and Eq. 15.20 may not be significantly different, but for spectra with overlapping peaks it is highly demanded to use Eq. 15.18. Figure 15.10 shows an SFG spectrum of sodium dodecyl sulfate in D_2O in the CH stretching region. The black line is the measured SFG spectrum that is normalized to the visible and IR intensities. The smooth line is the corresponding SFG fit of the spectrum using Eq. 15.18. The dotted line represents the best fit obtained with Lorentzian line shapes from Eq. 15.20 and the dashed line is a fit that is using the peak positions from the SFG fit as fixed parameters for a fit with Lorentzian line shapes. This plot illustrates, that the fit results from Eqs. 15.18 and 15.20 are different, i.e., they yield different peak positions.

While the presence of phase effects severely complicates SFG spectral analysis, it is also a source of information. First of all, let us consider a centrosymmetric arrangement, say two molecular groups that are oriented opposite to each other. Exciting both vibrations results in an opposite phase for each group, i.e., the transition dipole moments are oriented antithetic to each other. In simple pictorial

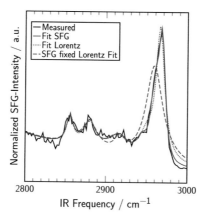

Fig. 15.10 Different fitting curves for an SFG spectrum of sodium dodecyl sulfate in D_2O in the CH stretch; the smooth line corresponds to a fit using the SFG equation, the *dotted line* to a fit with Lorentzian line shapes, and the *dashed line* to a Lorentzian fit that uses the peak positions from the SFG fit. This plot illustrates that different results are obtained when analyzing overlapping peaks with Lorentzian line shapes and the fit function for SFG (Eq. 15.18)

terms, when one molecules moves up, the other one moves down. The respective phases for these vibrations would be opposite and, therefore, destructive interference would take place that can only result in a vanishing SFG signal. This is the reason why molecules that are arranged in a centrosymmetric fashion, in other words, in arrangements that possess inversion symmetry, cannot produce SF signals. So the absence of a signal in fact provides information about the conformation of molecules. Prominent examples in this context are alkyl chains of surfactants: In an all-trans conformation the methylene groups have a center of inversion in between them. Therefore, a highly ordered alkane-chain does not produce a significant CH_2 signal, but the terminal methyl groups do because they are embedded in a symmetry that does not have inversion symmetry. For example, a self-assembled monolayer (SAM) of alkanethiols on a gold substrate does show CH_2 vibrations for a non-ordered layer (*gauche*-defects do not have inversion symmetry) and less CH_3 contributions that are outnumbered by CH_2 groups. However, a highly ordered SAM only shows CH_3 contributions because the CH_2 groups are in an all-trans conformation that cannot produce SFG signals.

Similar relations can be transferred to surfactants at air/water interfaces. Figure 15.11 shows SFG spectra that are recorded at aqueous solutions of sodium dodecyl sulfate at concentrations of 0.5 mmol/L (line) and 8 mmol/L (dashed) in the CH stretching region. The peak at around 2855 cm^{-1} is associated to CH_2 vibrations and the peak at around 2880 cm^{-1} includes CH_3 vibrations. At the lower SDS concentration, both contributions are similar, while at the higher concentration, around the critical micelle concentration, a stronger spectral contribution of the methyl groups can be detected.

Fig. 15.11 Comparison of spectra of aqueous solutions of sodium dodecyl sulfate at concentrations of 0.5 mmol/L (*line*) and 8.0 mmol/L (*dotted*) obtained under *ppp*-polarization; the increasing intensity of the CH_3 mode indicates a transition to a more ordered conformation within the alkyl chain

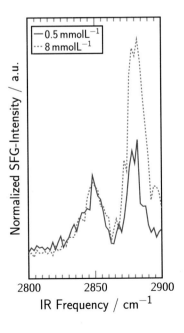

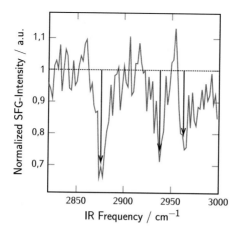

Fig. 15.12 SFG spectrum of a alkanethiol-SAM on Au. The spectrum is recorded in *ppp* polarization and normalized to IR and visible intensities. The dips around 2875, 2940 and 2965 cm^{-1} in the non-resonant background are related to terminal methyl groups within the SAM that are oriented away from the Au substrate

As mentioned earlier, phase and orientation are closely related. For a qualitative assessment of orientations, the phase analysis often helps to determine directions of transition dipole moments. For example, if the phase of the non-resonant contribution, ϕ in Eq. 15.18, is known, the relative phase to a resonant vibrations can be related to orientations of this group. Coming back to the example of the alkanethiol-SAM on Au, it can be experimentally shown that an orientation of the methyl group away from the Au surface results in a destructive interference with the non-resonant signal from the Au ($\phi_{Au} = -\pi/2$), so a dip in the non-resonant signal occurs (Fig. 15.12). If a group does have an orientation towards the Au substrate, its contribution to the spectra results in constructive interference (a peak).

15.4   Summary

This chapter summarizes basic concepts of SFG data analysis. It includes a discussion about the structure of $\chi^{(2)}$ including symmetry considerations which help identifying non-zero tensor elements. A mathematical translation between the molecular coordinate system and the laboratory coordinate system is provided as well as the description between $\chi^{(2)}$ and $\chi^{(2)}_{\text{eff}}$, which is accessible in the experiment. Examples are provided for several common molecular symmetries and practical aspects of data analysis have been presented. The purpose of this chapter is (i) to illustrate the opportunities and limitations of SFG spectroscopy, and (ii) to help researchers using the basics presented here to follow up on more complex data analysis routines that are presented in literature or that they use in their own work.

Acknowledgments The authors thank Hongfei Wang for helpful discussions. PK acknowledges funding from NIH Grant No. EB-002027 to the National ESCA and Surface Analysis Center for Biomedical Problems.

References

1. H.-F. Wang, W. Gan, R. Lu, Y. Rao, B.-H. Wu, Quantitative spectral and orientational analysis in surface sum frequency generation vibrational spectroscopy (sfg-vs). Int. Rev. Phys. Chem. **24**(2), 191–256 (2005)
2. F. Vidal, A. Tadjeddine, Sum-frequency generation spectroscopy of interfaces. Rep. Prog. Phys. **68**(5), 1095 (2005)
3. M.J. Shultz, C. Schnitzer, D. Simonelli, S. Baldelli, Sum frequency generation spectroscopy of the aqueous interface. ionic and soluble molecular solutions. Int. Rev. Phys. Chem. **19**(1), 123–153 (2000)
4. Y.R. Shen, Phase-sensitive sum-frequency spectroscopy. Annu. Rev. Phys. Chem. **64**, 129–150 (2013)
5. S. Roy, P.A. Covert, W.R. FitzGerald, D.K. Hore, Biomolecular structure at solid liquid interfaces as revealed by nonlinear optical spectroscopy. Chem. Rev. **114**(17), 8388–8415 (2014)
6. S. Roke, Nonlinear optical spectroscopy of soft matter interfaces. ChemPhysChem **10**(9–10), 1380–1388 (2009)
7. G.L. Richmond, Structure and bonding of molecules at aqueous surfaces. Annu. Rev. Phys. Chem. **52**, 357–389 (2001)
8. D. Liu, G. Ma, L.M. Levering, Heather C. Allen, Vibrational spectroscopy of aqueous sodium halide solutions and air liquid interfaces: observation of increased interfacial depth. J. Phys. Chem. B **108**(7), 2252–2260 (2004)
9. A.G. Lambert, P.B. Davies, D.J. Neivandt, Implementing the theory of sum frequency generation vibrational spectroscopy: a tutorial review. Appl. Spectrosc. Rev. **40**(2), 103–145 (2005)
10. S.A. Hall, K.C. Jena, P.A. Covert, S. Roy, T.G. Trudeau, D.K. Hore, Molecular-level surface structure from nonlinear vibrational spectroscopy combined with simulations. J. Phys. Chem. B **118**(21), 5617–5636 (2014)
11. L. Fu, Z. Wang, E.C.Y. Yan, Chiral vibrational structures of proteins at interfaces probed by sum frequency generation spectroscopy. Int. J. Mol. Sci. **12**(12), 9404–9425 (2011)
12. X. Chen, M.L. Clarke, J. Wang, Z. Chen, Sum frequency generation vibrational spectroscopy studies on molecular conformation and orientation of biological molecules at interfaces. Int. J. Mod. Phys. B **19**(04), 691–713 (2005)
13. J. Bredenbeck, A. Ghosh, H.-K. Nienhuys, M. Bonn, Interface-specific ultrafast two-dimensional vibrational spectroscopy. Acc. Chem. Res. **42**(9), 1332–1342 (2009)
14. C.D. Bain, Sum-frequency vibrational spectroscopy of the solid/liquid interface. J. Chem. Soc. Faraday Trans. **91**, 1281–1296 (1995)
15. S. Ye, F. Wei, H. Li, K. Tian, Y. Luo, Structure and orientation of interfacial proteins determined by sum frequency generation vibrational spectroscopy: method and application. Adv. Protein Chem. Struct. Biol. **93**(Biomolecular Spectroscopy), 213–255 (2013)
16. S. Gopalakrishnan, D. Liu, H.C. Allen, M. Kuo, M.J. Shultz, Vibrational spectroscopic studies of aqueous interfaces: salts, acids, bases, and nanodrops. Chem. Rev. (Washington, DC, United States) **106**(4), 1155–1175 (2006)
17. Y.R. Shen, V. Ostroverkhov, Sum-frequency vibrational spectroscopy on water interfaces: Polar orientation of water molecules at interfaces. Chem. Rev. (Washington, DC, United States) **106**(4), 1140–1154 (2006)
18. H. Goldstein, C.P. Poole, J.L. Safko, *Classical Mechanics* (Pearson Education, Limited, 2013)

19. E.C.Y. Yan, L. Fu, Z. Wang, W. Liu, Biological macromolecules at interfaces probed by chiral vibrational sum frequency generation spectroscopy. Chem. Rev. (Washington, DC, United States) **114**(17), 8471–8498 (2014)

20. X. Zhuang, P.B. Miranda, D. Kim, Y.R. Shen, Mapping molecular orientation and conformation at interfaces by surface nonlinear optics. Phys. Rev. B Condens. Matter Mater. Phys. **59**(19), 12632–12640 (1999)

21. S.A. Hall, A.D. Hickey, D.K. Hore, Structure of phenylalanine adsorbed on polystyrene from nonlinear vibrational spectroscopy measurements and electronic structure calculations. J. Phys. Chem. C **114**(21), 9748–9757 (2010)

22. Y. Liu, T.L. Ogorzalek, P. Yang, M.M. Schroeder, E.N.G. Marsh, Z. Chen. Molecular orientation of enzymes attached to surfaces through defined chemical linkages at the solid-liquid interface. J. Am. Chem. Soc. **135**(34), 12660–12669 (2013). PMID: 23883344

23. X. Wei, S.-C. Hong, X. Zhuang, T. Goto, Y.R. Shen, Nonlinear optical studies of liquid crystal alignment on a rubbed poly(vinyl alcohol) surface. Phys. Rev. E Stat. Phys. Plasmas Fluids Relat. Interdisc. Top. **62**(4-A), 5160–5172 (2000)

24. M.B. Feller, W. Chen, Y.R. Shen, Investigation of surface-induced alignment of liquid-crystal molecules by optical second-harmonic generation. Phys. Rev. A At. Mol. Opt. Phys. **43**(12), 6778–6792 (1991)

25. P. Ye, Y.R. Shen, Local-field effect on linear and nonlinear optical properties of adsorbed molecules. Phys. Rev. B Condens. Matter Mater. Phys. **28**(8), 4288-4294 (1983)

26. C. Hirose, N. Akamatsu, K. Domen, Formulas for the analysis of surface sum-frequency generation spectrum by ch stretching modes of methyl and methylene groups. J. Chem. Phys. **96**(2), 997–1004 (1992)

27. R. Lu, W. Gan, B. Wu, Z. Zhang, Y. Guo, H. Wang, C-h stretching vibrations of methyl, methylene and methine groups at the vapor/alcohol (n = 1-8) interfaces. J. Phys. Chem. B **109**(29), 14118–14129 (2005)

28. H.-F. Wang, L. Velarde, W. Gan, L. Fu, Quantitative sum-frequency generation vibrational spectroscopy of molecular surfaces and interfaces: lineshape, polarization, and orientation. Annu. Rev. Phys. Chem. **66**(1), 189–216 (2015). PMID: 25493712

29. L. Velarde, H.-F. Wang, Unified treatment and measurement of the spectral resolution and temporal effects in frequency-resolved sum-frequency generation vibrational spectroscopy (sfg-vs). Phys. Chem. Chem. Phys. **15**, 19970–19984 (2013)

30. A. G. F. de Beer, J.-S. Samson, W. Hua, Z. Huang, C. Zishuai, X. Chen, H. C. Allen, S. Roke, Direct comparison of phase-sensitive vibrational sum frequency generation with maximum entropy method: case study of water. J. Chem. Phys. **135**(22), 224701/1–224701/9 (2011)

31. M. Sovago, E. Vartiainen, and M. Bonn, Determining absolute molecular orientation at interfaces: a phase retrieval approach for sum frequency generation spectroscopy. J. Phys. Chem. C **113**, 6100–6106 (2009)

32. P.-K. Yang, J. Y. Huang, Phase-retrieval problems in infrared-visible sum-frequency generation spectroscopy by the maximum-entropy method. J. Opt. Soc. Am. B **14**(10), 2443–2448 (1997)

33. P.-K. Yang, J. Y. Huang, Model-independent maximum-entropy method for the analysis of sum-frequency vibrational spectroscopy. J. Opt. Soc. Am. B *17*(7), 1216–1222 (2000)

34. M. J. Hofmann, P. Koelsch, Retrieval of complex χ^2 parts for quantitative analysis of sum-frequency generation intensity spectra. J. Chem. Phys. **143**(14), 134112–134118 (2015)

Chapter 16
Microfluidics: From Basic Principles to Applications

Florent Malloggi

Abstract Microfluidics is the science and technology of systems that process or manipulate small amounts of fluids using channels with dimensions of one to hundreds of micrometers. This field is mainly driven by technological applications where the aim is to develop entire laboratories inside chips. Introduced more than a decade ago, microfluidics has quickly become an important tool in several fields including new technologies as well as basic research. One reason for its fast development is based on the predictability of the flows at such scale and the exquisite control of interfaces in microchannels. Nowadays microfluidics has a place in many scientific fields. More often it is seen as a tool for the development of various topics related to chemistry, biology or physics. The list of possible applications and developed systems is very long and it is not the purpose of this chapter. In the following, we focus on the physics foundations on which this discipline relies. After a brief introduction on lab on chip technology, we introduce the basis of fluid mechanics with the governing equation for a fluid in motion. We also introduce diffusion transport and capillary effects which are dominant in microfluidic systems. Throughout the chapter we will illustrate the basic principles with practical examples.

16.1 Introduction

16.1.1 Lab-on-Chip Technology

Microfluidics is the science and technology of systems that process or manipulate small amounts of fluids (microliter to attoliter), using channels with dimensions of one to hundreds of micrometers. This field is mainly driven by technological applications where the aim is to develop entire laboratories inside chips. In the 1990s, miniaturization of electronic integrated circuits and microelectromechanical

F. Malloggi (✉)
CEA Saclay, LIONS, Gif-sur-Yvette, France
e-mail: florent.malloggi@cea.fr

© Springer International Publishing Switzerland 2016
P.R. Lang and Y. Liu (eds.), *Soft Matter at Aqueous Interfaces*,
Lecture Notes in Physics 917, DOI 10.1007/978-3-319-24502-7_16

systems (MEMS) [8, 12, 16] coupled to fluid gave birth to microfluidics: a new discipline based on the theory of flow of fluids (fluid mechanics) and of suspensions in submillimeter-sized systems influenced by external forces (colloidal science). Although the theory involved is not new, interestingly, manipulating fluid at the micrometric scale changed our intuition and relative importance between forces (see Table 16.1). The fluid transport occurs as either a continuous flow or as a finite volume, i.e. droplet based, which is known as digital microfluidics. Fluid motion is achieved via mechanical or pressurized pumps as well as surface tension effects, electrostatic or magnetohydrodynamic [19]. The scientific and technological renewed interest has been motivated by the emerging and rapidly evolving field of lab-on-chip (LOC) systems [9, 17, 18]. Originally built on silicon based technology developed by the semiconductor industry, the field had a great expansion with the emergence of soft lithography [22] i.e. polymer devices in analogy to hard lithography used in microelectronics [15]. Main components are microchannels and micropumps as well as associated sensors, heaters or actuators. There are several advantages of scaling down standard laboratories. One obvious advantage is the reduction in the amount of required sample. A linear reduction by a factor of 10^3 amounts to a volume reduction by a factor of 10^9, so instead of handling 1 mL a lab-on-a-chip system could easily deal with 1 nL or even smaller down to 1 pL. Such small volumes allow for very fast analysis, efficient detection schemes, and analysis, even when large amounts of sample are not available. Moreover, the small volumes make it possible to develop compact and portable systems that might ease the use of bio/chemical handling and analysis systems tremendously. Nowadays lab-on-a-chip systems have great impact in biotechnology, pharmacology, medical diagnostics, forensics, environmental monitoring and basic research.

Table 16.1 Comparison of flows in macrochannels and microchannels

Phenomenon	Macrochannels	Microchannels
Gravity	Dominant (typical size $\gg$ capillary length: l_c	Negligible (channel size $\ll l_c$)
Continuum mechanics	Valid	Valid if lengthscale: fluid > 10 nm; gas > 10 μm
Reynolds number	Laminar and turbulent flow (Re $\approx$ 2000)	laminar flow (Re < 1) in most cases, Stokes flow approximation
Surface roughness	Negligible	To be considered; roughness may be comparable to dimensions of the system
Diffusion	Negligible	Important and used for separation and mixing
Surface tension	Negligible	Important and major contributing force
Viscous heating	Negligible	Major player due to high velocity gradient
Electrohydrodynamic effects	Negligible	Important

16.1.2 Scaling Laws

When analyzing the physical properties of microsystems, it is helpful to introduce the concept of scaling laws or size reduction effect. A scaling law expresses the variation of physical quantities with the size l of the given system or object, while keeping other quantities such as time, pressure, temperature, etc. constant. As an example, consider volume forces, such as gravity and inertia, and surface forces, such as surface tension and viscosity. The basic scaling law for the ratio of these two classes of forces can generally be expressed by

$$\frac{surfaceforces}{volumeforces} \propto \frac{l^2}{l^3} = l^{-1} \tag{16.1}$$

This scaling law implies that when scaling down to the microscale in lab-on-a-chip systems, the volume forces, which are very prominent in our daily life, become largely unimportant. Instead, the surface forces become dominant, and as a consequence, we must revise our intuition [7, 21]. The most common types of forces and their scaling with size are listed in Table 16.2.

Examples

At macroscopic scale the weight (which scales as l^3) of an object is predominant and it falls down under gravity. When an object of the same mass density, ρ, is very small, the weight becomes insignificant compared to the air friction force which depends on the surface area scaling as l^2. As a consequence even small air currents can keep the object floating. One way to see this is to consider the terminal velocity, v, of a particle in air. This velocity is simply determined by the balance between gravity force and viscous drag and writes:

$$\eta v l \sim mg \tag{16.2}$$

where η is the shear viscosity. The velocity v writes

$$v \sim \frac{\rho g l^2}{\eta} \propto l^2 \tag{16.3}$$

As a result the terminal velocity is very small for microparticles and this causes the particles to float with the moving air instead of falling.

Table 16.2 Scaling of the most common forces with system size l		
Surface tension	l^1	
Fluid force/electrostatic force	l^2	
Weight/inertia/electromagnetic	l^3	

In the same manner, let us consider the volume flow rate Q (an important quantity in fluid mechanics of microflows):

$$Q \sim \frac{l^4 \Delta P}{\eta l} \tag{16.4}$$

For a given pressure drop rate (i.e. $\Delta P/l$) $Q \propto l^4$ meaning a reduction of 10 in channel size induced a reduction of 10^4 in volume flow.

16.1.3 Dimensional Analysis

Dimensional analysis is a powerful analytic technique based on the Buckingam π-theorem [5]. In a system described by m dimensional variables containing n different physical dimensions (length, time, mass, temperature, etc.): the Buckingham π-theorem states that there are m–n independent non-dimensional groups that can be formed from these governing variables. When forming the dimensionless groups, we try to keep the dependent variable (the one we want to predict) in only one of the dimensionless groups. Once we have the m–n dimensionless variables, the Buckingham π-theorem further tells us that the variables can be related according to

$$\pi_1 = f(\pi_2, \pi_i, \ldots, \pi_{m-n}) \tag{16.5}$$

where π_i is the ith dimensionless variable.

Example: mixing scale.

In microfluidics systems, we will see that convection-diffusion mechanisms are important for mixing chemical species. In this problem we have three parameters: L is the distance over which the chemical spreads, D is a measure of the rate of diffusion (dimension L^2/t), and t is the time. Thus, we have three variables and two dimensions (L and t), the dimensionless group writes:

$$\pi_1 = L^a D^b t^c$$

and we want all dimensions to cancel out, giving us two equations

t gives: $0 = -b + c$
L gives: $0 = a + 2b$

Thus, we have

$$\pi_1 = \frac{L^2}{Dt} \tag{16.6}$$

Later, we will see that this dimensionless quantity is called the Peclet number.

Example: Reynolds number.
We consider here a famous number in fluid mechanics, the Reynolds number Re, which describes the transition between different flow regimes (laminar vs turbulent). This number is dimensionless and the variables it depends on are the flow velocity u, the flow disturbances characterized by a typical length scale L and the fluid properties (density ρ, viscosity η and temperature T). Since ρ and η are functions of T the most compact approach is to retain ρ and η in the form of the kinematic viscosity $v = \eta/\rho$ in m^2/s unit. Thus, we have three variables (u, L and v) and two dimensions (L and t), yielding one non-dimensional number:

$$\pi_1 = \frac{uL}{v} = Re$$

which is the Reynolds number defined later.

16.2 Fluid Motion: Governing Equations

16.2.1 Fluid Definition

The two main classes of fluids, the liquids and the gases, differ by the densities and by the degree of interaction between the constituent molecules. At ambient temperature, the density of a real gas is so low ($\lesssim 1\,kg/m^3$), that the molecules move largely as free particles that only interact by direct collisions at atomic distances, ≈ 0.1 nm. The relatively large distance between the gas molecules, ≈ 3 nm, makes the gas compressible. The density of a liquid ($\approx 10^3 kg/m^3$) is comparable to that of a solid, i.e. the molecules are packed as densely as possible with a typical average intermolecular distance of 0.3 nm, and a liquid can be considered incompressible in most practical cases. As opposed to a solid body, a fluid is a substance that deforms continuously when being acted on an arbitrarily small shearing stress. Whatever small this action may be, the fluid responses with a time-dependent deformation e.g. a fluid motion, a flow.

We will focus on liquids only in this chapter. One of the most important property of the liquid is its viscosity. In order to understand what is the viscosity let us consider two parallel planes of surface A with the top one sliding at a velocity u_0 (Fig. 16.1). The motion of the top plane gives rise to the motion of the fluid below: this flow is called Couette flow. The velocity profile is linear along the z-direction (supposing the flow stationary e.g. long time after the plane motion). This kind of flow is similar to what is obtained with the heat between two plates of different temperatures. The friction force F on the surface plane with area A (opposite to

Fig. 16.1 Couette flow

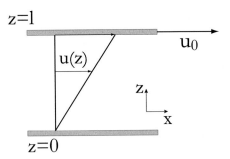

plane motion) is proportional to the velocity variation from one plane to the other and writes:

$$\frac{F}{A} = \frac{\eta u_0}{l} = -\eta \frac{\partial u}{\partial z} \tag{16.7}$$

The ratio $F/A = \sigma$ is called shear stress and it has a dimension of a pressure. The material constant η is called dynamic viscosity and it is expressed in Pascal $\times$ second (Pas). If the viscosity is low as for example in the case of gases the motion of the top plate induces only a very small or even no fluid motion. Accordingly, the shear stress is very small or vanishes completely.

16.2.2 Continuum Approximation

One possible approach to describe a fluid in motion is to examine what happens at the microscopic level where the stochastic motions of individual molecules can be distinguished. However, the resulting many-body problem of molecular dynamics is very complex for a liquid since it contains an enormous number of molecules. The standard way to solve this problem is to introduce the continuum approximation. According to this hypothesis, the fluid is infinitely divisible without change of character. This implies that all quantities such as density, viscosity as well as variables such as pressure, velocity and temperature can be defined without ambiguities. The motivation of this approach is that in many applications, we are concerned with fluid motions in the vicinity of bodies such as channel walls and not with the instantaneous forces of interaction between the surface and the molecules: at such macroscopic level we are interested in the time average of the forces of interactions between the fluid and the surface. Nowadays with the growing trend of using microfluidics and nanofluidics it is legitimate to ask whether this assumption is still valid. A general rule of thumb is that the continuum postulate holds for length scales greater than 1 μm for gases and larger than 10 nm for liquids [12]. In other words the continuum approximation is valid for a liquid in microchannels [3].

16.2.3 Acceleration of a Fluid Particle

Consider a fluid particle of volume V in the geometry as sketched in Fig. 16.2. The velocity depends on time and position as $\mathbf{v} = \mathbf{v}(x(t), y(t), z(t), t)$. For an infinitesimal volume V the acceleration is simply:

$$a = \frac{dv}{dt} \tag{16.8}$$

To facilitate our understanding we consider the acceleration along the x-axis a_x first. Given the functional dependency, it is derived as:

$$\begin{aligned}
a_x &= \frac{dv_x}{dt} = \frac{d}{dt} v_x(x(t), y(t), z(t), t) \\
&= \frac{\partial v_x}{\partial x}\frac{dx}{dt} + \frac{\partial v_x}{\partial y}\frac{dy}{dt} + \frac{\partial v_x}{\partial z}\frac{dz}{dt} + \frac{\partial v_x}{\partial t} \\
&= \frac{\partial v_x}{\partial x} v_x + \frac{\partial v_x}{\partial y} v_y + \frac{\partial v_x}{\partial z} v_z + \frac{\partial v_x}{\partial t}
\end{aligned} \tag{16.9}$$

If we perform the corresponding derivation along the y and z axes we obtain the 3-D expression for the acceleration which writes in a more compact form:

$$\frac{d\mathbf{v}}{dt} = \frac{\partial \mathbf{v}}{\partial t} + (\mathbf{v} \cdot \nabla)\mathbf{v} \tag{16.10}$$

The symbolic form in the second term involves a scalar product between the vector $\mathbf{v}$ and the operator ∇ with the components $\partial/\partial x$, $\partial/\partial y$ and $\partial/\partial z$. Equation (16.10) is a generic expression where (*i*) the first term $\partial v/\partial t$ relates to the acceleration of the fluid due to an increase of velocity over time and (*ii*) the second term $(\mathbf{v} \cdot \nabla)\mathbf{v}$ describes the increase of the fluid velocity due to mass conservation. We will use this expression later when introducing the fundamental equation of motion in fluid mechanics.

Fig. 16.2 Decomposition of the acceleration of a fluid particle in an unsteady flow

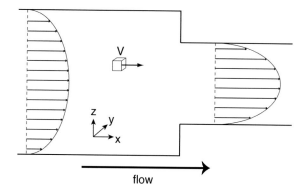

Note.

Equation (16.10) can also be used to express the variation of other quantities with position and time. For instance it is possible to look at the temperature variation of a particle $T(\mathbf{r}, t)$ along its trajectory.

$$\frac{dT}{dt} = \frac{\partial T}{\partial t} + (\mathbf{v} \cdot \nabla)\mathbf{T}$$

where $\partial T/\partial t$ is the temporal derivative of the temperature of the fluid at a given fixed point. The second term reflects the variation of T due to fluid flow in the direction of the temperature gradient.

16.2.4 Mass Conservation in a Fluid Flow

We consider an arbitrary volume (V) bounded by a closed surface (S), which is stationary with respect to the reference used to describe fluid flow (Fig. 16.3). At every moment, the fluid enters and exits the volume; the change of the total mass m therein is opposite to the outgoing flow through the surface.

Therefore

$$\frac{dm}{dt} = \frac{d}{dt}\left(\int\int\int_V \rho dV\right) = -\int\int_S \rho v \cdot n dS \qquad (16.11)$$

where $\mathbf{n}$ is the unit vector normal to the interface.

Applying Ostrogradski's theorem to the second term of the Eq. (16.11), we get

$$\int\int\int_V \left(\frac{\partial \rho}{\partial t} + \nabla \cdot (\rho v)\right) dV = 0 \qquad (16.12)$$

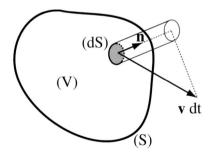

Fig. 16.3 Mass balance of the fluid within the fixed volume (V). The outgoing flow of mass per unit time is $\rho v \cdot n dS$

This equation is correct regardless of the volume V and hence we obtain the continuity equation

$$\frac{\partial \rho}{\partial t} + \nabla \cdot (\rho v) = 0 \tag{16.13}$$

In the case of incompressible fluids, i.e. the density remains constant over time, the equation of mass conservation (16.13) becomes

$$\nabla \cdot v = 0 \tag{16.14}$$

The conditions under which a fluid can be considered incompressible can be reduced to the inequality

$$U \ll c \Leftrightarrow M \ll 1$$

where U represents the scale of characteristic flow velocity and c the sound celerity of pressure waves in the fluid of interest. We defined M as the Mach number. This condition is clearly not satisfied in studies of gas dynamics at high speeds.

Conversely, this case is almost always encountered in microfluidics dealing with liquids.

Remark For small Reynolds number the previous condition is even more restrictive

$$M \ll \sqrt{Re} \Leftrightarrow fluid\ incompressible$$

16.2.5   Surface Forces

General expression of surface forces

Considering an element of fluid surface dS, the shear is the value of the force acting on this surface element.

For a fluid at rest, the force is normal to the surface and it is the hydrostatic pressure. For a fluid in motion, there are stresses appearing parallel to the surface element dS. Stresses are due to the friction between fluid layers. They come from the fluid viscosity. In other words there is momentum transport from region of high velocities to region with lower velocities.

Considering an element of fluid volume dV there are 9 stress components on dV.

These components represent the so-called stress tensor σ which writes (Figs. 16.4 and 16.5):

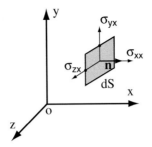

Fig. 16.4 Components of the stress exerted on a surface whose normal is oriented along the x-axis

Fig. 16.5 Components of the stress exerted on a volume element

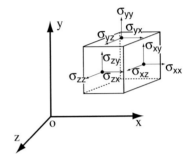

$$\boldsymbol{\sigma} = \begin{pmatrix} \sigma_{xx} & \sigma_{xy} & \sigma_{xz} \\ \sigma_{yx} & \sigma_{yy} & \sigma_{yz} \\ \sigma_{zx} & \sigma_{zy} & \sigma_{zz} \end{pmatrix}$$

where diagonal terms are normal stress and non-diagonal ones are the parallel or shear stress.

From the stress tensor $\boldsymbol{\sigma}$ one can extract the part corresponding to pressure stresses (fluid at rest or in global translation). This compound is purely diagonal (normal stresses) and isotropic (same value for the 3 coefficients of the diagonal). Hence the stress tensor re-writes:

$$\sigma_{ij} = \tau_{ij} - p\,\delta_{ij} \tag{16.15}$$

where p is the pressure, δ_{ij} is an element of the Kronecker tensor ($\delta_{ij} = 1$ if i = j and $\delta = 0$ if i ≠ j). Note that the negative sign before p means that the fluid at rest is in general pressurized: the stress is opposite to the unit normal vector **n** of the surface. The term τ_{ij} is the generic form of the viscous stress: this part is linked to the fluid deformation.

Viscous stresses τ_{ij} tensor for a Newtonian fluid
For a Newtonian fluid, the viscous stress tensor, τ, can be represented as a combination of the rate of strain tensor **d**, fluid velocity vector, shear viscosity η and the bulk coefficient of viscosity κ as

$$\tau = 2\eta\mathbf{d} + \left[\left(\kappa - \frac{2}{3}\eta\right)\nabla \cdot \mathbf{v}\right]\mathbf{I} \tag{16.16}$$

where $\mathbf{I}$ is the identity matrix. The rate of strain tensor can be calculated from the velocity vector as

$$d_{ij} = \frac{1}{2}\left(\frac{\partial v_i}{\partial x_j} + \frac{\partial v_j}{\partial x_i}\right) \tag{16.17}$$

The continuity equation, a statement of conservation of mass, states

$$\frac{\partial \rho}{\partial t} + \nabla \cdot (\rho v) = 0 \tag{16.18}$$

which for an incompressible fluid reduces to

$$\nabla \cdot (v) = 0 \Rightarrow \frac{\partial v_i}{\partial x_i} = 0 \tag{16.19}$$

In case of an incompressible newtonian fluid, the viscous stress tensor τ simplifies to the following expression:

$$\tau_{ij} = \eta\left(\frac{\partial v_i}{\partial x_j} + \frac{\partial v_j}{\partial x_i}\right) \tag{16.20}$$

16.2.6 The Navier-Stokes Equation

The equation of motion (Newton's Second Law) for a liquid in vector notation is:

$$\rho\frac{dv}{dt} = \rho g + \nabla \cdot \boldsymbol{\sigma} \tag{16.21}$$

where $\mathbf{g}$ is the acceleration of gravity and $\boldsymbol{\sigma}$ is the stress tensor introduced before and defined as

$$\boldsymbol{\sigma} = -p\mathbf{I} + \tau \tag{16.22}$$

and p, the pressure, is the stress experienced by the fluid at rest which always acts along the outwardly directed normal unit vector, $\mathbf{I}$ is the identity matrix, and τ the viscous stress tensor due to the motion of the fluid. Substituting Eqs. (16.21) into (16.22) and applying the definition of $\frac{d}{dt}$ [1] we get

$$\rho\left(\frac{\partial \mathbf{v}}{\partial t} + \mathbf{v} \cdot \nabla\right)\mathbf{v} = -\nabla p + \rho \mathbf{g} + \nabla \cdot \tau \qquad (16.23)$$

Introducing expression (16.20) into (16.23), and assuming a constant viscosity, we eventually obtain the Navier-Stokes equation

$$\rho\left(\frac{\partial \mathbf{v}}{\partial t} + \mathbf{v} \cdot \nabla\right)\mathbf{v} = -\nabla p + \rho \mathbf{g} + \eta \varDelta \mathbf{v} \qquad (16.24)$$

where the Laplace-operator $\varDelta = \nabla^2$.

The Navier-Stokes equation simply states that the time rate of change in linear momentum (left-hand side) must equal the applied body and surface forces (right-hand side).

16.2.7 Boundary Conditions in Fluid Flows

To have a complete description of the fluid motion, it is necessary to determine the boundary conditions i.e. the value of the velocity and stress fields at the fluid boundary. Two cases are possible depending on whether the medium limiting the fluid is a solid or another fluid.

16.2.7.1 Boundary Conditions at the Surface of a Solid Body

The fluid cannot penetrate the solid and this condition imposes the equality of the normal velocity at the interface solid/liquid

$$\mathbf{v}_{solid} \cdot \mathbf{n} = \mathbf{v}_{fluid} \cdot \mathbf{n}$$

Moreover due to the viscous stress, there is no-slip condition of the fluid at body surface. The tangential components of the fluid velocities and the solid must be equal, which, added to the condition of equality of normal components, leads to the relation

$$v_{fluid} = v_{solid} \qquad no-slip\ condition \qquad (16.25)$$

---

[1]

$$\frac{d}{dt} = \frac{\partial}{\partial t} + \mathbf{v} \cdot \nabla.$$

16.2.7.2 Boundary Conditions Between Two Fluids

In addition to the requirement of continuous velocities, a second condition demands the continuity of the stress (force per unit area) across the surface of separation between two fluids. Indeed there must be a balance between the internal stresses in each of the two fluids and the localized stresses at the surface.

At the fluid interface, the condition on the normal stress (pressure) is expressed by the Laplace law. The pressures p_1 and p_2 in the two fluids are related by

$$p_1 - p_2 = \gamma \left(\frac{1}{R} + \frac{1}{R'} \right) \tag{16.26}$$

where γ is the surface tension between fluid 1 and 2, and R and R' are the principal radii of curvature of the interface.[2]

Moreover, the balance of the tangential stresses at the interface is expressed by

$$\left(\boldsymbol{\sigma}^{(1)} \cdot \mathbf{n} \right) \cdot \mathbf{t} = \left(\boldsymbol{\sigma}^{(2)} \cdot \mathbf{n} \right) \cdot \mathbf{t} \tag{16.27}$$

This equation expresses the equilibrium at the interface between each action exerted by one fluid on the other, and the received reaction.

For an incompressible Newtonian fluid, the stress tensor can be simply written $\sigma_{ij} = \eta (\partial v_i / \partial x_j + \partial v_j / \partial x_i)$. The continuity condition (16.27) becomes

$$\eta_1 \left(\left(\frac{\partial v_i^{(1)}}{\partial x_j} + \frac{\partial v_j^{(1)}}{\partial x_i} \right) n_i \right) t_j = \eta_2 \left(\left(\frac{\partial v_i^{(2)}}{\partial x_j} + \frac{\partial v_j^{(2)}}{\partial x_i} \right) n_i \right) t_j \tag{16.28}$$

where t_i (resp. n_i) represents the components of a unit vector $\mathbf{t}$ (resp. $\mathbf{n}$) tangent (resp. normal) to the interface.

Let's consider the case of the Fig. 16.6 where locally the interface is extrapolated to a plane and the velocity $\mathbf{v}$ is parallel to the abscissa and a function of the y-coordinate, $v_x(y)$.

The tangential stress is reduced to the term $\sigma_{xy} = \eta \partial v_x / \partial y$. The condition of equality of tangential stress writes

$$\eta_1 \frac{\partial v_x^{(1)}}{\partial y} = \eta_2 \frac{\partial v_x^{(2)}}{\partial y} \tag{16.29}$$

which simply means that the velocity gradients at the interface are in the inverse ratio of the dynamic viscosities.

---

[2] The pressure is higher inside the concavity of the interface.

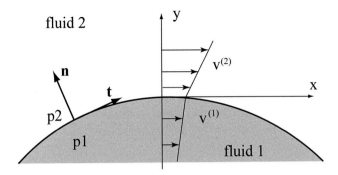

Fig. 16.6 Conditions at viscous fluid interface

The equation is even simpler when one of the fluids is a gas (then the interface is called free surface). The low viscosity of the gas allows to neglect the shear stress in the liquid at the interface and then

$$\left(\sigma^{(1)} \cdot \mathbf{n}\right) \cdot \mathbf{t} = 0 \qquad (16.30)$$

or, in the example described above $\partial v_x / \partial y = 0$

16.2.8 Stokes Flow

The non-linear term $\rho(\mathbf{v} \cdot \nabla)\mathbf{v}$ in the Navier-Stokes equation is responsible for making the mathematical treatment of the equation complex and difficult. However, in microfluidics systems, it is possible to neglect the non-linear term since the Reynolds number is small (Re < 1). In this condition and in the assumption of a stationary state the Navier-Stokes (16.24) equation becomes the Stokes equation where analytic solutions can be found:

$$0 = -\nabla p + \eta \Delta \mathbf{v} \qquad (16.31)$$

In deriving this approximation, we assumed that the time derivative $\partial \mathbf{v} / \partial t$ was controlled by the intrinsic constant time scale $T_0 = L_0 / U_0$. In case of non-stationary flows we must keep the temporal evolution of the velocity

$$\frac{\partial v}{\partial t} = -\nabla p + \eta \Delta \mathbf{v} \qquad (16.32)$$

16.2.9   Flows in Microchannels: Some Examples

The Navier-Stokes equation is notoriously difficult to solve due to its non-linearity. However, analytical solutions can be found in a few, but important cases. In the following we will develop some of these solutions. We will focus on steady-state problems and the so-called Poiseuille-flow problem, i.e. pressure induced steady-state fluid flow in infinitely long (translation invariant) channels. Such idealized flows are important since they provide us with a basic understanding of the behavior of liquids flowing in the microchannels.

16.2.9.1   Flow in a Cylindrical Microchannel

We consider a horizontal cylinder of radius R and we study the flow induced by a pressure difference Δp over the tube length L along the z axis (see Fig. 16.7).
Field of flow velocity
   Due to the cylindrical geometry of the problem we adopt the cylindrical coordinate to solve the Stokes equation. We can already make several intuitive arguments about the flow in this geometry. The tube is axisymetric which means that the velocity is independent of θ and infinite along the z-axis which means that the velocity is independent of z i.e. $\mathbf{v}(r, \theta, z) = \mathbf{v}(r)$. The mass conservation with the incompressible fluid condition, expressed in cylindrical coordinates, gives

$$\frac{1}{r}\frac{\partial r v_r}{\partial r} = 0 \Rightarrow v_r = 0$$

Moreover, as illustrated in Fig. 16.7 the velocity is along the z-axis only i.e. $v(v_r, v_\theta, v_z) = v(0, 0, v_z)$. We have simplified the velocity field by simple intuitive arguments and it writes

$$\mathbf{v}(r, \theta, z) = v_z(r)e_z$$

The Stokes equation can be reduced to

$$\frac{\partial p}{\partial z} = \frac{p_{z=0} - p_{z=L}}{L} = -\frac{\Delta p}{L} = \eta\frac{1}{r}\frac{\partial}{\partial r}\left(r\frac{\partial v_z}{\partial r}\right) \qquad (16.33)$$

Fig. 16.7 Laminar flow in a cylindrical tube

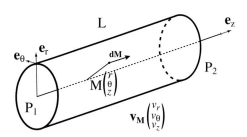

The no-slip boundary condition becomes

$$v_z(r = R) = 0 \tag{16.34}$$

This partial differential equation is easily solvable with a double integration over r as

$$v_z = -\left(\frac{\Delta p}{L}\right)\frac{r^2}{4\eta} + A\ln(r) + B \tag{16.35}$$

Obviously A = 0 since v_z cannot be infinite when $r = 0$. Now applying the boundary condition of (16.34), we find

$$B = \left(\frac{\Delta p}{L}\right)\frac{R^2}{4\eta}$$

The velocity profile is now fully determined

$$v_z = \left(\frac{\Delta p}{L}\right)\frac{R^2}{4\eta}\left[1 - \left(\frac{r}{R}\right)^2\right] \tag{16.36}$$

The functional form of the velocity profile indicates that the fluid front has a parabolic shape. From this velocity profile, several important parameters can be calculated. The maximum linear flow velocity, V_{max} which occurs at $r = 0$ is

$$V_{max} = \left(\frac{\Delta p}{L}\right)\frac{R^2}{4\eta} \tag{16.37}$$

The volumetric flow rate Q can be calculated

$$Q = \int_0^{2\pi}\int_0^R v_z r\,dr\,d\theta = \frac{\pi R^4}{8\eta}\left(\frac{\Delta p}{L}\right) \tag{16.38}$$

This equation is known as the Hagen-Poiseuille law and it shows that the flow rate and the pressure drop are linked. In microfluidics, fluids can be controlled either by syringe pump (controlled flow rate) or by pressure controller. According

Table 16.3 Analogy fluid mechanics/electromagnetism

Microchannel network	Electrical equivalent
Pressure	Potential
Flow rate	Electrical current
Hydrodynamic resistance	Electrical resistance

Table 16.4 Hydraulic resistance for straight channels with different cross-sectional shapes

Shape	Expression
circle, radius a	$\frac{8\eta L}{\pi a^4}$
triangle, length a	$\frac{320\eta L}{\sqrt{3}a^4}$
square, length a	$28.4\eta L \frac{1}{a^4}$
rectangle, height h width w	$\frac{12\eta L}{1-0.63(h/w)} \frac{1}{h^3 w}$

to Eq. (16.38) it is then easy to know the local pressure (resp. flow rate) when using constant flow rate (resp. pressure).

Hydraulic resistance

From the Hagen-Poiseuille law Eq. (16.38) we found that a constant pressure drop Δp leads to a constant flow rate Q. We can re-write this equation by introducing R_h the hydraulic resistance, in analogy of Ohm's law in electricity (see Table 16.3).

$$\Delta p = R_h Q; \qquad [R_h] = \frac{Pas}{m^3} = \frac{kg}{m^4 s} \qquad (16.39)$$

Due to this analogy, we can apply Kirchoff's laws: in series $R_h = \sum R_i$ and in parallel $1/R_h = \sum 1/R_i$.

In the particular case studied above (flow in a cylindrical microchannel) the hydraulic resistance writes

$$R_h = \frac{8\eta L}{\pi R^4}$$

Table 16.4 lists the hydraulic resistance for a number of different cross section [4]

Friction force on the walls

The friction force exerted by the fluid on the tube is calculated from the stress on the walls

$$\mathbf{F} = \int\int_{(wall)} \boldsymbol{\sigma} \cdot \mathbf{n} dS \qquad (16.40)$$

where $\mathbf{n}$ is the unit vector normal to the wall. The component F_z along the z- axis writes

$$F_z = \int\int_{(wall)} [\sigma_{zr} r]_{r=R} d\theta dz = \int_0^L dz \int_0^{2\pi} [\sigma_{zr} r]_{r=R} d\theta \qquad (16.41)$$

Using the expression of the stress $\sigma_{zr} = \eta(\partial v_z / \partial r)$, one obtains the force f_z per unit length of tube

$$f_z = F_z/L = 4\pi\eta V_{max} \tag{16.42}$$

16.2.9.2 Laminar Flow in a Rectangular Microchannel

For lab-on-a-chip systems many fabrication methods lead to rectangular cross-section microchannels instead of circular cross-sections as we discussed above. For such a geometry no analytical solution is known to the Poiseuille flow problem. The best that can be done analyticaly is to find a Fourier sum representing the solution [4].

$$v_x(y, z) = \frac{4\,h^2\,\Delta p}{\pi^3\eta L} \sum_{n,odd}^{\infty} \frac{1}{n^3} \left[1 - \frac{cosh(n\pi\frac{y}{h})}{cosh(n\pi\frac{w}{2h})}\right] sin\left(n\pi\frac{z}{h}\right) \tag{16.43}$$

The flow rate is found by integration

$$
\begin{aligned}
Q &= 2 \int_0^{\frac{1}{2}w} dy \int_0^h dz v_x(y, z) \\
&= \frac{4\,h^2\,\Delta p}{\pi^3\eta L} \sum_{n,odd}^{\infty} \frac{1}{n^3} \frac{2\,h}{n\pi} \left[w - \frac{2\,h}{n\pi} tanh\left(n\pi\frac{w}{2\,h}\right)\right] \\
&= \frac{h^3 w \Delta p}{12\eta L} \left[1 - \sum_{n,odd}^{\infty} \frac{1}{n^5} \frac{192}{\pi^5} \frac{h}{w} tanh\left(n\pi\frac{w}{2\,h}\right)\right]
\end{aligned} \tag{16.44}
$$

where we have used $\sum_{n,odd}^{\infty} \frac{1}{n^4} = \frac{\pi^4}{96}$.

In the limit $h/w \to 0$ i.e. a flat and very wide channel, we can approximate (16.22) with $h/w\,tanh(n\pi\frac{w}{2h}) \to h/w\,tanh(\infty) = h/w$ and Q becomes

$$Q \approx \frac{h^3 w \Delta p}{12\eta L} \left[1 - 0.63\frac{h}{w}\right] \tag{16.45}$$

This approximation is rather good. For the worst case, the square $h = w$, the error is just 13 %.

16.3 Diffusive Transport

A fundamental transport process in fluid mechanics and in microfluidics is diffusion. Diffusion differs from advection in that it is random in nature and the related transport is from regions of high concentration to regions of low concentration with an equilibrium state of uniform concentration.

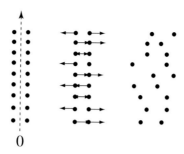

Fig. 16.8 Schematic of the one-dimensional molecular motion of a group of molecules illustrating Fickian diffusion

16.3.1 Diffusion

To derive a diffusive flux equation, it is somehow easier to begin by studying the simple 1D constant-step random-walk model of diffusion.

We consider two rows of molecules side-by-side and centered at $x = 0$ (Fig. 16.8). Each of these molecules moves randomly. Here we will consider only the component in one direction: motion right or left along the x-axis. We further define the mass of the particles on the left (right) as M_l (M_r), and the probability that a particle moves across $x = 0$ as k. After some time δt on average half of the particles have taken steps to the right and half of them have taken steps to the left. Mathematically, the average flux of particles from the left-hand column to the right is kM_l, and the average flux of particles from the right-hand column to the left is $-kM_r$, where the minus sign is used to distinguish direction. Thus, the net flux of particles j_x is

$$j_x = k(M_l - Mr) \tag{16.46}$$

For the one-dimensional case we can re-write in terms of concentrations using

$$c_l = M_l/\delta x \qquad c_r = M_r/\delta x \tag{16.47}$$

where δx is the average step along the x-axis taken by a molecule in the time δt. Next, we note that a finite difference approximation for dc/dx is

$$\frac{dc}{dx} = \frac{c_r - c_l}{x_r - x_l} = \frac{M_r - M_l}{\delta x(x_r - x_l)} \tag{16.48}$$

which gives

$$j_x = -D\frac{dc}{dx}, \qquad D = k(\delta x)^2 \tag{16.49}$$

Generalizing to three dimensions, we can write the diffusive flux vector at a point by adding the other two dimensions, yielding Fick's law

$$\mathbf{J} = -D\nabla c \tag{16.50}$$

where it is assumed that diffusion is isotropic, and the diffusion tensor $\mathbf{D}$ can be replaced by the scalar diffusion coefficient D.

16.3.2 Diffusion Coefficient

Since we derived Fick's law for molecules moving in Brownian motion, D is a molecular diffusion constant. The intensity of this Brownian motion controls the value of D. Thus, D depends on the state (solid, liquid, gas), temperature and molecule size. For dilute molecular species in water, D is generally of order $10^{-9} \mathrm{m}^2/\mathrm{s}$. For dispersed gases in air, D is of order $10^{-5} \mathrm{m}^2/\mathrm{s}$.

In the case of particles small enough (less than a micron), the effects of thermal agitation are sufficiently important to enable calculating the corresponding molecular diffusion constant. Let us consider a spherical particle of radius R, moving at velocity $\mathbf{U}$ relative to the liquid, the interaction force with the latter is the Stokes force $\mathbf{F} = -6\pi\eta R\mathbf{U}$ which can be expressed as

$$F = -\frac{U}{m}; \qquad m = \frac{1}{6\pi\eta R} \tag{16.51}$$

where m is the mobility coefficient. In addition, the mobility is linked to the diffusion constant D through the Einstein relation

$$D = mk_B T \tag{16.52}$$

which leads to

$$D = \frac{k_B T}{6\pi\eta R} \tag{16.53}$$

where k_B is the Boltzmann constant. This relation is valid for particle radii decreasing down to molecular scales.

16.3.3   Diffusion Equation

Although the Fick's law gives us an expression for the flux of mass due to the process of diffusion, we still require an equation that predicts the change in concentration of the diffusing mass over time at a point. This equation is achieved by considering mass conservation and combining with Fick's law we obtain the diffusion equation

$$\frac{\partial c}{\partial t} = D \Delta c \tag{16.54}$$

or written out fully in cartesian coordinates,

$$\frac{\partial c}{\partial t} = D \left(\frac{\partial^2 c}{\partial x^2} + \frac{\partial^2 c}{\partial y^2} + \frac{\partial^2 c}{\partial z^2} \right) \tag{16.55}$$

This equation describes the spreading of mass in a fluid with no mean velocity.

The most fundamental solution to the diffusion equation is that which describes the spreading of an initial slug of mass M introduced at time zero at $x = 0$. Since the equation is linear, this solution may be seen as building block to construct solutions to problems with more complex initial or boundary conditions.

Point source solution.

We consider the simple case where at the point $x = 0$ and at time $t = 0$, a mass of tracer, M, is injected uniformly across the cross-section of area A (see Fig. 16.9).

In one dimension, the governing Eq. (16.55) writes

$$\frac{\partial c}{\partial t} = D \frac{\partial^2 c}{\partial x^2} \tag{16.56}$$

which requires two boundary conditions and an initial condition.

As boundary conditions, we impose that the concentration at $\pm\infty$ remain zero:

$$c(\pm\infty, t) = 0 \tag{16.57}$$

The initial condition is that the dye tracer is injected instantaneously and uniformly across the cross-section over an infinitesimal width in the x-direction:

$$c(x, 0) = M/A \delta(x) \tag{16.58}$$

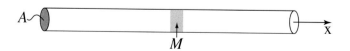

Fig. 16.9 Definition sketch for one-dimensional pure diffusion in an infinite pipe

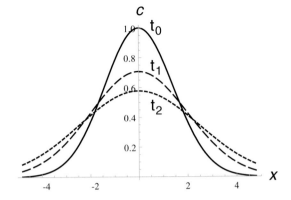

Fig. 16.10 Solution of the diffusion equation in one dimension for 3 different times ($t_0 < t_1 < t_2$) after the injection

where $\delta(x)$ is zero everywhere accept at $x = 0$, where it is infinite, but the integral of the delta function from $-\infty$ to ∞ is 1.

The fundamental solution of this problem is:

$$c(x) = \frac{M}{A\sqrt{4\pi Dt}} exp - (x^2/4Dt) \tag{16.59}$$

The solution is a Gaussian curve (Fig. 16.10). Its width increases as the square root of the time: a characteristic of diffusive phenomena. At the same time, the amplitude decreases as $1/\sqrt{t}$ so as to keep the area constant (the area corresponds to the initial mass).

16.4 Convective-Diffusive Transport

The derivation of the advective diffusion equation relies on the principle of superposition: advection and diffusion can be added together if they are linearly independent. In the previous section, diffusion was shown to be a random process due to molecular motion: each molecule during a time δt will move either one step to the left or one step to the right (i.e. $\pm\delta x$). Due to advection, each molecule will also move $u\delta t$ in the cross-flow direction. These processes are clearly additive and independent: the net movement of the molecule is $u\delta t \pm \delta x$. The total flux in the x-direction J_x, including the advective transport and a Fickian diffusion term, must be

$$J_x = uc + q_x = uc - D\frac{\partial c}{\partial x} \tag{16.60}$$

where q_x is the diffusive flux along the x-axis. Using this flux law and the conservation of mass, we can express the convection-diffusion equation for the concentration c of solutes in solutions having a weak velocity field $\mathbf{u}$

$$\frac{\partial c}{\partial t} + \mathbf{u} \cdot \nabla c = D \Delta c \tag{16.61}$$

Point source solution.

As we did previously, we study the simple case of a point source problem but this time the diffusion is coupled with advection. We substitute the coordinate transformation for the moving reference frame into the one-dimensional version of Eq. 16.61. In the 1-D case, $\mathbf{u} = (u, 0, 0)$, and there are no concentration gradients in the $y-$ or $z-$ directions, leaving us with

$$\frac{\partial c}{\partial t} + \frac{\partial (uc)}{\partial x} = D \frac{\partial^2 c}{\partial x^2} \tag{16.62}$$

The coordinate transformation for the moving system is

$$\phi = x - (x_0 + ut)$$

$$\tau = t$$

which substituted into Eq. 16.62 using the chain rule yields

$$\frac{\partial c}{\partial \tau} \frac{\partial \tau}{\partial t} + \frac{\partial c}{\partial \phi} \frac{\partial \phi}{\partial t} + u \left(\frac{\partial c}{\partial \phi} \frac{\partial \phi}{\partial x} + \frac{\partial c}{\partial \tau} \frac{\partial \tau}{\partial x} \right) \tag{16.63}$$

$$= D \left(\frac{\partial}{\partial \phi} \frac{\partial \phi}{\partial x} + \frac{\partial}{\partial \tau} \frac{\partial \tau}{\partial x} \right) \left(\frac{\partial c}{\partial \phi} \frac{\partial \phi}{\partial x} + \frac{\partial c}{\partial \tau} \frac{\partial \tau}{\partial x} \right) \tag{16.64}$$

which reduces to

Fig. 16.11 Solution of the advection-diffusion equation in one dimension

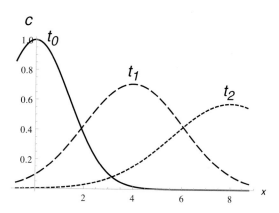

$$\frac{\partial c}{\partial \tau} = D \frac{\partial^2 c}{\partial \phi^2} \tag{16.65}$$

We recover the diffusion Eq. (16.54) expressed in the moving coordinates ϕ and τ with the solution

$$c = \frac{M}{A\sqrt{4\pi D\tau}} exp - (\phi^2/4D\tau) \tag{16.66}$$

Figure 16.11 shows the solution of the point source problem in the absolute system. Note that expressed in the moving coordinates we have exactly the solutions as shown in Fig. 16.10 with $\phi = x - (x_0 + ut)$.

This example is interesting since it highlights the importance of both diffusive and advective transport. If the cross flow was stronger, the cloud would have less time to spread out and would be narrower at each t_i. Conversely, if the diffusion was faster (larger D), the cloud would spread out more between the different t_i and the profiles would overlap. Thus, we see that diffusion versus advection dominance is a function of t, D and u, and we express this property through the non-dimensionnal Peclet number

$$Pe = \frac{u^2 t}{D} \tag{16.67}$$

or for a given downstream location $L = ut$

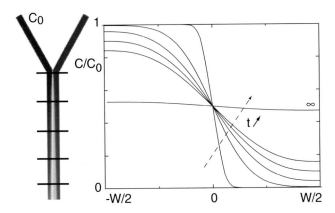

Fig. 16.12 Evolution of the concentration profile along the microfluidics channel

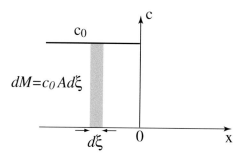

Fig. 16.13 Schematic of an instantaneous initial concentration distribution showing the differential element dM at a point ξ

$$Pe = \frac{uL}{D} \tag{16.68}$$

For $Pe \gg 1$ ($Pe \ll 1$), advection (diffusion) is dominant.

Advection-diffusion problem in microfluidics

One of the most popular mixers in microfluidics is based on the simple advection-diffusion problem. The geometry used is a channel with a Y-junction where a solute (concentration c_0) is diluted at the junction. Due to the laminarity of the flow there is an equivalence between time and length. In other words, in the steady state regime, measuring the concentration profile at a given length is equivalent of measuring the concentration at a given time.

Figure 16.12 shows the results of a numerical simulation of such a Y-junction. The dilution occurs at the interface and spreads along the flow (direction top to down). The right part on Fig. 16.12 shows the concentration profile across the channel at 5 different times.

The previous example is equivalent to solve the problem of an initial spatial concentration distribution as sketched in Fig. 16.13.

We consider the homogeneous initial distribution, given by

$$c(x, t) = c_0 \quad \text{if } x \le 0$$
$$= 0 \quad \text{if } x > 0$$

where $t = 0$ and c_0 is the uniform initial concentration. Since advection can always be included by changing the frame of reference, we will consider the one-dimensional stagnant case. At a point $x = \xi < 0$, there is an infinitesimal mass $dM = c_0 A d\xi$, where A is the cross-section area $\delta y \delta z$. For $t > 0$, the concentration at any point x is determined by the diffusion of mass from all the differential element

dM. The contribution dc for a single element dM is just the solution of Eq. 16.54 for an instantaneous point source

$$dc(x,t) = \frac{dM}{A\sqrt{4\pi Dt}} exp\left(-\frac{(x-\xi)^2}{4Dt}\right) \tag{16.69}$$

and by virtue of superposition, we can sum up all the contribution dM to obtain

$$c(x,t) = \int_{-\infty}^{0} \frac{c_0 d\xi}{\sqrt{4\pi Dt}} exp\left(-\frac{(x-\xi)^2}{4Dt}\right) \tag{16.70}$$

which is the superposition solution. To compute the integral we must make a change of variables as follows

$$\zeta = \frac{x-\xi}{\sqrt{4Dt}}; \qquad d\zeta = -\frac{d\xi}{\sqrt{4Dt}} \tag{16.71}$$

Substituting ζ into the integral gives

$$c(x,t) = \frac{c_0}{\sqrt{\pi}} \int_{+\infty}^{x/\sqrt{4Dt}} -exp(-\zeta^2)d\zeta \tag{16.72}$$

$$= \frac{c_0}{\sqrt{\pi}} \int_{x/\sqrt{4Dt}}^{+\infty} exp(-\zeta^2)d\zeta \tag{16.73}$$

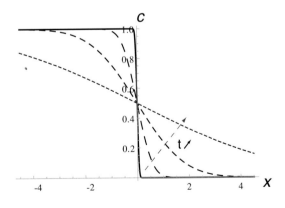

Fig. 16.14 Solutions of Eq. (16.76) for an instantaneous initial concentration distribution with $c_0 = 1$ and various time

$$= \frac{c_0}{\sqrt{\pi}} \left[\int_0^\infty exp(-\zeta^2)d\zeta - \int_0^{x/\sqrt{4Dt}} exp(\zeta^2)d\zeta \right] \qquad (16.74)$$

The first integral can be solved analytically (from a table of integrals) and its solution is $\sqrt{\pi}/2$. The second term of the integral is the so-called error function, defined as

$$erf(x) = \frac{2}{\sqrt{\pi}} \int_0^x exp(-\zeta^2)d\zeta \qquad (16.75)$$

Hence, the solution can be written

$$c(x,t) = \frac{c_0}{2}\left(1 - erf\left(\frac{x}{\sqrt{4Dt}}\right)\right) \qquad (16.76)$$

We plot the solution of Eq. (16.76) in Fig. 16.14 for 4 different times. Note that this solution is also valid if we use the moving coordinate system.

16.5   Capillary Effect

16.5.1   Notation of Interface

An interface is the geometrical surface that delimits two fluid domains. This definition implies that an interface has no thickness and is smooth. The reality is more complex and the separation of two immiscible fluids (water/air, water/oil) depends on molecular interactions between the molecules of each fluid and on Brownian diffusion (thermal agitation). However for macroscopic problems encountered in microfluidics we can consider an interface as a mathematical surface without thickness and the contact angle θ is uniquely defined by the tangent to the surface at the contact line (see Fig. 16.15).

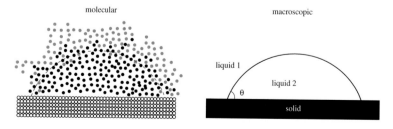

Fig. 16.15 Molecular and macroscopic sketch of the interfaces of a drop sitting on a solid

Fig. 16.16 Geometry of the
surface separation between
two liquids (1) and (2) and the
definition of the radius of
curvature R and R'

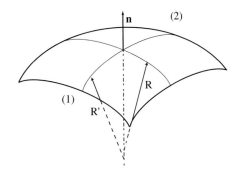

16.5.2 Surface Tension

Surface tension is related to the internal cohesion forces that act between the
molecules of a fluid: van der Waals forces, hydrogen bonding, ionic bonds. In the
bulk fluid, the forces exerted by each molecule are balanced by those exerted by
neighboring molecules. If an interface is introduced the forces are not balanced
anymore: this is the origin of surface forces and the corresponding energy. The
interfacial tension $\gamma_{A,B}$ between the two compounds (16.1) and (16.2) of free surface
tension γ_1 and γ_2 may be estimated from the approximate expression

$$\gamma_{1,2} = \gamma_1 + \gamma_2 - 2\sqrt{\gamma_1 \gamma_2} \tag{16.77}$$

16.5.3 Laplace Law

The force differential over the surface area of the interface, $\partial F/\partial A$, induces both a
surface tension and a pressure drop across the interface.

For the general case depicted in Fig. 16.16, the pressure difference can be written
as

$$\Delta p = p_2 - p_1 = \gamma_{1,2}\left(\frac{1}{R} + \frac{1}{R'}\right) \tag{16.78}$$

This equation is known as Laplace law or Laplace pressure. In the particular case
of a spherical drop ($R = R'$), the Laplace pressure is simply

$$\Delta p = p_2 - p_1 = \gamma_{1,2}\left(\frac{2}{R}\right) \tag{16.79}$$

16.5.4   *Wetting*

So far, we have dealt with interfaces between two fluids. Triple contact lines are the intersections of three interfaces involving three different materials: for example, a droplet of water sitting on a solid substrate in contact with air has a triple contact line. Liquids spread differently on a horizontal plate according to the nature of the solid surface and that of the liquid. In reality, it depends also on the third constituent, which is the gas or the fluid surrounding the drop. At the liquid-solid interface, the liquid interface molecules have strong interactions to both the bulk liquid and the solid. If the intermolecular force balance favors the solid, the liquid will seek to minimize the total surface energy by spreading out over the surface of the solid: this process is called *wetting*. If water wets the solid surface, the solid is called *hydrophilic*. Conversely, if the intermolecular force balance favors the bulk liquid, the liquid will not wet the surface but instead form a spherical cap in order to minimize the surface energy. The free surface of a solid on which water forms a spherical cap is *hydrophobic*.

A liquid spreads on a substrate in a film if the energy of the system is lowered by the presence of the liquid film. The spreading parameter S determines the type of spreading

$$S = \gamma_{SG} - (\gamma_{SL} + \gamma_{LG}) \qquad (16.80)$$

where γ_{ij} denotes the surface tension between the phases i and j, with S, L and G representing the solid, the liquid and the gas phase, respectively. If $S > 0$, the liquid spreads on the solid surface; if $S < 0$ the liquid forms a droplet.

16.5.5   *Contact Angle*

The angle the spherical cap of liquid makes with the solid surface is called the contact angle and is related to the surface tension of the various interfaces by the following equation:

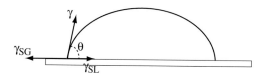

Fig. 16.17 Schematic of the tensions at the triple contact line

$$cos\theta = \frac{\gamma_{SG} - \gamma_{SL}}{\gamma_{LG}} \qquad (16.81)$$

In Fig. 16.17 we draw the different tensions that are exerted by the presence of a fluid on the triple contact line. At equilibrium, the sum of the tensions must be zero. The projection of the tensions on the horizontal line gives Eq. (16.81)

16.5.6 Capillary Length and Dimensionless Numbers

The effects of capillary, which are directly related to the curvature of the interface, are significant when considering phenomena at small length scales. For large objects, they are outweighed by volume forces like the gravity. In other words, it exists a length l_c, called capillary length, beyond which gravity becomes dominant. A manifestation of this phenomenon is the well known ascent of the liquid meniscus near the walls of a tube. An estimate is given by considering the Laplace pressure γ/l_c and the hydrostatic pressure $\rho g l_c$. The equality of the two pressures defines the capillary length

$$l_c = \sqrt{\frac{\gamma}{\rho g}} \qquad (16.82)$$

It is also possible to define a dimensionless number related to the gravity/capillary ratio. This number is called the Bond number and writes

$$Bo = \frac{gravitational force}{surface tension force} = \frac{\rho g l^2}{\gamma} = \frac{l^2}{l_c} \qquad (16.83)$$

where l is a characteristic length of the system. For water at room temperature the capillary length is around 3 mm which means that in microfluidics gravity is negligible.

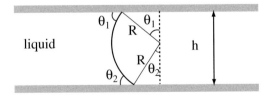

Fig. 16.18 Geometry of the liquid meniscus between two plates with different wettability

Laplace pressure in a microchannel.

We want to calculate the Laplace pressure in a flat channel with different contact angles θ_1 and θ_2. This example is observed where the bottom and the top of the channel are from different material i.e. there is a difference in wettability. In Fig. 16.18 we sketch such a system. The distance between the bottom and the top plate is denoted h. Let θ_1 and θ_2 be the contact angle at the bottom and top plates, respectively, then we have $h = R(cos\theta_1 + cos\theta_2)$. The Young-Laplace pressure is given by

$$\Delta p = \frac{\gamma}{R} = \frac{\gamma}{h}(cos\theta_1 + cos\theta_2) \tag{16.84}$$

This expression is validated in one dimension. If we consider the case where the channel is 2-D with a width w, Eq. (16.84) re-writes

$$\Delta p = \gamma \left(\frac{2}{w} + \frac{cos\theta_1 + cos\theta_2}{h} \right) \tag{16.85}$$

16.5.6.1 References

This chapter has been inspired by several books. Readers who wish to learn more about microfluidics and fundamental physics behind it are referred to the following monographs:

- *Introduction to fluid dynamics*, Batchelor [1]
- *Fluid mechanics*, Landau and Lifshitz [14]
- *Introduction to microfluidics*, Tabeling [20]
- *Theoritical microfluidics*, Bruus [4]
- *Elements of random walk and diffusion processes*, Ibe [13]
- *Mixing in inland and coastal waters*, Fischer [10]
- *Microdrops and digital microfluidics*, Berthier [2]
- french readers will appreciate also *Hydrodynamique physique*, Guyon et al. [11] and *Gouttes, bulles, perles et ondes*, de Gennes et al. [6]

References

1. G.K. Batchelor, *An Introduction to Fluid Dynamics* (Cambridge University Press, Cambridge, 1967)
2. J. Berthier, *Microdrops and Digital Microfluidics* (William Andrew Inc., Norwich, 2008)
3. L. Bocquet, E. Charlaix, Nanofluidics, from bulk to interfaces. Chem. Soc. Rev. **39**, 1073–1095 (2010)
4. H. Bruus, *Theoretical Microfluidics* (Oxford Master Series in Physics, 2007)

5. E. Buckingham, On physically similar systems; illustrations of the use of dimensional equations. Phys. Rev. **4**, 345–376 (1914)
6. F. Brochard-Wyart, D. Quéré, de P-G. Gennes, *Gouttes, bulles, perles et ondes* (Belin, Paris, 2005)
7. D.B. Dusenbery, *Living at Micro Scale* (Harvard University Press, USA, 2009)
8. M. Gad-el-Hak, *The MEMS Handbook* (CRC Press, Boca Raton, 2005)
9. D. Figeys, D. Pinto, Lab-on-a-chip: a revolution in biological and medical sciences. Anal. Chem. **72**(9), 330 A–335 A (2000)
10. H.B. Fischer, *Mixing in Inland and Coastal Waters* (Academic Press, New York, 1979) (Includes indexes)
11. L. Petit, E. Guyon, J-P. Hulin, *Hydrodynamique Physique* (EDP sciences, Paris, 2001)
12. C.M. Ho, Y.C. Tai, Micro-electro-mechanical-systems (MEMS) and fluid flows. Annu. Rev. Fluid Mech. **30**, 579–612 (1998)
13. O.C. Ibe, *Elements of Random Walk and Diffusion Processes* (Wiley Publishing, 1st edition, 2013)
14. E.M. Lifshitz, L.D. Landau, *Fluid Mechanics* (Pergamon press, London, 1959)
15. V. Lindroos, M. Tilli, A. Lehto, T. Motooka (eds.), *Front-Matter* (Micro and Nano Technologies. William Andrew Publishing, Boston, 2010)
16. M.J. Madou, *Fundamentals of Microfabrication* (CRC Press, Boca Raton, 1997)
17. A. Manz, E. Verpoorte, CS. Effenhauser, N. Burggraf, D.E. Raymond, H. Michael Widmer, Planar chip technology for capillary electrophoresis. Fresen. J. Anal. Chem. **348**(8–9), 567–571 (1994)
18. D.R. Reyes, D. Iossifidis, P.-A. Auroux, A. Manz, Micro total analysis systems. 1. Introduction, theory, and technology. Anal. Chem. **74**(12), 2623–2636 (2002)
19. T.M. Squires, S.R. Quake, Microfluidics: fluid physics at the nanoliter scale. Rev. Mod. Phys. **77**, 977–1026 (2005)
20. P. Tabeling, *Introduction to Microfluidics* (Oxford University Press, Oxford, 2009)
21. W.S.N. Trimmer, Microrobots and micromechanical systems. Sens. Actuators **19**(3), 267–287 (1989)
22. Y. Xia, G.M. Whitesides, Soft lithography. Angew. Chem. Int. Ed. **37**(5), 550–575 (1998)

Index

© Springer International Publishing Switzerland 2016
P.R. Lang and Y. Liu (eds.), *Soft Matter at Aqueous Interfaces*,
Lecture Notes in Physics 917, DOI 10.1007/978-3-319-24502-7

Printed in the United States
By Bookmasters